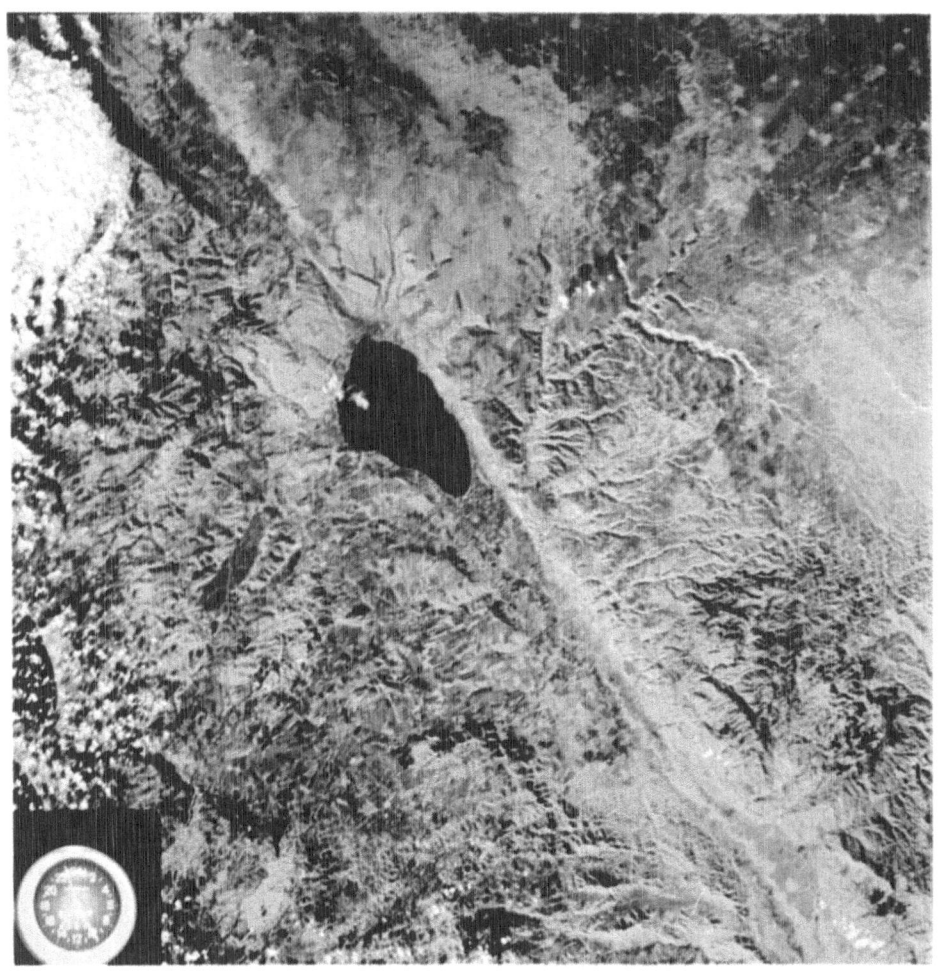

'SKYLAB picture of the lake Kinneret area, by courtesy of NASA'.

LAKE KINNERET

MONOGRAPHIAE BIOLOGICAE

Editor

J. ILLIES

Schlitz

VOLUME 32

Dr W. Junk bv Publishers The Hague–Boston–London 1978

LAKE KINNERET

Edited by

C. SERRUYA

Dr W. Junk bv Publishers The Hague–Boston–London 1978

ISBN-13: 978-94-009-9956-5   e-ISBN-13: 978-94-009-9954-1
DOI: 10.1007/978-94-009-9954-1

© Dr W. Junk bv Publishers The Hague
1978
Softcover reprint of the hardcover 1st edition 1978

Cover design Max Velthuijs, The Hague
1978
Zuid–Nederlandsche Drukkerij N.V.,
's–Hertogenbosch

Contents

viii

Foreword

Fauna and flora of lakes are an integrative result of regional past history and present environmental factors. In the Lake Kinneret area where Prehistoric Man witnessed the last tectonic readjustments of the Rift Valley, geological events do not belong only to the remote past but still strongly affect the lacustrine environment. It is therefore necessary to give a detailed picture of the regional background and limnological features of the lake (Parts I and II) before describing its planktic and benthic communities (Parts III and IV) and the Vertebrate fauna of the lake and its surroundings (Part V).

The trophic relationships between communities are beyond the scope of a Monograph and have consequently not been studied in detail but only mentioned occasionally.

It is intentional that Man and his penetration into the Kinneret area have been treated on a purely zoological basis. It underlines the fact that Man, as any other living organism, is part of the ecosystem and ruled by its laws and that his activities have an automatic feed back on his environment. However, in contrast with other living organisms, Man is able to 'utilize' the lakes and their watersheds for his benefit if, by appropriate management, he minimizes the damaging influence of his activities. This is the main purpose of the research carried out presently on Lake Kinneret and its watershed and briefly described in Part VI.

This Monograph is the result of the cooperation of many foreign and Israeli scientists who agreed to join me in this venture. To all I wish to express my deep gratitude. I am greatly indebted to all my colleagues of the Kinneret Limnological Laboratory who, besides their own contribution, have helped me in so many different ways. I am especially grateful to Mrs. U. Pollingher and Dr. M. Gophen for their constant interest and for providing me with valuable sources of information. I wish to thank Professor F. D. Por, of the Hebrew University, Jerusalem, for putting at my disposal new information concerning various groups of benthic fauna and for reading the chapters related to this subject. I express my gratitude to Dr. F. E. Round for reviewing the English of the manuscript. I am grateful to Mrs. L. Oved for drawing most of the illustrations, to Miss R. Herschovitch and Mrs. K. Diskin for typing the manuscript.

I am quite aware that, in spite of my search through recent works, I might have overlooked important contributions. Moreover, I had to coordinate the various chapters of independant co-authors and adjust them to the general plan of the book. I am then responsible for errors and omissions.

Colette Serruya

N.B. Permission has been granted to reproduce graphs and tables already published in scientific journals. This is gratefully acknowledged. In each case, detailed reference has been added to the legend.

Authors' Addresses

Avnimelech, Y., Department of Soil engineering, Technion, Haifa.

Bar Yosef, O., Department of Prehistory, Archeological Institute, The Hebrew University, Mount Scopus, Jerusalem, Israel.

Ben Tuvia, A., Department of Zoology, The Hebrew University, Jerusalem, Israel.

Berman, T., Kinneret Limnological Laboratory, P.O.B. 345, Tiberias, Israel.

Cavari, B. Z., Kinneret Limnological Laboratory, P.O.B. 345, Tiberias, Israel.

Freund, R., Department of Geology, The Hebrew University, Jerusalem, Israel.

Gophen, M., Kinneret Limnological Laboratory, P.O.B. 345, Tiberias, Israel.

Harpaz, A., Lake Kinneret Authority, Minhelet Ha Kinneret, Zemah, Emek Ha Yarden, Israel.

Horowitz, A., Institute of Archaeology, Tel Aviv University, Ramat Aviv, Israel.

Kugler, J., Department of Zoology, The George S. Wise Center for Life Sciences, Tel Aviv University, Ramat Aviv, Israel.

Lulav, S., The A. D. Gordon Agriculture & Nature Study Institute, Kibbutz Degania A, Jordan Valley, Israel.

Margalit, Y., Department of Zoology, The Hebrew University, Jerusalem, Israel.

Mazor, E., Isotope Department, Weitzmann Institute of Science, Rehovot, Israel.

Mero, F., Department of Hydrology, 'TAHAL', 54 Ibn Gabirol, Tel Aviv, Israel.

Michelson, H., Department of Hydrology, 'TAHAL', 54 Ibn Gabirol, Tel Aviv, Israel.

Neuman, Y., Department of Atmospheric Sciences, The Hebrew University, Jerusalem, Israel.

Pollingher, U., Institute of Oceanographic & Limnological Research Ltd., Tel Shikmona, P.O.B. 1793, Haifa, Israel.

Rodhe, W., Institute of Limnology, University of Uppsala, P.O.B. 557, S-751-22 Uppsala, Sweden.

Round, F. E., Department of Botany, The University, Bristol BS8 1UG, United Kingdom.

Rubin, S., Meteorological Service, P.O.B. 25, Beth Dagan 50200, Israel.

Serruya, C., Kinneret Limnological Laboratory, P.O.B. 345, Tiberias, Israel.

Serruya, S., Kinneret Limnological Laboratory, P.O.B. 345, Tiberias, Israel.

Stanhill, G., Institute of Soils & Water, The Volcani Center, Agriculture Research Organization, Ministry of Agriculture, Israel.

Tsurnamal, M., Department of Zoology, The Hebrew University, Jerusalem, Israel.

Introduction

C. Serruya

A. Lake Kinneret: an Old Lake

Lakes are relatively short-lived water bodies and their existence is permanently threatened by silting, evaporation, changes of hydrographic regime or climatological conditions. For example, the ice cover which, in the Pleistocene, extended all over northern Europe and America did not, in general, allow the persistence of tertiary water bodies and their fauna. It follows that most of the lakes of the northern Hemisphere are not older than 20,000 years. This represents a very brief period for the post-glacial fauna of these lakes to have produced new species, and these lakes generally show a very low level of species differentiation.

A few lakes in the world are exceptions to this general pattern; lakes Baikal, Tanganika, Ochrid and the Caspian Sea, various lakes in the Celebes, in China and Lake Kinneret can be considered as relict lakes from the Tertiary era. These lakes have in common two important characteristics: the presence of relict fauna of archaic features and a large number of endemic species. Lake Kinneret occupies a special position among the relict lakes of the world. It acquired its present configuration only approximately 20,000 years ago and, in this respect, may be considered as a 'post-glacial lake'. However, inland water bodies connected with the developing Jordan Valley existed during the Neogene and Pleistocene in this area. A certain number of species were preserved in these environments and invaded the lakes of the Jordan Valley when these developed. Even the hypersaline Lisan episode did not destroy the relict fauna which was maintained in the Hula Lake and in the rivers and springs. The long period of time involved allowed also a certain endemism. However, the eventful history of the Jordan Valley and the considerable variations of water level and salinity which took place mainly in the Central Jordan Valley did not allow the development of these two characters, reliction and endemism, to the high degree which is found in Lake Baikal and Lake Ochrid for example.

B. Lake Kinneret: A new Role

For many people, Lake Kinneret (Lake of Tiberias, Sea of Galilee) has a deep religious significance; for millions of Christians, the shores of the lake represent, together with Bethlehem, Nazareth and Jerusalem, the very place where Christianity began. Every year, numerous pilgrims visit the region where Jesus used to perform miracles and address the people of the towns of Chorazim, Capernaum and Betsaida, which were relatively large settlements.

Various archaeological remnants testify to the economic wealth of this region in the past, and there is no doubt that the presence of a large body of freshwater in this warm area was one of the determining factors which permitted flourishing agriculture and fisheries.

The lake and its water reserve of 4,000 million cubic meters has lost none of its importance in the modern State of Israel. In the same manner as thousands of years ago, the lake water contributes largely to the welfare of the country; in contrast to the situation in historical times, the beneficial influence of the lake is no longer restricted to the lake vicinity but has been extended to a large part of the country by the construction of the National Water Carrier.

1. *The conception of the National Water Carrier*

The period which preceded the creation of the State of Israel was characterized by the rapid development of agricultural settlements, and this highlighted a natural feature of the geography of Israel: two-thirds of the arable land are located in the central and southern part of Israel whereas most of the water resources are found in the northern, more mountainous part of the country. This geographical feature determined the guidelines of water policy. It was clear that part of the northern waters had to be directed to the southern, more fertile areas. The idea of a 'National Water Carrier' connecting both areas was first developed in 1944 by W. Lowdermilk in his book 'Palestine: Land of Promise'. Lowdermilk proposed to utilize the Jordan sources for both water supply and production of electricity. In his book 'Tennessee Valley Authority on the Jordan' (1948), the American engineer J. B. Hays presented a technical version of Lowdermilk's general conception. The project included the diversion of the source of the rivers Litani, Hatzbani and Jordan with the objective of hydroelectric production and irrigation of southern Palestine and of the Negev. It was clear that the diversion of the Upper Jordan River would rapidly transform Lake Kinneret into a lake without outlet and increase its salinity. Therefore, Hays proposed to divert part of the Yarmouk River, a tributary reaching the Jordan some 7 km south of the lake, into Lake Kinneret.

The War of Independence made obsolete these conceptions which were based on the cooperation of the different people of the region. Tahal, the government company responsible for the overall planning of water in Israel, together with the American engineer J. Cotton, revised the project and decided that the water of the Jordan River would be diverted in Israeli territory at Gesher Benot Ya'akov. Two-thirds of the diverted water were to be sent southwards for irrigation purposes and one-third was to be utilized for hydroelectric production. The Cotton–Tahal project was initiated in 1951 but border incidents which caused severe casualties together with diplomatic pressure from various countries led Israel to abandon the project.

However, the intensive immigration and the relative drought of the early fifties convinced the late Israeli Prime Minister Levi Eshkol that a new project should be elaborated and carried out without delay.

The major modifications introduced mainly concerned the location of the diversion point. In the previous projects, the waters were to be diverted upstream and flow by gravity; this idea was abandoned in the final project for security reasons. In the final scheme, the lake itself became the main reservoir of the system and the northwestern area of the lake was chosen as the location of the intake area of the National Water Carrier. However, as the lake is located at 209 m below mean sea level, pumps had to replace gravity.

In 1956, the Water Company 'Mekorot' was commissioned to implement the project. The construction of the National Water Carrier commenced in 1959 and was

2

completed in 1964. On June 10, 1964, the National Water Carrier was inaugurated, and for the first time the Kinneret water flowed to irrigate the fields of the Negev.

2. *The role of the lake in the water supply of Israel*

The total water consumption of Israel amounts to 1,500 million cubic meters per year; agriculture is the main consumer with 82% of the total amount, whereas domestic and industrial activities require respectively 13% and 5% of the yearly consumption.

Lake Kinneret has an interannual average water income of 650 million cubic meters out of which 300 to 400 million cubic meters are pumped through the National Water Carrier. It appears then that Lake Kinneret supplies from one-fifth to one-fourth of the total freshwater requirements of Israel.

Although its quantitative contribution to the water system is not negligible, the most important role of the lake in the water supply is that of a regulation reservoir. Its considerable capacity permits the storage of the winter flood waters of the Jordan River, and water from rainy years to dry years. In other words, the lake minimizes the effect of the dry season and even of the dry years. Moreover, the National Water Carrier, by connecting local and regional supply networks into an interconnected grid, proved to be a very flexible system offering considerable manipulation facilities. This permitted a regular and permanent water supply to the different regions of the country, even to those where the annual amount of rainfall is less than 50 millimeters.

A typical example of the flexibility of the system is the possibility of utilizing the lake water for reinjection in the coastal aquifer. Before the completion of the National Water Carrier, the coastal aquifer had been overpumped. The consequent decrease of water level had caused intrusion of sea water into the aquifer, which could no longer be used for irrigation purposes. In the winter of 1964/1965, forty million cubic meters of water mainly withdrawn from the lake through the National Water Carrier were reinjected in order to push back the salty interface. This amount increased in the following years up to approximately 100 million cubic meters. The reinjection of freshwater into the coastal aquifer was followed by a significant increase in the water level and drop in salinity. At present, the yearly injections are aimed at storing enough water in winter to allow a regular pumping in summer. This operation increases the storage capacity of the water system and diminishes losses by evaporation. It could be carried out only when Lake Kinneret became the main reservoir of the water system.

The juxtaposition, in this introduction, of the ancient faunistic characteristics of this old lake and of the problems related to the present water supply is intentional. It reflects the double purpose of the scientific research carried out on the lake: to preserve the existing faunistic communities and to elaborate lake management programs allowing an optimal water supply. Such a task requires a deep understanding of the trophic relationships which regulate the lake and watershed community. It also requires the inclusion of Man in this community and the consideration of his activities as an integral part of the transformations of the ecosystem. Man is not an outsider who 'utilizes' ecosystems; he is a member – a powerful one – of the ecosystem community.

3

References

Lowdermilk, W. 1944. Palestine: Land of Promise.
Hays, J. B. 1948. Tennessee Valley Authority on the Jordan.

Part one

General background

I Geography

C. Serruya

Lake Kinneret is located in the central part of the Jordan Rift Valley but the lake's watershed (2,730 km²) extends over the geographical units of Upper Galilee in northeastern Israel and southern Anti-Lebanon in Lebanon. The whole watershed is situated between 32° 40′ and 33° 38′N which corresponds approximately to the latitude of San Diego, California. The watershed is 110 km long in the N–S direction. Its maximum width is 50 km in the lake area and 15 km in its northernmost portion (Fig. 1).

The watershed of the Litani River occupies the northern border of the Kinneret drainage area and the Mounts of Meron limit its extension to the west. The Mounts of Meron are also the water divide between the drainage systems of the Mediterranean Sea and of the Rift Valley. The water divide with the northern tributaries of the Yarmouk River constitutes the eastern frontier of the Jordan watershed on the Golan Heights, and the northern limit of the watershed of the southern Jordan River is the southern limit of the Lake Kinneret drainage area. This relatively small area has an unusual variety of landscapes and vegetation which is partly due to the large range of altitudes encountered (+2,814 m to −210 m).

The mountainous area of the Hermon in the north (Djebel Ha-Sheikh in Arabic), an uplifted massif of Jurassic and Lower Cretaceous limestones, is the highest part of the watershed and culminates at 2,814 m. It covers an area of 650 km² and has a very high annual average precipitation. Above 1,500 m, most of the precipitation falls as snow and the area located above 2,000 m remains under snow cover from February to May. The karstic erosion led to the formation of numerous dolines and the main sources of the Jordan River spring out of the mountain karst.

In the Hermon Mountain, the altitude governs the distribution of plants, and three main vegetal belts have been described (Schmida, *et al.*, 1972; Peri, 1967).

(1) Up to 1,400 m, the oak forest dominates with *Quercus calliprinos* as the main species. The forest also harbours *Quercus infectoria*, *Spartium junceum*, *Crataegus azarolus* and *Rhamnus palaestinus*. In the understory grow *Poterium spinosum*, *Cestus villosus* and *Larandula stocchas*.

(2) From 1,400 to 1,800 m, the destructive influence of Man has reduced the arborescent vegetation to sparse *Spartium junceum*, *Prunus ursina*, *Quercus calliprinos*, *Styrax officinalis* and *Crataegus monogyna*; these are dwarf trees not exceeding 1 meter high. Most of the area is covered by thorny plants.

(3) Above 1,800 m, a type of Alpine vegetation develops belonging to the Irano–Turanian phytogeographical zone. The vegetation consists of thorny plants such as *Acantholimon* which are well adapted to the dry summer conditions. This type dominates on the slopes, whereas green meadows cover the wet and rich soils of the dolines.

The growth of trees and bush is limited in the Hermon by the long dry summers when the combined effect of low barometric pressure and strong winds generates a

7

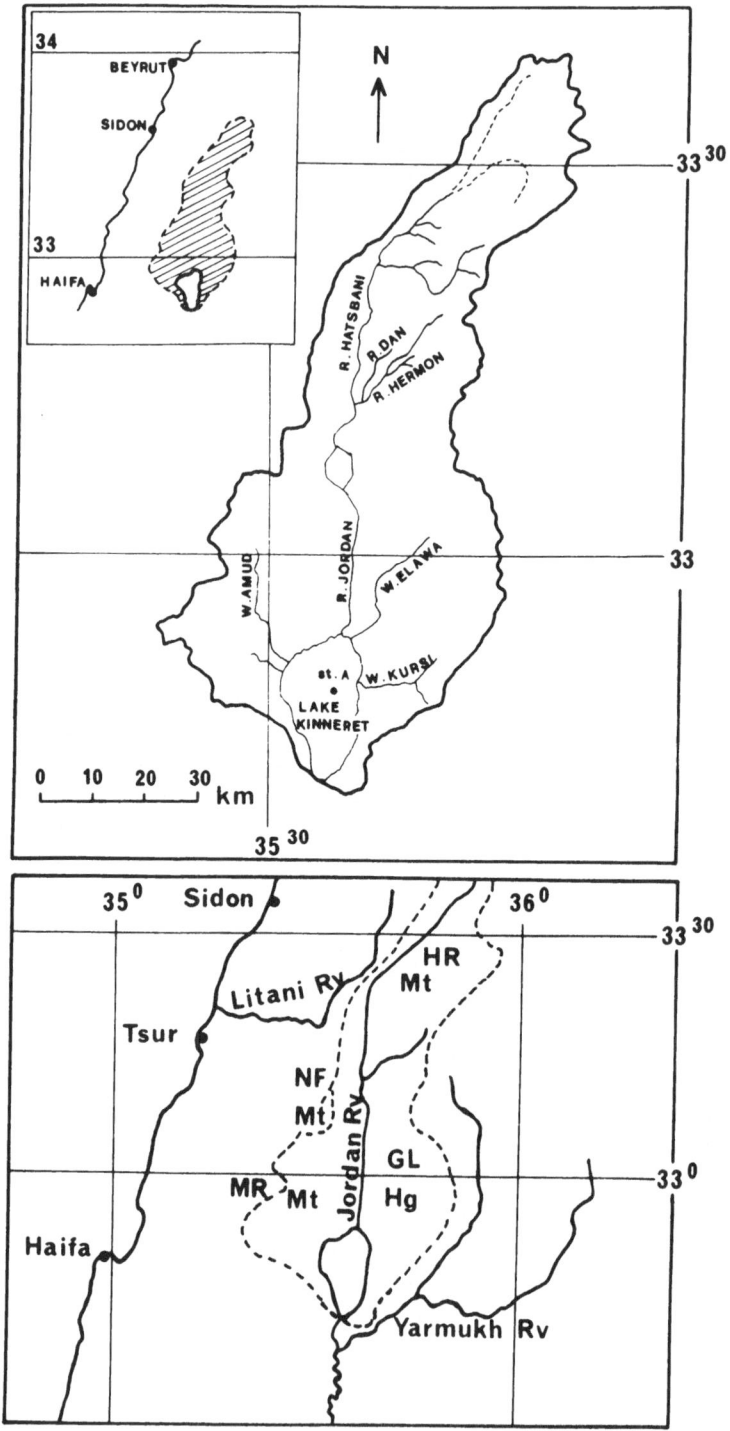

Fig. 1. Lake Kinneret and its watershed. HR Mt = Hermon Mountain, NF Mt = Naftali Mountain, MR Mt = Meron Mountain, GL Hg = Golan Heights.

considerable evaporation. Therefore, the Hermon 'forest' is not dense and covers only a small portion of the area. Besides unfavourable climatological conditions, excessive clearing of trees and overgrazing are additional factors which account for this degenerated forest where the average size of trees seldom exceeds two meters.

The Golan Heights in the east are covered with basalt. The thickness of the basaltic cover varies from 100–200 m in the western part of the plateau to 300–400 m in the east. In the highest part of the Golan Heights, a chain of about 60 cinder cones of non-active volcanoes dominates the grey or blackish rocky landscape. The basalt flows filled the thalweg of rivers and streams, and many a stream has since carved canyons and waterfalls in the lava. In certain areas, the slow solidification of lavas has produced prismatic columns of great beauty (plates 1 and 2).

According to the description of Schumacher (1888), one century ago, dense forests covered the slopes of Tel Abu Khanzin, but these are barren today. Remnants of the original forest are preserved near holy graves, for example, *Quercus boissieri* near Mansura. Natural forests are found in two areas (Danin, 1968): (1) Near Massada–Moumasia, the forest is composed of *Quercus calliprinos* and of *Quercus boissieri* (Fig. 2). Associated species are *Pistacia palaestina*, *Prunus ursina*, *Rhus coriaria*, two species of *Crataegus* and *Spartium junceum* as well as creepers such as *Lonicera etrusca* and *Rubia tenuifolia*. Among the non-arborescent plants are *Dianthus polycladus*, more abundant here than in other regions of Israel, and *Dryopteris villarii*. (2) The area of Yehuda, and in general the southern Golan below 300 m altitude, is the domain of *Quercus ithaburensis*. Associated species are *Pistacia atlantica*, *Ziziphus spina-Christi*, *Ziziphus lotus*,

Plate 1. The Meshushim pool, Meshushim River, Golan Heights. Note the prismatic columns of basalt (photo Y. Kaplan).

9

Plate 2. The Meshushim pool. Detail of the prismatic columns and the waterfall (photo Y. Kaplan).

Scolymus maculatus, Carlina corymbosa, Hordeum bulbosum, Echinops sp. and *Gundelia tournefortii.* It seems that in this area, *Ziziphus* spp. are secondary and have invaded the domain of *Quercus ithaburensis* when the forest was cleared.

In the streams flowing towards the Buteiha plain, the following species are common: *Nerium oleander, Vitex agnus castus, Myrtus communis* and *Salix acmophylla.* In the lower parts of the plain, *Vitex castus agnus* dominates accompanied by various species of reeds, *Trifolium fragiferum* and *Trifolium repens. Butomus umbellatus, Apium nodiflorum* and *Ludwigia stolonifera* are found, particularly in spring.

In the west, the chain of the Naftali mountains of Cretaceous and Eocene limestones is faulted in its northern part and forms a steep escarpment reaching an altitude of 900 m. The central part is occupied by karstic depressions at 500–600 m altitude covered with a thin cover of basalt and reddish-brown terra rosa soils suitable for plantations. The southern part of this area consists of the Sfad and Meron mountains which reach an altitude of 1,200 m.

The northern part of the Jordan Valley is occupied by the 175 km² of the Hula plain, a flat area located at an altitude of 70 m. This area of permanent subsidence has long been occupied by a lake where 1,000–1,500 m of sediments have been deposited. The Hula Lake was drained during the period 1951–1959. The scheme was worked out by the Jewish National Fund in two stages. In the first stage, the portion of the Jordan River south of the Hula plain was straightened, widened and deepened in order to carry the surplus water from the northern area. In the second stage, three main canals and a network of drainage canals were dug to drain the swamps and eliminate the yearly floods. Only a small area of the swamps and former lake has been kept as a nature reserve where the water buffaloes, water

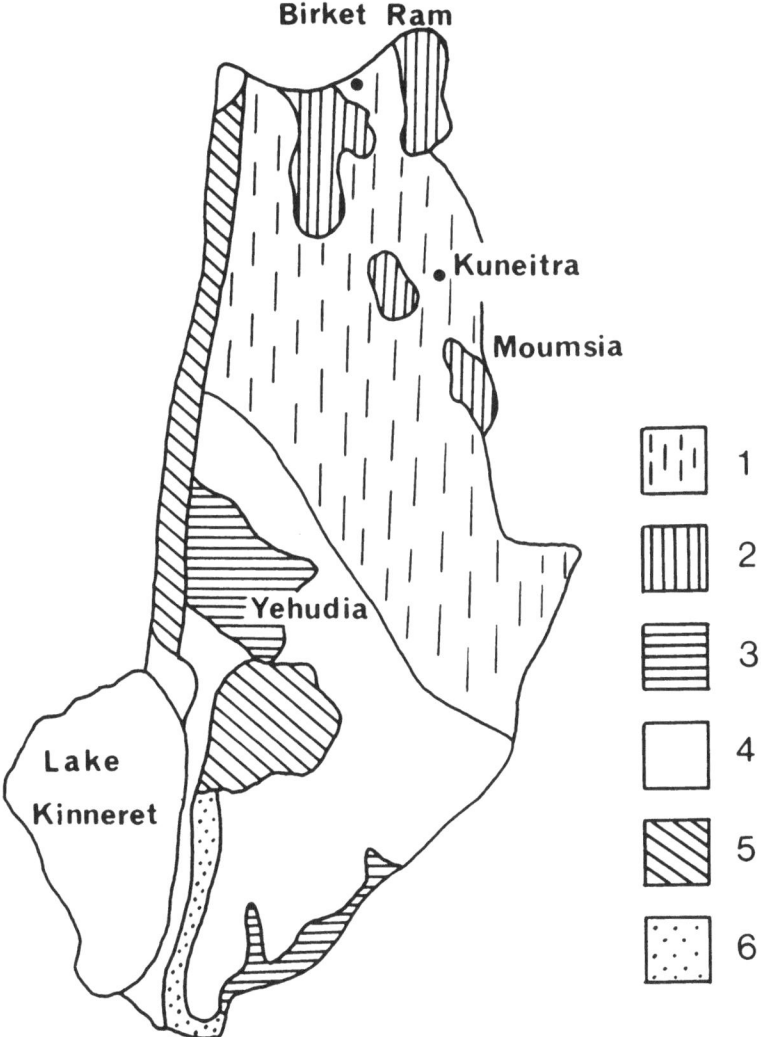

Fig. 2. Distribution of vegetation on the Golan Heights. 1, *Quercus calliprinos.* 2, *Quercus calliprinos* and *Quercus infectoria.* 3, *Quercus ithaburensis.* 4, *Zizyphus replacing Quercus ithaburensis.* 5, *Ziziphus lotus.* 6, *Poterium spinosum, Salsola vermiculata, Retama raetam.* (From A. Danin, 1968, Teva Vaaretz, 10).

fowl and papyrus reeds, the common fauna and flora of the swamps, are preserved.

An area of about 15 km², corresponding to the area of the drained swamp, is now covered by organic peat soils (60–80% of organic matter). The exposure to atmospheric oxygen of this organic material has created many problems of mineralization. Soils of the dried lake are more calcareous and contain no more than 10% of organic matter.

The region which was formerly classified by the British Army as a 'highly malarious area' has now become well irrigated and cultivated. The reclamation of 60 km² increased the irrigated area of the Hula plain to 126 km².

This area is utilized as follows (Elan, 1973): (1) Irrigated field crops occupy 76 km² and cotton is the most important of them (52% of the area). In 1972, the average yield was 1,170 kg/ha, requiring a water supply of 3,000–4,500 m³/ha/year. Approximately 22% of the area produces vegetables and flowers, 19% produces alfalfa and 7% is reserved for forage irrigated pasture. (2) Orchards represent an important source of income in the Hula Valley and occupy 36 km². Apples are the main product (50% of the area) but citrus, pears and avocados are also cultivated. (3) Fishponds, a typical branch of Israeli agriculture, occupy an area of 14 km² in the Hula. The ponds were introduced by fish breeders from Yugoslavia. The size of the ponds varies from 3 to 10 ha and the size of fish farms ranges from 50 to 200 ha. The ponds are generally fertilized with nitrogenous and phosphate fertilizers and the fish are fed with cereals and pellets. Very high yields are obtained in these ponds: 2,500 kg/ha is an average and 4,000 kg/ha in special cases. These ponds produce about 15% of the total fish production of the country. Carp represent 80% of the fish grown in the Hula ponds. The other 20% consists of Saint Peter fish (*Tilapia*) and grey mullet (*Mugil*).

In the Dan area, trout cultivation in running water has proved to be successful. The cool, oxygen-rich water of one of the sources of the Jordan River passes through artificial basins, and on a relatively small area, high yields of trout are produced and marketed to the tourist industry.

The drainage of Lake Hula has led to a complete modification of the landscape of the Valley. In particular, the natural vegetation has been replaced by agricultural crops. If we now follow the Jordan River and reach the lake, we find again a more natural environment, although the local economic development, especially shore building, already affects the natural vegetation.

In the immediate surroundings of the lake, the vegetation belongs to the semiarid Irano–Turanian zone (Lulav, 1973). The winter rainfall allows the growth of an abundant flora of annuals and bulb plants such as *Urginea maritima* but no forest vegetation. In summer, conditions prevailing in this depression, lying at −210 m, turn it into an arid plain, the dominant plant of which is *Ziziphus lotus*. A Mediterranean flora develops only on the northern shores of the lake from Ginosar to Buteiha and in the surrounding wadis. On the higher slopes, remnants of a marginal forest, including the Atlantic pistache (*Pistacia atlantica*), wild almonds (*Prunus amygdalus*) and in the north, the Tabor oak (*Quercus ithaburensis*), are found. In the lower areas, there are traces of tropical Sudano–Decanian flora (*Ziziphus spina-Christi* and *Acacia albida*).

The lake shore vegetation which will be described in more detail in a later chapter is not very rich: Abraham's bush (*Vitex agnus castus*), oleander (*Nerium oleander*), reeds (*Phragmites, Arundo*), bullrushes (*Cyperus articulatus*) and raspberry bushes (*Rubus sanctus*) occur.

The Buteiha plain is a complex area resulting from the deposition, in winter, of the alluvions brought by the three rivers, Meshushim, Yehudia and Daliot. This supports one of the rare remnants of the indigenous shore vegetation, and may be the only remnant in Israel of a natural hydrophylic vegetation in an area which is large enough to allow the development of a stable ecosystem.

The fields of bananas, grapefruit, oranges, avocados, grapes, dates, etc., harbour a rich wild life. Egyptian mongoose (*Herpestes ichneumon*), otters (*Lutra seistanica*), gazelles (*Gazella gazella*), wild boars (*Sus scropha*) and jungle cats (*Felis catolynx chaus*) were indigenous here and are still present. Recently, a desert lynx (*Felis caracal schmitzi*) has been spotted on the hills west of the lake.

About two hundred birds have been noted; the most numerous are the winter water fowl. The most outstanding local species are francolin (*Francolinus francolinus*), pied kingfisher (*Ceryle rudis*), Palestine sunbird (*Nectarinia osea*), Dead Sea sparrow (*Passer moabiticus*), brown fish owl (*Ketupa ceylonensis*), black headed gull (*Larus ridubundus*), and the herring gull (*Larus agentatus*).

Whereas in the areas located north of Israel and Lebanon the alpine folding has produced the landscape's east–west geomorphic features, the Kinneret area is characterized by a clear north–south topographic pattern. In other words, the tectonic lines of the Levant cross at right angles to the general east–west lines of the orogenic belts of the earth. This interesting fact, which affected significantly the development of the hydrographic net and the distribution of the flora and fauna, is the result of a complicated and still unclear geological history of this region.

References

Schmida, A., M. Zohari & A. Danin, 1972. The vegetation of the Hermon: New species. Sal'it 1.3: 100–111 (in Hebrew).

Peri, D. 1967. General characteristics of the vegetation of the Hermon. Teva Vaaretz 10, 1:4–8 (in Hebrew).

Schumacher, G. 1888. The Jaulan. London.

Danin, A. 1968. The vegetation of the Golan. Teva Vaaretz 10, 3:162–167 (in Hebrew).

Elan, S. 1973. Upper Galilee Region: Past, present, perspectives. Ministry of Agriculture, 54 p.

Lulav, S. 1973. The Lake Kinneret region: Lake Kinneret, general background. Israel Scientific Research Conference, Sept. 1973. T. Berman (ed.).

II Geology

A. The general frame

C. Serruya

The geological history of Israel was determined by the position of the area between the Precambrian Arabo–Nubian crystalline massif in the south and the pre-Mediterranean Sea in the north and west.

The Precambrian massif outcrops in the region of Eilat. It is composed of a thick sedimentary formation metamorphized into gneiss and micaschists and injected with magmatic rocks. The general tendency to uplift of the massif which prevailed in Paleozoic times caused its erosion and the accumulation of continental series, the Nubian Rocks, in the area extending north of the massif.

The Mesozoic Seas transgressed over the continental series. In the Triassic and Jurassic periods, the coastline was located in the present northern Negev where shallow marine rocks of this period outcrop. Jurassic rocks of a much deeper facies constitute the core of Mount Hermon in the northern part of the Lake Kinneret watershed. The transgression which took place from early Cretaceous until Middle Eocene covered the entire Middle East.

The tectonic movements which occurred in the Eocene period brought to an end the main marine phase in Israel. The Neogene transgressions were much more limited in space and duration and were separated by freshwater episodes. After the last Pliocene regression, lakes became a permanent feature, especially in the Rift Valley, as will be described in the following sections. We see that although rocks belonging to the most ancient geological series of the Earth are found in Israel, the stratigraphy of the Lake Kinneret area concerns relatively recent formations.

B. Stratigraphy of the lake area

H. Michelson

The watershed of Lake Kinneret can be divided into three areas: the northern area, mainly formed by the core of Jurassic limestone of Mount Hermon and the Cretaceous–Eocene series of Mount Naphtali; the western area, composed of Cretaceous and Eocene rocks; and the eastern area, almost completely covered with basalts (Fig. 3).

In the following, we have deliberately limited our stratigraphic description to the formations which outcrop in the vicinity of the lake (western and eastern areas (Figs. 4 and 5).

1. Upper Cenoman (Ce_3): Sakhnin formation

Grey, karstic dolomite, 200 m thick.

2. Turon (t): Bina formation

Detrital limestone, 50–80 m thick; facies from finely crystallized to lithographic, locally dolomitic. The limestone is well stratified; its high density and hardness make it valuable to the marble industry.

3. Senon (S): Mount Scopus formation

Marl, chalk, shales and bituminous shales with local horizons of flint. This group may reach a thickness of 290 m and is found in the Sfad and Ein Zeitim areas.

4. Lower–Middle Eocene (e):

a. Bar Kochba formation

Limestone with flint horizons, especially abundant in the lower part of the formation. The limestone is finely crystallized, rich in nummulites and may be karstic. It reaches a thickness of 400 m and outcrops in the area below Sfad and in the Arbel and Cana'an mountains.

These four formations outcrop only on the western side of the lake.

b. Zora formation

Outcrops on the eastern side of the lake and is the equivalent of the Bar Kochba formation. It is divided into the lower Adulam unit (Zad), composed of chalky limestone, well stratified and including flint horizons, and the upper Maresha unit (Zma), composed of unbedded marl and chalk. The Zora formation reaches a thickness of 400 m.

17

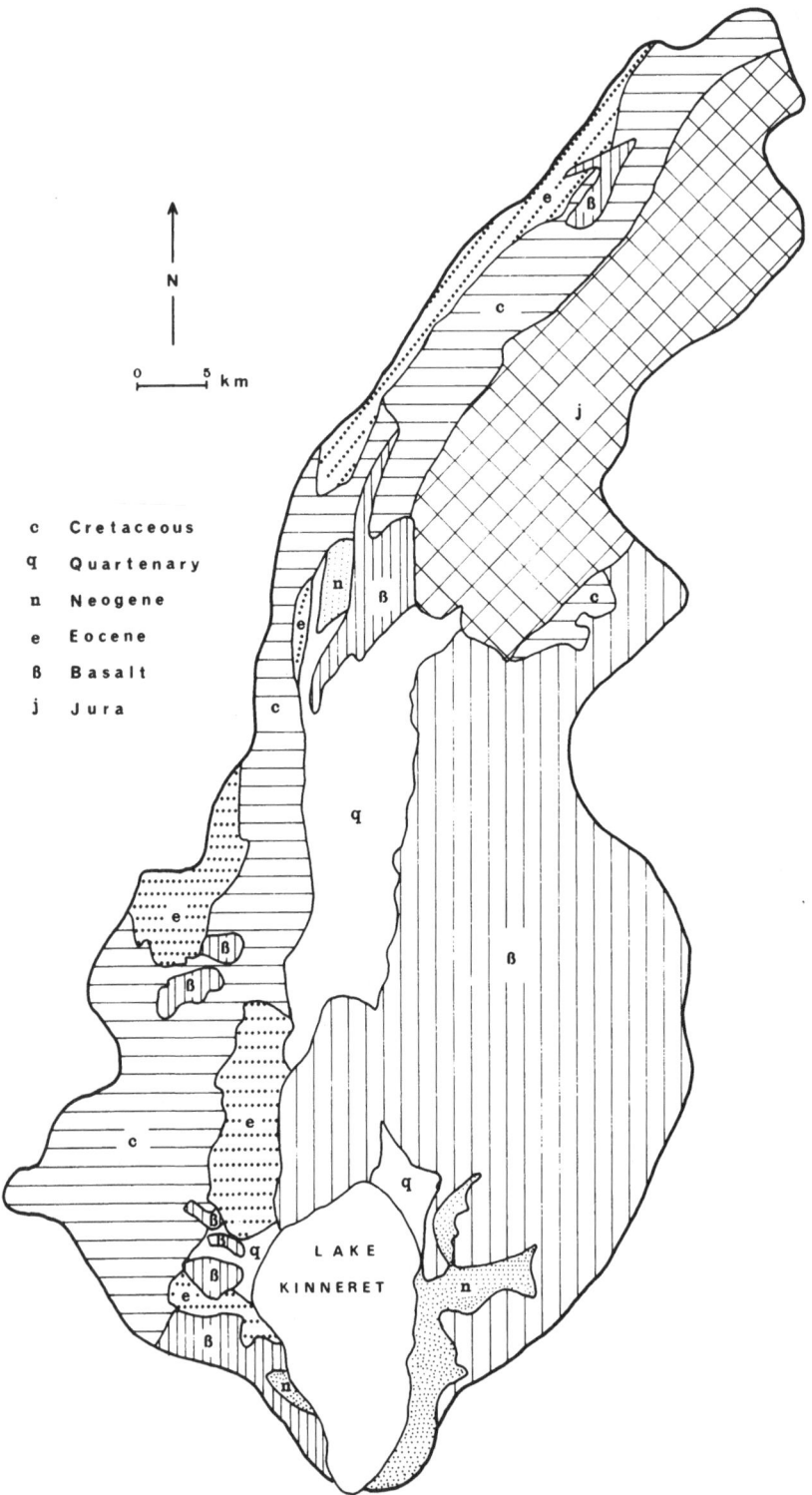

N

0 ——— 5 km

c Cretaceous
q Quartenary
n Neogene
e Eocene
ß Basalt
j Jura

LAKE
KINNERET

Fig. 3. Different geological formations of the Kinneret watershed.

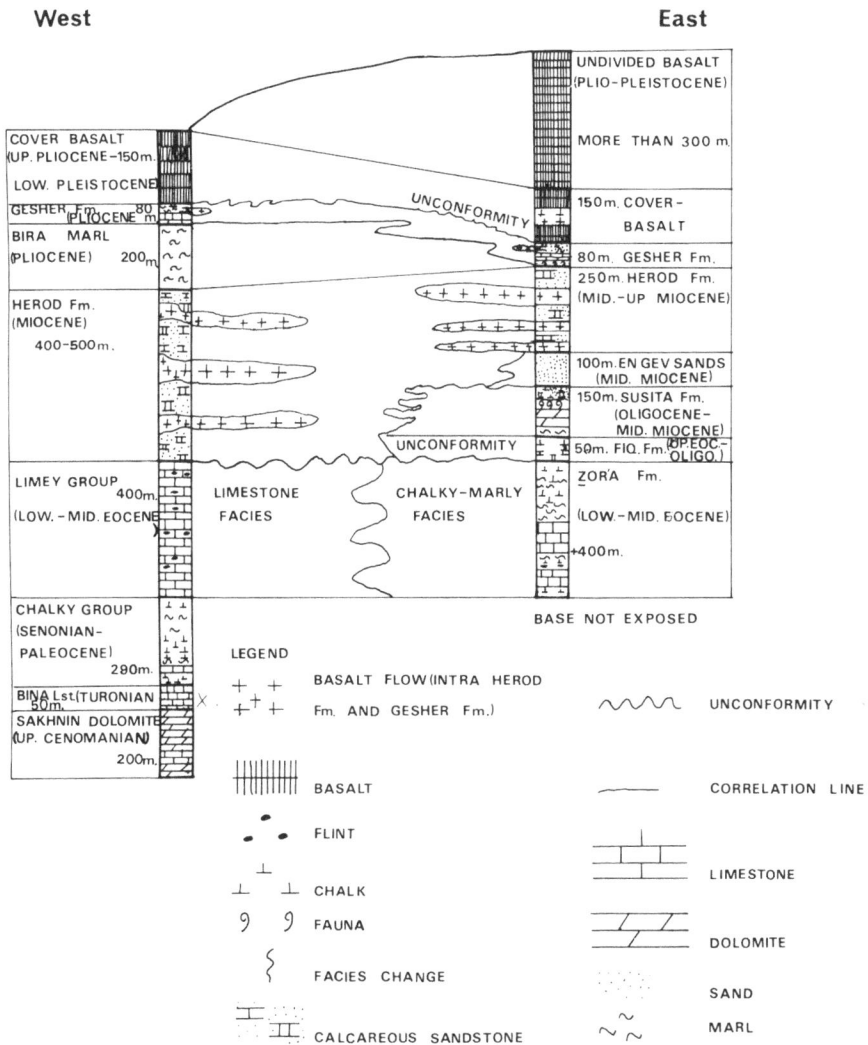

STRATIGRAPHIC RELATIONSHIPS BETWEEN
THE WESTERN AND EASTERN SIDES OF KINNERET LAKE (SCHEMATIC)

Fig. 4. Stratigraphic succession on the western and eastern sides of the lake (see text).

19

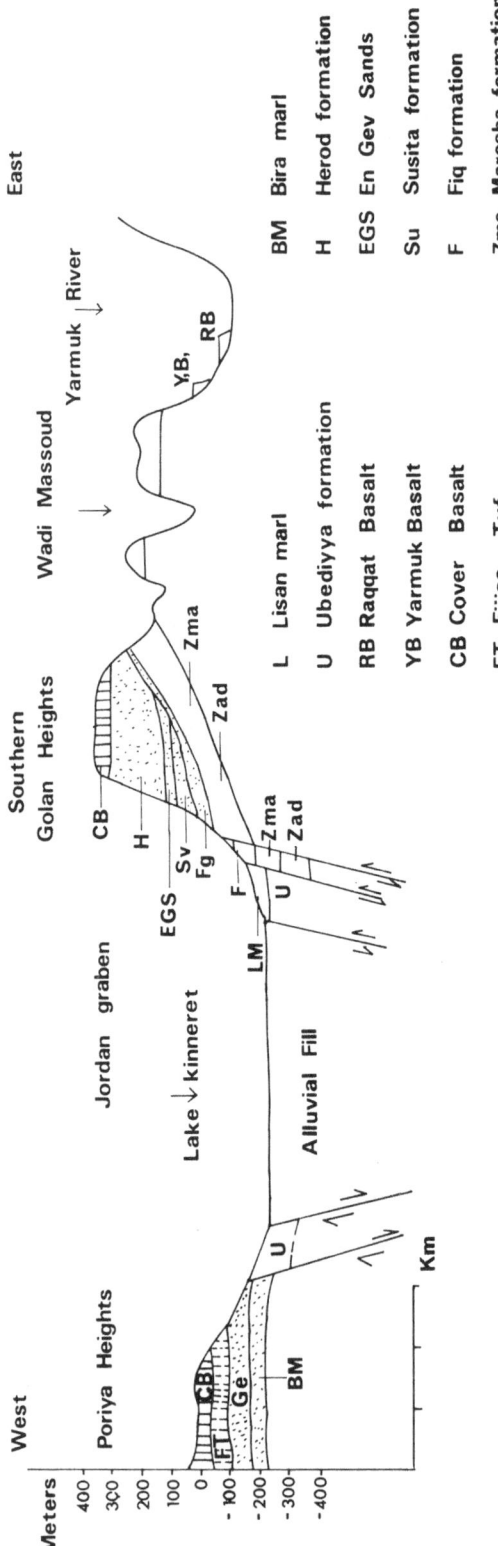

Fig. 5. Cross-section of the Lake Kinneret area.

West

Meters

Poriya Heights

Southern
Golan Heights

Wadi Massoud

Yarmuk River

East

Jordan graben

Lake ↓ kinneret

Alluvial Fill

Km

BM   Bira marl

H   Herod formation

EGS En Gev Sands

Su   Susita formation

F   Fiq formation

Zma Maresha formation

Zad Adulam formation

L   Lisan marl

U   Ubediyya formation

RB   Raqqat Basalt

YB   Yarmuk Basalt

CB   Cover Basalt

FT   Fijjas Tuf

Ge   Gesher formation

20

5. Upper Eocene–Oligocene (d): Fiq formation

Shales, detrital limestone, marls and glauconitic limestone showing traces of erosion. This formation is 50 m thick and rests in unconformity on the Zora beds; it is known only on the eastern side of the lake.

6. Miocene (m):

a. Susita formation (Su) (Oligocene–Middle Miocene)

Sandy dolomite, detrital limestone, lumashels, quartzite and shales which were deposited by a marine transgression intruding from the Persian Gulf into the Jordan depression. This formation, which reaches 150 m, has been found also in South Syria but is not known in areas west of the Jordan depression.

b. Ein Gev formation (E.G.S.) (Middle Miocene)

Yellow quartz sand with a few beds of sandstone which were deposited during rainy periods in limited areas of active erosion and may reach 90 m. The Ein Gev sands are known in the southeastern bank of the lake and rest in unconformity on the Susita formation.

c. Horodus formation (H) (Middle–Upper Miocene)

Sandstone, conglomerates, limestone and reddish, yellow, green and white clays. This formation constitutes the cliffs on both sides of the southern part of the lake. The environment of deposition was fluvio lacustrine. On the eastern side of the lake, on the Golan Heights, the Horodus formation is 250 m thick. On the western side, in the Poriya area, it exceeds 400 m. Three to four basalt flows (lower basalt) are intercalated in the sediments of the Horodus formation, which rests in unconformity on older layers.

7. Pliocene–Pleistocene (p–q):

a. Bira marl formation (BM)

The marls and gypsum of this formation give a landscape of smooth hills. These sediments were deposited in lagoons or shallow seas which resulted from a transgression of the Mediterranean Sea through the Qishon Bay and the Jezreel Valley, and supplied marine fossils (Ostrea). This formation is known in the area southwest of the lake. On the eastern side of the lake, only relics of Ostrea lumashels are known.

b. Gesher formation (ge)

This formation consists of oolithic limestone and marls with freshwater fauna

(Ostracoda, Melanopsis and Unio). These sediments were deposited in lakes on both sides of the present lake and reach 80 m in thickness.

c. Fijjas tuff (FT)

This formation corresponds to the stratified tuff outcropping between Poriya and the lake. Its thickness reaches 50–70 m.

d. Cover basalt (CB) (Upper Pliocene–Lower Pleistocene)

These basalts, which rest in unconformity on previous formations, cover the Golan Heights and the Poriya area. They reach a thickness of 200 m near Yavniel but are only a few meters thick on the hills near Huqqoq.

e. Young basalt (YB) (Middle and Upper Pleistocene)

These younger basalts are a few hundred meters thick and are found in the Yarmouk Valley and in the central and northern part of the Golan Heights.

f. Ubediyya formation (Ub) (Middle Pleistocene)

This formation, composed of layers of chalk and marls containing freshwater fauna, is restricted to the Jordan depression, where it reaches 200 m. This is the youngest layer which has been influenced by the tectonic activity of the graben.

g. Lisan marls (LM) (Upper Pleistocene: 70,000 to 20,000 years before present)

The Lisan beds are composed of varved marls and clays. The layers are horizontal and lay in unconformity on the Ubediyya formation. The Lisan formation is restricted to the Jordan depression; its northern limit is the Tiberias area.

8. Alluvium (from Pleistocene to recent) (see Section D)

C. Tectonics

More than a century of controversy has accompanied the tectonical interpretation of the Jordan–Dead Sea area. In 1869, Lartet interpreted the Jordan Valley as a sinistral wrench, an idea which was developed later by Dubertret (1932) and Freund (1968, 1970, 1973). Conversely, Picard (1943, 1970) sees in the Jordan Valley a tensional rift valley. It is clear that the paleogeography of the area is very different in both conceptions.

1. The paleogeography of the Kinneret area based on the concept of a tensional Rift valley

H. Michelson

The limestone and dolomites of the Cenomanian–Turonian period were deposited in a shallow marine environment. The late Turonian tectonic movements uplifted the sediments and restricted the extension of the sea. The Senonian–Paleocene chalks and marls were deposited in a deeper sea on folded structures (Saltzman, 1964). The folding and uplifting movements which took place at the end of the Senonian–Paleocene period induced the formation of synclinal depressions where the Eocene sediments were deposited (depressions of Safed–Migdal and Irbid–Southern Golan Heights). The anticlinorium of Galilee, extending in a NNE–SSW direction, was also formed during this folding phase. One of the anticlines of this complex extends in a N15E direction from Mount Herodus (near the Hot Springs of Tiberias), Tabgha, Kfar Hanassi and Rosh Pina to the 'windows' of Shamir in the north. Part of this anticline is at present buried under the Kinneret. The Cenoman–Turonian formation outcrops along its axis, for example, at Tel Rakat and at Mount Herodus. In the Korazim hills, the Cenomanian formation was found at a depth of 200–300 m. This anticline, which was probably formed during the Upper Senonian, had a considerable importance in the later history of the Kinneret basin. It divided the area into two sedimentation basins: the relatively deep eastern basin, where marls, marly limestone and flint were deposited (Irbid–Southern Golan), and the shallower western basin, where mainly fine grained and hard limestone was deposited (Tiberias, Migdal, Rosh Pina). The folding and uplifting which took place in the Middle Eocene brought the depressed areas of Irbid and Arbel–Sfad almost to their definitive forms.

This orogenic phase brought to an end the large regional marine transgressions which declined in space and time. In the Upper Eocene and Oligocene, a transgression from the Indian Ocean invaded the Southern Golan and the eastern side of the Jordan Rift Valley and deposited the marine sediments of shallow facies of the Fiq and Susita formations. This transgression was limited to the west by the Herodus–Shamir anticline; more limited folding and uplifting movements took place also during the Oligocene and early Neogene times, when the formations of Susita, Ein–Gev and Herodus were deposited. These movements were marked by erosive and angular unconformities and by faults.

During the Neogene, the area was partly covered by inland lakes, in which large amounts of clastic material were deposited by powerful streams draining the uplifted hinterland. The Neogene was a period of intensive volcanic activity. It is possible to distinguish two Neogene basins: the eastern basin (southern Golan Heights) and the deeper western Neogene basin (Tiberias, western Jordan Valley). These two basins were separated by the previously-mentioned anticline. The last marine transgression was episodic and reached the region during Pliocene times. It penetrated the area from the Qishon Bay through the Jezreel Valley. After the connection was cut off, brackish lagoons developed and the Bira marls were deposited. Tongues of this last marine transgression reached as far as the Ein Gev region. By the end of the Pliocene, the depressions were filled up and the Cenomanian ridge ceased to be a topographic barrier. Freshwater lakes covered the area and the Gesher formation, rich in freshwater fauna, was deposited.

In the late Pliocene and early Pleistocene, the first stages of tectonic movement were accompanied by intense volcanic activity. Basalt flows covered the Golan, the Hauran and eastern Galilee. A new phase of volcanic activity is known to have occurred in the Middle and Upper Pleistocene, accounting for the numerous volcanoes on the Golan and the Hauran.

The main tectonic phase which produced the Jordan graben in its present form took place during the Pleistocene. The faults of the graben, whose stratigraphic throw may reach several hundred meters, formed uplifted areas high above the narrow, deep faulted graben. The deep graben was filled during the Pleistocene with the material eroded from the upfaulted areas (pebbles, gravels, marls and clay) to a thickness of several hundred meters. The formation of the Jordan graben, and the consequent subsidence of the base level by several hundred meters, increased the erosive action of the rivers to such an extent that in a few cases, real canyons were formed. The last phases of tectonic activity in the Jordan Rift took place in the Middle Pleistocene. The renewal of the fault activity resulted in the tilting and faulting of the Ubediyya formation. The salty Lisan Lake covered the depression from Hazeva in the south (south of the Dead Sea) to Tiberias in the north. This lake was formed approximately 70,000 years ago, and dried up 20,000 years ago. In the Lisan Lake, evaporites, clays and marls were deposited. The horizontal position of the Lisan marl formation testifies that quiescent tectonic conditions prevailed in the Upper Pleistocene.

The freshwater Lake Kinneret is relatively very young (no more than 20,000 years) and is mainly of tectonic origin.

In this concept, the outcrops of the Cenomanian–Turonian of Mount Herodus and Tabgha are interpreted as an anticlinal structure which separated two basins of sedimentation. Moreover, the formation of the graben is a recent event which occurred during the Quaternary period. From a structural point of view, Picard (1970) sees in the trench-like feature of the Jordan Valley, its low seismicity, and the absence of areas of mylonitization, arguments which justify its origin as a graben limited by normal vertical faults.

2. The concept of a sinistral megashear

R. Freund

a. Evidence for the shear along the Lake Kinneret section of the rift

The Jurassic and Cretaceous sequences of Mount Hermon differ from those of Lebanon and Galilee and resemble those of the Judean mountains, opposite which Mount Hermon is placed by the restoration of the 105 km left-hand shear along the Dead Sea Rift (Freund *et al.*, 1970). It is, therefore, clear that this 105 km, post-Cretaceous shear affects also the two sides of Lake Kinneret. The post-Miocene stage of the shear is marked by a 40–45 km left-hand offset of several Miocene rock bodies (Fig. 6), such as the lake deposits and puddingstone of Zahle from Kefar Gil'adi to Kh. Kanafar, the Tanur red beds from the Hordos forma-tion at Hukok, the southern boundary of the Lower Basalt from northeast Kinneret to Beit Shean, and the Hordos formation from Ein Gev to Wadi Malih (Freund *et al.*, 1968). Several formations of Oligocene and Miocene age, which occur only on the east side of Lake Kinneret (Michelson, 1972) and have not been found so far anywhere on the west side, neither support nor disprove the 43 km post-Miocene shear.

The 15–20 km left-hand offset of the Cover Basalt (Fig. 7d) marks the last recorded stage of the shear movement in the region. This shear movement took place during the last two million years according to K-Ar dating (Siedner, 1973). This last part of the shear movement is supposed to have caused the formation of the fault trough Lake Kinneret, which is younger than the Cover Basalt.

b. Statement of the structural problems

It has been suggested (Freund *et al.*, 1968) that the grabens of Lake Kinneret–Beit Shean, Hula and Marj Ayun are rhomb-shaped grabens formed by shear move-ment along the 'en-echelon' faults-segments of this part of the Dead Sea–Jordan Rift (Fig. 7a). The 'en-echelon' offsets of these three grabens, and their separation by saddles of Rosh Pinna and Kefar Gil'adi, are explained by this simple model. However, careful examination reveals that while the Hula Valley is a rhomb-shaped graben whose length and width are compatible with the amount of shear and the offset of the 'en-echelon' fault segments, Marj Ayun is not rhombic but triangular, and its dimensions are far too small, while the Lake Kinneret–Beit Shean graben is several times too long and has an extra widening at Ginosar. Moreover, this model does not explain the occurrence of compressional features such as the flexure on the northeast side of the lake and the monocline on the slopes of Mount Cna'an. It does not account for the structural contrast between the undeformed Golan Heights and the intensely faulted Galilee, nor for the locations of the volcanic eruption centers in the Golan Heights and Galilee.

Furthermore, there are no indications of older troughs of the same nature, which should have been formed if the c. 80 km pre-Cover Basalt shear had taken

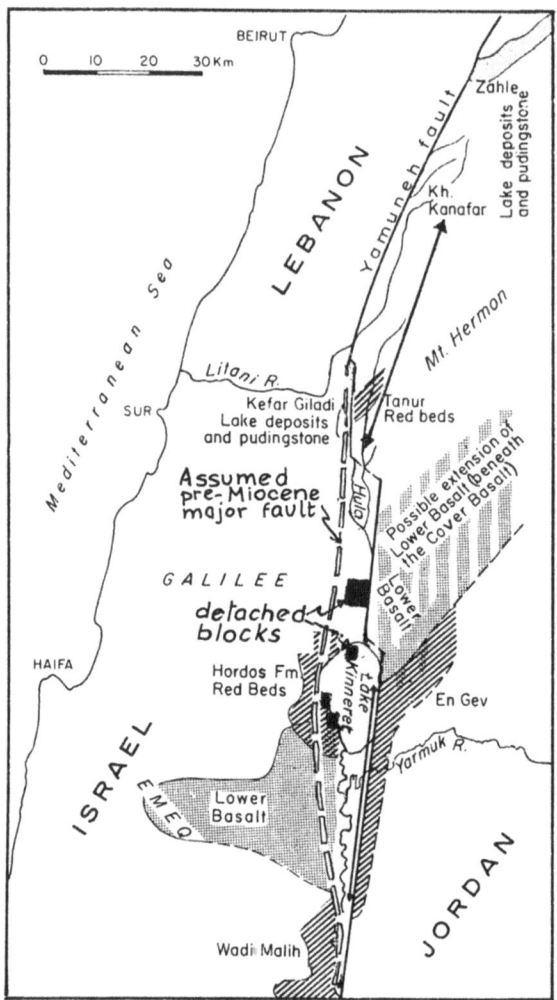

Fig. 6. Post-Miocene 40–45 km left hand shear indicated by the following pairs: Wadi Malih–En Gev red beds, Lower basalt, Hordos–Tanur red beds and Kefar Gil'adi–Kh. Kanafar Lake deposits and puddingstone. The assumed pre-Miocene fault is shown in dashed line, west of the detached blocks of Har Hordos, Fulliya, Barbutim and Rosh Pinna (black squares).

place along the same configuration as the rift faults. This model does not explain the pre-Neogene deformation of the blocks of Mount Hordos, Fulliya, Barbutim and Rosh Pinna. These blocks are supposed to constitute an anticline (Saltzman, 1964), but their fragmented distribution and their low topographical position compared to their structural height casts some doubt on this opinion.

c. Post-cover basalt structure of Lake Kinneret

The different shapes and sizes of the three grabens can be accounted for by rotating the eastern side 3° anticlockwise around a pivot situated between Hula

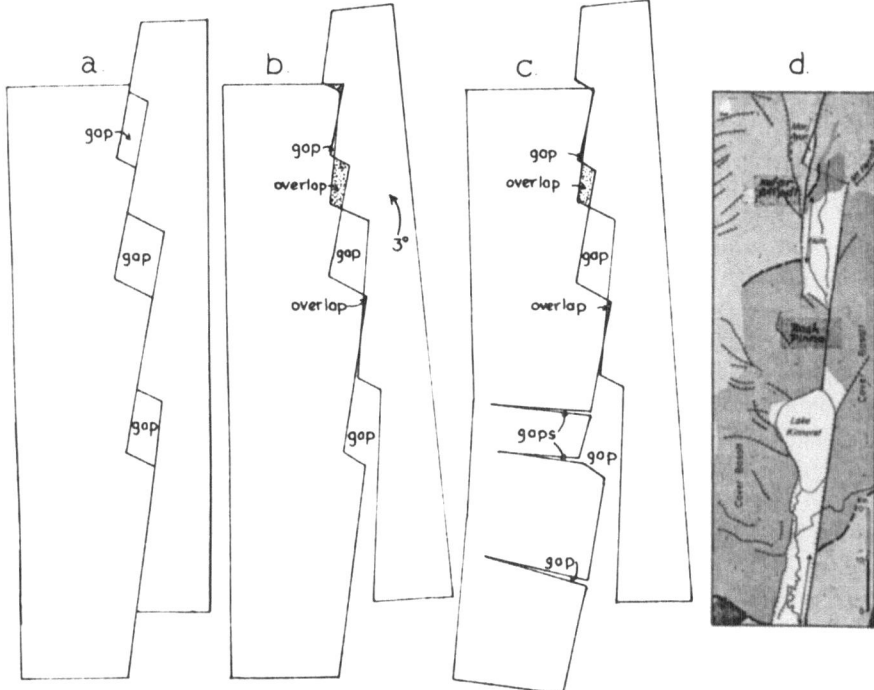

Fig. 7. a. Shear movement along 'en echelon' fault movement forms rhomb-shaped grabens of equal length. b. The same combined with rotation reduces the northern gap and expands the southern one with overlaps occurring between the gaps. c. The same with westwards bending of the west (Galilee) side by transversal normal faults, makes the southern gap wider. d. Simplified geological map compares with model C: 20 km offsets of the southern and northern boundaries of Cover Basalt are shown.

and Kinneret, in addition to the left-hand shear (Fig. 7b). Thereby, the northern gap is reduced to a small triangle resembling Marj Ayun, and the southern gap becomes much larger with an excess widening at its northern end. This gap is not unlike the shape of the Lake Kinneret–Beit Shean graben, although slightly too narrow. The width of the southern gap can be increased by bending the Galilean side westward (Fig. 7c). This westward bending is formed by the westward tapering narrow gaps, whose counterparts in nature are the large transversal normal faults in Galilee (Freund, 1970). The northward extension and westward bending of Galilee is attributed to the compression caused by the shear movement along the oblique Yamuneh Fault in Lebanon (Freund *et al.*, 1970). In this way, the structure of Galilee and that of the rift beside it are linked.

This model (Fig. 7b, c) shows overlaps between the gaps, implying that compressional features should occur in these places. This is indeed the case in the Rosh Pinna and Kefar Gil'adi saddles. In the Rosh Pinna saddle, the compression is reflected by the monocline of Mount Cna'an and by the tilt of the young Pleistocene beds at Gesher Bnot Ya'akov (Picard, 1963; Horowitz, 1973). The compression in the Kefar Gil'adi saddle is indicated by the tight folds of the Neogene lake deposits (Rosenberg, 1960) and the intense shear of the Ghareb–Taqiya formations near the Tanur waterfalls. This model does not

29

explain, however, the flexure on the northeast side of Lake Kinneret. The explanation for this structure has yet to be sought.

d. The location of the volcanic eruptions

Although it may be premature to tackle the problem of the volcanic eruptions outside the rift, an attempt has been made (Freund, 1973), following Gass (1973). The N–S shear movement of Transjordan (Arabian block) can be replaced by northeast extension (or northwest compression), where the shear movement is blocked by an obstacle. The deviation of the Dead Sea Rift fault to the northeast along the Yamuneh Fault in Lebanon provides the required obstacle, and the ensuing extension should lead to the formation of northwest trending fissures originating at the said obstacle. These fissures could provide the conduits for the volcanic eruptions. The northwest trending lines of eruption centers from Mount Hermon to Wadi Sirhan may have originated in this way.

e. Pre-cover basalt structure

It has been shown above that the last 10–20% of the shear movement after the Cover Basalt are supposed to have formed the three 'en-echelon' grabens of Lake Kinneret, Hula and Marj Ayun. Since the shear movement during earlier stages is much larger (about 80–90% of the total shear movement), it should be expected that large pre-cover basalt depressions occur along this stretch of the rift, and that they are recorded by thick sequences of sediments of these ages. The length of the rift from the northern end of Marj Ayun to the southern end of the Lake Kinneret–Beit Shean graben (northern end of Wadi Malih anticline) is 110 km, so that this part may be considered, on purely geometrical grounds, as a rhomb-shaped graben formed by the entire 105 km shear. However, it must be admitted that the stratigraphic record does not provide any solid evidence in this direction. The thickest sequences of the Miocene Lower Basalt and Hordos formation red beds, as well as the Pliocene Um Sabune, Bira and Gesher formations and the Plio-Pleistocene Cover Basalt, occur in the eastern Galilee, Emeq Yizre'el and Golan Heights, far away from the rift as well as at the rift or near it (Schulman, 1962). Most investigators (Picard, 1943, 1963, 1965; Schulman, 1959, 1962) concluded, therefore, that this section of the rift did not exist prior to the Pleistocene.

There is, thus, an apparent contradiction between the evidence for a pre-Cover Basalt shear and the absence of stratigraphic evidence for fault troughs of that age, along the northern Jordan Rift Valley. It may, of course, be claimed that the required thick sequences occur exactly underneath the present-day depression, which have never been drilled through deep enough, but this is rather unlikely. It is more probable that the location and orientation of the major fault, as well as the direction of the shear movement during pre-Cover Basalt times and pre-Miocene times, were different from the later ones.

Another argument about this fault concerns the Jurassic and Lower Cretaceous sequence in the Rosh Pinna 1 borehole, which resembles that of the western slope of Mount Hermon (Hasbaya), 40 km north on the east side of the rift, and that of Wadi Malih, 60 km south on the west side, more than the adjacent Galilean sequence. This has led to the suggestion that the Rosh Pinna block is detached from both sides (Freund *et al.*, 1970), with a strike slip movement of 60 km between the Rosh Pinna block and Galilee on a conjectured fault which is probably covered by the Neogene strata. The 60 km pre-Miocene shear thus occurred on the western side of the rift, whereas the later one took place on the eastern side of the rift.

It is proposed that the pre-Neogene fault has followed the thick dashed line on Fig. 6. In this way, there were no 'en-echelon' offsets of the major fault at that time, and hence, there was no cause for the formation of rhomb-shaped grabens. If the shear movement followed the general direction of this conjectured fault, the direction of the shear during the early stages deviated by several degrees anticlockwise from the young movement (the same clockwise deviation of the direction of the shear movement in time is observed in the Gulf of Eilat and the Dead Sea section of the Rift).

According to this interpretation, the blocks of Rosh Pinna, Barbutim, Fulliya and Mount Hordos (black squares on Fig. 6) are detached from both sides. This can be the reason for the contrast between the high structural elevation and their low topographic position.

3. Conclusion (note of the editor)

We have presented the two main concepts concerning the tectonic structure of the Jordan Valley and of the lake area. The intense geological work which has been done in the recent past has not supplied any new facts which would allow us to adopt one theory and reject the other. In 1973, Vroman presented a tentative compromise between the theories of tension and of shear, but it is clear that surface geology will not be able to provide a solution to the problem. A geophysical survey of the Rift was therefore initiated in 1974, and it is hoped that the complementary information obtained from different techniques will help our understanding of the structure of one of the most fascinating regions of the Earth.

D. The quaternary evolution of the Jordan Valley

A. Horowitz

The tectonic activity and climatic fluctuations which deeply modified the Jordan Valley during the Quaternary played a major role in the distribution and evolution of the flora and fauna of the Kinneret area.

In this account, the term 'Quaternary' includes both the Preglacial Pleistocene, which began about 2.8 and continued to about 1.8 million years ago, and the succeeding Glacial Pleistocene (Horowitz, 1973, 1974; Siedner & Horowitz, 1974). Four glacial phases have been identified in the latter, and the Alpine terminology is used here, save for the replacement of the term 'Glacial' by 'Pluvial'. A discussion of the problem of correlation of the European Glacial and the Mediterranean Pluvial phases is given in Horowitz (1971). It must be stated also that the term 'Preglacial' is used here only to denote the time interval stated above, and does not have any climatic implication. In fact, at least two well-defined Glacial phases, the Biber and the Donau, have been recorded within this period.

1. The Preglacial Pleistocene

After the regression of the Late Pliocene Mediterranean, the landscape of the entire country was of a rather flat topography, crossed by several wide rivers carrying sediments from Transjordan to the Mediterranean, eroding also to some extent the Pliocene deposits. The Calabro–Sicilian ingressions caused the deposition of conglomerates in these river systems – the HaMeshar Conglomerate in the Negev and the Bethlehem Conglomerate in Judea.

The period of erosion was much shorter in the central and northern Jordan Valley, which was covered with a 200 m thick basalt sheet (Cover Basalt). The main source area for these eruptions is southwestern Syria, but dykes and vents connected with this system can also be seen in Israel, such as those on the main road to Tiberias, near Gazit and Lavi. A volcanic neck, Karne Hittin, is also thought to be part of this eruption system.

The Cover Basalt filled and flattened the pre-existing erosional and to some extent tectonic relief, so that by the end of the eruption phases, the landscape became a wide plain. In particular, the series of eruptions filled the Pliocene Jordan Valley with basalts and consequently the depression ceased to function as an erosion base level. The only erosion base level in Preglacial Pleistocene times remained the Mediterranean.

The age of the Cover Basalt is considered Preglacial Pleistocene because of its stratigraphic position, overlying unconformably the Late Pliocene Gesher Formation sediments and underlying the Günzian sediments–Erk el Ahmar Formation in the central Jordan Valley and Gadot Formation in the Hula Valley. The Cover Basalt sequence comprises, in places where it is best developed, 10–12 flows, separated by red paleosols. The lower 2–3 flows cooled under normal magnetic polarity, during the Gauss Normal Epoch that terminated 2.5 million years ago. The rest of the flows cooled under reverse magnetic polarity of the Matuyama Reverse Epoch, lasting until about 1.7 million years ago (Siedner & Horowitz, 1974).

The Preglacial Pleistocene landscape of the Hula and Kinneret areas was a broad, shallow basin, in which flows of the Cover Basalt, separated by paleosols, have accumulated. The area, it should be noted, was a single basin, as the elevated Gadot–Korazim block was not uplifted yet. No sedimentary rocks are known to have been deposited in the central-northern Jordan Valley during Preglacial Pleistocene times.

2. The Glacial Pleistocene

The Glacial Pleistocene period corresponds to the time during which the four major Alpine glaciations have occurred in southern Europe and other areas. According to the radiometric datings recently carried out in Israel (Siedner & Horowitz, 1974), and supported also by datings of deep sea cores (Smith & Beard, 1973), this period began some 1.6 million years ago.

The influence of the Alpine glaciations on the Levant climate was discussed in Horowitz (1971) and Butzer (1973), and it seems that the glacial periods resulted in a pluvial climate. The hypothesis that the glacials correspond to an arid, periglacial climate in Israel is rejected, as untenable according to the results obtained by pollen analyses. In general, sediments correlated chronologically with the Alpine glacial periods yield in Israel rich arboreal spectra that indicate a well-developed oak forest (Horowitz, 1971, 1973a, b), while the arid periglacial belt lay to the north, e.g. in Italy (Bonatti, 1966), Greece (Wijmstra, 1969) and Iran (Wright et al., 1967).

The transition from the Preglacial to the Glacial Pleistocene was accompanied in Israel by a pronounced tectonic activity (Horowitz, 1974) in a general north–south direction. Two kinds of tectonic movements occurred. The Jordan–Arava–Bay of Elat depression was greatly accentuated by means of differential subsidence, from at least the southern tip of the Bay of Elat up to northern Syria, and this resulted in the formation of a number of deep tectonic depressions. These depressions are separated by relatively elevated blocks (though these are downfaulted as well) which form either watersheds, such as the central Arava and the Metulla blocks, or areas in which the Jordan River flows, almost never flooded to form lakes except for short periods during their maximal extensions, such as the Gadot–Korazim block or the southern Jordan Valley.

The second kind of tectonic activity took place contemporaneously with the downfaulting of the Jordan–Arava rift valley and resulted in the upwarping of the central areas of Israel along a north–south line that affected the entire country, from the Galilee (or in fact, from the Lebanon) down to Sinai. This upwarping resulted in the disconnection of the entire Jordan Valley area from the Mediterranean which, from that date onwards, does not play any role in the erosional and fluviatile processes of the Jordan Valley.

Within the Jordan Valley proper, three basins can be distinguished from the accumulations of lacustrine sediments during the Glacial Pleistocene: the Hula, the

Central Jordan Valley and the Dead Sea Basin.

Within the Dead Sea Basin, about 2,000 m of Glacial Pleistocene sediments have been deposited – the Upper Dead Sea Group – in which four pluvial phases were recorded in the pollen spectra, separated by interpluvials. Within the Central Jordan Valley Basin, 600–700 m of sediments, representing the four pluvial phases, are recorded, termed the Jordan Group. In the Hula Basin, about 1,700 m of sediments and basalts were encountered, probably representing the entire Glacial Pleistocene, named the Hula Group (Horowitz, 1973a, 1974). The correlations of these formations are given in Fig. 8.

3. The Hula Basin

Since its formation at the beginning of the Glacial Pleistocene until the present day, the shape and style of the Hula Valley have changed only slightly. The Hula Group accumulated during this period in a shallow lake or in marshes, and the valley floor subsided to compensate for the burden of the sediments, so that the shallow characteristics of the Hula Basin have been maintained during almost the entire time of its existence. This is proved by analysis of the deposits comprising the Hula Group (Horowitz, 1973a) which alternate between shallow lacustrine chalk and clays in pluvial periods and organic paludine sediments of the interpluvials. The deposition basin always existed as an intermediate link en route to the Jordan River, and was never blocked by any barrier; this is proved by the nature of the sediments and their fossil faunal assemblages which indicate a freshwater environment (Tchernov, 1973). Several weak phases of intragraben tectonic movements (Picard, 1963) and limited volcanism on the rims of the Hula Valley did not affect these characteristics.

The sediments of the Hula Valley have been studied in detail both from boreholes and from outcrops (Picard, 1963; Horowitz, 1971, 1973a). Because of the continuous subsidence of the Basin's floor, the outcrops are rather limited, and only sediments deposited during periods of greater extension of the lake, namely during the pluvials, are preserved on the valley rims (Fig. 9). The pluvial characteristics of these sediments are also confirmed by their pollen spectra which indicate a well-developed oak forest, in contrast to the interpluvial spectra of sediments encountered only in boreholes which comprise a mixed maquis spectrum, poor in arboreal species. In the boreholes, sediments with pluvial and interpluvial characteristics alternate. The best outcrop of the Hula Group sediments appears on the elevated block of Gesher Benot Ya'akov, discussed in Picard (1963) and later in Horowitz (1973a). In this area, the entire pluvial sequence is represented, while the interpluvials are represented by erosional surfaces, paleosols or occasionally by conglomerates. Another area in which outcrops occur is the northern Hula Valley, where lateral equivalents of the upper part of the Hula Group are exposed as a series of spring deposits that were formed during times of stronger spring activity, in pluvial times. In general, therefore, most of the study of the Hula Group and the geological and climatic history of the Hula Valley is based on pollen analytical studies of borehole material and is summed up in Horowitz (1973a).

The oldest deposit of the Hula Group is the Gadot formation. The Formation

K-Ar ages (×10^6 yr)	Chrono-stratigraphy	Hula Basin	Central Jordan Valley
	Holocene	Malaha Formation	Tabgha Formation
	Würm	Dan Travertine / Ashmura Formation	Lisan Formation
0.064 → 0.073 → 0.079 →	R-W	Hasbani Basalt and Hulata Formation	Raqqad (Naharayim) Basalt
	Riss	Benot Ya'akov Formation and Kefar Yuval Travertine	Naharayim Formation
	M-R	Ayelet Hashahar Formation	
0.56 0.64 → 0.68 →		Yarda Basalt	Yarmuk Basalt
	Mindel	Mishmar Hayarden Formation	Ubeidiya Formation
	G-M	Palaeosol	
1.7 → 2.0 →	Günz	Gadot-Hazor Fm.	Erk el. Ahmar Fm.
	Preglacial Pleistocene	Cover Basalt	

Fig. 8. Correlation chart for the Quaternary formations of the central and northern Jordan Valley.

outcrops south of the present Hula Valley, on the Gadot elevated block. It comprises white soft limnic chalk, bearing some scarce *Melanopsis* and *Melania* shells. The chalk interfingers westwards with gravel horizons, called in Picard (1963) Hazor Gravel. The Gadot and Hazor sediments overlie unconformably the taphrogenic and erosional relief of the Cover Basalt and are overlain in the Gesher Benot Ya'akov area by a paleosol, of grey-brown color. The formations are also overlain by the Mishmar HaYarden Formation and the Yarda Basalt, discussed below. The Gadot Formation, and the laterally interfingering Hazor Gravel, attain

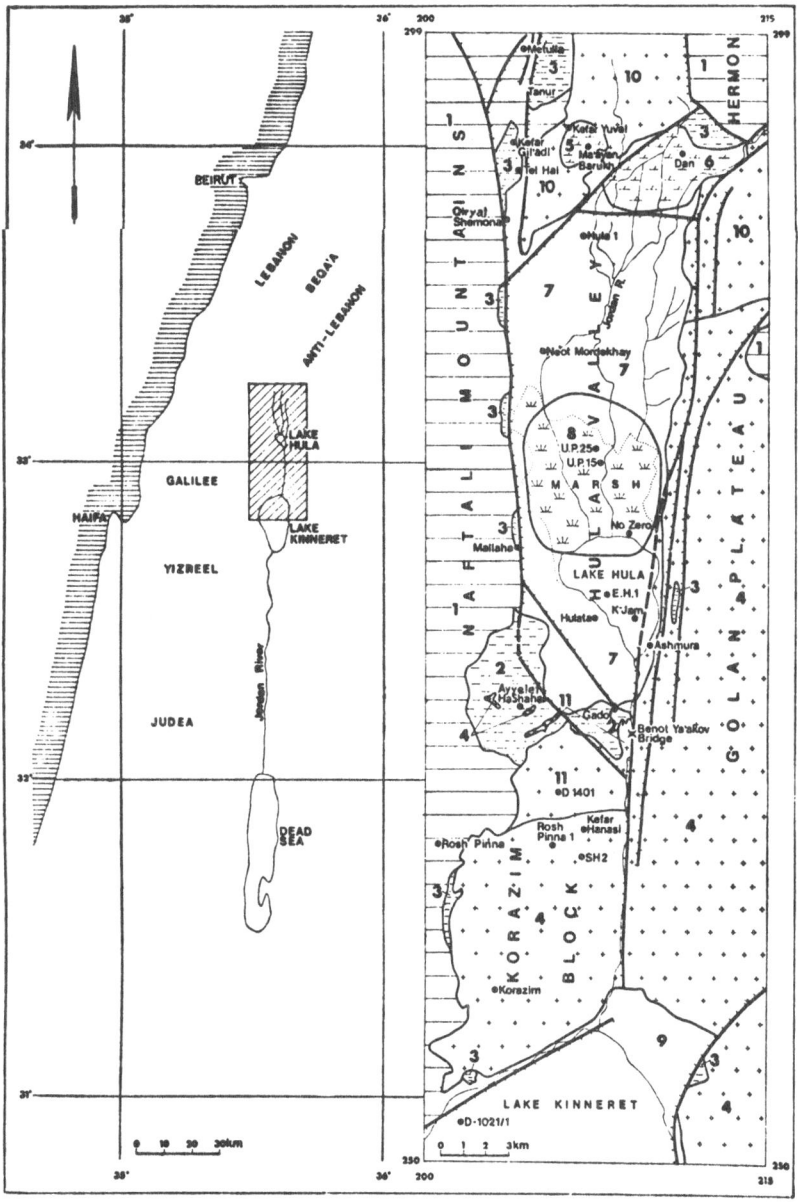

Fig. 9. General geological map of the Hula Valley. 1. Pre-Neogene formations, 2. Gadot Formation, 3. Bira Series, 4. Cover Basalt, 5. Kefar-Yuval Travertine, 6. Dan Travertine, 7. Hashmura Formation, 8. Approximate extension of the Mallaha Formation, 9. Tabgha Formation, 10. Hasbani Basalt, 11. Yarda Basalt. (from Horowitz, A. 1973, Isr. J. Earth Sci. 22, 2).

20–30 m in thickness, but it is suggested that this is only a marginal development, while the entire subsurface thickness is estimated as approximately ten times greater.

The Gadot and Hazor formations seem to represent the greatest extension of the Hula Lake in Günzian times. The unnamed paleosol which covers these sediments in places is thought to represent the erosion period during the lower lake level in Günz–Mindel Interpluvial times. The Gadot Formation was never reached by boreholes in the center of the Hula Basin.

The next pluvial formation in the Hula Group is the Mishmar HaYarden Formation. It outcrops near Gesher Benot Ya'akov, where it displays a littoral lake facies, very rich in molluscs, discussed in Tchernov (1973). The most common are melanopsids, *Melania* and *Theodoxus*, which resemble, although are not identical to, the malacological assemblage of the Ubeidiya Formation of the central Jordan Valley. Bones of fossil vertebrates were encountered in this formation, but have not yet been identified. Some artifacts, vaguely suggesting the Ubeidiya types, were also found (Horowitz *et al.*, 1973).

The pollen spectra of the Mishmar HaYarden Formation are typical of a pluvial climate and vegetation, oak pollen forming more than 70% of the total spectrum and being almost the sole constituent of the arboreal pollen.

The age of the Mishmar HaYarden Formation is considered as Mindel Pluvial, based on its being the third pluvial cycle before the present and its correlation with the Ubeidiya Formation. The Formation is covered by a basaltic sheet of the Yarda Basalt that was potassium/argon dated (Horowitz *et al.*, 1973) to an age of 640,000 years.

The Yarda Basalt (Picard, 1963, 1965) covers the elevated block of Gadot and was poured off over the Gadot and Mishmar HaYarden Formations. This basalt is characterized by its normal magnetic polarity and it underlies uncomformably the Benot Ya'akov Formation. Near Gesher Benot Ya'akov, the Yarda Basalt is strongly faulted and tilted together with the underlying formations. The stratigraphic age attributed to the Yarda Basalt is latest Mindel or earliest Mindel–Riss. The radiometric age obtained helps to date this climatic event.

The next sedimentary unit is the Ayyelet HaShahar Formation, of Mindel–Riss Interpluvial age, which is defined in the Emek Hula 1 borehole from its total depth – 455 m up to 340 m (Horowitz, 1973a). The base of the Formation was not penetrated and the structural relations with the tilted underlying formations are obscure. The sediments comprise highly organic chalk, frequently grading into peat, characterized by a very low share of arboreal pollen within the spectrum. The arboreal pollen is mainly derived from oaks, olives and some pistachios. No outcrops of this formation are known, and it seems that its distribution is confined to the central part of the Hula Basin, following the shrinkage of the lake under interpluvial influences.

The Riss Pluvial is much better represented in the Hula Basin by two sedimentary formations which outcrop in the northern and southern areas. In the north, the Kefar Yuval Travertine forms a sheet underlying the Riss–Würm Hasbani Basalt. In the south, the Benot Ya'akov Formation (known also as the '*Viviparus* Beds' from its unique and rich malacological assemblage) outcrops near Gesher Benot Ya'akov, overlying a coarse base conglomerate and underlying a top conglomerate. The sediments comprise grey to brown chalk, very rich in the mollusc *Viviparus apameae*, together with other species discussed in Tchernov (1973). Vertebrate remains (Hooijer, 1959, 1960) and prehistoric industries (Stekelis, 1960) were also uncovered from this formation. The pollen spectrum is extremely rich in oak grains and indicates deposition of the Benot Ya'akov Formation under pluvial climatic conditions. The formation attains 5–10 m thickness in the outcrops, but is known from many boreholes in the Hula Basin, attaining up to 200 m in the central part of the Basin. Pollen analyses of the sequence in the Emek Hula 1 borehole indicate two pluvial phases within the Benot Ya'akov Formation separated by an interstadial, during which the climate was warmer than the pluvial but the humidity remained high, as compared with the present day.

The Benot Ya'akov Formation is overlain in the subsurface of the Hula Basin by the Hulata Formation, known also as the 'Main Peat' (Picard, 1963). The Kefar Yuval Travertine is overlain, in the northern Hula Valley, by the Hasbani Basalt, both of Riss–Würm Interpluvial age. The Hulata Formation comprises mainly thick accumulations of peat, very poor in arboreal pollen, which were deposited in the restricted area of the Hula marshes, as a result of the dry and warm interpluvial climate. Several flows of the Hasbani Basalt were potassium/argon dated, yielding ages of 70–80,000 years. This figure is in full accord with estimates for the absolute age of the late Riss–Würm, as calculated from rates of sedimentation in the Hula (Horowitz, 1971) and the Dead Sea Basin (Neev & Emery, 1967).

The Würmian is represented in the Hula Basin by the Ashmura Formation, comprising mostly white to buff limnic chalk with some peat and conglomerate intercalations, as detailed in Horowitz (1971) and other publications. In general, the pollen spectra indicate a pluvial climate, separated by two main interstadials (Table 1). In the central area of the Hula Basin, where the lake persisted until the present day, the Ashmura Formation continued to be deposited also during the Holocene. In other localities, the deposition of the Ashmura ceased, either due to a minor tectonic movement, as in the Gesher Benot Ya'akov area (Horowitz, 1973a), where it occurred some 4,500 years ago, or, as in the northern Hula Valley, due to the shrinkage of the lake following the dry post-Atlantic climate.

The shrinkage of the lake in the north gave way to the formation of extensive marshes, in which the Mallaha Formation peat (Horowitz, 1973a) was deposited.

In the northernmost areas of the Hula Valley, extensive spring activity deposited the Dan Travertine in Würmian times, overlying the Hasbani Basalt.

4. The Central Jordan Valley Basin

The Glacial Pleistocene formations of the central Jordan Valley were grouped together under the term 'Jordan Group' (Horowitz, 1974). The Group comprises, from the oldest, the Erk el Ahmar Formation, the Ubeidiya Formation, The Yarmuk Basalt, the Naharayim Formation (or Naharayim Gravel), the Raqqad (or Naharayim) Basalt, the Lisan, Tabgha and Fatza'el formations. No complete, detailed study of the Jordan Group formations is yet available. A general discussion on the sequence is given in Horowitz (1974) where the stratigraphy, paleogeography, paleoecology and correlations with other Glacial Pleistocene sequences in Israel are suggested.

The Erk el Ahmar Formation overlies the Cover Basalt and is restricted in areal distribution. The formation comprises limnic to paludine sediments, mostly brown or grey clays, with occasional sandstone and some peat. Some horizons are very rich in molluscs. The sequence attains approximately 100 m thickness, but exact figures cannot be obtained, since the base is exposed only at a single locality and the top is always truncated. The sequence is strongly faulted and tilted. Studies on

Table 1. The main paleoclimatic conclusions drawn from the pollen diagram of borehole K-Jam, Lake Hula (modified from Horowitz A., 1971, Pollen and Spores, XIII).

Depth (m)	Main palynological features	Climate	Absolute age (based on C^{14})	Stage	Extension of Lake Hula
	A. P. low O.F. medium Gram. + Cyp. medium mixed A.P. spectrum	Warm and dry	4,500–5,000		Small
8				Holocene	
	A.P. medium O.F. medium Gram. + cyp. increasing *Quercus* prevailing	Warm and humid	11,500		Medium
18					
	A.P. high, *Quercus* prevailing O.F. high, Gram. + Cyp. low	Cool and humid pluvial	16–18,000	Late Würm	Great
25					
	A.P. medium, various components O.F. medium Gram. + Cyp. high *Thelyp.* pal. medium	Warm and humid– interstadial	20–22,000	Mid-Late Würm Interstadial	Small
35					
	A.P. high O.F. high Gram. + Cyp. low *Quercus* prevailing	Cool and humid– pluvial	28–32,000	Mid-Würm	Great
50					
	A.P. medium O.F. medium Gram. + Cyp. high *Thelyp.* pal medium *Quercus* and *Olea* prevailing	Warm and humid– Interstadial	45–50,000	Early-Mid Würm Interstadial	Medium
82					
	A.P. high O.F. high Gram. + Cyp. low *Quercus* prevailing	Cool and humid– pluvial	60–70,000	Early Würm	Great
95					
	A.P. very low, various components O.F. very low Gram. + Cyp. prevailing *Thelyp.* pal. high	Warm and dry– interpluvial		Riss–Würm Interglacial	Small
123.5					

the lithology and areal distribution (U. Baida, Tahal, Water Planning for Israel, Tel-Aviv), the fossil charophytes (Y. Lipkin, Department of Botany, Tel-Aviv

University) and the fossil molluscs (E. Tchernov, Department of Zoology, Hebrew University of Jerusalem) are in preparation. A pollen spectrum of the Erk el Ahmar Formation, from a peat horizon, is given.in Horowitz (1973b) and indicates a pluvial climate. The exact relationship of the Erk el Ahmar to the Ubeidiya Formation is not clear in the field, since the two outcrop at different localities. It is, however, clear that the Erk el Ahmar is older, as indicated both by the mollusc and charophyte assemblages. Since the Erk el Ahmar is older than the Ubeidiya Formation, to which a Mindel age was attributed (*cf.* below), and overlies the Cover Basalt, a Günzian age was tentatively implied (see also discussions in Horowitz, 1973b, 1974).

The Ubeidiya Formation has been dealt with by many authors, mainly because of the enormous amounts of human implements uncovered from its sediments. The lithology and type-sections are given in Picard & Baida (1966), who indicate the existence of two major cycles within the formation, alternating from lacustrine to fluviatile. The sediments comprise mainly chalk in the lacustrine members and conglomerates in the fluviatile. The topmost horizons seem to comprise paleosols. The faunal remains and the paleoecological implications, as well as some leaf impressions and a pollen spectrum, are discussed in Bar–Yosef & Tchernov (1972). The Middle age for the Ubeidiya Formation is based on its pluvial charactristics, being the third pluvial phase down from present (Horowitz, 1973a, b, 1974), on the occurrence of some Middle Pleistocene horses, *Equus stenonis* (Bar–Yosef & Tchernov, 1972), and on the palynological correlations with the Mishmar HaYarden Formation (Horowitz, 1973a).

The Yarmuk Basalt occurs as terraces in the Yarmuk gorge. It is magnetically normal and yielded a potassium/argon age of 690,000 years (Horowitz *et al.*, 1973). Geomorphologically, this basalt occurs higher than the Ubeidiya Formation sediments, although no direct connection between the two can be observed in the outcrops. It was suggested (Horowitz, 1973a) that the Yarmuk Basalt is younger than the Ubeidiya Formation and should be correlated with the Yarda Basalt. The latter, as mentioned above, overlies the Mishmar HaYarden Formation which is regarded as contemporaneous with the Ubeidiya. The radiometric dates prove this assumption.

The Naharayim Formation comprises conglomerates and paleosols which overlie the Ubeidiya and the Erk el Ahmar formations. These comprise a series of riverfills indicating stronger activity of the paleo-Yarmuk, most probably as a result of higher humidity. The Naharayim Formation is attributed a Riss age, based on its pluvial characteristics and its underlying the Würmian Lisan Formation.

The Naharayim Formation is at most localities overlain by the Lisan Formation, but in the vicinity of Naharayim the two are separated by a basalt sheet, called in Picard (1943) the Naharayim Basalt. Michelson (1973) identified this basalt as the same as that which, much earlier was called the Raqqad Basalt and indicated more outcrops, upstream of the Yarmuk and its tributary. the Wadi Raqqad. The basalt is magnetically normal, and is attributed a Riss–Würm Interpluvial age, corresponding to the Hasbani Basalt of the northern Hula Valley.

The Lisan Formation is described in detail in Neev & Emery (1967) from the Dead Sea area, but maintains almost the same characteristics elsewhere. It occurs

41

both in the central and southern Jordan Valley, continuously from the southern end of the present Lake Kinneret down to Hazeva in the Arava. During the maximal extension of the Lisan Lake, in early Würm, tongues of the formation were deposited up to the central part of the present day Kinneret. The Lisan Formation comprises finely laminated chalk and clays, grading laterally into conglomerates, which were deposited in a rather salty lake. Fossils are rare, but pollen assemblages indicate a pluvial climate of the surroundings. Absolute ages indicate that the Lisan Formation was deposited from about 60,000 up to 18,000 years ago, and its deposition ceased due to tectonic activity which resulted in a considerable deepening of the Dead Sea Basin and the formation of the Kinneret Basin, into which the Lisan Lake waters drained. Due to the dryer conditions in post-Würmian times, the Lisan Lake never came back to its original configuration.

Sediments deposited since post-Würmian time in the Kinneret Basin were designated in Horowitz (1971) as the Tabgha Formation (Table 2). Alluvial fans, in which red loams and conglomerates have been accumulated in the central Jordan Valley during the late Würm humid phase, were designated (Horowitz, 1973b, 1974) as the Fatza'el Formation. The Tabgha Formation comprises grey to black clays and silts and was continuously deposited during the last 18,000 years, while the Fatza'el Formation clastics accumulated from about 16,000 up to 12,000 years ago.

Table 2. The main conclusions drawn from the pollen diagram of borehole D-1016/2, Lake Kinneret (modified from Horowitz A., 1971, Pollen and Spores, XIII).

Depth (m)	Main palynological features	Climate	Absolute age	Stage
	A.P. medium O.F. medium Gram. + Cup. low Comp. medium	Warm, humidity increases		
4				Holocene
	A.P. low O.F. high Gram. + Cyp. medium Comp. high	Warm and dry	11,500	
9				
	A.P. high O.F. low Gram. + Cyp. high Comp. low	Cool and humid	16,000	Late Würm
13				
	A.P. low O.F. medium Gram. + Cyp. medium Comp. high	Warm and dry		Mid-Late Würm Interstadial

5. Conclusion

The northern and central sectors of the Jordan Valley display somewhat different lines of development during the Glacial Pleistocene. The Hula Basin was always covered by water: either lakes in pluvial times, or marshes during the interpluvials, in which sediments of the Hula Group have been continuously deposited. The Basin's floor subsided in response to the burden of the sediments, giving way to shallow water bodies during almost the entire period. The center of the Hula Basin is always located at approximately the same place, resulting in the accumulation of 1,600–1,700 m of sediments.

The Central Jordan Valley Basin, on the contrary, was occupied by lakes only during the humid pluvials, which resulted in a much more restricted sedimentary column, of the order of several hundred meters thickness. This Basin was also much more influenced by the tectonic activity during the Glacial Pleistocene. The strong faulting of Mindel–Riss times, which affected so profoundly the Erk el Ahmar and Ubeidiya formations, probably opened the Basin towards the Dead Sea, so that during the succeeding Riss Pluvial, no lake was formed in the area, and only fluviatile deposits accumulated – the Naharayim Formation. The Late Würmian faulting phase changed totally the location of the Basin's center, so that Lake Kinneret is in fact situated north of the position of most of its preceding lakes, and the Tabgha Formation is deposited over Preglacial Pleistocene and older formations.

Many uncertainties remain in our knowledge of the formation of the Rift area. However, it is clear that very recent tectonic events have modified considerably the regional hydrologic regime and the salinity of the waters. These events had considerable effects on both the organisms which developed in the successive water bodies of the Rift and the present freshwater organisms of Lake Kinneret.

References

Bar–Yosef, O. & E. Tchernov. 1972. On the paleoecological history of the site of Ubeidiya. Israel Acad. Sci. Hum. 35 p.
Bonatti, E. 1966. North Mediterranean climate during the last wurm glaciation. Nature. 209:984–985.
Butzer, K. W. 1973. Patterns of environmental change in the Near-East during Late Pleistocene and Early Holocene times. Proc. Intern. Conf. On NE African and Levantine Pleistocene prehistory, Dallas. 63 p.
Dubertret, L. 1932. Les formes structurales de la Syrie et de la Palestine; leur origine. C. R. Acad. Sci. Colon. 195:65–67.
Freund, R. 1970. The geometry of faulting in the Galilee. Isr. J. Earth Sci. 19:117–140.
Freund, R. 1973. Comments to 'The Red Sea depression, causes and consequences' by I. G. Gass. In: Implication of Continental Drift to the Earth Sciences. D. H. Tartling & S. K. Runcorn (eds.) 2: p. 786.
Freund, R., I. Zak & Z. Garfunkel. 1968. Age and rate of the sinistral movement along the Dead Sea Rift. Nature. 220:253–255.
Freund, R., Z. Garfunkel, I. Zak, M. Goldberg, T. Weissbrod & B. Derin. 1970. The sheer along the Dead Sea Rift. Phil. Trans. Roy. Soc. London. A, 267:107–130.
Gass, I. G. 1973. The Red Sea depression, causes and consequences. In: Implication of Continental Drift to the Earth Sciences. D. H. Tartling & S. K. Runcorn (eds.).
Hooijer, D. A. 1959. Fossil mammals from Jisr Banat Yaquv, south of Lake Huleh, Israel. Bull. Res. Counc. Isr. 8G:177–199.
Hooijer, D. A. 1960. A Stegodon from Israel. Bull. Res. Counc. Isr. 9G: 104–108.
Horowitz, A. 1971. Climatic and vegetational developments in northeastern Israel during Upper Pleistocene–Holocene times. Pollen and Spores XIII:255–278.

Horowitz, A. 1973a. Development of the Hula Basin, Israel. J. Earth Sci. 22: 107–139.

Horowitz, A. 1973b. The pleistocene paleoenvironments of Israel. Intern. Conf. NE African and Levantine Pleistocene prehistory. Dallas. 65 p.

Horowitz, A. 1974. The Late Cenozoic stratigraphy and paleogeography of Israel. Institute of Archaeology, Tel Aviv University. 187 p.

Horowitz, A., G. Siedner & O. Bar Yosef. 1973. Radiometric dating of the Ubeidiya formation, Jordan Valley, Israel. Nature. 242 : 186–187.

Lartet, L. 1869. La Geologie de la Palestine (Thèse) Paris. 292 p.

Michelson, H. 1972. The hydrology of southern Golan Heights. Water Planning for Israel Ldt. (TAHAL) Tel Aviv. Intern. Report.

Michelson, H. 1973. Yarmuk basalt and Roqqat Basalt. Two volcanic phases which flowed through preexisting gorges. Isr. J. Earth Sci. 22 : 51–58.

Neev, D. & K. O. Emery. 1967. The Dead Sea. Bull. Geol. Surv. Isr. 41 : 1–147.

Picard, L. 1943. Structure and evolution of Palestine. Bull. Geol. Dept. Hebrew Univ. 4 : 1–187.

Picard, L. 1963. The quaternary in the northern Jordan Valley. Proc. Isr. Acad. Sci. Hum. 1(4) : 1–34.

Picard, L. 1965. The geological evolution of the quarternary in central-northern Jordan Graben, Israel. Geol. Soc. Amer. Sp. paper 84 : 337–366.

Picard, L. 1970. Further reflections on graben tectonics in the Levant, Int. Upper Mantle proj. Sci. Rep. 7 : 249–267.

Picard, L. & U. Baida. 1966. Stratigraphic position of the Ubeidiya formation. Proc. Isr. Acad. Sci. Hum. 4 : 1–8.

Rosenberg, E. 1960. Geologische Untersuchungen in den Naftali–Bergen. Mitt. E. T. H. Zurich C. 80.

Saltzman, U. 1964. Geology of the Tabgha–Huquq–Migdal region. M. Sc. thesis Hebrew Univ. Jerusalem. (in Hebrew).

Schulman, N. 1959. Geology of central Jordan Valley. Bull. Res. Counc. Israel. 8G : 63–90.

Schulman, N. 1962. The geology of the central Jordan Valley. Ph.D. thesis, Hebrew Univ. Jerusalem. (in Hebrew).

Siedner, G. 1973. K-Ar chronology of cenozoic volcanics from northern Israel and Sinai. Fortsch. d. Mineralogie 50 : 129–130.

Siedner, G. & A. Horowitz. 1974. Radiometric ages of late Cenozoic basalts from northern Israel: chronostratigraphic implications. Nature. 250 : 23–26.

Smith, L. A. & Beard, J. H. 1973. The late Neogene of the Gulf of Mexico. In: Worzel, J. L. et al. Initial reports of the deep sea drilling project, X : 643–677.

Stekelis, M. 1960. The Paleolithic deposits of Jisr. Banat Yaquv. Bull. Res. Counc. Isr. 96 : 61–87.

Tchernov, E. 1973. On the fossil molluscs of the Jordan Valley. Proc. Isr. Acad. Sci. Hum. 1–47.

Vroman, A. J. 1973. Is a compromise between the theories of tension and of shear for the origin of the Jordan Dead Sea trench possible? Isr. J. Earth Sci. 22 : 141–156.

Wijmstra, T. A. 1969. Palynology of the first 30 m. of a 120 m. deep section in northern Greece. Acta Bot. Neerl. 18 : 511–527.

Wright, H. E. Jr., J. H. McAndrews & W. van Zeist. 1967. Modern pollen rain in western Iran and its relation to plant geography and Quaternary vegetational history. J. Ecol. 55 : 415–443.

III Meteorology

45

A. History of climatological measurements

G. Stanhill & J. Neumann

The first series of climatological measurements at Lake Kinneret were made at the Scottish Mission Hospital in Tiberias on the shores of the lake (32° 48'N, 35° 32'E, −205 m MSL) from 1890 to 1911. They were followed in 1910 by a series of measurements at the German Templar colony of Tabgha on the northern shore (32° 52'N, 35° 32'E, −200 m MSL) which continued until 1916. In the same decade, rainfall measurements were started in the Jewish agricultural settlements of Kinneret (32° 43'N, 35° 34'E, −170 m MSL) and Migdal (32° 49'N, 35° 34'E, −206 m MSL).

Routine and standard climatological measurements were started at Deganya A near the southern shore of the lake close to the River Jordan (32° 43'N, 35° 34'E, −200 m MSL) in 1933 and continue currently.

During this same period, D. Ashbel of the Meteorological Department, Hebrew University of Jerusalem, was instrumental in starting routine climatological measurements at Kinneret, at Ein Gev (32° 47'N, 35° 38'E, −210 m MSL), on the eastern bank of the lake, and at Migdal. The wind and temperature measurements have been analyzed by Ashbel (1936, 1937) and also summarized in his review of the climate of the Rift Valley (1964).

Standard climatological measurements at Tiberias were resumed in 1939 and continued at a new site in the upper town (−110 m MSL) from 1942 to 1962. Routine measurements at Ein Gev were resumed in 1951. Measurements at Ginosar on the western shore (32° 51'N, 35° 31'E, −205 m MSL) are available from 1958.

Rainfall measurements are also available from some twenty lakeside sites, a few of which also record evaporation from class A pans and windrun.

In 1968, the Kinneret Limnological Laboratory set up a meteorological station at Tabgha where air temperature and barometric pressure are recorded. The Kinneret Limnological Laboratory also operates two stations for wind measurements, one on the lake at Barbutim and another on the wharf of Ein Gev (plates 3 and 4).

Plate 3, 4. The meteorological stations of the **Kinneret** Limnological **Laboratory**:
(3) The Ein Gev Station; the pier of Ein Gev and a tourist boat.
(4) The meteorological station anchored at station A in the central part of the lake. (photos., Y. Edelstein).

B. The general meteorological background

J. Neumann & G. Stanhill

The meteorology of Lake Kinneret (between 32° 42′ and 32° 55′N, 35° 31′ and 35° 39′E, about −210 m MSL) is largely determined by two major factors: its position with respect to the climatic belts of the hemisphere and the topography of its environs.

1. The shift of the climatic belts

a. Winter

Lake Kinneret, like the whole of the area of the eastern Mediterranean, is located on the poleward side of the northern hemisphere SHPB (= Subtropical high-pressure belt). In winter, when the SHPB shifts equatorward, along with the other climatic belts of the hemisphere, the area is, in effect, part of the belt of westerlies with their characteristic migratory cyclones and anticyclones and the variable weather associated with these pressure systems. It is during that season, and primarily during cyclonic weather, at the time of the passage of cold fronts, and in the often unstable cold air advancing behind these fronts, that upward motions in the air are, apparently, strong enough to outweigh the downward component of the air mass forced on it by the fall-off of topography, from about 300 to 500 m MSL just west and southwest of the lake (the storms usually move in from the southwest) to −210 m at the bottom of the Jordan Rift. As was pointed out earlier, the upcurrents of air must be stronger than the downward component due to the topography in order to sustain the development of clouds and precipitation. The latter is estimated to amount to about 400 mm yearly over the lake, whereas over the surrounding hills, both to the west and to the east, the annual rainfall is about 500 mm. It is almost certain that the lower rainfall over the lake is a consequence of the reduction in the intensity of upcurrents in the air brought about by the fall-off of the topography.

b. Summer

In summer, when the climatic belts of the hemisphere undergo a northward displacement, the area of Lake Kinneret, like that of the eastern Mediterranean, is well within the SHPB. The major anticyclones of the belt are characterized by slow downward motion, a subsidence of the air aloft, especially on the eastern sides of the oceans (in this respect, the eastern Mediterranean is on the eastern side of the North Atlantic). The air of the middle and upper troposphere sinks, and as the 'upper' air is brought to lower altitudes, where the environmental pressure is relatively high, it is compressed. There is very little doubt that this sinking motion is an adiabatic process, that is, no heat is passed to, or taken from, the environment, and since the compression releases internal energy, the air warms up during its subsidence. The rate of adiabatic heating is 1°C per 100 m of sinking and, since outside the region of subsiding motion the 'environmental' temperature lapse rate normally is but 0.65°C per 100 m, the air within the subsidence layer is, at all heights, at higher temperatures than in the surrounding 'environment' where no important vertical motions take place.

In contrast to the above, heating of the ground by the solar radiation during the summer months convects the surface layer of air upward. At a height ($\sim$1 km) where the subsiding motion of the 'upper' air and the generally upward motion of the surface air peter out, a temperature inversion (elevated temperature inversion) forms and this, like all inversions, inhibits the upward transfer of water vapour from the humid lower layers. The virtual suppression of upcurrents (and of significant amounts of water vapour) is tantamount to the absence of clouds, or, at least, of clouds of significant depth, and precipitation cannot develop.

The sparsity of clouds during the five rainless summer months, May to September, when the midday cloud cover observed at the lakeside averages less than two tenths, means that the usual depletion of insolation caused by the reflection, scattering (mainly back to outer space) and absorption of this radiation by clouds is virtually absent from the Lake Kinneret area during the summer. As a result, global radiation levels are high, leading to enhanced evaporation from the lake.

Although there are no upper-air soundings of temperature available for the general area of the lake, such radiosonde stations that operate at moderate distances from the lake (e.g. Bet Dagan, near Tel-Aviv, Israel, at a distance of about 110 km to the south-southwest, and Beirut, Lebanon, at approximately the same distance to the north-northwest) show the subsidence inversion as a persistent feature of the summer season. Moreover, the dryness of the air at hill stations, situated at heights usually above the level of the subsidence inversion, is a clear indication for the existence of the inversion over the general area of the lake. An example for a nearby hill station is Mt. Kena'an (32° 59′N, 35° 30′E, 934 m MSL), which is about 15 km from the lake and whose data show clearly the dryness of the air in summer.

2. Pressure distribution: surface and aloft

Mention was made of the fact that the subsidence inversion discussed above prevents the formation of a deep or extensive cover by clouds in the summer and that related to this, the summer solar radiation rates at lake level and at ground level of the environs are high. Thus at 'Amir in the Upper Jordan Rift (33° 11′N, 35° 37′E, 68 m MSL), global radiation from May to September inclusive averaged 720 cal cm^{-2} day^{-1}. True, the comparatively deep atmosphere over the lake (about $2\frac{1}{2}$% more mass over the lake than what is 'normal' at MSL) causes an extra attenuation of solar radiation, but this attenuation is considerably less than that due to thick clouds. The intense solar heating of the surface (especially that of the environs of the lake) causes, as was pointed out earlier, a heat convection in the lower air layers. In the framework of that heating, an air circulation develops that 'evacuates' air horizontally, with the result that the surface (and the whole surface layer) indicates a (relatively) low atmospheric pressure. To be specific, the

50

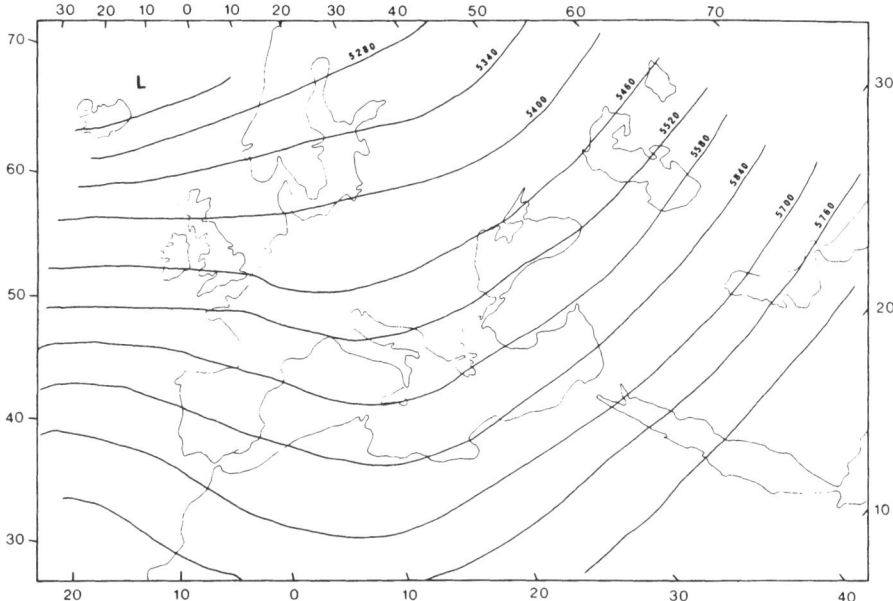

Fig. 10. Average contours of the 500 mb pressure surface in (geo-dynamic) meters in January for the years 1945–1953. The isopleths of height are, to an approximation, equivalent, with approximate pressure values, to isobars. The winds will blow along the isopleths; in the present case, more or less, from the west. Note that circulation conditions over the Eastern Mediterranean are essentially similar to those over the middle latitudes. The anticyclones are displaced toward the equator.

MSL pressure for the general area of the lake is estimated to be 1,017 mb in winter but only about 1,007 mb in summer (see Fig. 12 below). Thus, in summer, we have the seemingly paradoxical situation of low pressure at the surface and high pressure aloft. The surface low is a 'warm low'; the low pressure loses 'intensity' with height relative to its environment until, eventually, the 'low' is superseded aloft by a pressure 'high'.

The great contrast between the characteristic seasonal pressure distributions is illustrated by Figs. 10 and 11 which show the 'topography' of the 500 mb surface in January and July, respectively. In winter (January), in the general area of the eastern Mediterranean, the isopleths of altitude of the 500 mb pressure level run nearly west–east about a pressure low that is centered about Iceland. In summer (July), the eastern Mediterranean is in a belt of high pressure with closed centers of high pressure over the Sahara and the Saudi Arabia–Persian Gulf area.

Figure 12 shows the pressure distribution at the surface in summer. Although the lowest pressures are found to the southeast of Israel, surface pressure over Israel is low compared with areas to the west. Pressure measurements made at a lakeside station are presented in Table 3.

3. Summer winds: 'amplification' of the Mediterranean Sea breezes over the lake (and rift)

It can be noted from Fig. 12 that in summer, over Israel, as over all the coastal

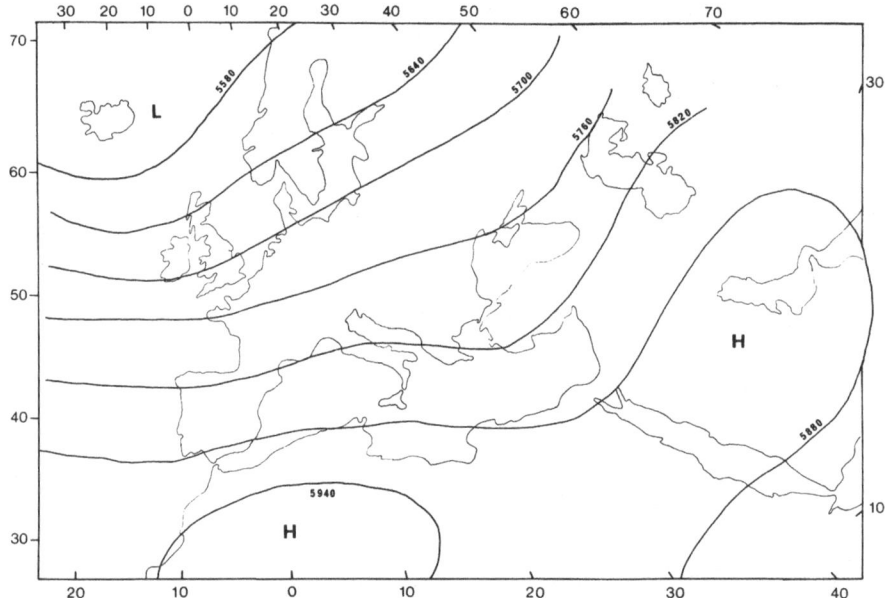

Fig. 11. Same as Fig. 10 but for July of the same years. Note that in the area of the Eastern Mediterranean the height values of the isopleths are about 200 (geo-dynamic) meters larger than for January (Fig. 10). This situation implies that the atmospheric pressure aloft in July is higher than in January. In the latitudes, and to the south, of Israel one finds centers of high pressure. Over the Eastern Mediterranean itself, there is a minor pressure trough but this fact does not alter the correctness of the statement in the text that the circulation is essentially anticyclonic with all its concomitant processes such as the subsidence of air etc.

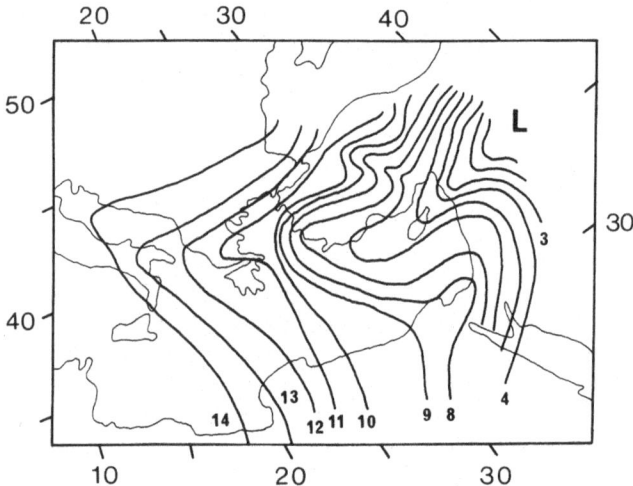

Fig. 12. Average pressure distribution at mean sea level in July 1951–1960 in mb (the 1,003 mb curve is referred to as 3 and so on). Note that over most of the Mediterranean the pressure distribution is slack so that winds connected with this distribution are sluggish. Under such conditions, local pressure gradients, which are not obvious in a map of the scale involved, become important and give rise to local wind circulations such as the sea and land breezes.

Table 3. Mean climatological values measured at Deganya A (32° 43′N, 35° 34′E, −200 m MSL) 1945–1970

	I	II	III	IV	V	VI	VII	VIII	IX	X	XI	XII	Yr.
Barometric pressure, adjusted to station height – mbars. (mean of 08, 14 and 20 hr observations, 1957–1970)	1042	1041	1038	1036	1035	1032	1029	1029	1033	1037	1040	1041	1036
Maximum air temperature, °C	18.4	19.5	22.2	26.8	32.0	35.0	36.5	37.0	34.9	31.5	26.1	20.4	28.4
Minimum air temperature, °C	9.0	9.0	10.5	13.2	17.0	20.2	22.5	23.4	21.2	18.2	14.7	11.2	15.8
Air temperature at 08 hr, °C	11.3	12.1	15.7	19.5	24.0	27.7	29.4	29.6	27.1	23.4	18.4	13.5	21.0
at 14 hr	17.2	19.4	20.9	25.4	30.4	33.9	35.7	36.1	33.9	30.3	25.2	19.6	27.3
at 20 hr	12.8	13.4	15.5	19.0	23.5	26.7	28.5	29.0	27.5	24.2	19.4	14.7	21.2
Rainfall monthly total, mms	96	74	58	23	6	0	0	0	1	14	47	89	408
Cloud cover at 08 hr, tenths	4.7	4.6	4.4	3.8	3.1	2.0	2.5	2.8	2.1	2.0	3.4	4.4	3.1
at 14 hr	4.7	4.7	4.6	3.7	2.8	0.9	0.7	0.7	1.3	3.7	3.8	4.6	2.9
at 20 hr	3.3	3.0	2.6	1.8	1.4	0.5	0.2	0.2	0.5	1.1	2.0	3.1	2.0
Wind force at 08 hr, Beaufort	1.4	1.2	1.2	1.1	1.0	0.9	1.0	0.9	0.7	0.8	1.1	1.5	1.1
at 14 hr Scale	1.8	1.9	2.4	2.5	3.0	3.5	3.6	3.5	2.8	2.1	1.3	1.5	2.5
at 20 hr (1945–1962)	1.2	1.2	1.2	1.5	1.7	2.4	2.7	2.5	2.2	1.5	1.0	1.3	1.7
Vapour pressure at 08 hr, mbs	10.4	11.0	10.9	14.6	21.8	22.7	25.5	26.9	23.1	16.9	13.0	11.2	16.2
at 14 hr	11.8	12.1	15.2	14.0	16.1	18.5	21.2	22.5	21.2	18.3	15.3	12.7	16.2
at 20 hr	11.0	11.7	14.3	12.9	15.9	19.4	22.6	24.3	21.4	17.8	14.1	11.8	15.9

areas of the eastern Mediterranean, the large-scale pressure distribution is slack, corresponding to a 'surface geostrophic wind speed' (a wind that would obtain in the absence of friction, etc., friction being, of course, of outstanding importance near the ground) of about 5 m s⁻¹.

In such conditions, air motions associated with the large-scale distribution of pressure alone are sluggish. Winds or breezes that develop are then primarily due to such factors as sea–land temperature difference and topography. We will discuss in the following paragraphs the case of Lake Kinneret's situation, and, in fact, that of the Jordan Rift.

One of the outstanding meteorological features of the Jordan Rift in summer is the arrival 'in force' in the afternoon hours of the Mediterranean sea breeze. We surmise that the combination of topography and the subsidence inversion brings about not only the increase in speed of the sea breeze (which is almost the only noteworthy wind of the summer) at some elevation especially over hill ridges athwart of the wind, but also a lowering of stream lines over the Rift bottom including the lake area. The Mediterranean sea breeze arrives at the lake with speeds of 10–15 m s⁻¹, that is twice and even two-and-a-half times its speed near the ground, west of the hills flanking the lake toward the Mediterranean. These comparatively strong winds give rise to almost daily storms on the lake in the afternoon hours of the summer.

The process of the strengthening of the Mediterranean sea breezes, as they arrive in the lake area, are at present being studied theoretically by E. Doron and J. Neumann, Department of Atmospheric Sciences, the Hebrew University of

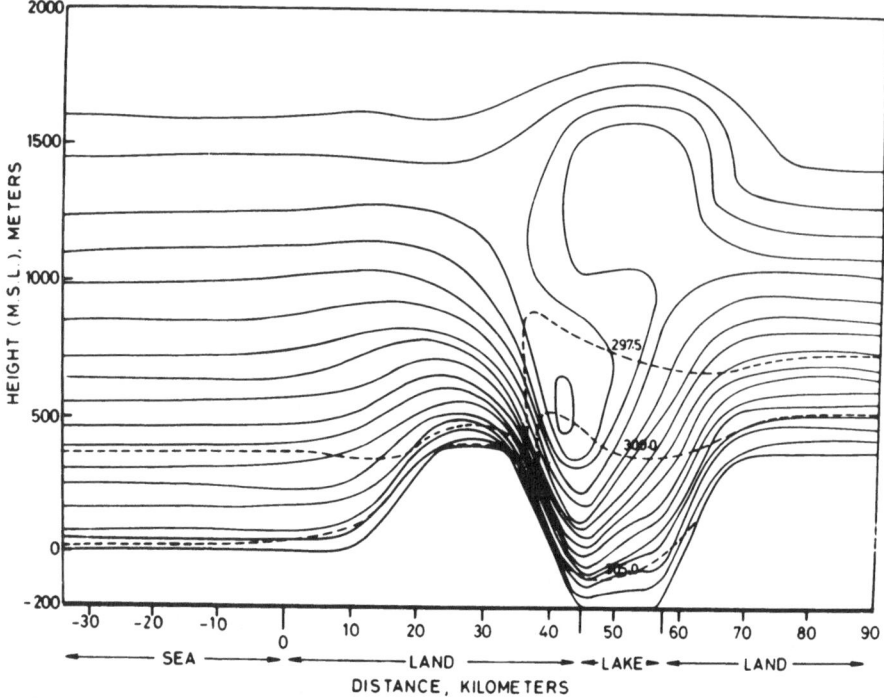

Fig. 13. Theoretically calculated winds (more precisely: resultants of the westerly and vertical components of the winds) and temperatures for a land area that is an idealized version of the topography in northern Israel from the Mediterranean to the Golan Heights via Lake Kinneret. The boundary conditions represent surface conditions of temperature in summer at 1700 hrs; the valley is taken to be covered with water at a temperature corresponding to that of Lake Kinneret in summer. The abscissa is distance from the Mediterranean coast while the ordinate is altitude above MSL. The full lines are streamlines for the resultants of the westerly and the vertical components of the winds. At any point the direction of the streamlines is that of the resultant and the spacing of the streamlines is inversely proportional to the speed. The dashed lines are isotherms.

Note the close packing of the streamlines near the surface just west of the 'lake', the spacing representing a speed of close to 15 m s⁻¹; the winds are weaker on the eastern shore. Note the frontlike structure of the flow in the lee of the western ridge: there is a sharp gradient of temperature across the front. Also note the whirl developing over the area representing the Jordan Valley and the lake. In parts of the whirl aloft the winds are easterly. The winds are fall winds on the western side and weaker, rising wind on the east side of the Valley.

This diagram has been taken from a paper by Doron & Neumann (1976) with permission of the authors.

Jerusalem. These authors have constructed a mesometeorological flow model involving: (a) a land surface representing an idealized version of the actual topography along a W–E cross-section of northern Israel, from the Mediterranean to the Golan Heights via Lake Kinneret – the adopted topography is shown in Fig. 13; (b) a special coordinate system ('σ-system') to be able to cope with a nonlevel ground surface (a nonlevel topography presents many difficulties both in the physics of the model and in the numerical handling of the equations of the model); (c) the equations of motion for a turbulent, stratified atmosphere, the equation of continuity and the equation for the turbulent conduction of (sensible) heat; (d) a large-scale pressure gradient, simulated by a geostrophic wind of 4 m s^{-1} from the WNW, portraying the large-scale flow in summer in the lower atmosphere over northern Israel; (e) a subgrid physics, that is, a parameterization of the turbulent transport of momentum and (sensible) heat adopted in earlier studies of the sea and land breeze circulations by Neumann & Mahrer (1975); (f) improved numerical methods for handling the nonlinear terms of the equations, as well as the terms representing the pressure gradient which present certain problems in the vicinity of a steep topography.

The horizontal grid distance used in the numerical work is 2 km and thus it affords about 5 grid intervals over the width of Lake Kinneret, offering a reasonably detailed resolution of the motion across the lake.

The results (Doron & Neumann, 1976) of the study for the daylight hours, when the stratification in the air is mostly unstable, are rather satisfactory; the results for the night hours, when the stratification is stable, still require improvement. The computed flow across northern Israel and the lake (and the topography adopted) for 1700 hours is shown in Fig. 13. The densely-packed streamlines (the continuous lines depict streamlines, see legend to Fig. 13) near the purported western shore of the lake indicate strong westerly winds near the surface; the less densely-packed streamlines on what represents the eastern shore symbolize somewhat weaker winds. A more detailed look at the results (not shown in Fig. 13) shows that the winds on the eastern shore have a small northern component relative to the winds on the western shore, in approximate agreement with observations (see Fig. 5 in Serruya, 1975). Another encouraging result of the theoretical study relates to the observed large pressure amplitude in the air near the surface in the lake area. The computations indicate a diurnal pressure drop of about 3 mb at the time of the strong winds, in close agreement with observations. For more details, reference is made to the manuscript by Doron & Neumann which is anticipated to be ready for publication in 1977.

Although the theoretical model topography differs considerably from that of the Lake Kinneret area, the results are relevant to the mode of generation of the strong afternoon winds of the lake area in summer which, as was pointed out above, bring about almost daily storms in the lake in the afternoon hours of the summer season.

It is natural to expect that the comparatively strong winds should produce a set-up of the lake waters. S. Serruya (1975) has recently described the magnitude and other aspects of the process. Waters of the epilimnion pile up on the eastern, downwind side of the lake, whereas the cold waters of the hypolimnion well up on the upwind side. The thermocline and even the unaerated waters of the hypolimnion rise up to the surface.

4. Air temperatures

The forced descent of air, from altitudes of up to 300–500 m MSL from the hills flanking the lake to the level of the lake, which is at about −210 m, causes a compression of this air. It is almost certain that the process is adiabatic (as is the case with the air subsiding in summer in the mid- and upper reaches of the troposphere,

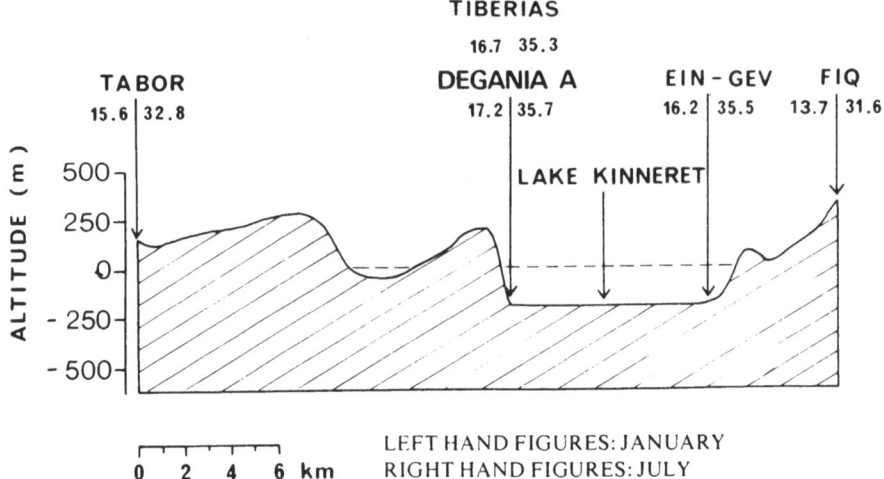

Fig. 14. Mean air temperatures at 1400 hrs LST in January and July in °C: Vertical cross-section across Lake Kinneret from Tabor (Agricultural School) in the west to Fiq to the east. Note that the air temperatures of the low-lying stations conform rather well with the picture of an adiabatic compression of the air as it crosses the Jordan Rift. The figures in Figs. 14 and 15 should be regarded as illustrative being based on unequal and in some cases relatively short periods of measurement. They were kindly prepared by the late N. Rosenan, on the basis of data available in the archives of the Israel Meteorological Service. Note from the coordinates of the stations that they do not actually lie along a straight line.

see Section 1(b) above) and, since the adiabatic heating rate of an air parcel is 1°C per 100 m, the air should arrive in the lake area at a temperature about 2°C higher than the temperature at MSL – other things being equal. Thus, at Tabor (Agricultural School: 32° 42′N, 35° 24′E, 144 m MSL), approximately 15 km west of the lake, the annual mean temperature is 20.5°C, and that of Degania A (32° 43′N, 35° 34′E, −200 m MSL), about ½ km south of the lake, is about 22.1°C, while the mean daily maximum in August at Tabor amounts to 33.8°C, and at Degania A to 37.0°C.

Fig. 14 is a cross-section of the topography and temperature in the environs of the lake, from Tabor (Agricultural School), in the west, to Fiq (32° 46′N, 35° 42′E, 330 m MSL) on the Golan Heights, to the east. The temperature conditions represented by this figure are for 1400 hours LST, for January and July, respectively, and illustrate the effect of topography on air temperatures described in the foregoing paragraph.

5. Some consequences of lake surface temperatures

The winter isotherms of the lakeside area follow closely the border of the lake in the winter season, suggesting a warming effect from the water surface.

Current theoretical calculations by E. Doron & J. Neumann, alluded to in Section 3, suggest that in the absence of the lake and assuming valley-bottom temperatures as observed at lakeside stations, the atmospheric whirl developing would be significantly different from that developing in the presence of the lake. Thus in the hypothetical absence of the lake, according to the computations over the Rift and the adjoining hills, the calculations indicate major upward motions on the slopes to the east when the air motion is from the west, as it is most of the time in the summer. It appears that the comparatively cool surface waters of the lake prevent the development of an upward motion to the east. That the topography of the Rift gives rise to a major whirl reaching up to some height was already surmised by L. Weickmann in 1915 as a result of his meteorological observations in the area* made on behalf of the German Air Force and published in 1922. The streamlines of the whirl, as conjectured by him, are reproduced by Ashbel (1950, p. 20). Several of the features of his suggested cross-section are certainly not verified by the observations. For example, his diagram suggests easterly winds on the east side of the lake and on the hills and plateau to the east, which is certainly not the case when the flow to the west of the lake is from the west. However, the calculations by Doron & Neumann do indicate the formation of an interesting vortex.

6. Humidity conditions

Figure 15 is a cross-section similar to that in Fig. 14, except that Fig. 15 shows the distribution of water-vapour pressure. In July, the increase in vapour pressure

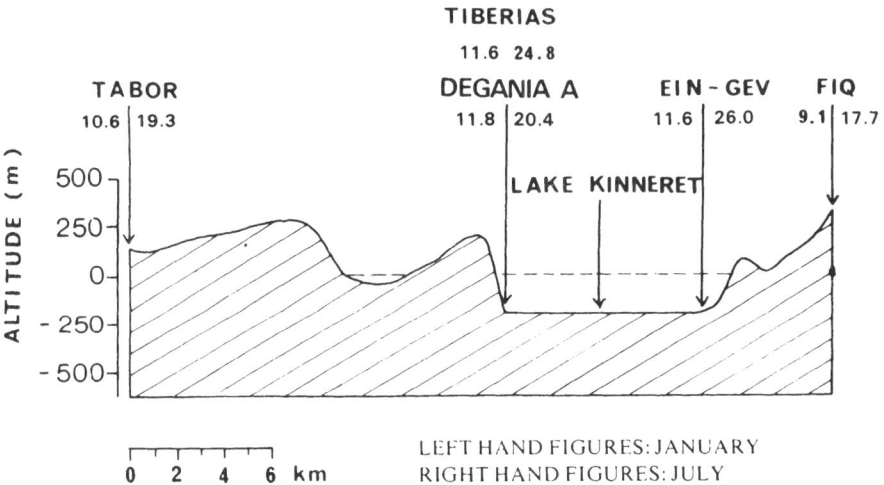

Fig. 15. Mean water-vapour pressure at 1400 hrs LST for January and July in mbar. Vertical cross section is the same as in Fig. 14. The small difference between the stations in January is mainly due to the greater variability of wind directions in winter (see Table 4 for wind directions at Fiq). In July winds are almost exclusively from the west, especially during daylight hours. Note the high vapour pressure at Ein Gev, on the east shore of the lake in summer, almost certainly due to the large evaporation rate from the water surface which enriches the air crossing the lake.

* Strictly, Weickmann's cross-section related to the Dead Sea area, about 100 km south of Lake Kinneret. As the hills about the lake are less high than those around the Dead Sea, and as Lake Kinneret is but 210 m below MSL while the Dead Sea is about 400 m below, one expects that the 'disturbance' in atmospheric flow set up by the topography about Lake Kinneret should be smaller than is the case for the Dead Sea. Nevertheless, a disturbance is expected, and this is indicated, as was pointed out above, by the theoretical calculations, see Fig. 13.

from Tiberias, the city on the western shore of the lake, to Ein Gev, a kibbutz on the eastern shore, is a conspicuous feature of the distribution. This increase is almost certainly the result of the enrichment in vapour content of the air through evaporation from the lake as the air crosses from a westerly direction, daylight-hour winds of the summer season being almost exclusively from the west. The lower vapour pressures at Tabor and Fiq are, of course, due in part to the relative elevation of these stations.

In January, the differences are small. We note, in this connection, that the vapour pressures shown in the cross-sections are monthly averages. The absence of any differences in the averages for Tiberias and Ein Gev is, in all probability, due more to the greater variability of wind directions in the cold season rather than to the lower evaporation rates from the lake at that time. In winter, the percentage of easterly winds coming from the comparatively dry areas of the Golan Heights and the Syrian Plateau is considerable, and thus Ein Gev is frequently affected by these somewhat dry winds. Table 4 is a summary of the

Table 4. Mean percentage frequency of easterly and westerly winds, at Fiq (32° 46'N, 35° 42'E, 330 m MSL)* 1961–1966

	I	II	III	IV	V	VI	VII	VIII	IX	X	XI	XII
Easterly (NE–SE)												
08 hr	70	54	59	41	26	11	–	1	14	69	83	72
14 hr	38	17	16	7	2	–	–	–	–	4	29	38
20 hr	44	24	27	15	5	1	–	–	1	11	47	31
Westerly (NW–SW)												
08 hr	13	17	29	40	50	64	83	63	31	6	3	10
14 hr	45	59	75	88	94	99	100	100	99	85	60	37
20 hr	23	38	51	59	56	71	85	75	55	20	12	15

* Source – N. Rosenan, Israel Meteorological Service.

frequency of easterly and westerly winds at Fiq, illustrating the wind-direction conditions just described.

7. Climatic classification

According to Köppen's classification (Köppen, 1931), the climate of the district is that of a hot, arid steppe with dry summers (BShS). Thornthwaite's more recent system of classification (Thornthwaite, 1948) designates the area as Megathermal (A') on his thermal efficiency index, i.e., it has a yearly potential evapotranspiration greater than 1.14 meters – the highest category in this scale. According to his moisture index (which is based on the moisture surplus and deficit in the annual water balance as calculated by Thornthwaite's formulae), the district is semi-arid with a winter moisture surplus (D).

C. Data concerning meteorological parameters measured on the lake or on the lake shores

S. Serruya

1. Solar radiation

Mero & Isaac from Water Planning for Israel measured the net surface radiation using the lake water as the reflecting surface at Ginosar Station. The net surface radiation is equal to the total incident energy (radiation from the sun and from the sky) minus the energy reflected by the water surface. Their unpublished data were summarized by Berman (1976). The daily net radiation ranges from a maximum of 5,800 kcal m^{-2} d^{-1} in June–July to a minimum of 430 kcal m^{-2} d^{-1} on an overcast day in December (Fig. 16).

High maximum values of net surface radiation are often measured in summer in Northern countries: 4,900 kcal m^{-2} d^{-1} were measured on a clear day of August

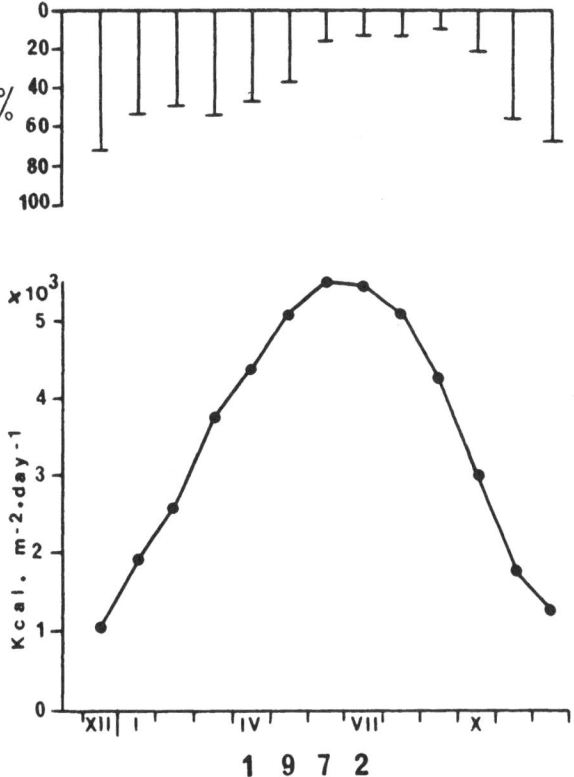

Fig. 16. Daily net solar radiation in the Lake Kinneret area and percentage of noontime cloud cover (from Berman, 1976, according to Tahal data, Hydrobiologia, 49, 1).

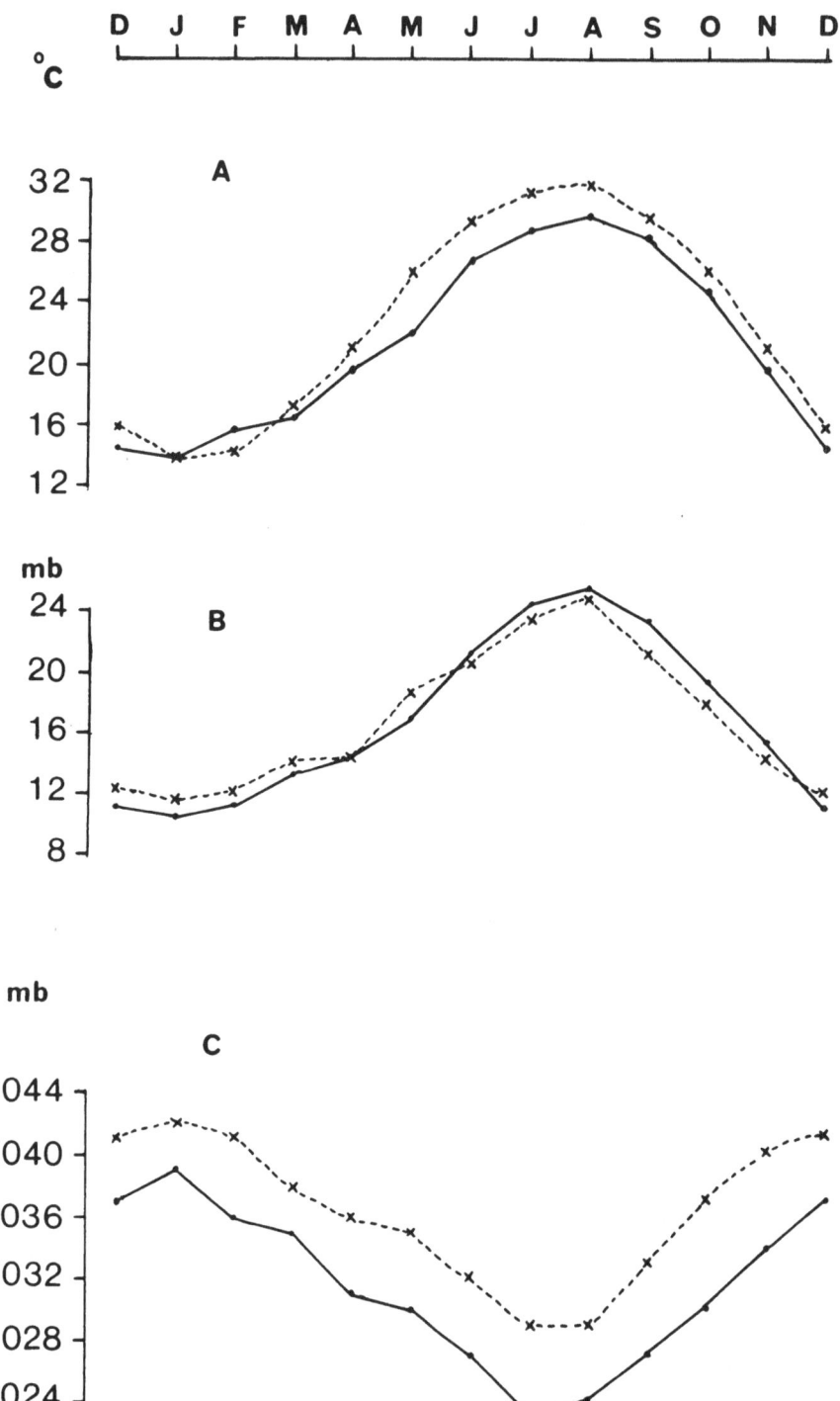

Fig. 17. Monthly average of air temperature (A), vapour pressure (B) and barometric pressure (C) at Tabgha station (1970–1974) (full line) and Deganya station (1945–1970) (dotted line).

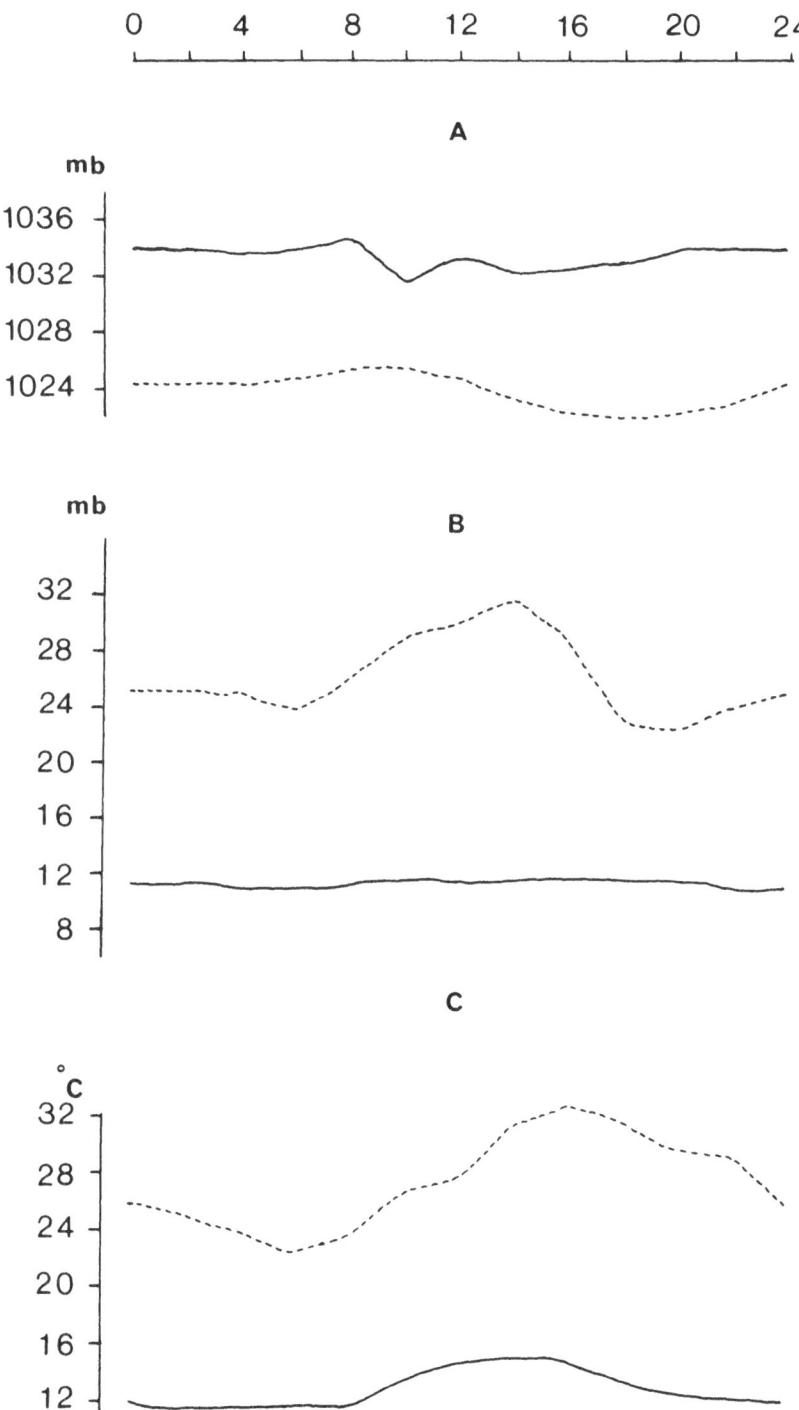

Fig. 18. Summer (dotted line) and winter (full line) hourly variation of barometric pressure (A), vapour pressure (B) and air temperature (C) at Tabgha station.

in Lake Klämmingen, southern Sweden (Johnsson, 1946 quoted by Hutchinson, 1957). However, in the temperate zone, the number of clear days is much smaller than in the Middle East where, as can be seen on Fig. 16, the cloud cover is important only from November through April.

The result is that while in the temperate zone the daily radiation averages about 2,200 kcal m^{-2} d^{-1} (Odum, 1971), it reaches an approximate value of 3,600 kcal m^{-2} d^{-1} in the Kinneret area.

2. Air temperature

a. Seasonal variations

The monthly average values obtained from the daily mean for the period 1970–1974 at the Tabgha Station show a maximum temperature of 30°C in August and a minimum of 14°C in January. In Fig. 17, these monthly averages have been compared with the monthly average values found by Stanhill at the Degania station for the period 1945–1970. (The slightly higher values of the Degania station are due to the more southerly location of this station.)

The highest maximum temperature recorded at Tiberias was 49.0°C (21 June 1942) and the lowest minimum was −3.3°C on three occasions (Stanhill, 1967).

b. Hourly and daily variations

Figure 18 shows a typical summer hourly variation of the air temperature at Tabgha. The temperature is minimum at 0600 hours (20–23°C) and maximum at 1400 (32–36°C); it shows a very regular daily pattern which is not observed in winter. The daily range variation reaches 10°C in July but does not exceed 3°C in January.

c. Interannual fluctuations

The monthly averages of temperature for 0800, 1400 and 2000 hours and maximum and minimum values are shown in Fig. 19 for the period 1970–1974. We can observe very similar yearly patterns with slight differences from year to year. We note that the 1972–1973 winter was the coldest one of the record whereas the summer of the same year was the warmest one.

3. Vapour pressure

The vapour pressure has been computed from the recorded relative humidity in percent and from the saturation pressure corresponding to the temperature measured simultaneously, for the period 1971–1974 at the Tabgha Station.

a. Seasonal variations

The vapour pressure was observed to vary from 10 millibars in January to 24 millibars in July–August. Very small differences exist between the Degania and Tabgha series of data (Fig. 17).

62

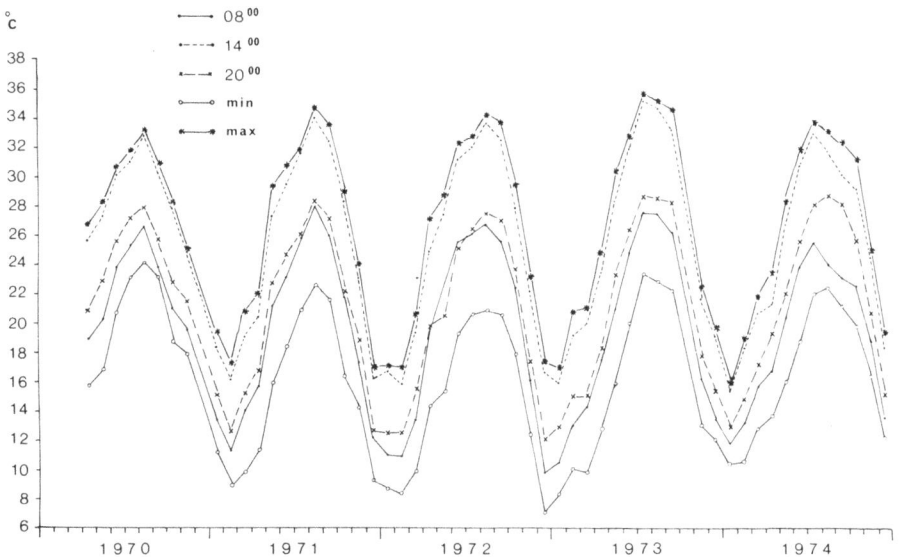

Fig. 19. Monthly averages of air temperature at 0800, 1400 and 2100 hours, minimum and maximum for the period 1970–1974 at Tabgha station.

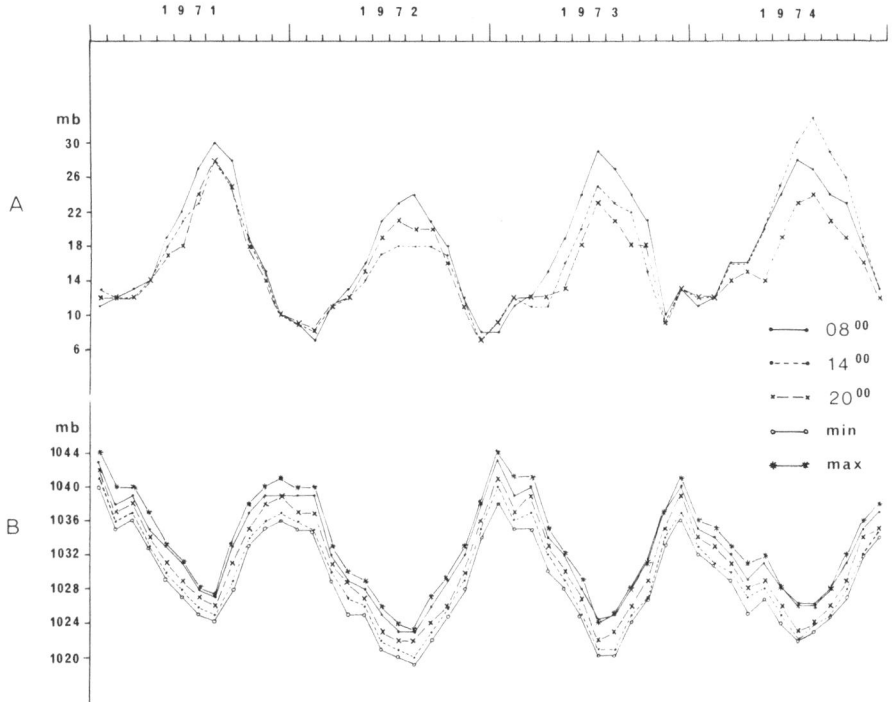

Fig. 20. Monthly averages of vapour pressure (A) and barometric pressure (B) for the period 1971–1974 at Tabgha station.

b. Hourly and daily variations

In summer, the minimum values are observed in the evening (18–20 millibars) and the maximum values at 1400 hours (32–36 millibars, Fig. 18). As expected, the increase of vapour pressure corresponds to the mid-day increase of temperature. The sharp drop of vapour pressure, observed at about 1400 hours when the daily temperature is at its maximum, must be related to the occurrence of the daily wind which starts blowing at this hour of the day.

In winter, the daily fluctuations range from 8 to 14 millibars and do not show any regular pattern.

c. Interannual fluctuations

The monthly mean values of vapour pressure for the period 1971–1974 are shown in Fig. 20. We note that the 1972 summer values were lower than the average. We also note that the spring increase of vapour pressure was delayed by approximately one month. This is related to the occurrence of higher than usual values of wind speed recorded in these months.

4. Barometric pressure

The location of the Kinneret area below sea level explains why the values of the barometric pressure are always above the 1,013 mb value of sea level.

a. Seasonal variations

The barometric pressure decreases regularly from January, with a monthly average value of 1,039 mb, to July, with a monthly average value of 1,024 mb. Fig. 17 shows that the Degania data present a similar pattern. The slightly higher values of the latter series may be explained by the different locations of the stations.

b. Hourly and daily variations

In summer, we observe a daily regular variation ranging from 1,020 to 1,028 mb. The maxima occur between 1000 and 1200 hours and the minima between 1800 and 2000 hours. In winter, variations of greater period (3 to 4 days) and greater amplitude are observed (Fig. 18).

c. Interannual fluctuations

The monthly average values for the period 1971–1974 (Fig. 20) show a very regular pattern. We note, however, that the summer minimum of 1971 was about 4 millibars higher than in other years. We also note that the spring values of 1974 were slightly lower than the average values for this period.

5. Winds

Like most meteorological parameters, the wind direction and velocity show a very regular seasonal pattern which can be divided into four periods: the winter period with storms of eastern winds; an intermediary period with very weak winds generally occurring in March; a summer period with dominant western winds; and a second 'weak wind' period in autumn. The short duration of the calm periods is a major climatological feature of the Kinneret area. The large amount of mechanical energy received by the lake surface has considerable bearing on hydromechanical, chemical and biological processes occurring in the lake.

The data presented in this section were obtained at two meteorological stations of the Kinneret Limnological Laboratory (Tabgha and Ein Gev stations).

a. Seasonal variations

Monthly averages are shown in Fig. 21 for Tabgha and Ein Gev stations for the period 1970–1974. This display shows the net dominance, at Tabgha, of the westerly wind in spring, summer and autumn. At Tabgha, the westerly winds have greater speed and blow longer than at Ein Gev. At this latter station, the winter eastern winds are stronger, and this direction prevails during a longer period. The southerly component is also stronger at Ein Gev than at Tabgha.

The monthly average speeds are higher in summer (3.8 m s^{-1}) than in winter (2.5 m s^{-1}). The instantaneous maximum velocities of eastern and western winds are comparable and reach 12 m s^{-1} (hourly integrated values), but the daily oc-

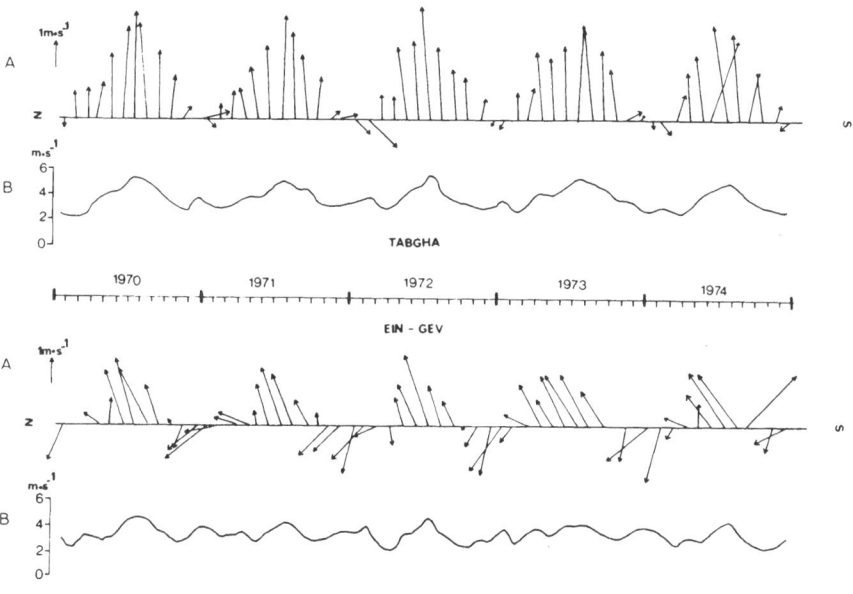

Fig. 21. Monthly averaged wind data at Tabgha and Ein Gev stations for the period 1970–1974. A = velocity as monthly vectorial average; B = speed as monthly mean.

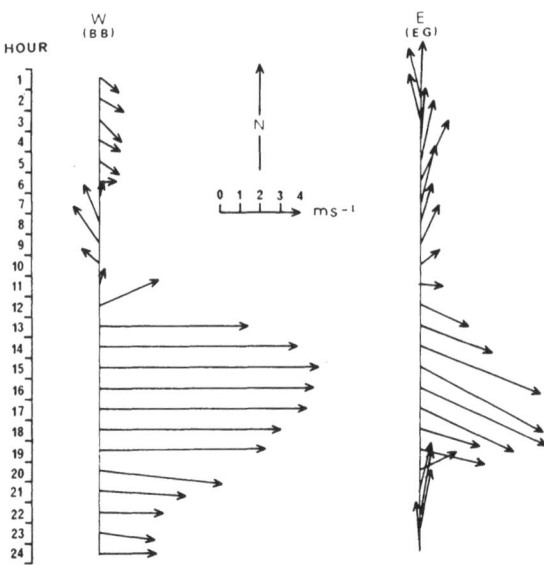

Fig. 22. 24 hour wind hourly vectorial distribution at the western Tabgha and eastern Ein Gev station. Period averaged: July 15 to 31, 1971. (From Serruya, S., 1975, Verh. Internat. Verein. Limnol., 19.)

currence of the westerly wind in summer, in contrast to the relatively low frequency of winter storms, accounts for the higher monthly averages.

b. Hourly and daily variations

The winter eastern storms have a duration of 2 to 4 days and a maximum speed of 12 m s⁻¹. During the four winter months (November to February), approximately fifteen such storms occur. Their role in the mixing processes of the lake will be discussed later.

Figure 22 represents the typical summer wind pattern. The periodic westerly wind starts blowing every day about noon, reaches its maximum speed at 1600 hours and drops down at 2200 hours approximately. This figure shows also that a very steady southerly component is observed at Ein Gev with the same regularity and daily frequency as the westerly component.

An hourly resultant vector, calculated by averaging the wind components of Tabgha and Ein Gev stations for the same hour of each day during a fortnight, is represented in Fig. 22. At night and in the early morning, winds of various directions show a breeze pattern; in the afternoon and evening, the westerly wind is recorded on both shores, lightly deflected to the right at the eastern station.

c. Interannual fluctuations

Table 5 shows the monthly average values of wind direction, velocity and speed from 1969 to 1974. We note the remarkably lower wind speed recorded during the winter 1969–1970. The yearly averages of the wind speed, resultant velocity and duration are shown in Fig. 23.

Table 5. Monthly wind averages—period 1969–1975

Tabgha	January			February			March			April			May			June		
	D	V	S	D	V	S	D	V	S	D	V	S	D	V	S	D	V	S
1969	–	–	–	267	0.9	1.9	270	1.1	2.1	279	1.4	2.8	271	1.7	3.0	269	2.3	3.3
1970	88	0.4	2.3	264	1.1	2.2	270	1.2	2.8	278	1.5	3.1	269	2.6	3.9	273	3.7	4.9
1971	60	0.4	3.0	269	0.6	2.8	283	1.2	3.1	257	1.4	3.5	261	2.2	3.5	266	3.1	4.6
1972	45	0.8	3.2	42	1.5	3.7	269	1.0	2.8	270	1.0	3.0	262	3.0	4.1	265	3.3	4.5
1973	118	0.4	3.3	269	1.2	2.5	277	1.4	3.4	264	2.8	4.1	265	2.6	4.2	267	3.2	4.9
1974	69	0.2	3.0	56	0.6	3.2	286	1.1	2.6	265	2.1	3.2	263	2.7	3.8	287	3.3	4.4
1975	3	0.4	2.9	267	0.6	3.3	269	1.2	2.9	273	1.5	3.4	270	2.7	4.2	263	3.5	4.6
Average	60	0.4	3.0	293	0.5	2.8	275	1.2	2.8	269	1.7	3.3	266	2.5	3.8	270	3.2	4.5
St. Dev ±		0.4	0.4			0.6			0.4			0.4			0.4			0.5

Tabgha	July			August			September			October			November			December		
	D	V	S	D	V	S	D	V	S	D	V	S	D	V	S	D	V	S
1969	272	2.8	4.1	271	1.9	3.9	277	1.7	3.0	293	0.8	2.7	327	0.3	2.1	276	0.3	2.4
1970	270	4.4	5.4	265	3.9	5.0	269	2.7	4.1	274	1.8	3.3	304	0.5	2.7	348	1.4	3.7
1971	271	4.1	5.2	268	3.5	4.5	264	2.6	4.2	274	1.8	3.1	311	0.4	3.1	351	0.5	3.1
1972	264	4.6	5.6	265	2.9	4.1	265	2.1	3.5	265	1.8	3.1	280	0.9	2.9	127	0.1	2.9
1973	271	4.0	5.3	264	3.7	5.1	259	2.8	4.2	262	2.1	3.5	332	0.7	3.5	302	0.2	2.7
1974	260	3.9	5.0	265	3.6	4.7	279	2.0	3.8	262	2.0	3.3	286	0.7	2.7	140	0.3	2.8
1975	261	4.0	5.1	264	4.0	5.1	266	2.5	4.1	275	1.8	3.0	292	0.7	2.5	67	1.2	3.9
Average	267	4.0	5.1	266	3.4	4.6	267	2.3	3.8	270	1.7	3.1	301	0.6	2.8	14	0.3	3.1
St. Dev ±			0.5			0.5			0.4			0.3			0.4			0.5

D = direction in degrees   V = velocity in m s⁻¹ (resultant vectorial speed)   S = scalar speed in m s⁻¹

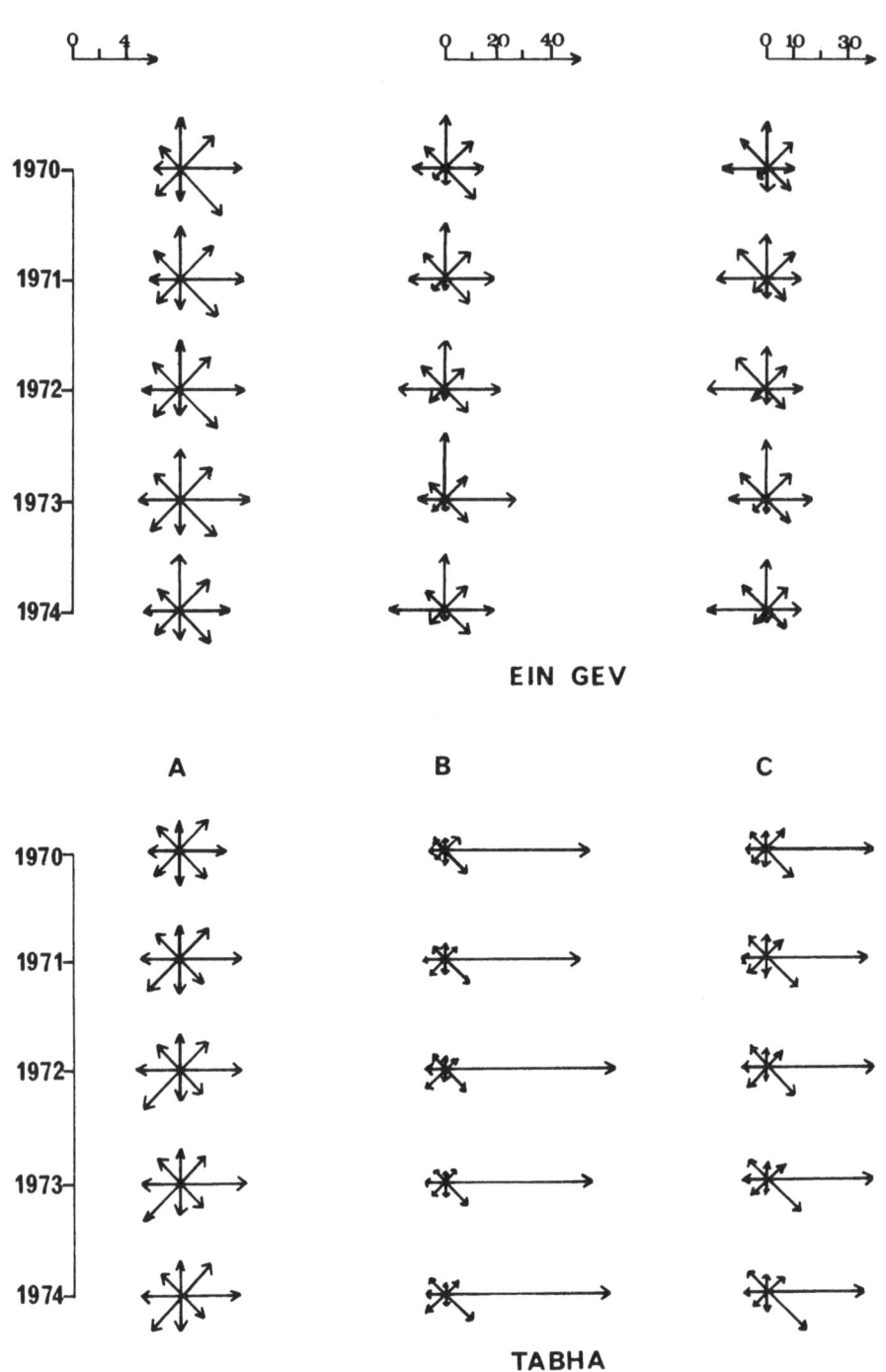

Fig. 23. Yearly distribution of wind at Ein Gev (upper) and Tabgha (lower)

(A) average speed in m.s^{-1}
(B) percentage of total run
(C) percentage of total duration.

D. Precipitation

S. Rubin

The location of Israel at the boundary of the semi-humid and semi-arid climatic zones accounts for the small amounts of rainfall in the southern parts of the country and the relatively abundant precipitation in northern Israel. The lake area is situated approximately at the boundary of these two zones; it is then clear that the rain regime in the lake watershed is a determining factor for the development of the vegetation and the fauna. That is why this section includes not only quantitative data concerning rainfall but also a detailed study of its spatial and temporal distribution and variability.

1. Rain regime

Details on the rain regime can be found in Harel & Nir (1963), Elbashan (1967), Katsnelson (1964, 1969), the Archives of the Meteorological Service of Israel, the Atlas of Israel (1970) and Gat & Paster (1974). The rain regime is governed by factors depending on the general circulation and by more local geographical factors.

a. Factors depending on the general circulation

In winter (October–April), the pressure systems in the area influencing the rain regime are:
- Depressions over the Balkans and the Caucasus which cause inflow of cold air into our area.
- A ridge over North Africa reaching as far as western Egypt.
- Low pressure over the Mediterranean which enables the passage of depressions.
- A depression over the Sudan and Ethiopia with an extension to the Red Sea.

According to these pressure systems, the rainfall in Israel is generally related to one of the following situations (Aelion, 1958; Shaia, 1964): (i) Our area is located on the southern border of the depressions moving from west to east. In winter, the depressions move southwards and enter our area. On their way to our area, the air masses carried by the depressions lose a great amount of their moisture and change their thermal characteristics compared with depressions in Europe. These lows are weaker and move faster in our area. (ii) The area is influenced by depressions moving from north-northwest. (iii) The area is affected by depressions formed over the Mediterranean, or intensified there, in case of meridional flow (low index). In winter, there is a tendency towards blocking ridges developing over western and northwestern Europe, which cause the formation of upper air troughs over eastern Europe or western Russia. This pattern induces the inflow of very cold air to the eastern Mediterranean, encouraging the formation or reinforcement of depressions in or near our area. Those depressions are generally very active and move slowly. They bring long rain periods with heavy precipitations to the eastern Mediterranean. The southern part of Israel is less influenced by these pressure systems in comparison to its northern part.

It is clear, then, that the annual variations of the duration of the rainy period, the seasonal amount of rain and the distribution of the rain are determined by these synoptic situations in the area.

The meteorological situation which develops in summer and prevents the formation of clouds has already been described (see Section B). Although the air mass which reaches our area in this period absorbs moisture while passing over the Mediterranean, this moisture is not able to produce rain. Another factor preventing precipitation in summer is the fact that the surface of the Mediterranean is cool in comparison to the land areas surrounding it, and this strengthens the stability of the air masses. Very rarely, a suitable synoptic situation develops and brings small amounts of rain to the area in this season.

b. Geographical factors

(i) Latitude: The general circulation factors previously described cause a gradual decrease of precipitation from north to south.
(ii) Distance from the sea: Israel lies at a boundary position between the Mediterranean Sea and the Asiatic continent. When moving from west to east, the moderate marine climate becomes gradually continental; the air becomes drier and the amount of rain decreases.
(iii) Topography and exposure: Rainfall increases with elevation above sea level (orographic rain). The Jordan Valley offers a typical example. The Valley, which is lower than the adjacent regions, causes air subsidence and hence gets less precipitation.

Exposure of the stations to the wind also affects the precipitation. For a given altitude, the stations facing the wind receive more rain than the others. We note that the rainiest area does not correspond generally to the crest of the mountain but is slightly displaced to the leeward side, in the case of a very steep leeward slope.

2. Area and period of this survey

The data used here are not homogeneous with regard to the period of measurements and to their reliability. There are about 100 long-term and reliable stations that can be reduced to a simultaneous period. These stations are located in the Jordan Valley and in Galilee (Elbashan, 1967). For the other regions, the data are sparse; therefore, the reduction is problematic and of a lower degree of reliability. Thus, the analysis of rainfall in these regions has a more general nature and is partly based on data, not always for simultaneous periods, from various sources (mainly from a survey by the Agrometeorological Division, Israel Meteorological Service, Gat & Paster, 1974).

The tables and graphs in this survey are based on data obtained from 9 base stations only, which have at least 25 years of records during the period 1931–1960 for the Galilee and the Jordan Valley, and from one base station (Kuneitra) with 6 years of systematic and reliable measurements, from 1967 to

1973, in the Golan Heights. In most of the tables and graphs, the data for this base station, Kuneitra, are not reduced. Only for the analysis of annual rain amounts was a reconstruction performed, for each year of the period 1935–1960 (Israel Meteorological Service, Gat & Paster, 1974). Information concerning the 10 base stations is given in Table 6 and their location is shown in Fig. 24.

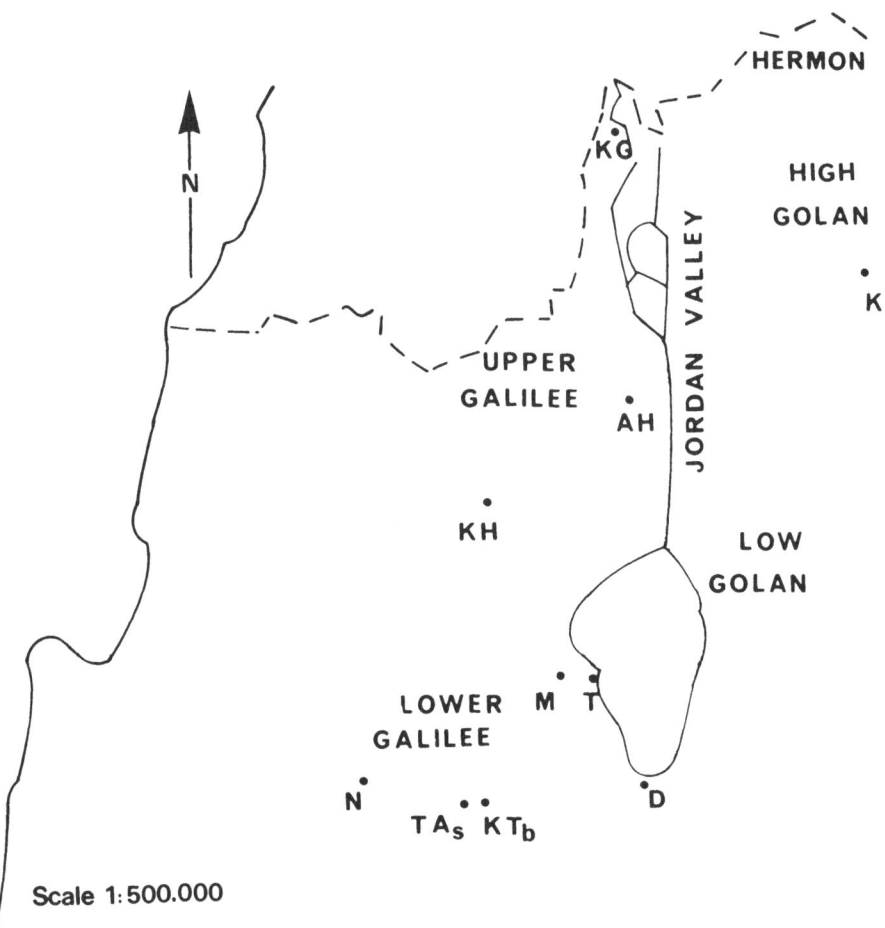

Fig. 24. Location of the meteorological base stations. KG = Kefar Giladi, K = Kuneitra, AH = Ayyelet Hashahar, KH = Kefar Hananya, M = Mizpa, T = Tiberias, D = Deganya, N = Nazareth, TAs = Tavor (Agricultural school), KTb = Kefar Tavor.

The reasons for choosing the period 1931–1960 as the homogeneous period are as follows: (i) In 1961, the Israeli cloud seeding experiment was extended also to our area. Therefore, it is preferable to survey the rainfall conditions in the area during a period which was less influenced by artificial rain experiments. (ii) The period 1931–1960 is the period for computation of climatological standard normals, according to W.M.O. regulations. (iii) A great part of the processed data concerns this standard period.

71

Table 6. Key Stations

Station	Co-ordinates		Israel grid		Elev. m	No. of years during period 1931–1960
	Long. E	Lat. N				
Galilee						
Kefar Gil'adi	35° 34'	33° 15'	204	294	340	30
Kefar Hananya	35° 25'	32° 56'	189	260	410	30
Nazareth	35° 18'	32° 43'	178	235	445	30
Tavor, Agr. School	35° 24'	32° 42'	188	234	145	25
Kefar Tavor	35° 25'	32° 41'	189	232	120	25
Golan						
Kuneitra	35° 49'	33° 07'	227	281	940	*
Jordan Valley						
Ayyelet HaShahar	35° 34'	33° 01'	204	269	175	30
Tiberias	35° 32'	32° 48'	200	244	−110	30
Mizpa	35° 30'	32° 47'	198	243	75	30
Deganya	35° 34'	32° 43'	204	235	−200	30

* Six years beginning in 1967/68.

3. Annual rain amount

Although this area is relatively small (2,760 km²), its diversified topography divides it into three main climatological units: Galilee Hills, Jordan Valley and Golan Heights–Mount Hermon.

Generally speaking, there are similar rain patterns in the Galilee, Golan Heights and Mount Hermon. However, in the Golan, because of its more homogeneous structure, the areal variability is smaller. The areal rain distribution is less variable in the Jordan Valley than in the other subregions except for short periods (less than one day), because of cloudburst phenomena.

Most of the rainfall differences among the subregions are due to topography, as indicated by the increases of amount of precipitation with altitude (Fig. 25).

Rain in the Kinneret area is mainly of cyclonic origin, but local conditions may produce local rain phenomena. In the mountainous part of the watershed, rainfall is mostly of cyclonic–orographic nature, whereas it is partly of convective origin in the Jordan Valley, which explains the variation of rain intensities between both areas.

a. Average (median) annual rainfall

Although it is recommended to base rainfall analysis on accumulative frequency distributions rather than on averages and standard deviations, we could use these latter parameters since, in the case of the Kinneret area, the deviation of the averages from the medians is small (Table 7). The reason is that the distribution of annual rainfall is well represented, even for extreme values, by the statistical parameters based on normal distribution. Thus, $\bar{P} \pm 1.28\sigma$ ($\bar{P}$ = average annual rainfall and σ = standard deviation) is 239–529 mm at Deganya while equivalent

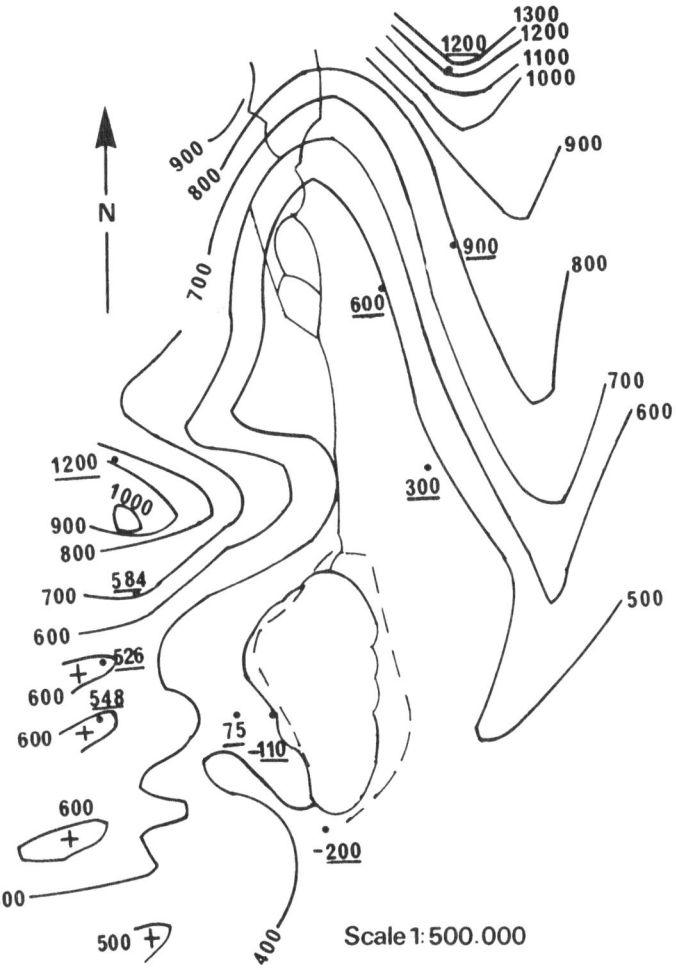

Fig. 25. Average annual rainfall (1931–1960). The underlined numbers correspond to the yearly amount of precipitation at the base stations. All results in mm per year.

percentage centered on the median (Med ± 40%) is 241–529 mm, and $\bar{P} \pm 1.65\sigma$ is 198–570 mm while the equivalent percentile interval (Med ± 45%) is 216–614 mm.

The variations of rainfall due to topography are shown in Fig. 25. In the Galilee, the annual rainfall ranges from 500 to 1,000 mm (rainfall in the mountainous Upper Galilee being usually above 800 mm). It varies from 400 to 700 mm in the Lower Golan and from 700 to 1,000 mm in the higher part of the plateau; reaching 1,000 to 1,300 mm in the Hermon area. The northern part of the Jordan Valley receives 1,000 mm; the decrease in altitude from north to south, superimposed on the general decrease of rainfall from north to south, explains the rapid drop of rainfall along the valley down to 350–400 mm in the southern part of lake Kinneret.

Table 7. Averages, cumulative frequency distributions and extremes of annual rainfall for the 1931–1960 period (mm)

Station	Average	Percentiles							Extremes		
		5%	10%	25%	50%	75%	90%	95%	max.	min.	1968/69
Galilee											
Kefar Gil'adi	769	421	478	638	746	922	1,030	1,082	1,097	383	1,362
Kefar Hananya	765	409	522	628	724	928	1,026	1,116	1,183	386	–
Nazareth	626	334	417	501	609	736	957	1,003	1,036	332	1,001
Tavor Agr. School	497	303	323	400	479	586	658	724	730	298	749
Jordan Valley											
Ayyelet HaShahar	434	234	250	331	423	514	676	681	683	233	852
Tiberias	431	193	240	348	421	523	646	665	670	179	–
Mizpa	451	240	283	321	431	537	587	733	754	228	628
Deganya	384	216	247	279	384	480	529	614	661	209	511
Golan											
Kuneitra	895	472	510	761	897	1,054	1,172	1,240	1,265	485	1,607

The topographical variety of the mountains located on the western side of the lake affects the rainfall distribution near the lake. For example, Hamat (32° 46′N and 35° 33′E; −200 m), bordered by the Poriya ridge (+240 m), receives more rainfall than Menorim (+190 m), located on the ridge itself.

Although our meteorological knowledge of the eastern shore of the lake is based on fewer data than in other areas, we can state that, on an average basis, the eastern shore receives 50–60 mm less than the western area (Stanhill, 1967; Archives of the Meteorological Service of Israel).

The north–south gradient of amounts of rain over the lake itself can be illustrated by comparing the annual amount of precipitation at Ein Sheva, on the northwest coast (436 mm), and at Zemah, on the south shore (332 mm). Data concerning the lake surface are only estimated, since there are no direct observations. The interpolations made from shore stations give an annual amount of rain slightly below 400 mm for the lake surface.

A section of the area discussed here is included in the northern seeding zone of the Israeli cloud seeding experiment (Gabriel, 1970). The experiment is still under operation, but according to a preliminary analysis of the data, the precipitation on seeded days is about 15–20% higher than on non-seeded days. A recent analysis (Brier *et al.*, n.d.) indicates that the effects found in the initial targets extend to downwind areas, though further experiments would be necessary to confirm this conclusion.

It is evident that any computation based on data collected after the date on which seeding started will be biased. It is also expected that the changes will not be uniform and the areal pattern might be distorted.

b. The frequency distribution of annual rainfall

Annual rainfall amounts are characterized by a great variability (Table 7). Although this variability is not uniform throughout the whole area, the trend of variation from year to year is usually similar (Fig. 26). This figure also indicates the absence of distinct periodicity or trend during the period under investigation, confirming the results of Gabriel & Kesten (1963).

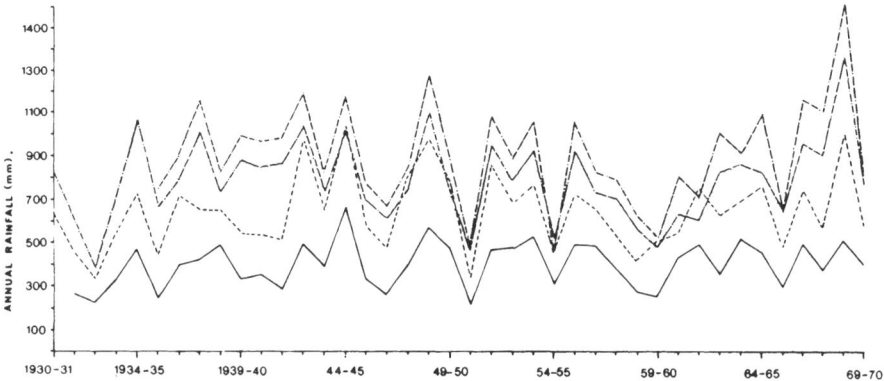

Fig. 26. Interannual fluctuations of rainfall. Period 1930–1970 at four base stations.
._ _.Kuneitra, ._ _. Kefar Giladi, _ _ _Nazareth, _____ Deganya.

The maximum annual rainfall in the greater part of the area was recorded in 1968/1969. Annual rainfall in this year generally exceeded 160% of the long-term average and reached 200% in some of the mountainous areas. A new absolute maxima for annual rainfall total (1,759 mm) was recorded in that year at Meron (941 m) in Upper Galilee.

From the data of Table 7, it is possible to determine the probability of specific annual amounts of rain. For example, in 50% of the seasons of the period 1931–1960, the annual rainfall at Deganya and at Kefar Gil'adi ranges respectively from 280 to 480 mm and from 639 to 922 mm. In 80% of the seasons, the precipitation range for the two locations is 247 to 529 mm and 478 to 1,030 mm.

c. Relative variability of annual rainfall

The present description is based on four different expressions of the relative variability:[*]

(i) Relative variability (%)

$$\frac{\frac{1}{n}\Sigma|P_i - \bar{P}|}{\bar{P}} = \frac{\Sigma|P_i - \bar{P}|}{\Sigma P_i} \tag{1}$$

(ii) Coefficient of variation (%)

$$\frac{\sigma}{\bar{P}} = \frac{\sqrt{n\,\Sigma(P_i - \bar{P})^2}}{\Sigma P_i} \tag{2}$$

(iii) Relative interannual variability (%)

$$\frac{\frac{1}{n-1}\Sigma|P_i - P_{i+1}|}{\bar{P}} = \frac{n}{n-1}\frac{\Sigma|P_i - P_{i+1}|}{\Sigma P_i} \tag{3}$$

(iv) Ratio between the interquartile range and the median (%)

$$\frac{Q_3 - Q_1}{\text{Med}} \tag{4}$$

The advantage of the expression (3) lies in the fact that it is determined by the rainfall amounts and the chronological order, whereas the other measures are determined by the annual amounts only. The measure (4) has the advantage that no normal distribution is assumed. These are only expressions of the average variability, and it is clear that the deviation from normal may be very high in certain years (Table 7).

In general, for a given amount of annual rainfall, the variability is higher in the Mediterranean climate than in temperate regimes. The variability measures of different locations of the investigated areas are presented in Table 8. The Golan and Galilee, being mountainous areas with rains of mainly cyclonic-orographic origin, have a more stable rain regime than the Jordan Valley, where the precipitation is partly of a convective nature.

[*] Variability measures (1) and (3) are based on the period 1921–1950 (Katsnelson, 1964).

76

Table 8. Relative variability of annual rainfall

Station	Average annual rainfall mm*	Coefficient of variation%*	Relative variability %†	Relative interannual variability %†	Relative interquartile range%*
Galilee					
Kefar Gil'adi	769	24.2	19.6	31.5	38
Kefar Hananya	765	25.0	–	–	41
Nazareth	626	29.4	22.1	32.9	39
Tavor, Agr. School	497	28.8	–	–	38
Kefar Tavor	502	27.1	–	–	39
Jordan Valley					
Ayelet HaShahar	434	30.2	24.4	33.4	43
Tiberias	431	30.6	26.8	42.9	42
Mizpa	451	29.0	23.9	39.1	50
Deganya	384	29.4	–	–	52
Golan					
Kuneitra	895	26.0	20.0	32.0	–

* For the period 1931–1960.
† For the period 1921–1950.

4. Description of the seasonal rainfall

a. General outline

The rainy season in Israel extends from October to May (Figs. 27 and 28). The first 5 mm of rain generally fall between October 5–20, and the last 5 mm fall between May 5–20. Two-thirds of the annual amount of rain fall during the period December–February. The remaining third is distributed almost equally between the beginning of the season (October–November) and its end (March–May). The eastern part of Israel receives higher amounts of rain at the end than at the beginning of the season (Elbashan, 1966), and this is well illustrated in the Golan Heights. The rainiest month is January (February in some northern parts of the Golan), with 24–30% of the annual amount, whereas 14–22% of the annual amount falls in December and February. In the Galilee and the Jordan Valley, November and March have approximately the same rainfall amount (7–13% of the yearly amount), but in the Golan, March receives more rain (12–16% of the yearly amount) than November (9–14%). April has a greater average (2–4%) than October (0.2–2%).

In general terms, in the area under investigation, a quarter, a half and three-quarters of the average annual amount are accumulated in the second half of December, January and February respectively (Atlas of Israel, 1970).

b. Frequency distribution of monthly amounts

Considerable yearly deviations from the average (median) situation previously described are observed. On some occasions, October, April and even May were

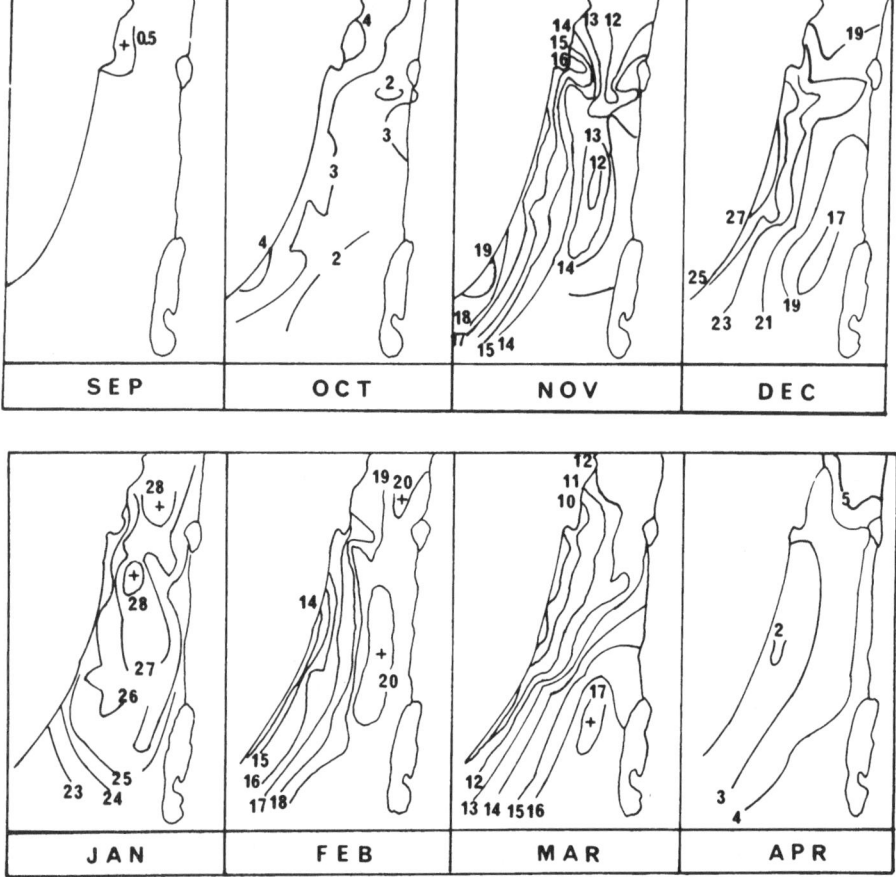

Fig. 27. Rainfall isomers† (1931–1960) (from Elbashan, 1966).
† percentage of total annual rainfall.

the rainiest months (109 mm were recorded in May 1923 at Mizpa near the lake and 114 mm fell in April 1949 at Deganya). Usually these rare cases are due to cloudbursts.

The relative variability of monthly rainfall is considerably higher than that of annual amounts, especially in the marginal months. At Deganya, for example, the relative interquartile range in January is 80% and in November, 144%. In September–October and April–May, this measure is meaningless because of the high number of zero cases.*

5. Rain days and daily rainfall

The areal distribution of number of rain days (≥0.1 mm) is more uniform and stable than the distribution of amounts of rain (Table 9). The number of rain days

* Other relative variability measures are based on the assumption that the distribution is normal and therefore could not be used for monthly rainfall which certainly deviates considerably from a normal distribution. The great disadvantage of all these measures is their dependence on the average or median value which varies considerably in time and space and approaches zero in some cases.

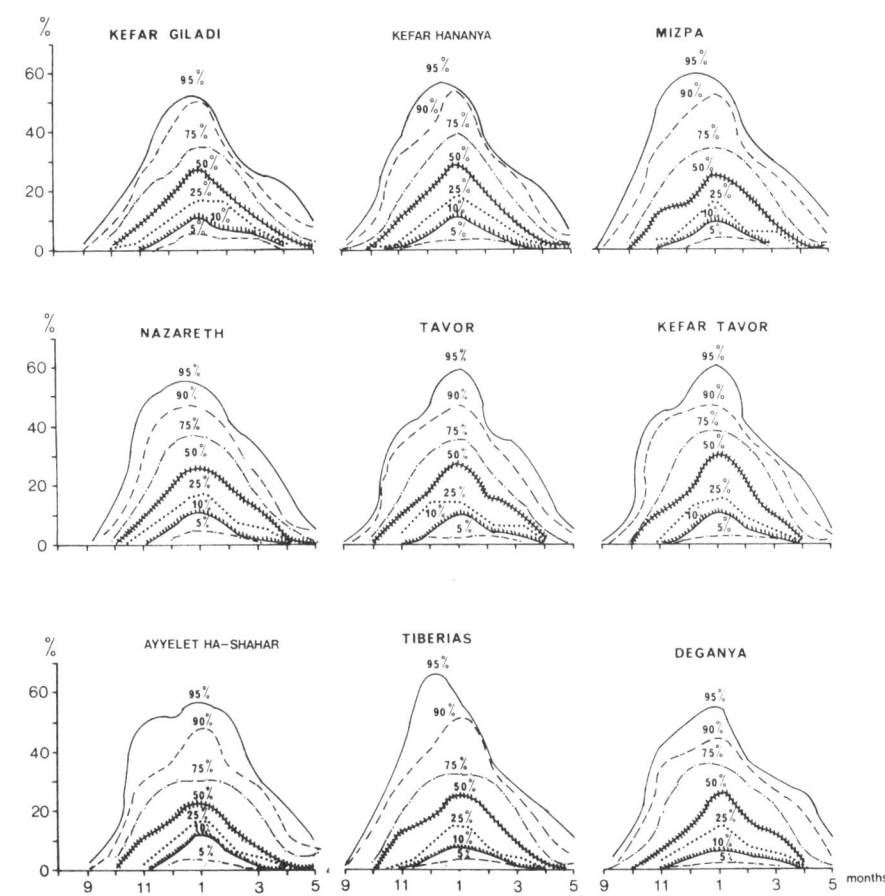

Fig. 28. Annual rainfall run, indicated by monthly percentiles (expressed as percentage of the median of annual amounts) for the period 1931–1960.

Table 9.   Cumulative frequency distribution of annual number of rain days (≥0.1 mm), 1931–1960

Station	5%	10%	25%	50%	75%	90%	95%
Galilee							
Kefar Gil'adi	45	52	57	62	75	88	89
Kefar Hananya	43	44	55	59	74	80	88
Nazareth	38	39	47	53	60	75	84
Tavor, Agr. School	38	45	52	55	69	75	82
Kefar Tavor	37	38	45	51	61	70	81
Jordan Valley							
Ayyelet HaShahar	31	37	43	50	58	65	71
Mizpa	28	33	43	50	64	75	80
Tiberias	31	34	47	49	66	80	89
Deganya	36	41	47	58	71	78	80

79

(more than 0.1 mm, 1 mm, 5 mm, etc.) decreases from north to south in approximately the same proportion as the amounts of rain. To a smaller extent, it decreases also from west to east but in a smaller proportion than the amounts of rain.

Topography is the main factor determining the areal distribution of the number of rain days. The variation in the number of rain days, due to topography, is smaller proportionally than the equivalent variation in amounts of rain. When comparing the stations Har Kena'an (+934 m) and Ayyelet HaShahar (+175 m), it appears that the former station has 165% greater annual amount of rain than the latter, but only 135% more rain days.

The median of the number of rain days (with 0·1 mm or more) ranges from 50 to 60 near the lake and from 60 to 80 in the higher mountainous regions. The corresponding values for the number of rain days with 1.0 mm or more are 40–50 and 60–70. The seasonal run of the medians of the total number of rain days with 0.1 mm or more is for the most of the area nearly symmetrical around January, with a slightly higher number of rain days in the second half of the season (Fig. 29).*

The highest number of rain days (≥0.1 mm) corresponding to 55–65% of the seasonal number of rain days is concentrated in December–February. This percentage is even higher for the number of rain days with greater daily amounts (Fig. 30). In the lower areas, this increase is noticed only up to daily amounts of 20–30 mm approximately, and for greater amounts, the percentage of number of rain days in these months tends to decrease.

January has the greatest number of rain days throughout the whole area for any daily amount up to 25 mm. For daily amounts of 25–50 mm, the peak is found, in some areas and especially the lower ones, in November, December and even October. However, in general, the distribution of the number of rain days with daily amounts smaller than 25 mm is almost symmetrical around January, with slightly higher values in the second half of the season (especially in the lower areas). The number of days with daily amounts of rain of 25 mm or more is usually greater in the first half of the season.

A most interesting phenomenon is observed in the frequency distribution of daily amounts in November–December. In November the values of the percentiles are relatively high compared to other months. This indicates that days with low daily rainfall are less frequent in November compared to other months, and the weight of higher daily rainfall is greater.

In December the percentiles up to 75% are relatively low compared with the annual run of the higher percentiles. Compared with other months the values of the percentiles up to 75% are as expected, or somewhat lower. For percentiles of 75% and more the values are relatively high compared to other months and the peak is at some stations in December (see the graph of Tavor in Fig. 30). This means that the percentage of days with rain amounts up to 15–20 mm in December is approximately as expected according to other months or somewhat lower, but the percentage of very high daily rain amounts is relatively higher than

---

*For other percentiles of number of rain days, the pattern, though basically the same, is somewhat different, especially in February–March. As February has 28–29 days, the number of rain days for the computation of percentiles was increased accordingly.

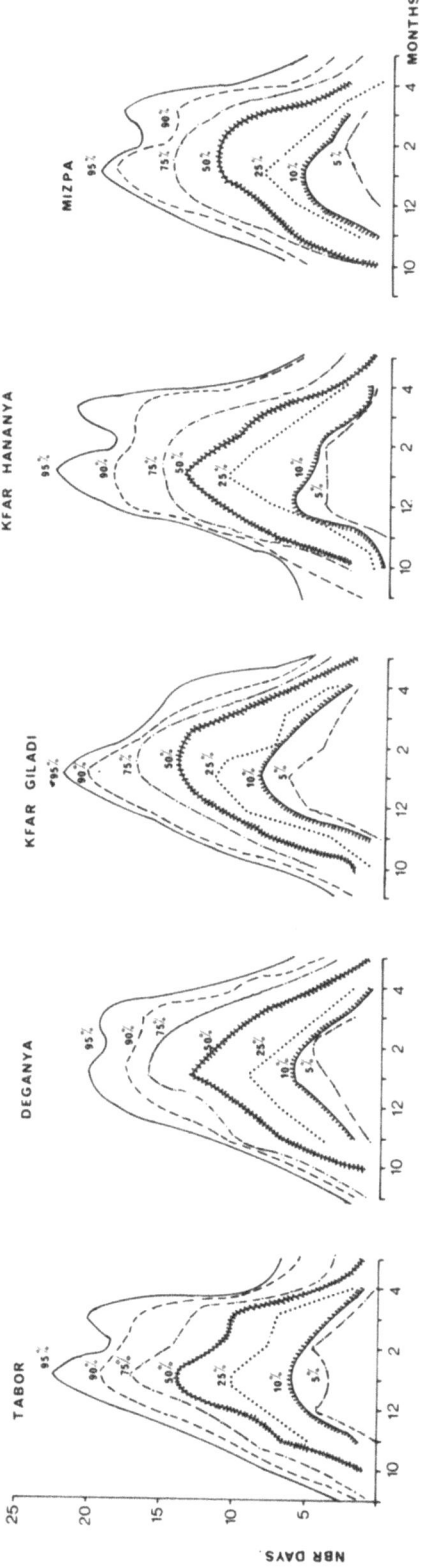

Fig. 29. Frequency distribution of number of rain days ($\geqslant 0.1$ mm) for the period 1931–1960.

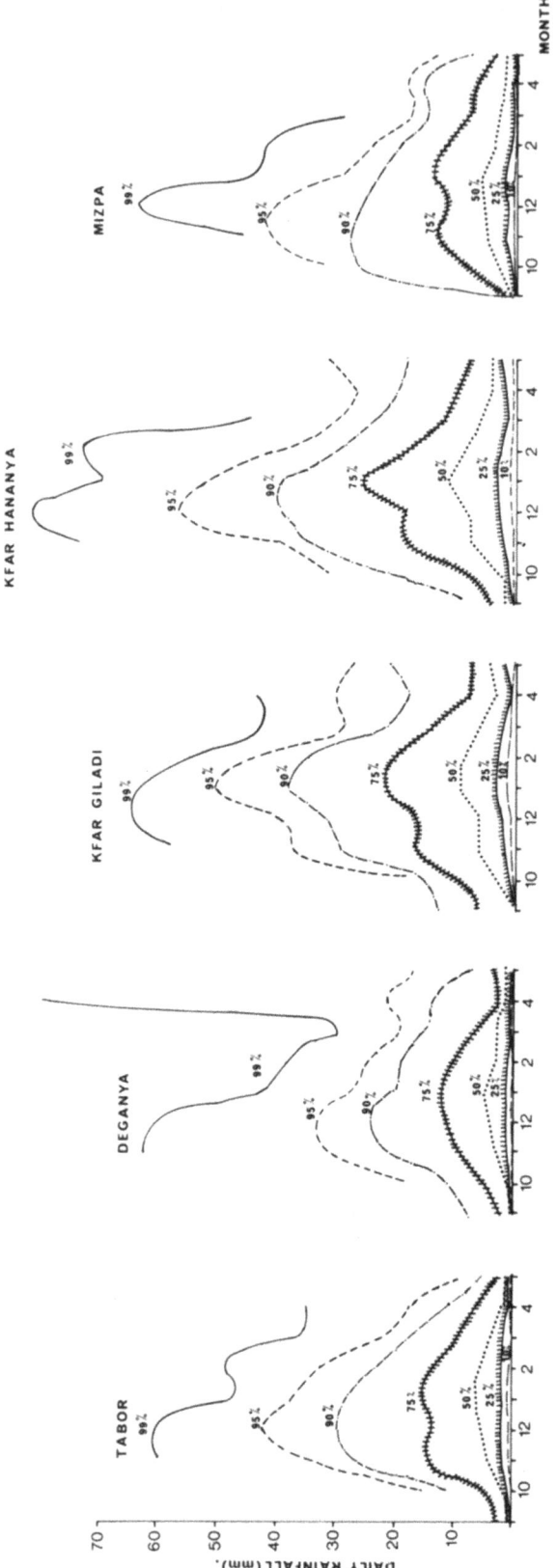

Fig. 30. Frequency distribution of daily rainfall amounts for the period 1931–1960.

it would be expected from the annual run. This means also that the proportion of daily rainfall between 15–20 mm and 30–40 mm is lower than expected.

Fig. 30 shows the high frequency of the number of days with small amounts of rain, in comparison with the small number of days with great daily rainfall.

Daily rainfall of 100 mm or more happens once in 5–10 years approximately in the higher mountainous area, but never in the Jordan Valley (the maximum daily amount at Tiberias was 80 mm in February 1929). Daily rainfall of 75 mm or more is recorded in the highest regions (Upper Galilee and high Golan Heights) once in 2–5 years. It is less frequent in the lower mountainous area and in the Jordan Valley, where such a case may happen once in 10–15 years and even less frequently. Daily rainfall of 50 mm or more is also more frequent in the mountainous area – 2–4% of all rain days – while in the Jordan Valley, these cases occur in less than 1% of the total annual number of rain days.

The variability of the annual number of rain days is much smaller than that of amounts of rain.

6. Rain intensity

This subject has been investigated by Katsnelson (1955, 1969) and Schein & Buras (1970).

High rain intensities during the short periods (sometimes called cloudbursts) are local and rare. They usually occur at the beginning and end of the season (October–November; April–May) when convection is more pronounced (see Table 10A). Cloudbursts are usually the product of a depression north of the Red Sea formed by the combination of suitable synoptic factors. The southeastern flow induces a great amount of moisture and heat into the surface layers, thus creating instability. This synoptic pattern is found in our area mainly in autumn and spring. As convection is particularly strong in the afternoon, most of the cloudbursts occur at this time of the day.

High intensities are relatively more frequent in the Jordan Valley, especially along its western edge (Table 10), and also, to a lesser degree, at the eastern edge of the mountainous area west of the Jordan Valley (Table 10). A lack of data prevents us from discussing rain intensities east of the Kinneret.

The high rain intensities, combined with the special topographical and pedological structure and the meager vegetation cover, can explain the higher frequency of floods in the area.

The general trend observed is that the longer, less extreme events are more frequent at the middle of the season, while the more extreme events, and especially those of short duration, concentrate in the marginal months, especially in the Jordan Valley.

7. Solid precipitation: hail

Lack of reliable information (because of the local and instantaneous nature of hail) prevents a thorough analysis of hail phenomena. We can merely point out that in the higher mountainous areas, hail is observed 7–9 days per year, and in the lower mountainous areas, 4–6 days or less. In the Jordan Valley, the average

Table 10.A. Some examples of extreme rain intensities*

Station	Co-ordinates Long. E	Lat. N	Israel grid		Elev. m	Duration	Rainfall mm	Date
Galilee								
Gazit	35° 27'	32° 38'	192	227	125	6 min	19.5	10 Oct. 1962
Bet Keshet	35° 24'	32° 43'	187	236	170	2 hrs	120.0	16 Apr. 1964
Jordan Valley								
Tiberias	35° 32'	32° 48'	200	244	−110	20 min	38.0	31 Oct. 1937
Tiberias	35° 32'	32° 48'	200	244	−110	25 min	44.6	25 Nov. 1947
Tiberias	35° 32'	32° 48'	200	244	−110	45 min	54.0	14 May 1934
Tiberias	35° 32'	32° 48'	200	244	−110	45 min	54.0	11 Nov. 1934
Mizpa	35° 30'	32° 47'	198	243	75	1 hr	75.0	11 May 1923
Mizpa	35° 30'	32° 47'	198	243	75	1½ hrs	109.0	11 May 1923
Afiqim	35° 34'	32° 41'	204	231	−200	30 min	51.0	7 Nov. 1957
Qiryat Shemona	35° 34'	33° 12'	203	289	90	1½ hrs	81.3	30 May 1958

* Source: Katsnelson, 1969.

Table 10.B. Computed rainfall for 4 durations and 4 annual probabilities (mm) (from Schein 1970).

Duration minutes	Probability %	Har Kenaan	Dafna	Deganya	Maoz Hayyim
15	1	24	16	30	41
	5	17	13	20	36
	20	13	10	13	31
	50	9	8	9	10
30	1	35	22	44	60
	5	24	17	28	51
	20	16	13	17	44
	50	12	11	11	13
60	1	47	32	60	70
	5	32	23	36	60
	20	21	17	22	51
	50	16	14	15	15
360	1	108	57	108	75
	5	73	42	65	65
	20	51	33	42	56
	50	42	27	30	26

number of days with hail is 2–3 per year in the north of the valley and 1–2 days in the areas near the Kinneret.

Hail phenomena are more frequent in January–March. The peak for most of the area is in March (in most stations about 25–30% of the hail events are recorded in this month). Hail occurs also in the marginal months, in which some of the most severe hail events were recorded.

8. Snow

Data sources concerning snow can be found in Katsnelson (1967) and in the Archives of the Meteorological Service of Israel.

Snow is influenced more by topography than are the other forms of precipitation. In the high Golan Heights (not including Mount Hermon) and Upper Galilee, snow occurs in 80% of the years measured, with an average of 4–5 days of snow per year.

Snow is most frequent in January and February; it occurs also in March, though to a lesser extent, still less in December, and it is very rare in November and April. The snow depth is usually not more than a few centimeters, but in extreme events, the snow depth reaches some tens of centimeters. Usually snow remains not more than 1–2 days, but sometimes it stays on the ground for several days.

In the lower mountainous areas, the average number of days of snow is only 1–2 per year, and the amount of snow is also smaller than that of the higher areas.

In the Jordan Valley, snow is very rare and the amount of snow is negligible.

Snow distribution in the Hermon is completely different. Up to a height of 1,000 m, most precipitation is rain; thus, snow cover does not remain more than 2–3 days. In the Hermon area, between 1,000–1,700 m, a considerable percentage of precipitation is snow, especially in January–February. The snow cover is maintained for several weeks at the middle of the season and for several days in the marginal months. In the high Hermon, above 1,700 m, most precipitation is snow. The snow season lasts from November to April, with a peak in January–February. On account of the strong winds on Mount Hermon, snow may accumulate up to tens of meters on leeside locations.

The snow melting begins in March, and influences, to a great extent, the inflow to the Kinneret, the runoff, ground water replenishing, etc. Towards the spring, most of the snow cover disappears. In summer, snowfall is very rare, and most of the snow cover is already melted by this time of the year.

References

Aelion, E. 1958. A report on weather types causing marked storms in Israel during the cold season. Isr. Meteorol, Serv. Series C, No. 10.

Ashbel, D. 1936. Wind conditions on the western and southern shores of Lake Kinneret. Teva Vaaretz (Nature & Country) 4(5):189–195 (in Hebrew).

Ashbel, D. 1937. The temperature of the shores of Lake Kinneret. Teva Vaaretz (Nature & Country) 5(3):316–321 (in Hebrew).

Ashbel, D. 1950. Bioclimatic Atlas of Israel. Central Press, Jerusalem.

Ashbel, D. 1964. The climate of the Kinneret Valley. 35–54. In: Climate of the Great Rift, Arava, Dead Sea, Jordan Valley. Hebrew Univ. Jerusalem. 55 p. + tables and data.

Atlas of Israel. 1970. ch. 4 – Climate. Survey of Israel, Ministry of Labour, Jerusalem.

Berman, T. 1976. Light penetrance in Lake Kinneret. Hydrobiologia. 49:41–48.

Brier, G. W., L. O. Grant & P. W. Mielke, Jr. n.d. An evaluation of extended area effects from attempts to modify local clouds and cloud systems. Colorado State Univ. Fort Collins, Colorado.

Doron, E. & J. Neumann. 1976. A mesometeorological model with topography. Dept. Atmosph. Sc. Hebrew Univ. Jerusalem. 29 p. and 14 diagrams.

Elbashan, D. 1966. Monthly rainfall isomers in Israel. Israel J. Earth Sci. 15:1–7,

Elbashan, D. 1967. Climatographical notes. In: Climatological standard normals of rainfall 1931–1960. Isr. Meteorol. Serv. Ser. A, No. 21.

Gabriel, K. R. 1970. The Israeli rainmaking experiment, 1961–1967. Final statistical tables and evaluation. The Hebrew Univ. Jerusalem.

Gabriel, K. R. & H. Kesten. 1963. Statistical analysis of annual rainfall in Jerusalem 1860–1960. Bull. Res. Counc. Israel 11G:3. 142–145.

Gat, Z. & Z. Paster. 1974. The Agroclimate of the Golan Heights (A rainfall analysis). Isr. Meteorol. Serv., Agromet. Rep. 2/74 (in Hebrew).

Harel, M. & D. Nir. 1963. Geography of the Land of Israel. Tel Aviv, Am Oved Publ. (in Hebrew).

Hutchinson, G. E. 1957. A treatise on Limnology. 1: John Wiley & Sons. New York, 1014 p.

Johnson, J. 1945. Termisk-hydrologiska studier: Sjön klammingen. Geogr. Ann. Stockh. 28:1–154.

Katsnelson, J. 1955. Rain intensities in Palestine. Isr. Meteorol. Serv. Ser. E. No. 3 (in Hebrew).

Katsnelson, J. 1964. The variability of annual precipitation in Palestine. Archiv. für Meteorologie, Geophysik und Bioklimatologie, Ser. B, Band 13, 2 Heft. Wien 163–172.

Katsnelson, J. 1967. Regional Climatology of Palestine. Israel Meteorol. Serv. Ser. A, No. 23 (in Hebrew).

Katsnelson, J. 1969. Rainfall in Israel as a basic factor in the water budget of the country. Isr. Meteorol. Serv. Ser. A, No. 24 (in Hebrew).

Koppen, W. 1931. Grundriss der Klimakunde, Berlin–Leipzig.

Neumann, J. & Y. Mahrer. 1975. A theoretical study of the lake and land breezes of circular lakes. Mo. Wea. Rev. 103:474–485.

Odum, E. P. 1971. Fundamentals of Ecology 3rd Edition. W. B. Saunders Co., Philadelphia. 574 p.

Schein, Z. & Buras, N. 1970. Analysis of rainfall intensities in Israel. Isr. Inst. Tech., Agricult. Engin. Fac. Haifa (in Hebrew).

Serruya, S. 1975. Wind, water temperature and motions in Lake Kinneret. Verh. Internat. Verein. Limnol. 19:73–87.

Shaia, J. 1964. Characteristics of the 500 millibar surface over Beer Ya'aqov on rain days. Isr. Meteorol. Serv. Ser. A, No. 20.

Stanhill, G. 1967. The climate of Lake Kinneret: A review of existing data. TAHAL Intern. Report.

Stanhill, G. 1969. The temperature of Lake Tiberias. Isr. J. Earth Sci. 18:83–100.

Thornthwaite, C. W. 1948. An approach towards a rational classification of climate. Geogr. Rev. 38:55–94.

Sources of climatological data

Publications of Israel Meteorological Service
Series B. Meteorological Notes

No. 3B. Climatological normals. Part One–B. Temperature and relative humidity (2nd edition), 1961 (reprinted 1964).

No. 21. Climatological standard normals of rainfall, 1931–1960 (1967).

Series B. Observational Data

Monthly weather report. Published since November 1947 (1948 omitted).

Annual weather report. Published since 1948.

Annual rainfall summary. Published since 1947/1948.

Agro–Meteorological Reports

No. 3. Gat, Z. & J. Lomas. 1968. Agroclimatic analysis of Lake Tiberias Valley (in Hebrew).

IV Hydrology

F. Mero

A. The watershed

The Jordan Valley is the inland drainage with the lowest base level in the world (400 m below sea level at the Dead Sea). Lake Hula and Lake Kinneret represent intermediary base levels. Lake Kinneret drains a catchment area of 2,730 km² from an altitude of 2,800 m (Mount Hermon) down to a base level of 209 m below MSL.

The Kinneret watershed can be subdivided into four sectors (Fig. 31): (i) The catchment area of the tributaries of the Jordan: Hermon, Dan, Snir and Ayun rivers. This sector represents an area of 820 km². (ii) The Hula Basin, collecting the waters of Sector (i) and the Golan Heights. These waters are collected in the canals of the presently regulated Jordan River. This sector is approximately 648 km². (iii) Small catchments along the banks of the Jordan River in the portion of the river between the Pardes Huri station and the lake. This sector includes a narrow strip of 122 km². (iv) Catchment areas draining their waters directly into the lake through wadis, ground water or shore and sublacustrine springs. This sector is located in the immediate vicinity of the lake and covers an area of 968 km².

Each of the hydrological sectors includes one or more superficial catchment units drained by a central river or wadi. The underground flow system is much more complex. The stratigraphical discontinuities due to the considerable faulting and block type tectonics explain why the underground aquifers are divided into numerous and discontinuous units. The natural outlets of these aquifers are the springs located on the fringes of the Rift Valley, along the faults and on the lake bottom.

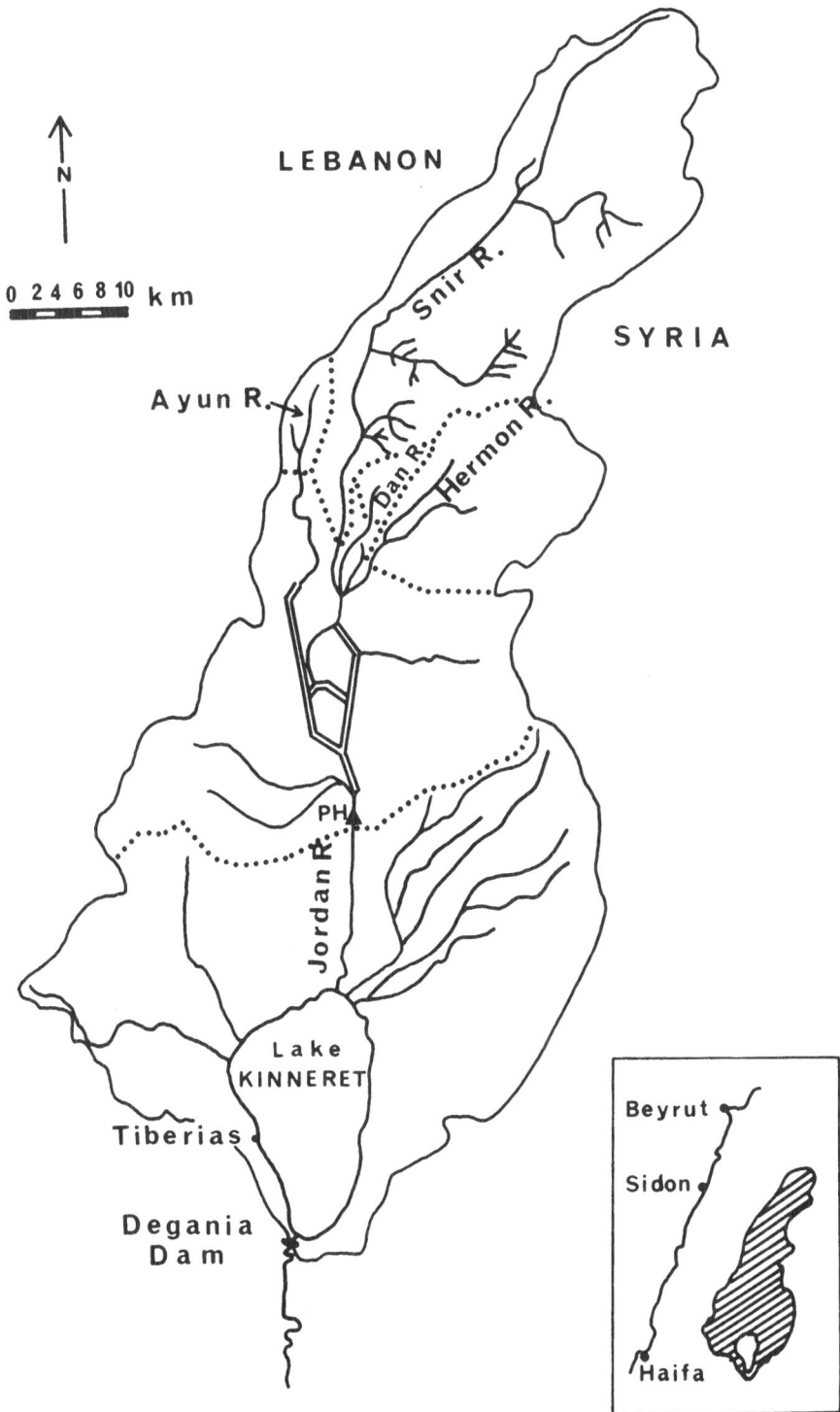

Fig. 31. The main hydrological units of the Kinneret watershed. PH = Pardes Huri, station of the hydrological service.

B. The superficial waters

1. The Jordan River

The Jordan River is the most important inlet to the lake. It results from the junction of three rivers: the Snir (Hatsbani River), the Dan River and the Hermon River (Banias River). Formerly, the Jordan River used to flow into the swamp area of Lake Hula. In 1957, the artificial drainage of the Hula plain was completed and the river regulated. It now flows into two canals (western and eastern canals) which drain the plain on both sides and join into a single canal which ends at the gauging station of Pardes Huri.

The average yield values (Pardes Huri station, period 1959–1960/1970–1971) range from 7.9 m³/sec in August to 29.6 m³/sec in February (Table 11). The maximum known discharge was 214 m³/sec on 23 January 1969, and the minimum was 0.80 m³/sec on 10 July 1973.

Table 11. Yields of the Jordan River at Pardes Huri station. Monthly averages for the 1959–1960/1970–1971 period.*

	Yields m³/sec		Yields m³/sec
October	9.5	April	24.6
November	11.6	May	17.6
December	15.9	June	11.6
January	23.5	July	8.1
February	29.6	August	7.9
March	29.0	September	9.0

* Data from the Hydrological Service.

The base flow of the Jordan River is not higher than a few cubic meters per second and its regime is of the flood type. Its yields are consequently very variable from year to year, as shown in Fig. 32 for the yield of the Jordan River during the period 1965–1974.

2. Other rivers and wadis

The western watershed is drained mainly by Wadis Amud and Tsalmon which are completely dry in summer. The central part of the Golan Heights, which forms the eastern watershed of the lake, is drained essentially by the Meshushim River. Its average yield is generally below 1 m³/sec, although the maximum known yield reached 115 m³/sec on 4 April 1971. The other wadis are of secondary importance.

91

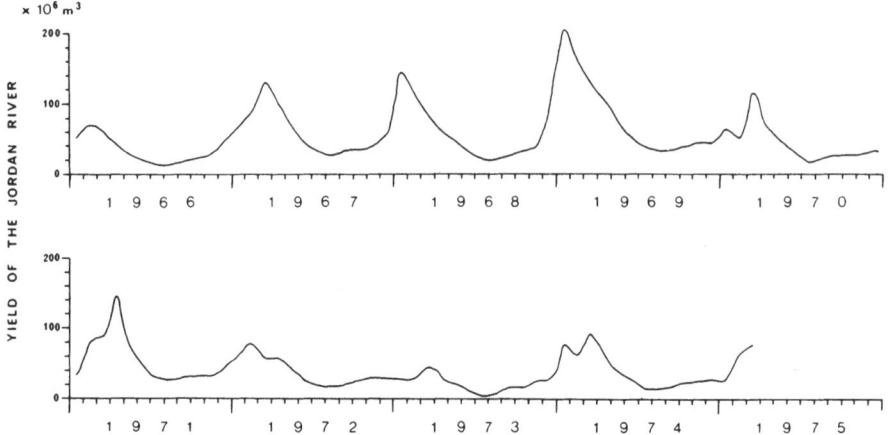

Fig. 32. Monthly yields of the River Jordan. (Data from the Hydrological Yearbook of Israel.)

3. The springs

The very recent tectonic disturbances have caused the discontinuity of water-bearing layers which find an outlet in numerous freshwater springs. In Table 12, a list of the main springs of the lake vicinity can be found, together with their main characteristics. Their location is shown in Fig. 33.

Table 12. List of freshwater springs in the lake vicinity. Serial numbers indicate location of spring in Fig. 33. Data from the Hydrological Yearbook of Israel.

No.	Name	Elevation m	*Average discharge l/sec	†Cl mg/l
3	Meron	+695	23	24
4	Bar-Yohay	+655	–	–
5	Taron	+530	4	22
6	Yaqim	+520	24	23
7	Zetim	+705	3	26
8	Amud	−100	–	–
9	Parod	+480	12	26
10	Ravid	−100	121	30
11	Nun	−192	70	54
12	Arbel	−140	17	45
13	Sanabir	+310	38	26
14	Sheikh-Husein	+260	4	57
15	Dardare	−130	3	30
16	Qutsbiya-el-Jedida	+445	44	29
17	Hushnie	+790	3	23
18	Fahem	+710	142	22
19	Tanuriya	+630	30	22
20	Umm-a-Dannir	+600	9	22
21	Umm-a-Dapun	+590	16	22
22	Mantsura	+560	20	18＼
23	Mujahiya	+100	26	37
24	Rapid 1	+700	–	–
25	Rapid 2	+730	9	28
26	Butmiya	+695	11	25
27	Beja	+660	26	30
28	Betset Juhader	+570	52	–
29	Sahina	−120	300	72
30	Magela	−160	128	532
31	Balzam	−150	182	330
33	Po'em	+515	–	–
34	Ramiel	+630	–	–
36	Re'ah	−100	–	–

* Interannual average.
† Values of October 1972.

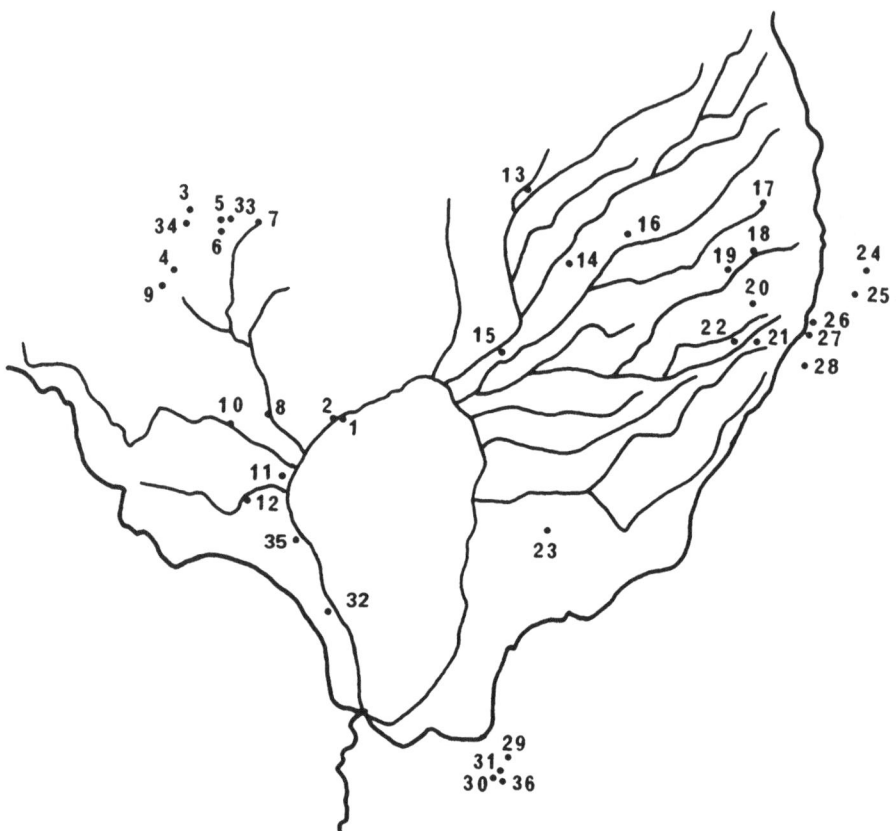

Fig. 33. Location of freshwater springs. The numbers correspond to the list of Table 12. (data from the Hydrological Yearbook of Israel).

C. The thermo-mineral springs

Although the main source of water of the lake, the Jordan River, presents the 'normal' ionic sequence of most of the inland freshwater bodies ($Ca > Mg > Na > K$ and $HCO_3 > SO_4 > Cl$), a completely different chemical pattern characterizes the lake water ($Na > Mg > Ca > K$ and $Cl > HCO_3 > SO_4$) as shown in Fig. 34.

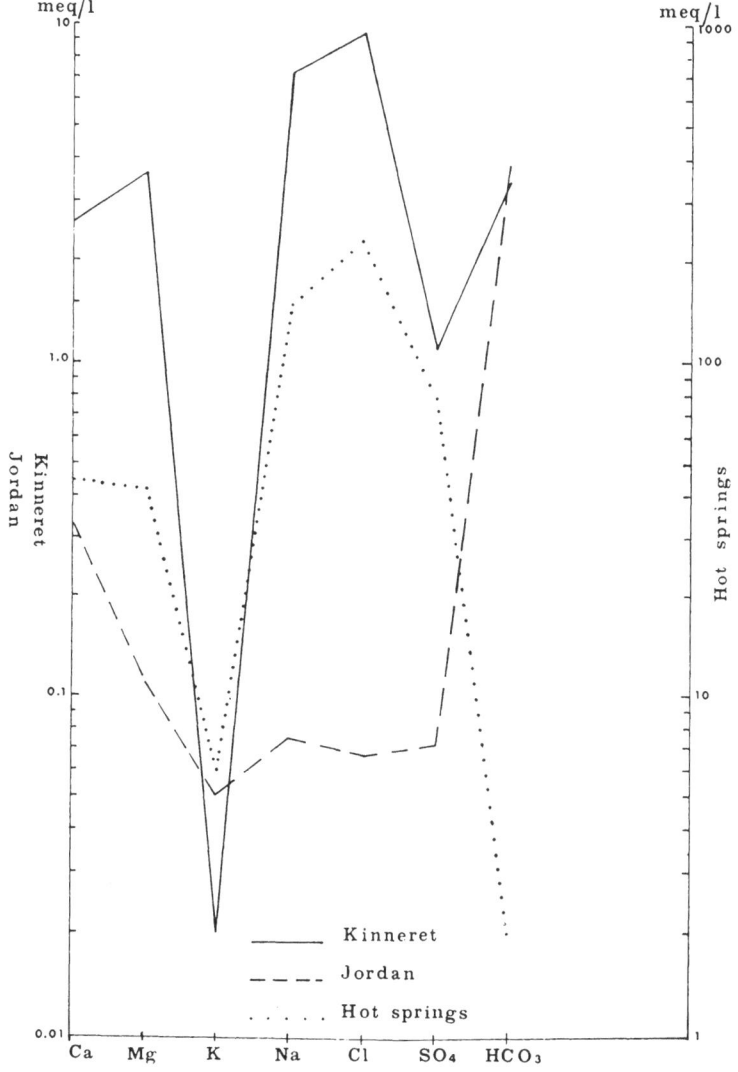

Fig. 34. Comparative chemical composition of the Kinneret, River Jordan and Hot Springs waters. (From C. Serruya & U. Pollingher, 1971, Mitt. Internat. Verein. Limnol., 19.)

95

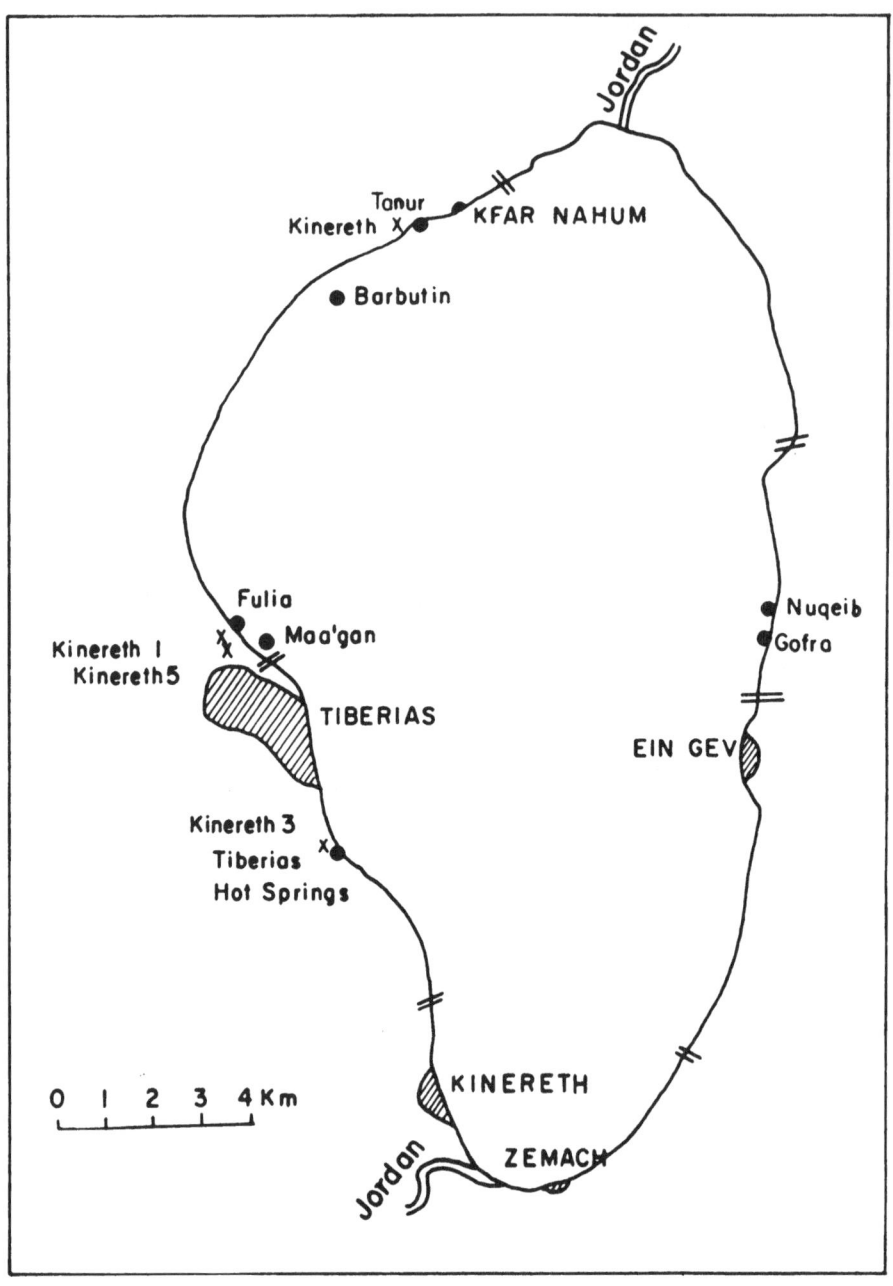

Fig. 35. Location of the thermo-mineral springs in the Kinneret vicinity.

Moreover, the chloride concentration of the Jordan water does not exceed 20 ppm, whereas in the lake, it amounts to 250 ppm. These considerable differences between the chemistry of the lake water and the water of its main inlet are due to the presence of thermo-mineral springs on the shores and on the bottom of the

Table 13. Main features of the thermo-mineral springs

Spring name and location	Discharges m³/sec			Chloride concentration ppm			Mean Cl⁻ discharge 10^3 tons/yr	Temp. °C	
	Max.	Min.	Ave.	Max.	Min.	Ave.		Max.	Min.
Western coast:									
Ein Sheva Group (201.87–253.11)	0.210	0.080	0.135	1,400	400	1,050	4,500	29	21
Ein Nur Group (201.79–253.16)	0.893	0.486	0.700	2,500	1,300	1,970	43,500	30	22
Ein Fuliya Group (199.63–246.00)	~0.500	~0.05	~0.350	1,300	~600	~1,100	~12,150	32	27
Hamei Tveria Group (201.83–241.40)	0.035	0.022	0.029	16,500	18,200	17,900	16,400	61	56
Total known springs on western coast	1.156	0.638	1.214	–	–	–	~76,550	–	–
Sublacustrine springs:									
Barbutim Group (201.40–251.70)	?	?	?	4,000	1,100	~2,500		31	26
Ma'agan Group				2,400	1,100	1,800		36	33
Saline seepages along eastern and western shores of lake	?	?	?	22,000	1,000	?		?	?
Estimated total	?	?	~1.60			~2,000	~71,000	–	–

lake. The location of these springs is shown in Fig. 35, and the characteristics of the main springs are listed in Table 13.

The exact location of three groups of sublacustrine mineral springs was determined during the bathymetric measurements carried out in the 1960's by TAHAL (Water Planning for Israel).

The Barbutim group in the NW sector of the lake consists of a series of 'craters' within a general NE–SW directed depression. The largest crater is located at the southern part of the depression and has the strongest discharge. The Ma'agan group, situated 800 m offshore of the Tiberias fisheries port, consists of six circular funnels. The largest one is approximately 40 m wide and 20 m deep (Fig. 36). The Fuliya group consists of various sublacustrine springs emerging at about 150 m from the shore or close to the shore. Although their discharge is modest in comparison with other sources of water, the chloride balance indicates that the springs bring to the lake an annual amount of 150,000 tons (±20%) of chloride. As can be seen from Table 13, only half of the saline influx originates from springs on or close to shore. The other half is contributed by sublacustrine springs and

97

seepages. The danger that this salt represents to the quality of the lake water explains the detailed investigations which have been conducted in order to clarify the origin of the salty water and the hydraulic mechanisms which bring it back to the surface. The results of these studies are reported in Chapter V.

D. The water balance of Lake Kinneret

In the Kinneret watershed, the amount of precipitation ranges from 400 mm/yr in the lake area to 1,200 mm/yr in the Hermon mountains, and reaches a total average volume of 2,160 MCM/yr. The deviation from this average value is considerable: 160 to 30% of this amount is received by the watershed in very wet and very dry years respectively.

Approximately 75% of the total superficial runoff is of underground origin, and corresponds to the rather uniform baseflow of rivers and springs. The remaining 25% comes from storm runoff and surface flow occurring during and after the rainy period.

The considerable interannual variations of the discharge are caused by the nearly direct response of the surface flow to the rainfall amounts and their time distribution.

The generalized water balance presented in Table 14 has been established according to the average precipitation and discharge data of the period 1959–1974. The water inflow of the drainage area of the Jordan River accounts for 66% of the total lake inflow. It is interesting to note that the average net Jordan inflow (558 MCM) is practically equal to the average lake outflow (546 MCM). This means that the 295 MCM of water lost annually by evaporation must be compensated for by an equivalent additional inflow coming from the immediate vicinity of the lake (the 282 MCM of the Kinneret Basin). If we subtract the amount of annual precipitation on the lake (70 MCM), we obtain an amount of water corresponding to the surface flow of the wadis and to the freshwater and saline springs (212 MCM). Since the saline flux is estimated to be 90×10^6 m^3/yr, it follows that the freshwater supply of the basin close to the lake amounts to 122 MCM.

An attempt has been made to determine the contribution of each geological formation of the Kinneret Basin, which was divided into four areas: (i) An area covered with basalts in the northern and western sectors of the basin. It covers 125 km^2 and includes minor aquifer units. (ii) An area of 85 km^2 where the dominant Eocene limestone forms confined aquifers of medium size. (iii) The Turonian–Cenomanian karstic formations extending over 180 km^2. The water of this aquifer seems to mix with the saline water and consequently plays a role in the final chemical composition of the mineral springs. (iv) An area of 205 km^2 along the eastern shore of the lake. It is covered with basalts and includes a minor aquifer giving rise to numerous springs.

The known hydrological features of each formation allowed the estimation of the superficial and underground annual influx through each formation for the period 1959–1960/1973–1974 (Table 15).

The average total influx (217 MCM) is in excellent agreement with the contribution of the Kinneret Basin (212 MCM) calculated in the overall water balance of Table 14. We note also that the Golan Heights contributes as much as 60% of

Table 14. Generalized water balance of Lake Kinneret

Name of watershed or river basin	Drainage area 10^6km^2	Aver. prec. mm/yr	Yearly discharges (MCM) winter	summer	total
(1) *Upper Jordan:*					
Hermon River	141.0	?	82.0	35.0	117.0
Dan River					
(mainly ground water)	24.0*	?	150.0	114.0	264.0
Snir River	623.0	?	98.0	32.0	130.0
Ayun River	32.0	?	9.0	0.3	9.3
Total:	820.0	~1,200	339.0	181.3	520.3
(2) *Hula Basin:*					
Eastern border catchments	237.0	881	~40.0	20.0	60.0
Western border catchments	254.0	720	35.0	14.0	49.0
Hula Valley	157.0	551	~16.0	8.0	24.0
Total:	648.0	703	91.0	42.0	133.0
Total Jordan at Pardes Huri gauging station	1468.0	980	430.0	223.3	653.3
(3) *Jordan–Korazim Valley*†	122.0	460	~10.0	~5.0	~15.0
Total Jordan into Lake Kinneret (undiverted)	1,590.0	955	440.0	228.3	668.3
Estimated upstream use†	–	–	~15.0	~95.0	~110.0
Total net Jordan inflow	1,590.0	955	425.0	133.3	558.0
(4) *Kinneret Basin:*					
Eastern catchments (S. Golan)†	583.0	615	91.0	12.0	103.0
Western basins and springs†	385.0	620	71.0	38.0	109.0
Lake surface (direct rain)†	169.0	415	70.0	–	70.0
Total Kinneret Basin contribution†	1,137.0	587	232.0	50.0	282.0
Total net lake inflow‡	2,727.0	793	662.0	188.0	840.0
(5) Evaporation losses from lake (calculated by energy balance)	169.0	1,740.0	−101.0	−193.0	−294.0
Total of average lake outflow					~546.0

* The indicated catchment area for the Dan springs refers to the surface catchment only. The extent of the ground water catchment area is unknown. Its extent is about 500 km^2, estimated from water balance considerations.

† Discharges are estimated averages, obtained by hydrometeorological and conventional water balances.

‡ Based on daily water balances, lake levels, outflow measurements, etc.

100

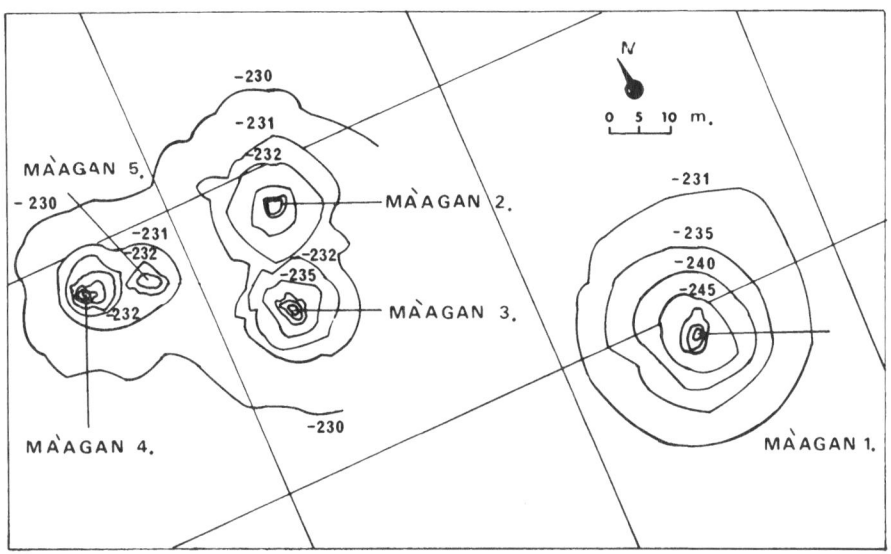

Fig. 36. Topography of the funnels of the Ma'agan sublacustrine springs.

the total influx. Although one can hardly speak of basaltic aquifers, large amounts of water pass through the basalts and find their way through underlying formations.

Table 15. Contribution of the different geological formations to the lake inflow. Results in MCM.
GW = ground water flow. SUR = superficial flow.

		West & North basalts	Eocene	Cenoman	Golan	Total
1959/60	GW	1.4	25.7	53.3	22.5	102.9
	SUR	1.8	1.9	3.9	10.5	18.1
	TOTAL	3.2	27.6	57.2	33.0	121.0
1960/61	GW	2.2	17.8	49.4	36.1	105.5
	SUR	2.8	1.4	3.0	27.5	34.7
	TOTAL	5.0	19.2	52.4	63.6	140.2
1961/62	GW	17.9	25.1	57.1	85.4	185.5
	SUR	10.2	4.4	7.9	55.1	77.6
	TOTAL	28.1	29.5	65.0	140.5	263.1
1962/63	GW	5.6	18.9	52.2	70.5	147.2
	SUR	3.7	2.3	4.1	40.5	50.6
	TOTAL	9.3	21.2	56.3	111.0	197.0
1963/64	GW	8.7	17.8	49.9	78.2	154.6
	SUR	6.3	3.1	4.4	46.8	60.6
	TOTAL	15.0	20.9	54.3	125.0	215.2
1964/65	GW	20.3	25.7	56.9	96.4	199.3
	SUR	10.3	4.7	8.1	49.9	72.5
	TOTAL	30.6	30.4	64.0	145.8	271.8
1965/66	GW	3.0	13.3	45.6	39.9	101.8
	SUR	3.1	1.3	1.9	19.9	26.2
	TOTAL	6.1	14.6	47.5	59.8	128.0
1966/67	GW	14.0	21.4	50.8	96.7	182.9
	SUR	7.8	3.9	5.6	59.4	76.7
	TOTAL	21.8	25.3	56.4	156.1	259.6
1967/68	GW	7.2	20.0	51.8	75.1	154.1
	SUR	4.4	3.5	5.5	41.9	55.3
	TOTAL	11.6	23.5	57.3	117.0	209.4
1968/69	GW	34.7	40.5	69.5	165.1	309.9
	SUR	23.5	13.6	21.1	131.7	189.9
	TOTAL	58.2	54.1	90.6	296.8	499.7
1969/70	GW	4.6	20.6	55.7	71.3	152.2
	SUR	3.5	2.4	3.7	35.2	44.8
	TOTAL	8.1	22.0	59.4	106.5	197.0
1970/71	GW	7.7	18.6	51.2	85.8	163.3
	SUR	7.2	2.9	3.8	63.3	76.4
	TOTAL	14.1	21.5	55.0	149.1	239.7
1971/72	GW	1.7	15.0	46.0	67.8	130.5
	SUR	1.9	1.6	1.6	32.2	37.3
	TOTAL	3.6	16.6	47.6	100.0	167.8
1972/73	GW	0.0	7.3	35.5	38.8	81.6
	SUR	0.6	0.5	0.5	21.8	23.4
	TOTAL	0.6	7.8	36.0	60.6	105.0
1973/74	GW	6.0	16.7	45.3	91.1	159.1
	SUR	3.7	3.5	5.5	62.7	75.4
	TOTAL	9.7	20.2	50.8	153.8	234.5
Average	GW	9.1	20.3	51.3	74.7	155.4
Average	SUR	5.9	3.4	5.4	46.5	61.2
Average	TOTAL	15.0	23.7	56.7	121.2	216.6

V Mineral waters of the Kinneret basin and possible origin

E. Mazor

A. General features

The thermo-mineral springs are all located in the deepest part of the depression at or below an altitude of 200 m below MSL. Nearly all the springs are located on the western shore and the western part of the lake bottom. The salinity and temperature of the springs vary in a broad range from 400 to 22,000 ppm chloride and from 27 to 61°C respectively; in general the waters having the highest salinity have also the highest temperature. Besides their high salinity and temperature, these springs are also radioactive.

The radioactivity of the mineral water was utilized by Braudo et al. (1970) to detect sublacustrine springs or seepage areas. A preliminary scanning of the lake bottom with a scintillation probe (Braudo et al., 1968) indicated two areas of high radioactivity which were subsequently investigated in detail. The first study concerned a lake portion of approximately 2,000 m long in the N-S direction and 600 m wide in the E-W direction, situated offshore of the group of Tiberias Hot Springs (Hamei Tveria). The radioactivity and salinity measurements indicate that the major saline water upwelling takes place along a line parallel to the shore at a distance of 150 m from the coast. A second line parallel to the coast and distant from it by 500 m represents another, but weaker, source. A rather large scattering was found between the salinity and radioactivity data. Since the radioactive material is strongly absorbed by clay minerals, areas of high radioactivity are those where the brine flow has taken place for a long time but not necessarily up to the present period. Conversely, salinity indicates present or very recent discharge. Other radioactive areas were located in the Tabgha area, at about 100 m offshore, Ein Sheva and Ein Nur springs, and along the eastern shore, between Haon and the Buteiha plain at distances of 50 to 150 m from the shore.

Approximately 200 wells have been drilled in and around the lake, in order to clarify the origin and emergence mechanism of the mineral water. In general, the hydraulic pressure of the saline aquifers increases from west towards east. The water pressure distribution in the wells follows the tectonic pattern of faulted blocks described by Saltzman (1964). The water pressure is then uniform in all the wells drilled in one block, but may differ by tens of meters from the pressure in the adjacent block. Similarly, dye tests with Rhodamine showed interconnections between the wells of each block but no connections at all between the blocks, even if the wells tested were very close.

B. Chemical composition

1. General composition:

The mineral waters of the Kinneret area (see location on Fig. 35) have a uniform composition but a very wide range of concentrations. The data of the Kinneret basin, obtained from 260 samples taken in different seasons and during different pumping regimes, plot on straight lines or bands in Na, Ca, Br and K diagrams (Fig. 37). However, they can be divided into four subgroups.

    a. The Fulya-Barbutim-Tabgha subgroup plots best on straight lines in the different diagrams and its lines have a somewhat different angle than the other subgroups, especially for Bromine. The Br line shows Br values lower by a factor of 2 than in the other subgroups and quite close to the oceanic value. The HCO_3 content is the highest among the Lake Tiberias mineral waters (Table 16).

    b. The Tiberias city subgroup is indistinguishable from the Zemach and Ein Gev subgroups on the diagrams of Na, K and Br. It is unique in the shape of the Ca and Mg contents. The lines of these elements change their slope at 700–800 meq/litre total solids. Above this concentration the Ca line bends. The excess of Ca over Mg characterizes the Tiberias city subgroup and its northern neighbour. The Tiberias city waters are lowest in HCO_3.

    c. The Zemach subgroup resembles its adjacent neighbour subgroups in the relative Na, K, Cl and Br contents. However its Mg exceeds its Ca and it has the highest SO_4 and HCO_3 values.

    d. The Ein Gev subgroup resembles the Tiberias city and Zemach waters in its relative Na, K, and Br contents. Its Mg exceeds also the Ca. The SO_4 and HCO_3 contents are low and that puts them aside from the Zemach subgroup.

    To sum up, the Fulya-Barbutim-Tabgha subgroup differs from the rest in its lower Bromine content. The two western subgroups have Ca > Mg whereas the two eastern ones have Mg > Ca. The SO_4 and HCO_3 relative concentration ranges are rather uniform in each subgroup but differ from one to the other.

    e. The Gofra and Nuqueib springs have Mg > Ca but otherwise resemble the Tabgha-Barbutim-Fulya groups.

    The geographical boundaries between the subgroups seem to be sharp and abrupt. This fact is explained by the existence of separate saline water reservoirs, varying slightly in composition one from the other because of variable reactions with the different host rocks.

    The lines found by the analyses of waters from the various subgroups extend to the origin of the coordinates in the diagrams of Fig. 37. This seems to indicate that in each area, one kind of saline water exists, which is diluted to various degrees by fresh water. If the sources should, for example, result from the mixture of two or more saline waters of different composition, the extrapolated lines would cross the vertical axis. If more than two types of saline water or dissolution of salts should be involved, no straight lines would be obtained.

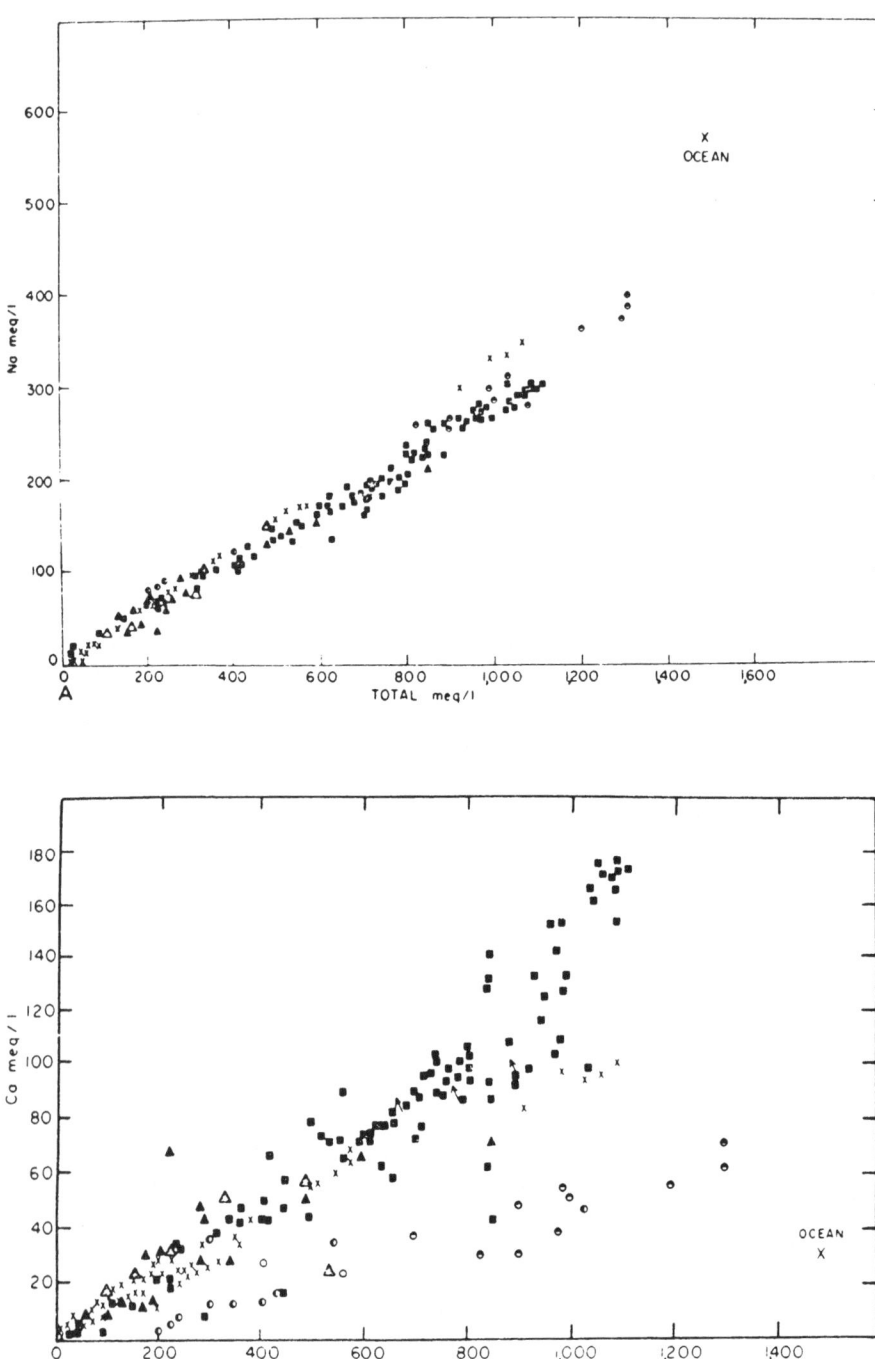

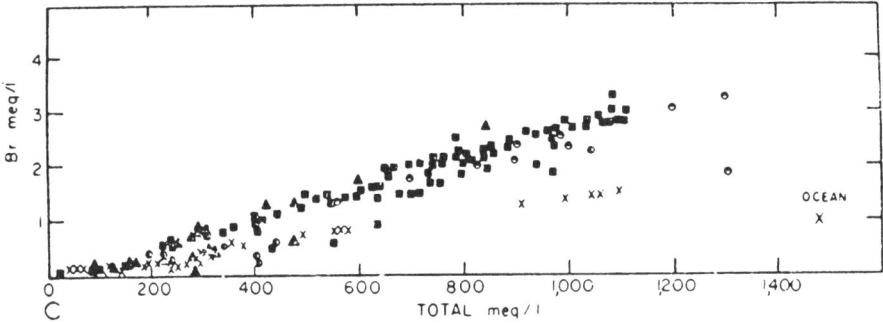

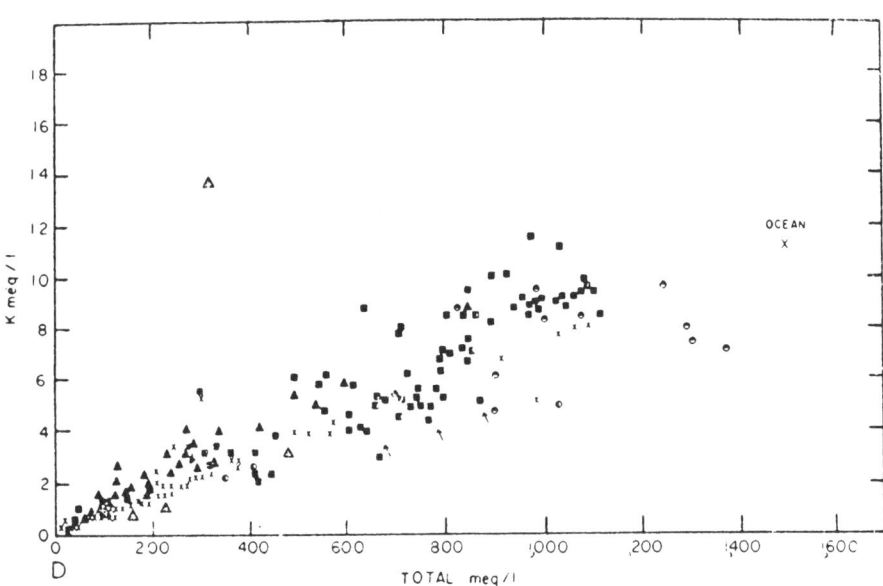

Fig. 37. Na (A), Ca (B), Br (C) and K (D) diagrams.

107

Table 16. Representative chemical analysis of springs and boreholes in the Kinneret Basin (mg/litre)

Spring	Coordinates	°C	date	Li	Sr	K	Na	Mg	Ca	Br	Cl	SO₄	HCO₃
Tannur 1 (Tabcha)	2007/2526	28	11.3.60			25	259	76	260	12	1,318	104	280
Tannus 1 (Tabcha)	2007/2526		9.8.60			44	1,024	113	398	19	2,325	186	305
Tannur 2 (Tabcha)	2007/2526		12.3.61			29	661	65	289	12	1,484	128	262
Tannur 3 (Tabcha)	2007/2526		12.3.61			55	1,300	119	499	25	2,915	230	317
Kinneret 4 borehole	2018/2532		12.3.61			96	2,302	207	770	38	5,089	416	323
Kinneret 4 borehole	2018/2532		22.3.61			154	4,038	379	1,317	69	9,046	749	293
Barbutim, 48″ casing (in lake)	2014/2017		16.2.64			56	1,301	143	546	29	3,015	267	363
Barbutim spring (in the lake)	2014/2017		30.11.60			31	758	108	347	18	1,807	132	366
Russian Garden Spring A (Fulya)	1997/2460	28	16.3.60				670	88	243		1,400	169	323
Russian Garden Spring B (Fulya)	1997/2460	26	16.3.60				404	63	176		851	102	317
Russian Garden Spring C (Fulya)	1997/2460	29	22.7.58				137	33	100		284	41	293
Ma'agan 1 (in lake)	2009/2456		22.11.63			36	848	126	300	17	1,942	167	286
Kinneret 5 borehole	1997/2457		8.6.61			21	504	94	156	6	1,061	139	305
Kinneret 1 borhole	1996/2458		12.7.60			19	436	55	178	5	912	101	274
Tiberias Hot Springs	2019/2412	62	28.8.60			296	5,322	560	2,562	176	13,927	732	152
Tiberias Hot Springs	2019/2412		3.6.64	0.30	58.0	335	6,211	576	3,028	198	16,270	770	159
Kinneret 3 borehole	2018/2415		9.3.64			355	6,388	652	3,280	226	17,096	782	171
Kinneret 2 borehole	2016/2417		26.8.60			358	6,505	634	3,240	218	17,171	774	152
Nuqueib spring	2106/2462	30		0.060	20.5	51.0	925	105	156	16.1	1,663	356	263
Gofra spring	2105/2457	33		0.072	15.8	68.5	1,100	172	265	27.5	2,431	205	322

The bicarbonate shows no dependence on the total salinity, but the values of the subgoups differ in their ranges of concentrations. The Tiberias city subgroup has the lowest HCO_3 content of 0–4 meq l^{-1}, the Zemach subgroup has the highest values of 5–13 meq l^{-1} (300–800 mg l^{-1}) and the rest have intermediate values. It is hoped that future work will provide an explanation for these differences, but as they are quite systematic, they are encouraging even at the present stages of the study. They strengthen the division into geographical-compositional subgroups, and raise confidence in the analytical data, which is important as bicarbonate is known to be somewhat difficult to sample and analyse.

The oceanic values were included in the diagrams for comparison. It is seen that the Tiberias-Noit subgroups are altogether close to the oceanic values, diverging from them occasionally. The Fulya-Barbutim-Tabgha subgroups is, for example, closest to the ocean composition in its Na, K, Br and also SO_4 and HCO_3 values, but is significantly different in its Ca and to some extent also its Mg contents. The Zemach and Ein Gev subgroups are, on the other hand, closest to the ocean in their Ca values.

The different subgroups have a wide range of concentrations of dissolved salts which are continuous and stop abruptly close to, but never above, the ocean (Gulf of Eilat) concentrations, as seen in Fig. 37.

2 Radium and radon content

Extensive measurements of radon and radium (Amiel, 1955; Mazor, 1962) revealed the mineral waters of the Kinneret Basin to be enriched in radon and radium. Maximum values for the Tiberias Hot Springs are 7,000 picocurie/litre (pc/l) radon and 140 pc/l radium (Table 17). Maximum values obtained for the Fulya springs were 9,700 pc/l radon and no detectable radium, and the corresponding values for the Tanur springs were 5,000 pc/l and 13 pc/l.

Table 17. Isotopic composition of the Tiberias Hot Springs [2, 3, 4]

Temperature	Cl		Ra	Rn
°C	mg/litre	10^{-8}cc STP/cc water	10^{-12} Ci/litre	10^{-12} Ci/litre
63	1,700	2,660	140	7,000
T	$\delta^{13}C$	^{14}C	Age I*	Age II†
T.U.	%	p.m.c.	Years B.P.	Years B.P.
0.0 ± 0.2	−5.7	4.8 ± 0.5	20,000	$12,500 \pm 900$

* Age calculated assuming initial content of 80 p.m.c.
† Taking into account the $\delta^{13}C$ value.

The radon occurs in a significant excess over the dissolved radium, and the radium, in turn, is in excess over the dissolved uranium. Hence, each species enters the water independently. The radium seems to be extracted from common rocks, a process that is possibly enhanced by the Cl content of waters, as radium chloride is highly soluble. The radon, being an inert gas, seems to be even more mobile and to enter the water with a remarkably high efficiency.

109

The radioactive elements originate from flushing of common rocks and not from uranium deposits, as some scientists postulate. This is indicated by the following observations: (i) The radium and radon-enriched waters issue in the Jordan Rift Valley from various types of rocks belong to various stratigraphic units and seem therefore to originate in common rocks, (ii) No uranium anomalies were encountered in the many drillings which took place in the area.

3. Radiogenic helium

The more saline and thermal water sources have a significant enrichment in radiogenic helium (Mazor, 1972), arising from the radioactivity decay of uranium and thorium, present in common rocks in the ppm range. In the Tiberias Hot Springs, up to $2,660 \times 10^{-8}$ cc STP He/cc water were found. Such radiogenic helium enrichments are common in thermal waters all over the world.

C. Comparison between Kinneret mineral waters and mineral waters in other sections of the Jordan-Dead Sea and the Suez Rift Valleys

Mineral waters also occur in other sections of the Rift Valleys (Fig. 38). The main springs include Hammam Gader in the Yarmouk Valley (Mazor, 1969; Mazor *et al.*,

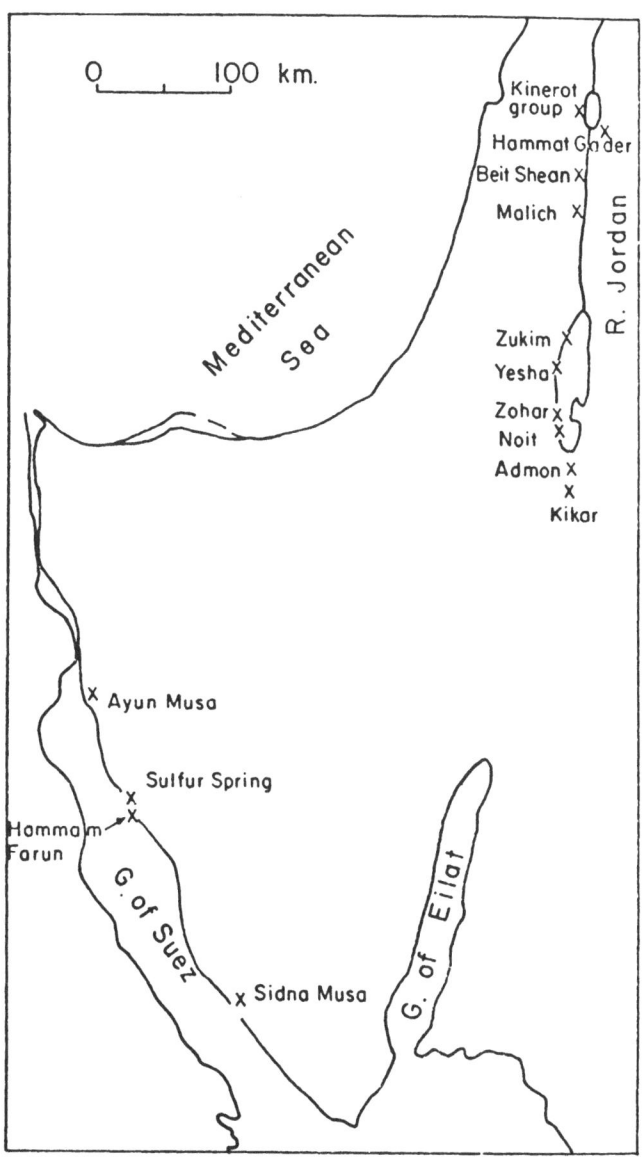

Fig. 38. Location of thermo-mineral waters in the Jordan–Dead Sea and Suez Rift Valley.

1973), Hammam Malik in the Beit Shean Valley, and Feshcha, Ein Noit, Nahal Boqeq, Tamar, Amiaz, Admon, Kikar springs and wells in the Dead Sea area (Mazor *et al.*, 1969; Mazor & Molcho, 1972). Data from the Beit Shean and Dead Sea groups have been included in Fig. 37 and fit the general pattern of the mineral waters in the Kinneret Basin. On the shores of the Gulf of Suez in western Sinai, mineral waters of the same type are also found.

The similarity of general composition of the mineral waters of the Rift system justifies the term "Tiberias-Noit association" which has been applied to them.

D. Origin of saline waters and mechanism of their upward movement

1. Previous theories

Numerous theories have been proposed to explain the origin of saline waters and the mechanism of their upward movement.

a. *Outflushing of formation waters*

The oil geologists defend the theory stating that meteoric waters, infiltrating in the mountains, flush out the original connate waters and then become saline. However, connate waters always deviate significantly from the sea water in composition and frequently have a higher concentration. The fact that the Tabgha-Fulya-Barbutim subgroup has a composition very close to that of oceanwater does not support this theory. Moreover, the flushing theory requires continuous hydraulic and salinity gradients from the mountain to the Rift Valley, which have not been observed so far.

b. *Dissolution of salt deposits*

This theory (Picard quoted by Yaron, 1952) explains the existence of mineral water as a result of dissolution of salt deposits by meteoric waters. However, no known salt would, by dissolution, give the Tiberias-Noit composition. In the case of a complex of different salts, the leaching would be differential, the more soluble salts would oe removed first, resulting in a gradual change of the composition of the dissolved salts. Considerable local variations would also be expected due to porosity, hydraulic transmissivity, etc., and it would be unlikely that the various water sources would plot on straight lines. Another serious objection is that the concentrations of ^{18}O and D of the saline waters change with salinity (Gat *et al.*, 1969), which should not be observed if the origin was by dissolution.

c. *Entrapment of highly evaporated brines*

One might argue that the proposed oceanic gulf underwent first an evaporation cycle and what was trapped was not original ocean water but a highly concentrated brine. This line of thought can be based on (a) the Cl:Br ratio which might be considered relative Cl depletion, as compared to the ocean values, and (b) the Na:(Ca + Mg) ratio which shows a relative depletion of Na, as compared to sea water (Goldschmidt *et al.*, 1967; Starinsky, 1974). Both these apparent depletions might be explained by the assumption that the postulated oceanic gulf reached the evaporation stage of NaCl precipitation.

However, evaporation of sea water would not give place to brines of the specific Tiberias-Noit composition. This could be explained only by assuming in addition intense exchanges with the country rocks, but then the model gets rather complicated. The hypothesis of entrapment of highly evaporated brines, where evaporation processes are called upon to explain all deviations from ocean composition, cannot explain the existence of the Lake Tiberias subgroups. All the Fulya-Barbutim-Tabgha waters have for example Cl:Br values that are quite close to the oceanic values, whereas the other subgroups have only half this ratio. The observed continuity in dissolved salt contents up to oceanic concentration, but not beyond it (Fig. 37) does not support the theory of entrapment of highly evaporated brines.

d. *Infiltration of Dead Sea water*

Another theory takes into account the possibility of infiltration of Dead Sea water inland. This process would generate the mineral waters on the Dead Sea shores and even northwards in the Tiberias area. This theory cannot explain the existence of the Tiberias-Noit waters which differ completely from the waters of the Dead Sea in their composition. It also does not explain the fact that 22 sources of this water association are found along the Dead Sea shores at a distance ranging from a few kilometers to a few hundred meters and yet, with their typical composition. No genetic connections can thus exist between the Dead Sea and the Tiberias-Noit waters although the opposite seems to be true.

e. *Magmatic exhalations*

Magmatic exhalations have been considered in the past as a possible source for the heat, salts and even part of the water of the Tiberias Hot Springs. These exhalations were, accordingly, described as last echoes of the volcanism that formed the last basalt flows in the surrounding region. This hypothesis seems, however, to be ruled out in light of the wide distribution of the Tiberias-Noit waters. A magmatic source could not give rise to water sources of an almost identical composition in an area over 200 km long. The concept of magmatic waters has been abandoned in the last decades for most mineral waters in the world, even where heat is still regarded as coming from a nearby hot magmatic body.

f. *Rising of the brines by hydrostatic pressure*

Goldschmidt *et al.*, (1967) think that the pressure of the freshwater on the edges of the saline body causes the upward movement of the saline water in the centre of this body. However, the artesian pressure of the Hot Springs of Tiberias, found among freshwater wells of lower hydrostatic pressure, indicates that the mechanism of ascension of the mineral waters is caused by forces that are independent of the local freshwater regime. The small variability of the Hot Springs' yield with precipitation (which modifies considerably the pressure of the freshwater) cannot be explained by this theory either.

g. *Infiltration of Mediterranean Sea water*

The short distance between the Mediterranean Sea and the Lake Kinneret basin (40 km) and the fact that the water level of the lake is 200 m below that of the Sea have led to the theory that the Tiberias-Nolt waters are directly fed by infiltrating seawater. However, the presence between both water bodies of the anti-clinorium of Galilee seems a serious obstacle to such an hydrological connection.

2. Model of entrapped sea water

Mazor Mero (1969) suggested that the brines originate from sea-water which has been trapped in the porous formations of the Rift Valley during the last marine transgression. When the Pliocene Sea covered the Rift Valley, the salty water penetrated into the freshwater aquifers, replacing local ground-water, intermixing with it at various dilution rates. The infiltration of the sea water must have been enhanced by the hydrostatic pressure of the sea water that filled up the newly formed gulf which during certain periods, had a depth of about 200 m.

When the connection with the open sea was interrupted, probably due to a tectonic rise in Esdraelon Valley, the sea water was trapped in the Rift Valley. The initial composition was modified by interaction with the host rocks, causing the formation of the observed geochemical-geographical water subgroups.

a. *Possible variations of entrapped sea water with aquifer rocks*

The composition of a 15% evaporated ocean water has been entered into the composition diagrams of Fig. 37. The value of 15% evaporation has been chosen because such an inland gulf must have been slightly saltier than the open sea, perhaps like the present day Gulf of Eilat. The following may be observed:– The Fulya-Barbutim-Tabgha subgroup plots in the diagrams of Fig. 37 in lines that point closer to the sea value than the lines of the other subgroups do. It shows a significant deviation from the ocean value in the Ca and Br diagrams and remarkable similarity to seawater in Na, Mg, K, Cl, HCO_3 and SO_4 (Mazor, E. & Mero, F. 1969a, 1969b). The Cl:SO_4 ratio is also very close to the ocean value. (Mazor & Rosenthal, 1967). This subgroup corresponds then to diluted but otherwise very slightly modified seawater. – The other subgroups agree well with seawater in the K diagram but show depletion in Na (that is more or less balanced by enrichments in Ca and enrichment in Br). Enrichment and depletion of Mg and SO_4 and to some extent HCO_3 are noticed (Mazor, E. & Mero, F. 1969a, 1969b). – The waters of the various subgroups are seen in Fig. 37 to occur in a wide range of concentrations, which are continuous but stop abruptly close to, but not above, the present day sea concentration. This agrees well with the model of entrapped sea water, diluted to various degrees with the then prevailing groundwater and intermixing with present day fresh water. – In Fig. 39 the composition of the most concentrated member observed in each subgroup is plotted along with seawater composition. The lines obtained vary from the oceanic line each in a different manner but altogether they follow its major trend.

b. *Laboratory experiments on the compositional change of sea water in contact with common rocks.*

We have checked in laboratory experiments the feasibility of the proposed chemical interactions with the aquifer rocks. Stirring of sea water with powdered chalk, mont-

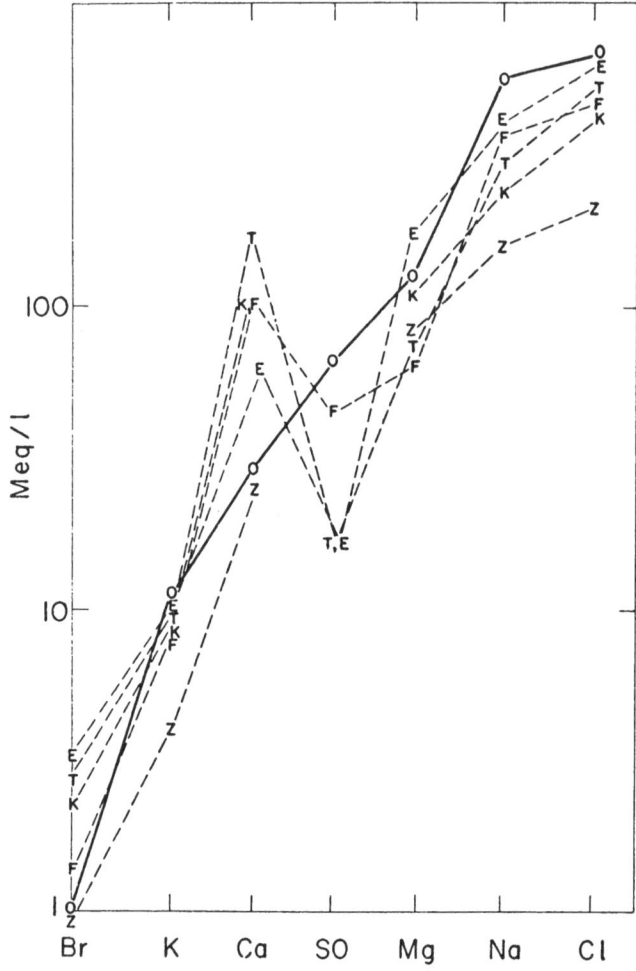

Fig. 39. Composition of the thermo-mineral waters of the Rift Valley compared to sea water: O = ocean, E = Ein Gev subgroup, F = Fulya subgroup, T = Tiberias subgroup, K = Kinneret group and Z = Zemah subgroup.

morillonite, dolomite, gypsum, bituminous shales, phosphorite and basalt resulted in "limitations" of the Tiberias-Noit-Farum water subgroups (Mazor, Nadler & Molcho, 1973,). The experimental results deviate from sea water in the same direction as the natural waters. Depletions are observed in the Na content, enrichments are seen in the Ca, Mg, Sr and Li concentrations and both depletions and enrichments are seen in the K and SO_4 values. Interestingly, even an enrichment in radium, $200 + 10^{-12}$ Ci/litre, took place by contact with the bituminous limestone.

The reported results are preliminary and the conclusion so far is that it is feasible to explain the deviations from sea water concentrations observed in the Rift Valley waters by reactions with local rocks. The only element which so far cannot be explained by such reactions is Bromine. Its enrichment in the Rift Valley water is still an open question. We would like to point out only that decrease in the Cl/Br ratio of sea waters undergoing diagenetic changes is common (e.g. in saline waters in oil fields) but the mechanism is unclear.

c. *Notes on the hydrological regime of the mineral waters in the Kinneret Basin.*

The Kinneret Basin seems to be underlain by saline water aquifers from which the water ascends in an endless array of cracks, faults and channels. This is manifested by the mineral springs, the frequency in which mineral water is found in boreholes, by the observation of radioactive anomalies on the bottom of the lake, in the areas where saline radioactive waters issue (Braudo *et al.*, 1970) and by the recent observation of a constant flux of saline water entering the lake through its bottom all over its area (Stiller, 1974).

An important complex of Talmudic and Roman bathing installations has been excavated about 1 km north of the present day Tiberias Hot Springs. No aquaduct or pipes that guided the water from the present spring was found. It might well be that at that time another spring was active north of the present complex.

E. Conclusions

Going through the various possible hypotheses for the genesis of the Tiberias-Noit water association, the model of trapped ocean water, controlled by tectonic forces, seems to best explain the host of available observations. The essential points of the present model are:

a. The ocean had a temporary connection with the Rift Valley and its water infiltrated into the faulted and tectonically shattered Rift Valley, forming traps of ocean water in porous rocks in various tectonic structures. This part of the model explains the vast distribution of the studied waters in the Rift Valley.

b. The deviations of the Tiberias-Noit waters from oceanic composition are attributed to exchange and dissolution reactions with the host rocks only, and not to any evaporation processes.

c. The constant changes in the tectonic settings of the Rift Valley blocks are suggested to have controlled the movement of the trapped brines. The waters of the suggested temporary ocean gulf later infiltrated into tensional blocks through faults and fissures. Part of these blocks became compressional and the infiltrated waters were squeezed out and forced to the surface. This explains the independence of the pressures of the Tiberias-Noit waters in the various tectonic blocks and the large pressure differences found.

d. The heat of the waters is suggested to result from a normal heat gradient. This would require the brines to be trapped at the depth of 1200 m in the case of the Tiberias Hot springs (assuming a gradient of 1°C per 30 m and a surface temperature of 20°C) and perhaps less in the other areas.

e. The saline waters are mixed with various amounts of fresh waters. This mixing takes place, at least in the Fulya-Barbutim-Tabgha region, close to the surface.

References

Amiel, S. 1955. Detection of disintegration and spontaneous fission products of uranium in natural sources and gases in Israel. Ph.D. thesis. Hebrew Univ. Jerusalem (in Hebrew).

Braudo, C. J., F. Mero & A. Mercado. 1968. Sub-marine spring discharge measurements using radioactive tracers. J. Hydraulic Div. ASCE 94: No. HY2, Proc. paper No. 5854. 399–409.

Braudo, C. J., F. Mero & A. Mercado. 1970. Detection of upward percolation of saline lake water. J. Hydraulic Div. ASCE 9: 7519–1802 Tel Aviv.

Gat, J. R., E. Mazor & Y. Tzur. 1969. The stable isotope composition of mineral waters in the Jordan Rift Valley, Israel. J. Hydrology 7: 334–352.

Goldschmidt, M. J., A. Arad & D. Neev. 1967. The mechanism of the saline springs in the Lake Tiberias depression. Israel Geol. Surv. Bull. 45.

Lovengart, S. 1961. Airborne salts – the major source of the salinity of waters in Israel. Bull. Res. Counc. Israel 10G.

Mazor, E. 1962. Radon and radium content of some Israeli water sources and a hypothesis on underground reservoirs of brines, oils and gases in the Rift Valley. Geochim. et Cosmochim Acta 26: 765–786.

Mazor, E. 1968. Compositional similarities between hot mineral springs in the Jordan and Suez Rift Valleys. Nature, 219: 477–478.

Mazor, E. 1968. Genesis of mineral waters in the Tiberias–Dead Sea–Arava Rift Valley, Israel. XXIII Int. Geol. Cong. 17:65–80.

Mazor, E. 1969. The springs of Hammat Gader, Nuqeib and Gofra. Teva V'aretz 11:260–263 (in Hebrew).

Mazor, E. 1972. Paleotemperatures and other hydrological parameters, deduced from noble gases dissolved in groundwaters; Jordan Rift Valley, Israel. Geochim. et Cosmochim. Acta, 36:1321–1336.

Mazor, E. & F. Mero. 1969. The origin of the Tiberias–Noit mineral water association in the Tiberias–Dead Sea Rift Valley, Israel. J. Hydrol. 7:318–333.

Mazor, E. & F. Mero. 1969. Geochemical tracing of mineral and fresh water sources in the Lake Tiberias Basin, Israel. J. Hydrol. 7:276–317.

Mazor, E., E. Rosenthal & J. Ekstein. 1969. Geochemical tracing of mineral water sources in the south-western Dead Sea Basin, Israel. J. Hydrol. 7:246–275.

Mazor, E. & M. Molcho. 1972. Geochemical studies on the Fescha springs, Dead Sea Basin. J. Hydrol. 15:37–47.

Mazor, E., A. Kaufman & I. Carmi. 1973. L Hammat Gader (Israel): Geochemistry of a mixed thermal spring complex. J. Hydrol. 18:289–303.

Mazor, E., A. Nadler & M. Molcho. 1973. Mineral springs in the Suez Rift Valley – comparison with waters in the Jordan Rift Valley and postulation of a marine origin. J. Hydrol. 20:289–309.

Saltzman, U. 1964. Geology of the Tabgha–Huquq–Migdal region. M.Sc. thesis, Hebrew Univ. Jerusalem (in Hebrew).

Starinsky, A. 1974. Relationship between Ca-chloride brines and sedimentary rocks in Israel. Ph.D thesis, Hebrew University of Jerusalem.

Stiller, M. 1974. Rates of transport and sedimentation in Lake Kinneret. Ph.D. Thesis, Feinberg graduate school, Weizmann Institute of Science, Rehovot.

Yaron, F. & M. Heintner. 1952. The chloride–bromide ratios of the water sources of eastern Emek Israel and Beit Shean Valley. Bull. Res. Counc. Israel. 2.

Part two

The Lacustrine environment

I. General background

C. Serruya

A. Names of the Lake

In the course of history, the lake has received many names. 'Kinneret' seems to be the most ancient and is frequently found in the Old Testament (Deut 3:17, Joshua 11:2, 12:3, 13:27). The ancient town Kinnarot is also mentioned (Joshua 19:35, I Kings 15:20). In the period of the Talmud (Third and Fourth Centuries A.D.), the lake is designated according to the names of the 'new' towns in the vicinity of the lake, that is, towns founded during the period of the Second Temple: Sea of Tiberias (founded in 18 A.D.), Sea of Ginosar, and Waters of Ginosar. In the New Testament, the lake is called Sea of Genesareth, which is simply one alteration of Ginosar, and Sea of Galilee. Nun (1977) reports that the Talmudists, investigating the origin of the name Kinneret, related it to the fruit of the *Ziziphus*, called in Hebrew 'kinar'. According to them, the fruit of the town Kinneret was as sweet as the kinar. However, another Hebrew word, 'kinor' (meaning violin), was thought by later Talmudists to be the origin of the name Kinneret. Their version was then that 'the fruit of Kinneret was as sweet as the voice of the violin'. The discovery, in 1928, of the ancient Canaanite town of Ugarit on the northern shore of Syria brought a new and unexpected solution to the problem of the origin of the name Kinneret. Letters dating back to 1600 B.C. tell the story of a righteous man who prayed to the gods to have a son. The gods promised him a son who would live forever if no evil could be found in him, but he sinned and was killed. The father decided to bury him in 'the fields of Kinnarot, famous for their fisheries'. In 1956, the excavations of Ugarit supplied a tablet written in Akadian; it was a list of the gods worshipped by the people of Ugarit, and among them the god Kinnar. The custom in those times was that a god was associated with a town where his temple was built; the town was then considered as his sacred wife and was called according to the feminine form of the god's name: Kinnar – Kinneret. It was later found that the cult of the god Kinnar included musical performances.

It then appears that our lake draws its name from a musical Canaanite god worshipped nearly 4,000 years ago in the town Kinnarot; the ruins of Kinnarot have been excavated in the late 1950's on the northwest shore of the lake, during the construction of the main pumping station of the National Water Carrier.

B. Location

Lake Kinneret occupies a rectangle defined by the latitudes 32° 42' 15"N and 32° 53' 44"N, and by the longitudes 35° 30' 52"E and 35° 38' 55"E.

Politically speaking, the lake is entirely located within the borders of the State of Israel. From an administrative point of view, the lake area falls under the jurisdiction of the Regional Council of the Jordan Valley District.

C. Morphometric characteristics

1. *Absolute altitude of the water level*

During the expedition led by the Duke of Luynes in 1864, Lieutenant Vignes found the absolute level of the lake to lie at 189 m below MSL. New measurements were carried out by the Palestine Exploration Fund (PEF). The first determinations were rather approximate, and Wilson & Warren (1871) reported that the level fluctuates between 600 and 700 feet below sea level. Lortet came to the area in 1875 and in 1880, and performed numerous measurements of air pressure with various types of instruments. He came to the conclusion that the level of the lake lay at 212 m below MSL. In 1877, Kitchener & Conder performed for the PEF the first exact leveling from the Mediterranean Sea to the Kinneret, and found a level of −208 m. This value became the official level of the lake during the whole period of the British Mandate. In 1937, the level of the lake was connected to the geodesic net by the Department of Measures. Since then, periodic measurements of the lake level have been carried out.

2. *Lake area*

The surface of the lake was reported by Barrois (1894) to be 2,000 km², probably as a result of a typographical error. The interesting point is that this error was copied by numerous authors and is still found in Gruvel (1931) and Bodenheimer (1935).

The most important morphometric data of Lake Kinneret are presented in Table 18 together with similar data for Lakes Victoria and Erie.

Special mention should be made of the development of the shoreline, D_L, which is the ratio of the length of the shore line to the length of the circumference of a circle of an area equal to that of the lake. This parameter quantifies in a rough manner, the potential effect of littoral processes on the lake. The very low value of D_L in the case of Lake Kinneret reflects the small number of bays and inlets on the coast line.

The hypsometric curve (Fig. 40) shows clearly that when the level decreases, the water volume drops very rapidly in comparison with the water surface: for

Table 18. Morphometric parameters of Lake Kinneret (level−209 m), Lake Victoria and Lake Erie

	Kinneret	Victoria	Erie
1. Surface A (m²)	167.87×10^6	$68,800 \times 10^6$	$25,820 \times 10^6$
2. Maximum depth Zm (m)	43	79	64
3. Mean depth $\bar{z}$ (m)	25.6	40	21
4. Depth of cryptodepressions z^c (m)	253		
5. $\bar{z}$:Zm	0.59	0.51	0.33
6. Volume V (m³)	$4,301 \times 10^9$	$2,700 \times 10^9$	540×10^9
7. Length l (m)	22×10^3 NS		
8. Width w (m)	12×10^3 EW		
9. Shore line L(m)	53×10^3	$3,440 \times 10^3$	$1,200 \times 10^3$
10. Development of shore line D^L	1.16	3.7	2.1

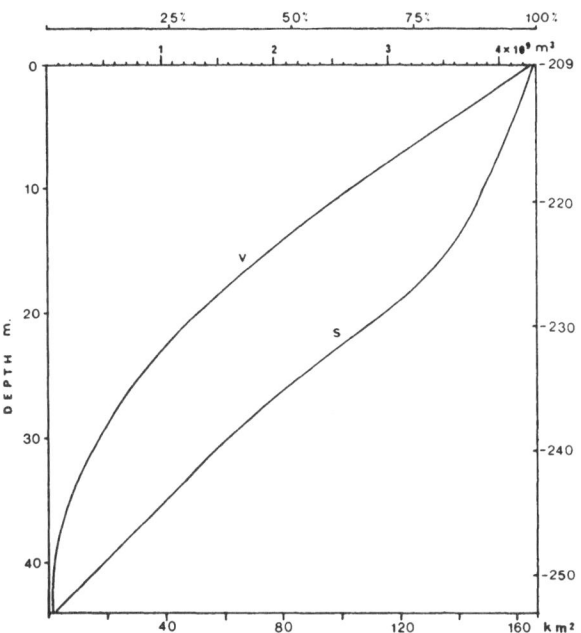

Fig. 40. Hypsometric curve of Lake Kinneret (from S. Serruya, 1975, Verh. Internat. Verein. Limnol., 19).

example, when the water level decreases from −208 down to −218 m, the surface diminishes by only 10% and the volume by 36%. This is a quantitative way of describing the morphology of the lake, which has steep shores and few shallows.

D. Bathymetry

1. *History*

The modern bathymetric investigation of Lake Kinneret began in August 1847 when Lieutenant Molyneux, from the British Navy, bringing a boat from Haifa to Tiberias, sailed and worked for two days on the lake. He was very eager to check whether the legendary unfathomed deeps of the Lake of Tiberias really existed. The numerous soundings that he performed demonstrated that the depth of the lake did not exceed 156 feet or 47.55 m.

In 1848, Lieutenant Lynch, leading an American mission (1849, 1852), crossed the lake from Tiberias to the lake outflow but did not perform any bathymetric measurements. However, he reported that the maximum depth of the lake was 165 feet or 50.30 m. He merely used Molyneux's value, but reported 165 feet instead of 156, and the lake deepened a few meters because of this typographical error!

In 1858, Van de Velde published a bathymetric map of the lake where, according to his own statement, the soundings were those of Molyneux (Fig. 41).

In his book, 'The Rob Roy on the Jordan' (1886), MacGregor presented a map

125

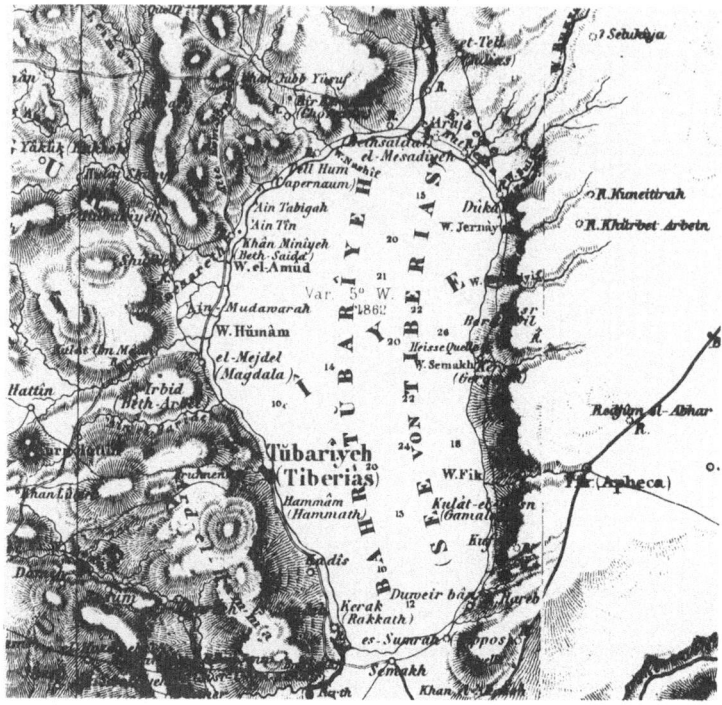

Fig. 41. Map of Van de Velde, amended edition of 1862. Soundings given in fathoms. Original scale 1:315,000.

of the lake and curiously enough, mentioned that he utilized the bathymetric points of Van de Velde 'who took them from Lynch' (Fig. 42)! Besides this gross inexactitude, MacGregor, reporting incorrectly the value of the maximum depth found by Molyneux (156 feet), states on page 424 of the 7th edition of his book (1886) that the maximum depth of the lake is 156 fathoms or 936 feet. This unfortunate mistake renewed the erroneous concept of the great depths of the lake.

In 1883, in his work on 'Fish and Reptiles of Lake Tiberias', Lortet wrote that he found in the middle of the northern basin a depth of 250 m.

In his study of the lakes of Syria published in 1894, the Frenchman Barrois reported on the same map the soundings of Molyneux and the numerous soundings he performed on six transepts. Barrois established definitively that the 'depth of the lake does not exceed 40 to 45 m according to the seasons' and the map published in 1931 by Gruvel differs very little from Barrois' rather exact document (Fig. 43).

In 1936, Captain Meininger prepared a new map for the Jewish Agency for Palestine but did not publish it. In 1945, Ashbel published the first bathymetric map with bathymetric contour lines (Fig. 44). This map was based on approximately 500 soundings partly carried out by himself, and partly from information collected from fishing boats utilizing echo-sounders.

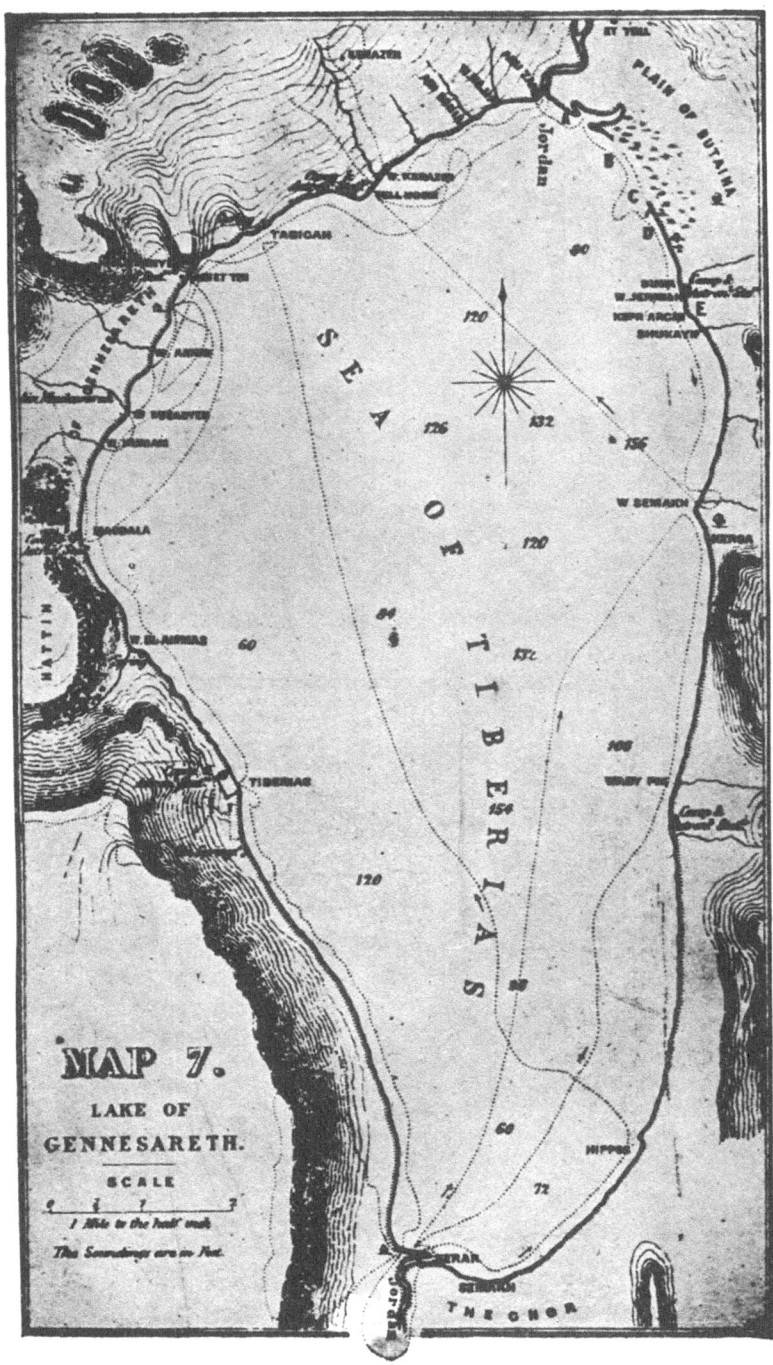

Fig. 42. Map of MacGregor. Contouring according to Anderson and Wilson. Soundings in feet according to Molyneux.

127

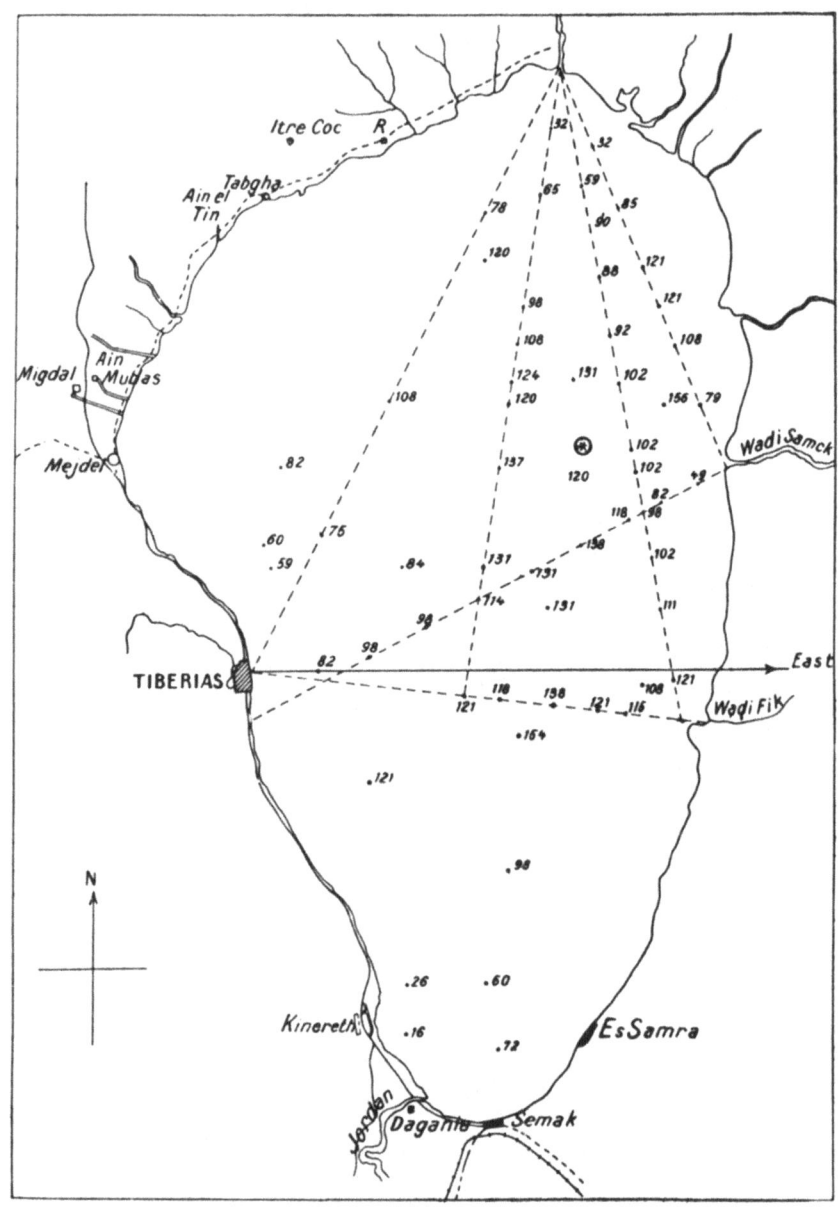

Fig. 43. Map of Grevel, 1931. Soundings in feet derived from Barrois.

In 1950–1951, Oren (1957), in his work for the National Planning Board of Israel Water Scheme, carried out 288 new soundings with a meter wheel; the location was determined by sextant measurements of three points on the shores (Fig. 45).

In 1961, Tahal (Water Planning for Israel) published a map based on echo-sounding and precise navigation with 1 m interval bathymetric lines (Fig. 46).

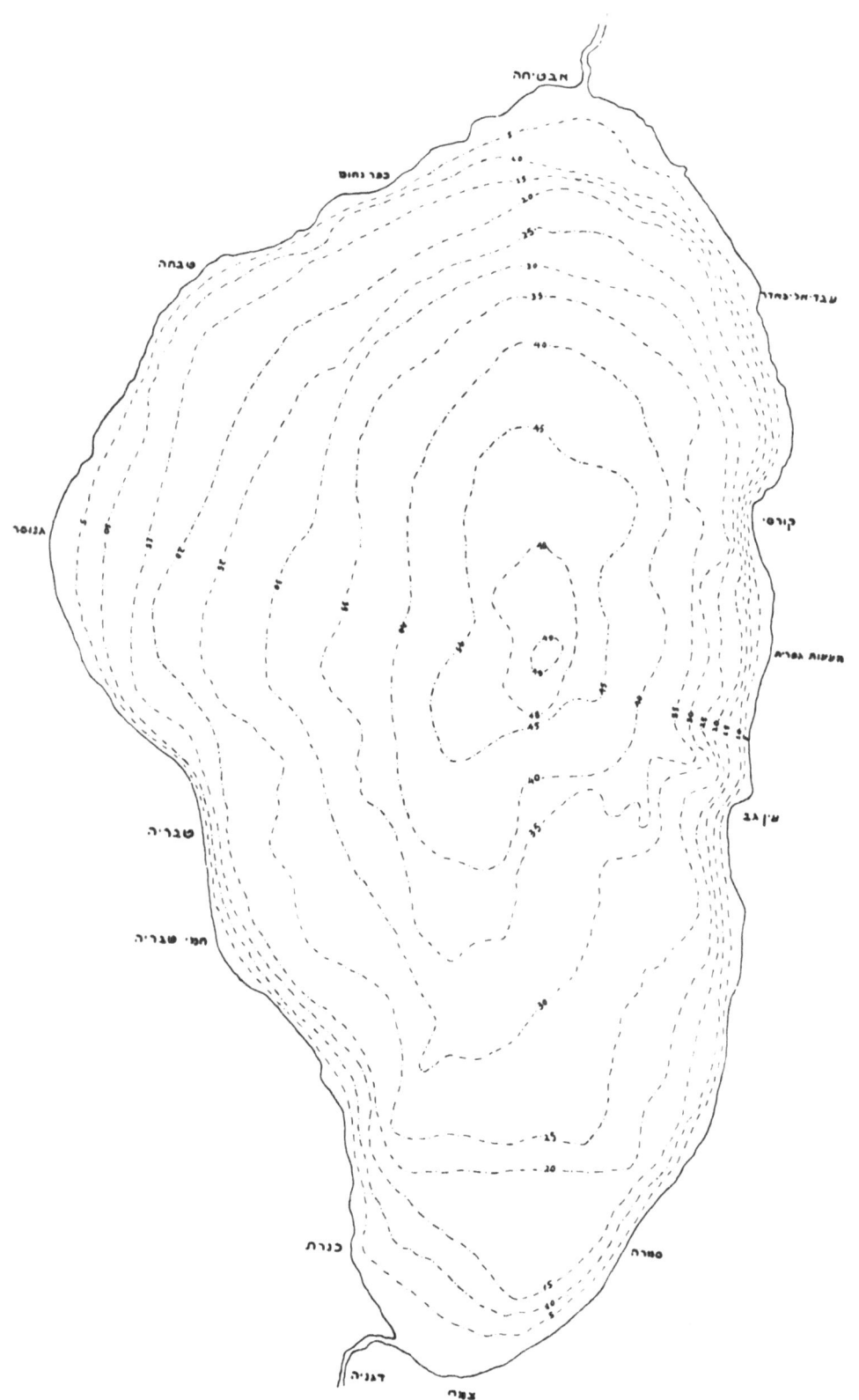

Fig. 44. Contour map of Ashbel, 1945. Original scale 1:115,000. Soundings in meters.

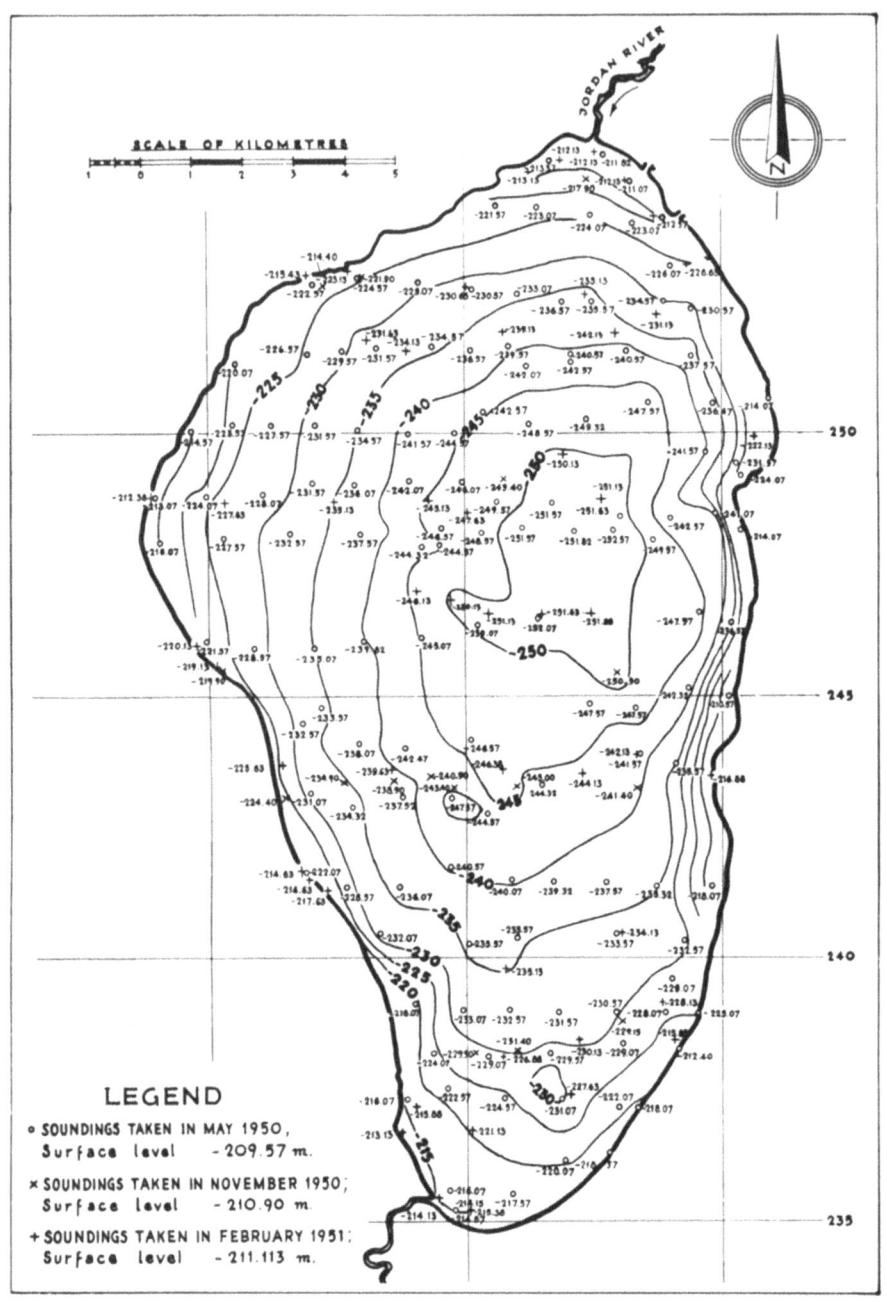

Fig. 45. Map of Oren, 1951. Original scale 1:100,000. Soundings in meters.

2. *Bathymetric features*

The bathymetric curves of the lake have the general form of an eccentric ellipse. The deepest part of the lake is located slightly NE of the center of the lake and approximately 8 km south of the Jordan mouth.

130

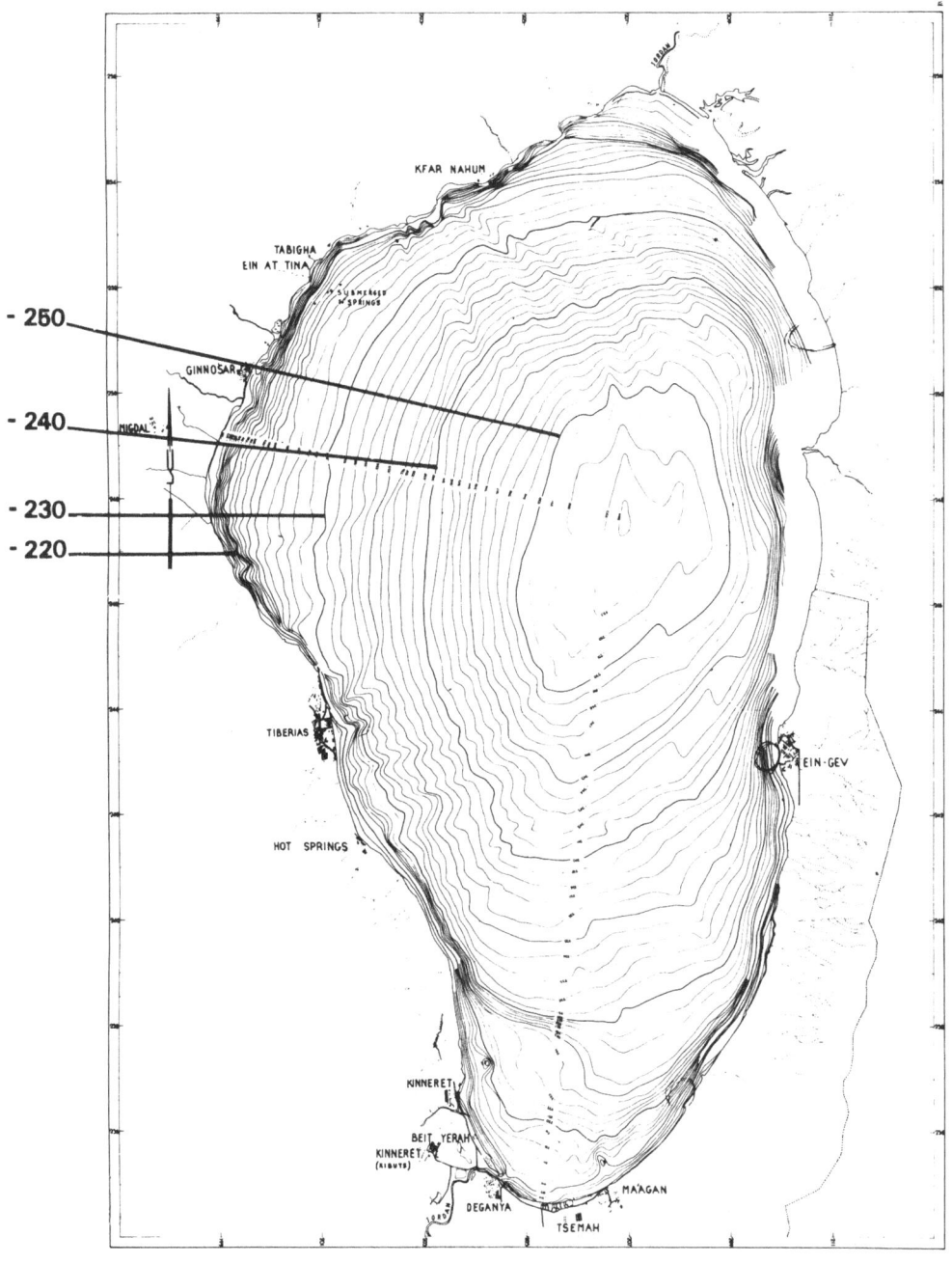

Fig. 46. Map of Tahal, 1961. Original scale 1:20,000. Contours given in meters.

The eastern coast of the lake – nearly a straight N–S line – runs along the major fault of the Rift Valley. On the western coast, the large Bay of Ginosar largely contributes to the lake's asymmetric shape. The western coast is formed of various NW–SE oriented blocks separated by faults. One of these faults forms the

131

coastline of the Hamei Tveria area.

The northwestern part of the lake is much shallower than the eastern one, where the slope is as high as 5%. Similar slopes are also found in the Hamei Tveria area. Conversely, the southern part of the lake does not exceed 15 m. However, at the absolute altitude of −224 m, the smooth slope of the southern area (0.1%) increases suddenly to 0.5%. This is probably the topographic expression of the fault Mount Herodus–Hamei Tveria.

The sublacustrine springs are generally found at the bottom of deep funnels. The spring Ma'agan 1, for example, occupies a circular depression of 40 m diameter and 20 m depth. The funnels are maintained open by the constant discharge of artesian water. At Ma'agan 1, the artesian pressure at the bottom of the funnel, located at −252 m, rose to −203.0 (Mero, personal communication).

E. Shores

1. *General*

The structure and composition of the shores, both on the shoreline and on the bottom at shallow depths, is very variable from place to place. In the northwestern area, the rather steep slopes of Kfar Nahum end in a few meters high cliffs. The area between the cliff and the water is covered with black basalt boulders (plate 5). Southwards on the same side of the lake, the flat background of the plain of Ginosar with its alluvial origin explain the existence of the large Bay of Tabgha–Ginosar. In this area, a very gentle slope leads to the water line and the shore is made of sand, gravel and pebbles, most of them brought by Wadi Amud. In the most sheltered areas, the shore is covered with a thick layer of shells (mainly of unionids). Southwards, the high cliffs of Arbell Mountain dominate the lake and nearly reach the shore line. In this area, the shore is very narrow and covered with limestone boulders.

In the southern part of the lake, the Lisan beds outcrop on the shores. They are composed of clayey material, including gravels and pebbles of basalt and of Eocene chalk and flint. In the southeastern part of the lake, erosion has cut vertical cliffs in the horizontal Lisan formations. North of these formations, a narrow coastal plain develops; it is limited in the east by the Neogene hills of the Golan Heights, deeply eroded by the river bed of Wadi El Kursi. In the northwestern corner of the lake, the coastal plain widens into a flat marshy area called the Buteiha plain. It covers an area of 25 km² and drains a watershed of 368 km². This plain is crossed by three streams, Meshushim, Yehudia and Daliot, which show a typical winter flood regime. The overgrazing caused by goats and sheep during the last millennium has destroyed the plant cover, and the winter floods bring considerable amounts of alluvion which are partly deposited in the lower part of the plain. This results in a complex deltaic formation where tongue-shaped alluvial formations isolate shallow lagoons (plate 6). This is one of the few areas in Israel where natural hydrophilic vegetation can be found on a large scale. The vegetal cover is dense and includes a large number of species; Gal (1972) mentions rare species belonging mainly to the Gramineae and Cyperaceae. Twenty-three species of fish populate the lagoons and represent the food source of a rich population of birds and otters. The lagoons represent also the spawning ground of a few species of lake

Plate 5. Lake Kinneret, northern area:
the bay of Sheikh Ibrahim; the shore is covered with basalt boulders. (photo, Y. Edelstein).

Plate 6. The lagoon area in the north eastern corner of the lake, one of the spawning areas of Tilapia galilaea. (photo, Y. Edelstein).

fish. Coypus, ichneumons, hogs, jungle cats, jackals and hyaenas form the mammalian population of these lowlands. Approximately 100 species of water fowl live in the Buteiha plain. Some of them, such as the flamingo and the goose, became very rare in Israel. Nests of the Jordan sparrow abound. The black partridge is also common.

Agriculture and tourism are endangering this area. Moreover, the rich fish population of the lagoons is a great attraction to fishermen, with the possible result of overfishing.

2. *Vegetation*

The eastern winds in winter and the daily western wind in summer generate strong and frequent storms, and the littoral area is submitted to a very severe wave action. This doubtlessly contributes to the wash-out of the fine-grained particles from the shore and explains the small percentage of muddy shore areas. The predominance of boulders and gravels, together with the limited development of the shore line and the bathymetric features of the lake, account for the poor development of the phanerogamous vegetation.

The investigations of Waisel (1967) showed that three main plant associations occupy the shore area. The *Myriophylletum spicati* association is restricted to the substrates rich in clay and organic matter. Moreover, it is a submerged association developing below 1 or 2 meters of water. It includes two species: *Myriophyllum spicatum* L. and *Najas marina* L. The conditions required by this association explain its distribution in the sheltered and relatively quiet areas of the Bay of Ginosar and the southern shore of the lake.

The *Phragmitetum communis* association develops only in shallow water and on clay-rich substrate. This association includes *Phragmites australis* (Cav.) Trin., *Arundo donax* L., *Juncus acutus* L., *Jussiaea repens* L., *Panicum repens* L., *Tamarix jordanis* Boiss. and *Typha australis* Schum. and Thon. In this association, *Typha australis*, although not submerged, occupies deeper water than *Phragmites*, whereas *Tamarix jordanis* is dominant on the slopes. This association is frequently found in the Bay of Ginosar and on the eastern shore north of Haon.

The *Viticetum* agri-caste association develops on shores covered with fine to coarse basaltic gravel. The components of this association are all terrestrial plants, such as *Alhagi maurorum* Medik, *Arundo donax* L., *Inula viscosa* (L.) Desf., *Juncus acutus* L., *Nerium oleander* L., *Panicum repens* L., *Phragmites communis* L., *Prosopis farcta* (Banks et Sol.) Eig., *Salix acmophylla* Boiss. and *Vitex agnuscastus* L.

F. Deltas of rivers

The morphology of the lake shores is considerably influenced by the presence and building up of deltas at the mouth of the main rivers. In Lake Kinneret, the River Jordan and Wadi El Kursi are the two main influent streams where delta formations could be observed.

Plate 7. The mouth of the River Jordan and the River Zaki. (photo Y. Edelstein).

Plate 8. The vegetation near the River Zaki. (photo Y. Edelstein).

1. *The delta of the River Jordan* (plates 7 and 8).

The relatively recent down faulting of the lake area accounts for the young relief of the watershed area and the strong slopes of the River Jordan. This, together with the poor vegetation cover of the mountains and the short but stormy rainy season, causes an intense erosion in the Upper Galilee and a high suspended load in the River Jordan (Inbar, 1976). However, the bathymetric map of the lake and earlier maps do not show any delta morphology. Prior to 1969, even at low water levels (−212 m), no direct sedimentation was observed at the river mouth.

S. Serruya (1974) studied the flow pattern of the Jordan water after its entrance into the lake. He traced the Jordan water in the lake by measuring such characteristics as temperature, current speed and direction, electrical conductivity and nitrate content, since these values are different in the river and in the lake (Table 19). Such parameters were measured on profiles located on four E–W

Table 19. Some characteristic differentiating Jordan and Kinneret waters. (from S. Serruya, 1974, Limn. Ocean. 19.)

	Conductivity (umho cm^{-1})	Total solid content (mg.l^{-1})	Chlorides (mg.l^{-1})	Nitrates NO$_3$-N (µg.l^{-1})
River	400–500	300–350	16–18	1,500
Lake	1,000–1,100	600–650	220–250	20–40

transects in the northern part of the lake. At the beginning of the flood season, the large differences of temperature, current and conductivity between the lake and the river make it very easy to identify the Jordan water in the lake. From December until early February, the observed flow pattern is as represented on Fig. 47A, B. The Jordan water, cooler than the lake water, flows in a N–S direction in the deep layers near the lake bottom; it can be identified over 5 to 6 km. The suspended matter of the early floods is then likely to be deposited in this tongue-shaped area, according to a granulometric gradient. As a matter of fact, a study of sediment distribution in the northern area of the lake (see section on sediments) confirmed this hypothesis.

In mid-February, the lake water becomes cooler than the river (13.7 and 15.2°C respectively on 21 February 1972). From late February until the end of the flood period, the flow pattern of Jordan water in the lake is according to Fig. 47C, D. During this period, the lighter Jordan waters float on the lake water. Like all the superficial water layers, the Jordan waters are affected by the general counter-clockwise circulation (see section on currents) and at this time, these waters can then be seen progressing along the NW coast. This means that the suspended material of later floods is deposited along the western coast.

Similar measurements, carried out in summer (Fig. 48), showed that the Jordan water flows on the thermocline. At this period, the solid yield of the river is at a minimum, but since, in the thermocline area, the frictional forces are low, the suspended matter can be transported over great distances.

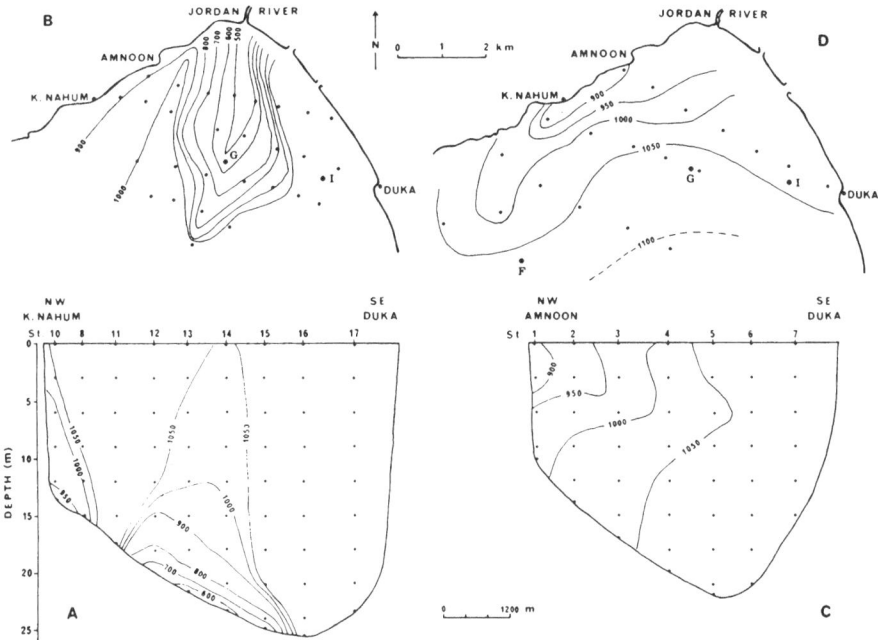

Fig. 47. A: Kfar Nahum–Duka conductivity on 7 February 1972. B: The front of the Jordan flood progressing on the lake bottom, 7 February 1972. C: Amnoon–Duka conductivity on 21 February 1972. D: Jordan flood front in the upper western layers. All results are in umhos cm⁻¹. (From S. Serruya, 1974, Limnol. Ocean., 19.)

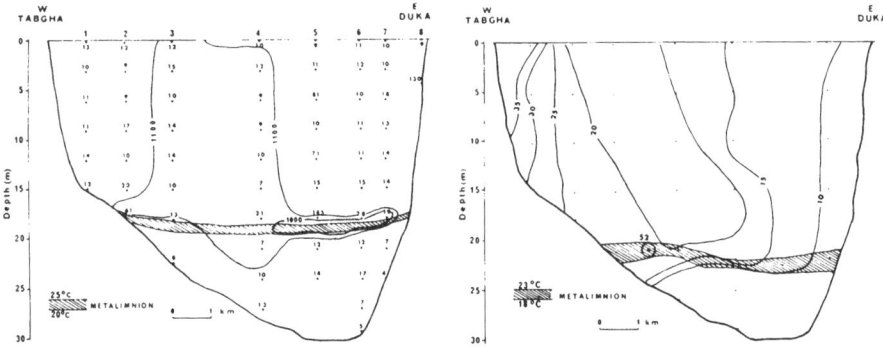

Fig. 48. A: Tabgha–Duka section 23 October 1972. Isoconductivity curves in umhos cm⁻¹. Numbers represent NO_3–N in ppb. B: Tabgha–Duka section 6 November 1972. Curves represent NO_3–N in ppb. (From S. Serruya, 1974, Limnol. Ocean., 19.)

137

The large differences of density, which make it possible for the Jordan water to keep its identity over long distances in the lake, also allow the flood load to be dispersed over wide areas and prevent the formation of deltas.

Considerable changes in shore morphology were caused by the exceptional flood of the 1968–1969 winter; these changes were studied in detail by Inbar (1974). The winter of 1968–1969 was unusually rainy, with more than 200% of the annual average being recorded. The Jordan River discharge reached unheard peaks of 225 m³/s and the lake level rose to −208.3 m (the highest level measured during the last 60 years), although the gates of the dam at the southern exit were fully opened. The enhanced erosion in the watershed and the numerous slides and rockfalls in the water channel increased considerably the solid yield of the river. In the period January–March 1969, several isles were formed at the mouth of the river, which was diverted eastwards. These isles then played the role of a sediment trap, and the space between them rapidly silted. In 1973, the newly-formed delta had a surface of 0.4 km² and was partly covered with vegetation.

The eastern diversion of the river remains until this day the only open channel; this new direction of flow towards a shallow and flat bottom area accelerates the development of the delta. Conversely, any significant decrease of the water level in future will cause the erosion of the delta and the release of considerable amounts of sediments.

2. *The delta of Wadi El Kursi*

Wadi El Kursi is the only stream which has built a delta in the lake. As the stream exits from the highlands, it divides into various channels, anastomosing into braids which have eroded previous alluvial deposits. Kay (1971), who studied the bed and delta material, found that sorting was better in the delta material than in the thalweg. He also found that the former samples contained more basaltic elements than the latter ones.

G. Water levels

Nun (1973) gathered archaeological evidence which enables us to deduce the level of the lake during the historical period.

On the way from the hot springs of Hamei Tveria to Tiberias, a Roman wall made of large stones juts out into the lake down to a level of −212 m. One has to assume that the wall was built to this level in order to maintain 1 to 1.5 m of water even during low lake levels and prevent potential enemies from landing near the city wall. It follows that the lake level did not drop below −211 m.

On the northwest shore at Capernaum, excavations were carried out in 1969, when the water level was at −209 m, and led to the discovery of a house dating to the 1st Century A.D. The floor of the house was covered with ground water. Since the Capernaum dwellers of this period certainly did not build their houses below the water table, it must be concluded that the lake level was below −209 m.

The remains of the ancient site of Bethsaida, mentioned in the New Testament, are found a few hundred meters east of the Jordan inflow. Investigators always wondered about the small area of a site which was described in the historical records as a large settlement. In January 1973, the water level dropped to −211 m, and previously unknown building stones, remains of walls and foundations of a round tower were uncovered.

Further south, the shore of the area of Wadi El Kursi is associated in the Christian tradition with the miracle of the healing of the unclean spirit. This site was called by the early Christians Gerasa or Gergasa. In 1970, a large basilica was excavated 0.5 km from the shore. In the submerged portion of the shore (the level in 1970 varied from −209 to −210.5 m), a fisherman's quarter was discovered. This quarter extends over an area 300 m long and 100 m wide. The fishing harbour has a surface of 1,500 m² and is open to the NW. The site also includes the fish market and a vast building which may have played a role in fishery administration. This is the most ancient fishery infrastructure found in the Kinneret area. The planning of the whole system was done in accordance with water levels varying from −209.5 to −211.0 m.

At Ein Gev, the ruins of a fortified settlement established in the time of King Solomon (1000 B.C.) were excavated 15 years ago. This site was inhabited until the Hellenistic period. In the lake shore buildings, the floors of private houses were below the −209 m level.

These archaeological data indicate that for a very long period, the maximum level of the Kinneret could not be higher than −209.5 m, and that the amplitude of the fluctuation did not exceed 1.5 to 2 m.

More recent information is found in the narratives of pioneer scientists who visited our area in the 19th Century and reported by Ben Arieh (1965). Burckhardt (1822), who came to Tiberias in 1812, Turner (1820) and Thomson (1860) reported that the lake level increased by a few feet in winter during the rainy period. Schumacher (1888) was the first to give a more precise, although relative, estimation of the yearly fluctuation of the lake level, by noting the water levels on the town wall at different periods. During eight years of observations (1881–1889), the largest difference that he found between winter and summer levels was 1.55 m. From 1900 until 1912, the Director of the Scottish Hospital in Tiberias made monthly measurements of the lake levels according to a constant reference point on the town wall (plates 9, 10, 11 and 12).

The measurements of the lake levels according to an absolute reference began only in 1925, but has been continuous since then (Fig. 49). The last fifty years of the lake level history can be divided into four periods:

1. *The 1925–1932 period*

A certain increase of the amplitude of the annual variation, due to the wider range of rainfall distribution, was noted in comparison with the 1900–1910 period; this reached 2.00 m in 1928–1929, a very rainy year.

2. *The 1932–1948 period*

In 1932, a dam was built near Deganya bridge at the southern exit of the lake. The idea was to store part of the winter flood of the Jordan and release it in summer, according to the requirements of the hydroelectric plant of the Electricity

Plates 9 and 10. The town of Tiberias seen from the lake:
The northern part of the town (9) and the area south of Tiberias (10). (photo Y. Edelstein).

Plate 11. The wharf of fishermen north of Tiberias: in the foreground a passenger boat, in the background fishing boats. (photo Y. Edelstein).

Plate 12. The town of Tiberias in the 19th century (from W. F. Lynch, Verlag der Deutschen Buch-hanlung, Leipzig, 1854).

141

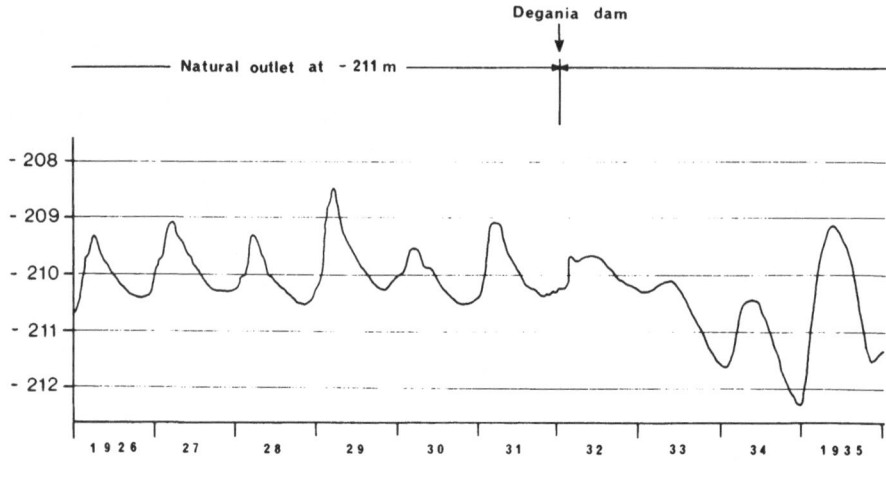

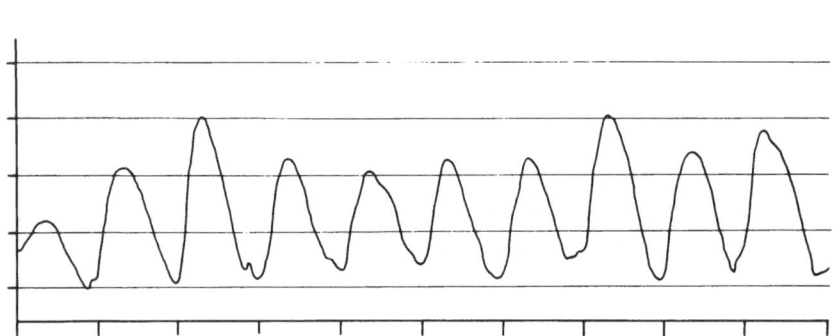

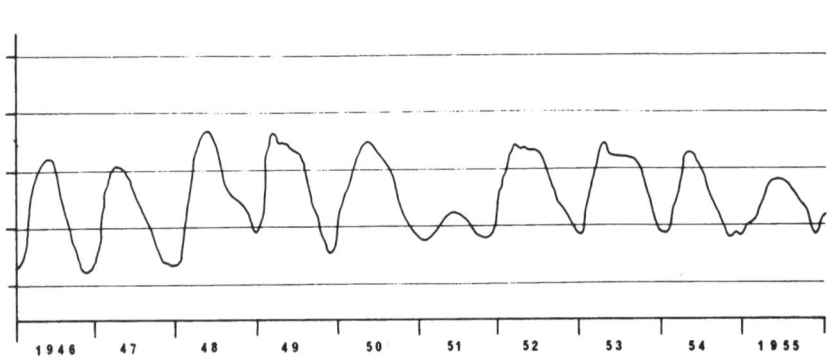

Fig. 49. The fluctuations of the lake level during the period 1925–1975. Data from Tahal.

Fig. 49. (contd.)

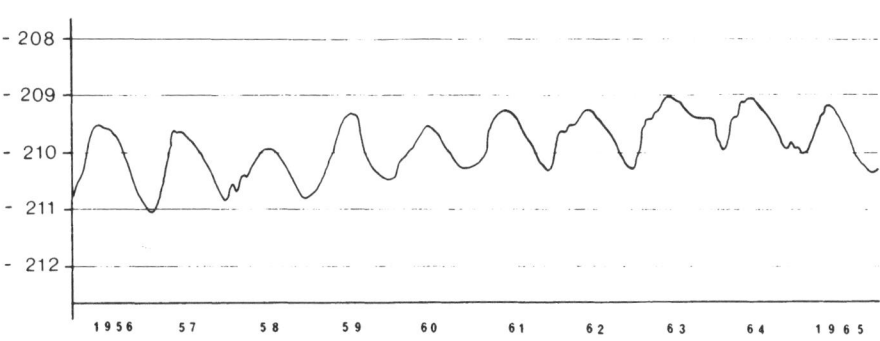

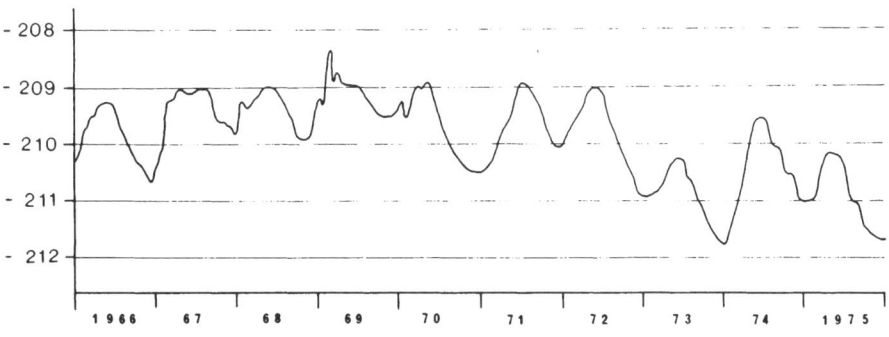

Company. The plant itself was built at the junction of the rivers Jordan and Yarmouk. Since the two rivers have their maximum flood period in winter, it was decided to operate the plant with the Yarmouk water in winter and with the Jordan–Kinneret water in summer. From March 1932 onwards, the lake level was no longer natural, and depended to a large extent on the amount of water allowed through the dam and the timing of the release. The operation of the dam resulted in an overall decrease of the average water level by one meter (from −210 to −211 m); the amplitude of the annual variation reached 3.2 m in 1933–1934 and the average value was about 2 m.

3. *The 1948–1964 period*

In 1948, with the War of Independence, the hydroelectric plant ceased to function and the Deganya dam served only to regulate the flood of the Jordan River and to supply water for the summer irrigation of the Jordan Valley. The amplitude of the annual variation of the lake level decreased, and its maximum value did not exceed 1.5 m. The average water level rose again, reaching −210 m in 1959–1960 and −209.5 m in 1963.

4. *The period 1964 until the present*

In 1964, the construction of the National Water Carrier was completed and the pumping station was put into operation on June 10. Since then, besides the flow of the Jordan and the amounts of water released through the Deganya dam, the intensity and timing of pumping through the National Water Carrier became a new and determining factor affecting the water levels. From 1964 to 1967, the water level dropped slightly (down to −209.8 approximately) but rose again in the rainy period extending from 1967 to 1970 (approximately −209.40 m). The maximum level was recorded in January 1969 (−208 m). From 1972 to 1975, the combined effect of relatively dry years and intensive pumping caused the level to drop from −209.0 m in spring 1972 down to −211.85 m in winter 1973 (plates 13, 14 and 15).

We also note that the annual peak of the water level, which occurred generally in January–February before the construction of the dam, took place later (April–May) after 1932.

H. Man-made modifications in the Lake vicinity

1. *The hydroelectric plant of Naharayim*

The construction of the hydroelectric plant of Naharayim was initiated in 1927; the plant was officially inaugurated in June 1932. Besides the Deganya dam, aimed at storing up to 600 million m^3 in Lake Kinneret, a second dam was built about 9 km south of the lake to retain the water of both the Jordan and Yarmouk rivers. In 1932, the plant included two turbines, and in 1933, a third one was added, each one of 8,500 HP. In 1948, the Naharayim area became Jordanian territory, and the plant ceased to function.

2. *The National Water Carrier*

The National Water Carrier was completed in 1964. The main exit of the lake was transferred from the southern part of the lake (Deganya area) to its northwest corner, with the exception of very rainy years, when large amounts of water had to be released through the dam to prevent flooding of the shores.

3. *The diversion of salty springs*

The high concentration of sodium chloride in the lake water has been a permanent problem, since a large amount of the water withdrawn from the lake was used for irrigation. The salty waters of the springs located on the shores of the lake were diverted into a canal specially built for this purpose. This canal starts at the salty spring of Ein Nur, located in the plain of Tabgha on the northwest shore. The canal runs southwards on the western side of the lake and brings the salty waters into the Jordan River south of the lake. The diversion of the salty water was completed in 1965.

144

Plate 13. The River Jordan at its outlet in summer 1967. Note the high water level. In the background the Golan Heights. Photo A. Belkind, Deganya A.

Plate 14. The 'century flood' of 1969. Beach houses and trees are surrounded with water. Photo A. Strud, Deganya A.)

Plate 15. The River Jordan at its outlet in winter 1974, after a dry hydrological cycle. Note the low water level. The exposed river bed is utilized by fishermen to collect the 'Kinneret sardine' from the nets. The seagulls are waiting for their share of fish. Photo A. Belkind, Deganya A.

References

Ashbel, D. 1945. The temperature of freshwater lakes in Palestine. Hateva Ve Ha'aretz. 2:72–79.

Barrois, M. J. 1894. On the depth and temperature of the Lake of Tiberias. PEF QST 211–220.

Ben Arieh, Y. 1965. The Central Jordan Valley – A regional geography (in Hebrew) Ha Kibbutz Ha Me'uchad publ. House ltd.

Bodenheimer, F. 1935. Animal life in Palestine. 506 p.

Burckhardt, J. L. 1822. Travels in Syria and the Holy Land. London.

Conder, C. R. & Kitchener, H. H. 1880. Map of Western Palestine from surveys conducted for the Committee of the Palestine Exploration Fund during the years 1872–1877. London.

Gal, I. 1972. The valley of Betsaida, Sal'it 1, 2:57–62.

Gruvel, A. 1931. Les Etats de Syrie, richesses marines et fluviales. Exploration actuelle – Avenir. Societe d'Editions geographiques, maritimes et Coloniales. Paris.

Inbar, M. 1976. Movements of suspended matter in the northern Jordan River. Haifa University. Report no. 3.

Kay, P. A. 1971. Geomorphology of Wadi Samak, Israel Dept. of Geography. Univ. Toronto, Canada.

Lortet, L. 1883. Poissons et Reptiles du lac de Tiberiade. Arch. Mus. Hist. Nat. Lyon. Tome III.

Lynch, W. F. 1852. Official report of the US Expedition to explore the Dead Sea and the River Jordan. Baltimore.

Mac Gregor, J. 1886. The 'Rob Roy' on the Jordan. 7th ed. London.

Molyneux, A. 1848. Expedition to the Jordan and the Dead Sea. J. of the Roy. Geogr. Soc. 18: part 2. London.

Nun, M. 1973. Water levels in Lake Kinneret in the historical period. Symposium on warm lakes. Nat. Counc. Res. Dev. 12–73. Berman, T. (ed.).

Nun, M. 1977. The Kinneret (in Hebrew). Kibbutz Ha Me'uchad Publ. House Ltd.

Oren, G. H. 1957. Physical and chemical characteristics of Lake Tiberias. Bull. Res. Counc. Israel 11 G:1–33.

Schumacher, G. 1888. The Jaulan, London.

Serruya, S. 1974. The mixing pattern of the Jordan River in Lake Kinneret. Limnol. Oceanogr. 19:175–181.

Thomson, W. M. 1860. The Land and the Book. 2 vol. New York.

Turner, W. 1820. Journal of a Tour in the Levant. London.

Van de Velde, C. W. H. 1852. Memoir to accompany the map of the Holy Land. Gotha.

Waisel, Y. 1967. A contribution to the knowledge of the Phanerogamous vegetation of Lake Tiberias. Bull. Sea Fish. Res. Stn. Haifa. 44:3–16.

Wilson, W. R. & Warren, E. 1871. The recovery of Jerusalem, narrative of exploration in the City and in the Holy Land. London.

II. The Physical environment

A. Light penetration

T. Berman

1. History

The first observations concerning light penetration date back to Barrois who, in April 1873, noted that the transparency of Kinneret water was considerably limited by numerous particles of algal origin.

Rechnitzer (1967), using a photometer equipped with four optical filters calibrated in energy units, was the first to collect data concerning the penetration of light at different depths and at various stations in the lake. His survey covered the period April 1966–May 1967, but unfortunately it is unavailable in published form.

From February to September 1969, Rodhe carried out weekly measurements of Secchi disc depths and of relative downwelling light with a photometer (Åberg & Rodhe, 1942) equipped with a single selenium cell covered with a diffuser plate. Standard Schott filters, BG12, VG9 and RG2, with optical midpoints at 460, 540 and 635 nm respectively were used with this instrument. These measurements, which were continued throughout 1970, did not permit the determination of absolute quantities of light energy but only of values relative to intensities measured at the water surface.

From August 1972, Berman used a Whitney LMT 8B illuminance meter equipped with a selenium cell giving a linear response in the visible light spectrum. This instrument, which has an on-deck sensor and two in-water sensors measuring both downwelling and upwelling irradiance, also gives only relative values of light intensity but, in addition, supplies information on the ratio of scattered to absorbed light. A summary of the characteristics of light penetrance in Lake Kinneret from 1970 to 1973 has been published (Berman, 1976).

Most recently, detailed studies of the spectral distribution of light energy at different depths were made by Dubinsky & Berman (1976) with an ISCO scanning radiometer, adapted for underwater work, and routine measurements with a Lambda Quantum Meter.

2. Secchi disc depths

The maximum Secchi disc depths (4.5 to 5 m) are generally found in August and November–December (Fig. 50). During the *Peridinium* bloom, Secchi depths are minimum, and when the algae are close to the surface, values as low as 0.6 to 0.7 m have been recorded. At other times, Secchi depths vary from 2 to 3.5 m.

3. Relative downwelling light penetrance

Figure 51 (Rodhe, 1969) shows the attenuation with depth of blue, green and red light in July 1969. The most penetrating component of the spectrum is, as in many

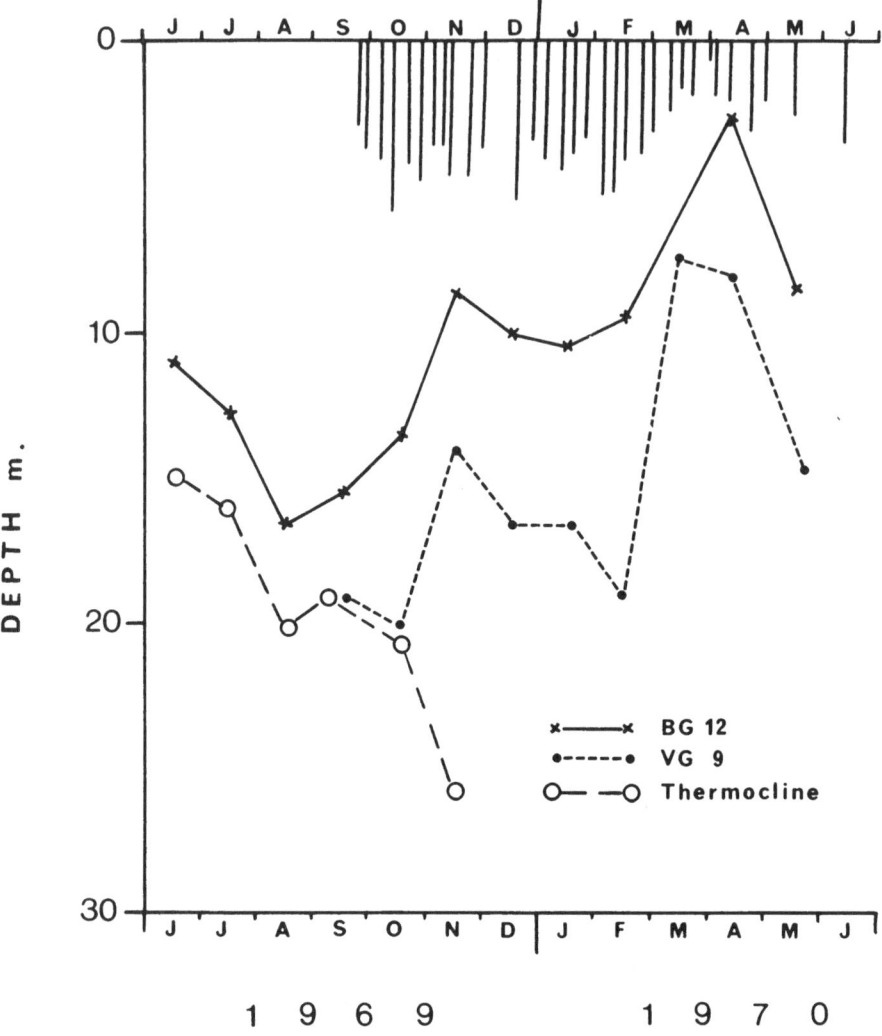

Fig. 50. Light penetration in Lake Kinneret – Secchi disc measurements are indicated by the vertical lines. The graphs show the depths at which the light intensity is 1% of the incident light (full line corresponds to blue light, BG 12 and dotted line to green light, VG 9). The position of the thermocline has been indicated for the stratified period.

other lakes, in the green range; blue light penetrates less and red the least. Occasionally, when very high concentrations of *Peridinium* cells are located near the surface, blue light is absorbed to a greater extent than red. The values of the photosynthetic assimilation rates at different depths, expressed as percentages of the maximum photosynthetic carbon fixation, sometimes give a curve which is most closely parallel to the curve of blue light (Rodhe, 1972). This is rather unusual; in general, the photosynthetic curve has been found to follow the most penetrating light component (Rodhe, 1965).

150

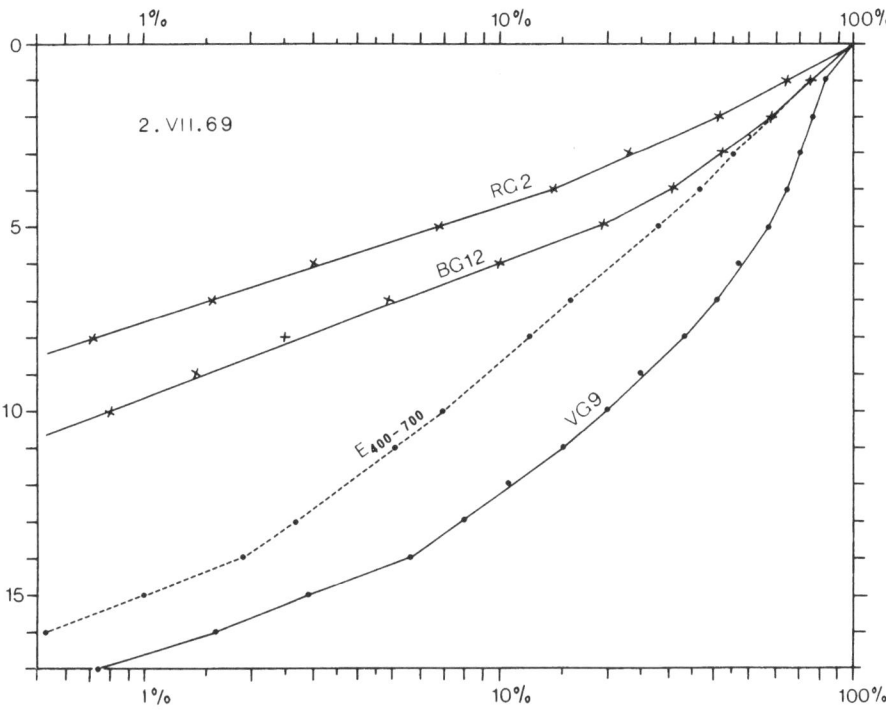

Fig. 51. Attenuation with depth of red, blue and green light. The dotted line corresponds to the visible spectrum between 400 and 700 nm (from Rodhe, 1969, mimeographed report).

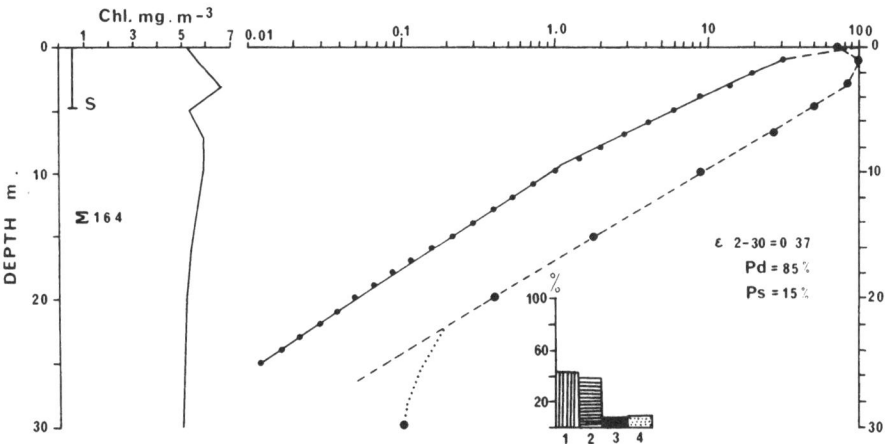

Fig. 52. Profiles of downwelling irradiation (full line), relative photosynthesis as percentage of maximum rate (interrupted line), chlorophyll a concentrations and percentage of algal biomass composition. (1 – Pyrrhophyta, 2 – Chlorophyta, 3 – Cyanophyta, 4 – Chrystophyta), Secchi disc reading (S), total chlorophyll standing crop in the trophogenic zone (Σ) in mg m^{-2}, extinction coefficints for indicated depths (ε), percentages of light energy absorbed (Pd) and scattered (Ps) by water. 10 January 1973 (from Berman, 1976, Hydrobiologia, 49, 1).

151

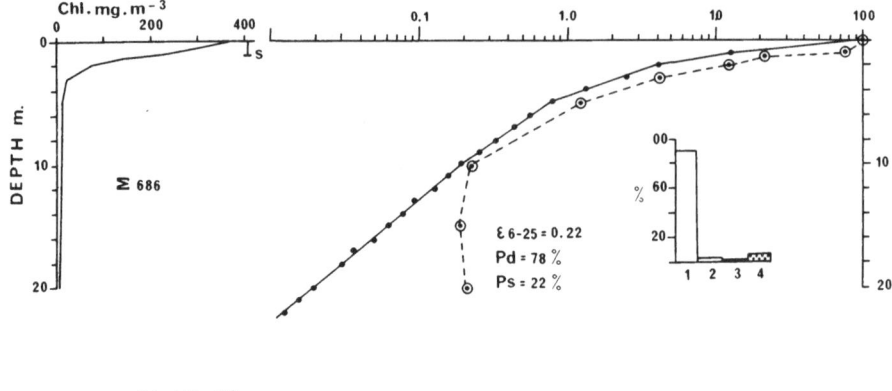

21.III.73

Fig. 53. As Fig. 52, 21 March 1973 (from Berman, 1976, Hydrobiologia, 49, 1).

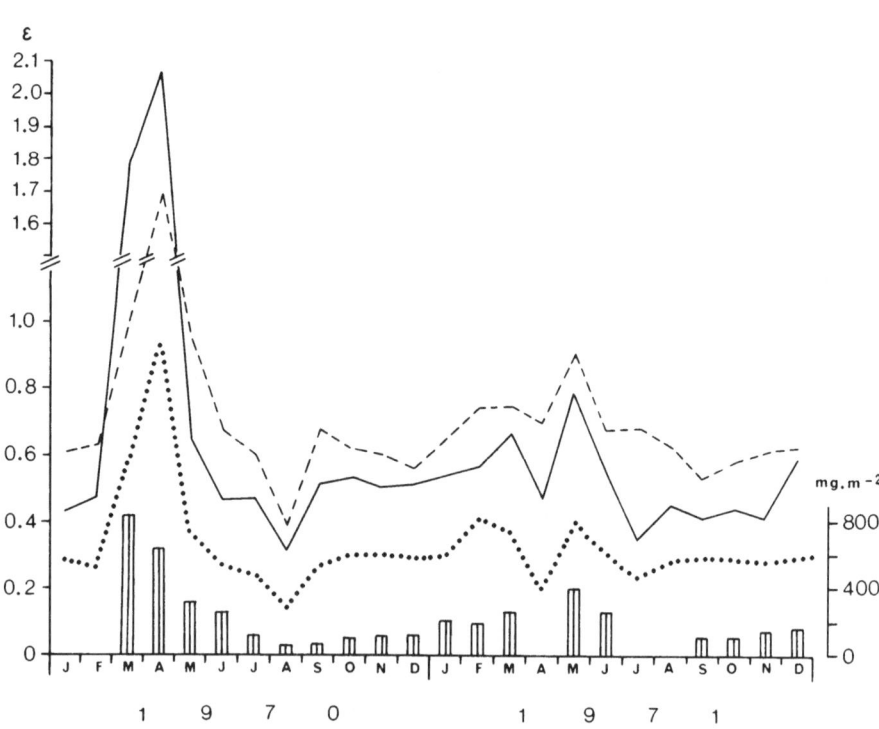

Fig. 54. Average monthly extinction coefficients measured with Aberg–Rodhe photometer with filters RG 2(interrupted line), BG 12(full line) and VG 9(dotted line). Average monthly chlorophyll a contents in trophogenic zone (mg.m^{-2}) are shown by histograms (from Berman, 1976, Hydrobiologia, 49, 1).

152

The dominant feature affecting light penetration characteristics in the lake waters is the presence or absence of the algal bloom of *Peridinium cinctum* (Figs. 52 and 53). When these organisms dominate, their patchy vertical distribution results in very uneven light penetrance, and relatively high extinction coefficients are obtained (maximum 2.1 ln units m^{-1}). The greatest range of monthly averaged values for $\mathcal{E}g$ (extinction coefficient of 'green' light, VG9) were observed in 1970, when this parameter varied from 0.15 (August) to 0.93 ln units m^{-1} (April) (Fig. 54). In 1971, when low standing crops of *Peridinium* developed, the fluctuations in extinction coefficients were more limited ($\mathcal{E}g$ maximum to minimum, 0.49 to 0.21 ln units m^{-1}). From August 1972 to December 1973, the average monthly extinction coefficient ranged from 0.31 (February 1973) to 0.88 ln units m^{-1} (May 1973) (Fig. 55). These observations, which were made at a central lake station, were probably only slightly affected by inflows of silt and detritus resulting from winter rains. The annual variations in minimum vertical extinction coefficients in Lake Kinneret are much greater than those recorded for lakes such as Lake Victoria, where the phytoplankton populations are generally stable and less dense throughout the year (Talling, 1965).

4. Ratios of absorbed to scattered light

For the main upper water column (0–15 m), the ratios of absorbed to scattered light ranged from 89:11 to 75:25. The highest ratio was observed in January 1973, even though algal standing crop was practically identical to that of August 1972, when the lowest value (75:25) was found. As yet, little is known about the non-living particulate matter or dissolved colored compounds in this lake, but clearly these, together with living particulate matter, will determine the above ratios (Hutchinson, 1957). For a number of Wisconsin lakes, Whitney (1938) determined ratios ranging from 96.8:3.2 to 76:24, and Åberg & Rodhe (1942) found ratios of 94.8:5.2 to 93.2:6.8 in two Swedish lakes. Thus, in comparison to other lakes examined, Kinneret waters show a greater degree of light scattering.

5. Spectral distribution of downwelling irradiance

An ISCO Scanning Radiometer adapted for underwater operation was used to measure downwelling light energies over the spectral range from 350 to 700 nm. The spectral response of the instrument was calibrated against a standard light source and corrected for the effects of the water-instrument window interface. Detailed profiles of downwelling irradiance were made from 1973 to 1974.

In Fig. 56 (Dubinsky & Berman, 1976), the pattern of attenuation of spectral irradiance typical for the non-bloom season (July to December) is shown. The wavelength of maximum light penetrance was at about 550 nm, and significant light readings were still obtained at 17.5 m. This is similar to the spectral distribution of downwelling light reported in Great Lake by Duval *et al.* (1969), but it contrasts with shorter maximum penetration wavelengths (440 to 500nm) measured in very oligotrophic lakes, Crater Lake and Lade Tahoe (Smith *et al.*, 1973). During the intense *Peridinium* bloom, light attenuation with depth is much more severe and the absorbance of the algal chlorophylls can be clearly noted with a peak at 650 nm.

6. Energy of downwelling irradiance

The total light energy penetrating into the water can be measured by the integration of the spectral distribution curves. Attenuation coefficients can also be

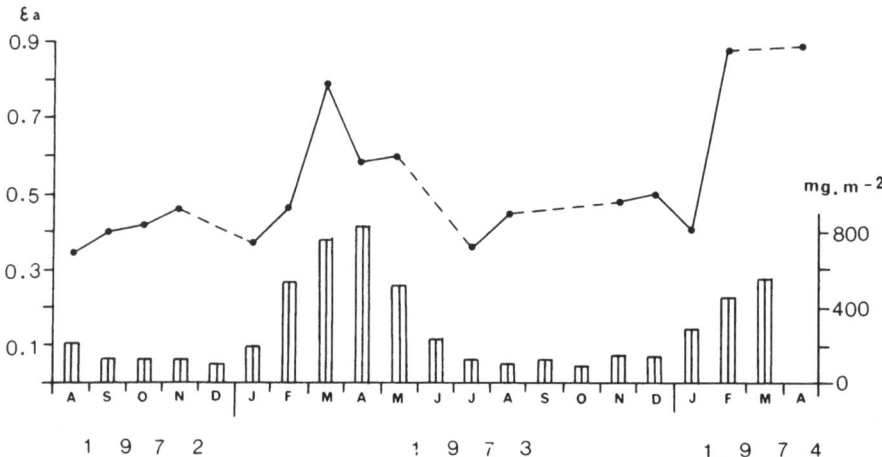

Fig. 55. Average monthly extinction coefficient measured with a Whitney photometer without filters. Histograms as in Fig. 56 (from Berman, 1976, Hydrobiologia, 49, 1).

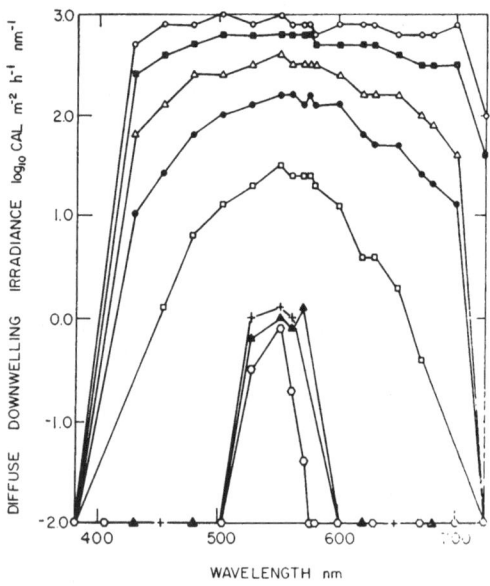

Fig. 56. Depth profiles of diffuse downwelling irradiance, energy assimilated in photosynthesis, light utilization efficiency and temperature measured at the central lake station (from Dubinsky, Z. & Berman, T., 1976, Limnol. Ocean., 21, 2).

154

calculated, either for the total penetrating light energy or within any selected spectral range. For the light profile measured in August 1973, photosynthetically available radiation varied from 3.63×10^5 cal m^{-2} h^{-1} at the surface to 1.2×10^2 cal m^{-2} h^{-1} at the thermocline (16.5 m depth), i.e. 0.03% of the surface energy. The attenuation coefficient of total downwelling light energy was 0.48 ln units m^{-1}.

For some purposes, it is desirable to express light energy in terms of photons, usually in units of micro-einsteins m^{-2} sec^{-1} (1 micro-einstein equals 6.02×10^{17} photons). The results of the ISCO radiometer can be converted using the approximation that one einstein of visible light has an energy of 52 k cal. Recently, routine measurements have been carried out with a Lambda Quantum Meter fitted with an underwater sensor calibrated directly in photons. Preliminary results appear compatible to those obtained earlier with the ISCO radiometer.

B. Water temperature

S. Serruya

1. History

The first measurements of temperature at different depths were made by Barrois in April–May 1893. This author related the large diurnal fluctuations of temperature in the surface layers to the contrasting effect of strong insolation and wind cooling. He also reported that below 20 m, the temperature was constant. Barrois explained the thermal features of Lake Tiberias by its latitude, its negative altitude and the thermal spring water entering the lake.

In a more recent period, Ashbel performed systematic temperature measurements on the lake water from December 1943 to February 1945. In 1945, Oren resumed these measurements in the framework of his work for the National Planning Board of Israel, and gave the first comprehensive description of the yearly thermal pattern of the lake. From 1963 until 1967, the Mekorot Water Company performed routine temperature measurements at discrete depths with a reversing thermometer. During this period, several other investigators from different institutions, studying the Kinneret environment, also measured temperature for their own purposes.

From 1968 until the present, the Kinneret Limnological Laboratory has been measuring temperature profiles at various stations on a weekly basis. The replacement of reversing thermometers by thermistor probes resulted in more detailed and frequent profiles. Moreover, automatic instruments, Aanderaa current meters equipped with temperature sensors and Aanderaa thermistor chains, were used to measure temperature every 10 or 20 minutes and obtain long term records. During various surveys, a bathythermograph was used to obtain almost synoptic pictures of the thermal pattern of the lake.

2. Thermal pattern

The isotherm variation based on the KLL weekly temperature profiles determined at the deepest central station A for the period 1971–1975 is shown in Fig. 57. The vertical axis represents the altitude (MSL) of the water layer instead of the depth referenced to a 0 m water level, in order to take into account the water level variation during the recorded period.

In Lake Kinneret, the temperature ranges from 14 to 28.5°C on an average basis. The minimum temperature measured in winter during the homogeneous period was 13.6°C in January 1973 and the maximum temperature measured in this season was 16.2°C in February 1970. Temperatures of 12.5°C at 40 m were reported by Oren in February 1950. During the stratification period, the maximum temperature measured in the upper layer was 30.2°C in August 1968 and the minimum was 28.2°C in August 1970 and September 1971.

The monthly temperature profile for the year 1971 is displayed in Fig. 58.

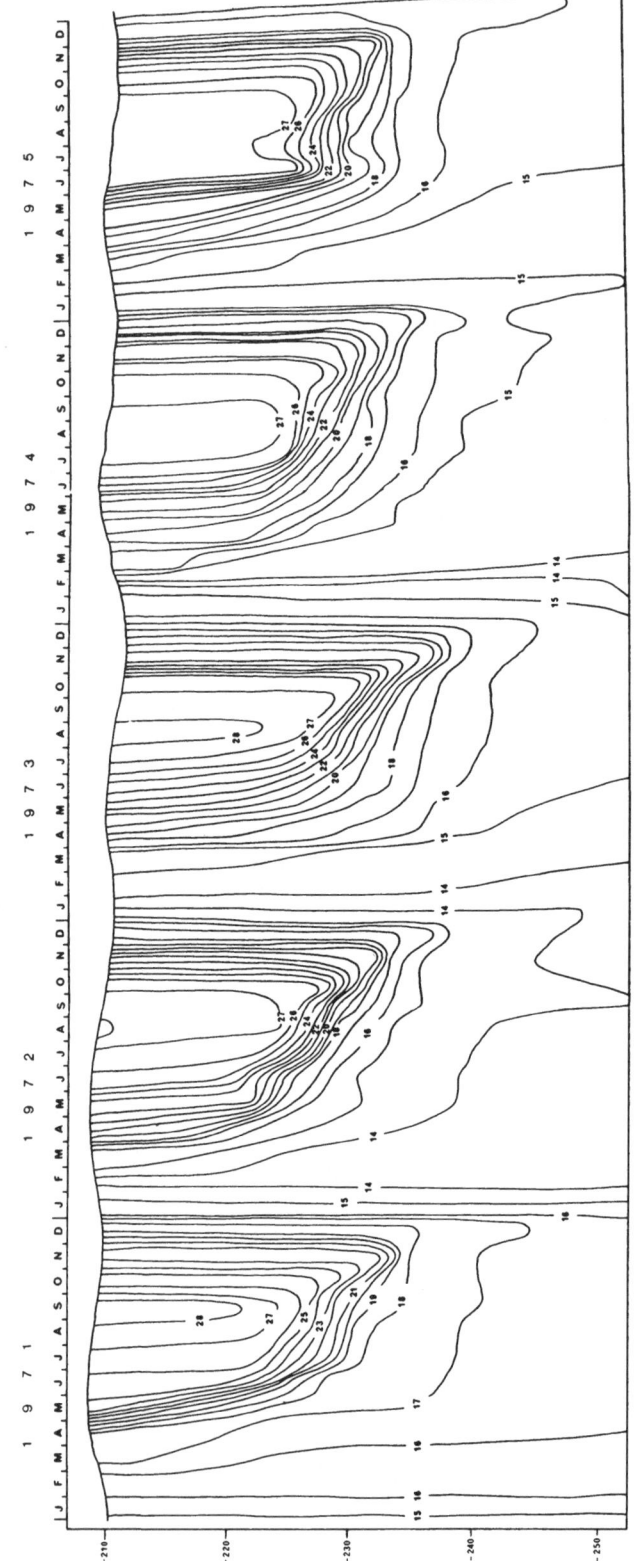

Fig. 57. Isotherm curves for the period 1971–1975 at station A. The isotherms are related to the actual level of the lake. Numbers are degrees centigrade.

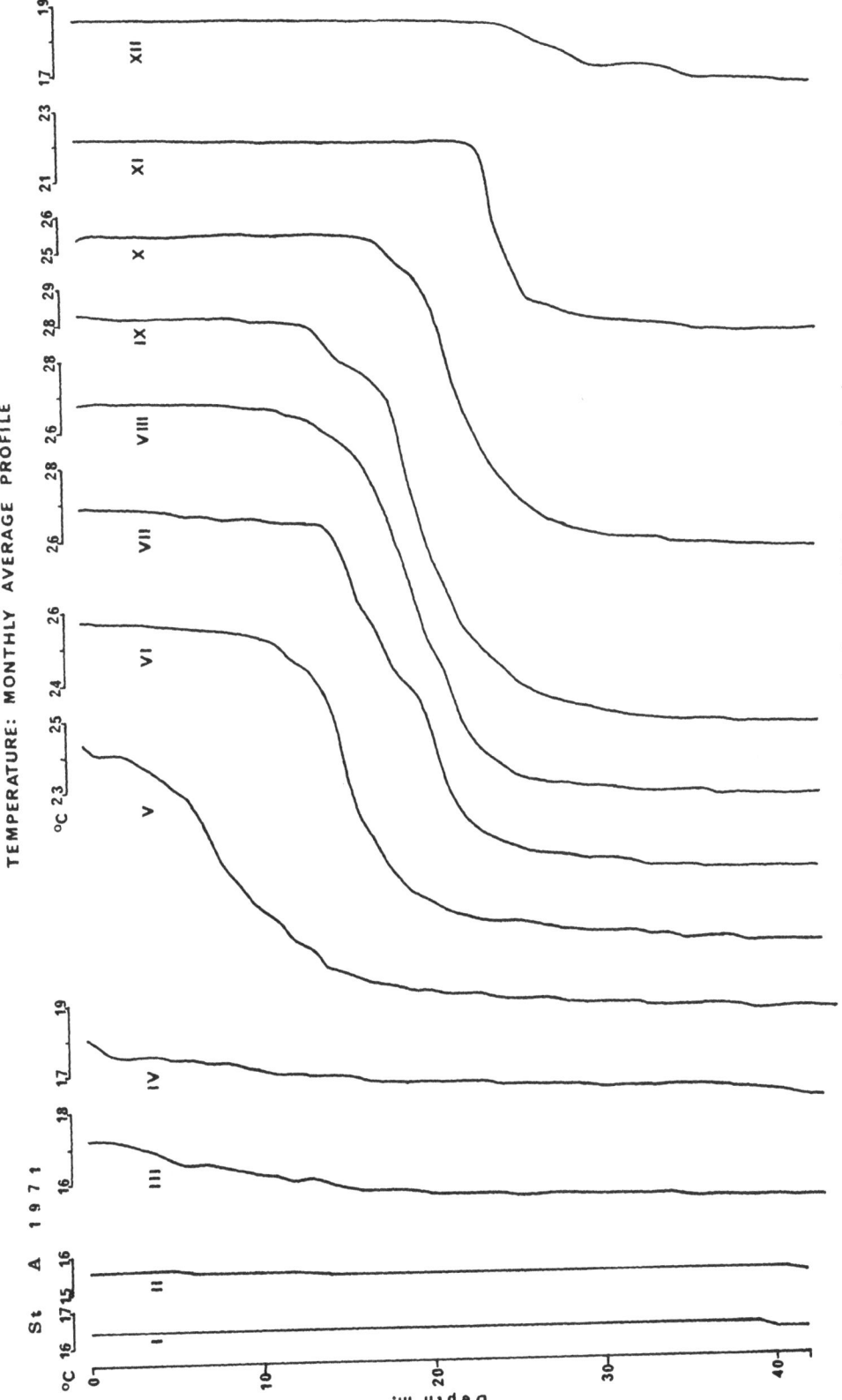

Fig. 58. Average monthly profile of temperature for the year 1971 in degrees centigrade.

Table 20. Mean monthly temperature of lake Kinneret at different depths for the period 1969–1975. Results in degrees C.

	JAN		FEB		MAR		APR		MAY		JUNE	
Depth	Av.	S.D.	Av.	S.D.	Av.	S.D.	Av.	S.D.	Av.	S.D.	Av.	S.D.
0	15.6	1.0	14.9	0.9	19.3	1.1	22.7	0.9	25.7	0.9	25.7	0.9
5	15.6	1.0	14.8	0.9	16.2	0.7	19.1	1.5	22.3	0.5	25.5	0.7
10	15.6	0.9	14.8	0.8	15.7	0.6	17.9	0.6	21.3	0.8	25.0	0.4
15	15.6	1.0	14.7	0.9	15.2	0.9	17.3	0.7	19.2	0.9	21.7	1.2
20	15.6	0.9	14.7	0.8	14.9	0.8	16.1	0.9	17.1	1.1	18.1	1.5
25	15.6	1.0	14.7	0.9	14.8	1.0	15.6	1.0	16.0	1.1	16.5	1.2
30	15.6	0.9	14.6	0.8	14.7	0.9	15.2	1.0	15.4	1.1	15.7	1.2
35	15.5	0.9	14.6	0.9	14.7	0.9	15.0	1.1	15.2	1.1	15.3	1.1
40	15.0	0.9	14.5	0.8	14.6	0.8	14.9	1.0	15.0	1.1	15.2	1.0

	JULY		AUG		SEP		OCT		NOV		DEC	
Depth	Av.	S.D.	Av.	S.D.	Av.	S.D.	Av.	S.D.	Av.	S.D.	Av.	S.D.
0	27.2	0.5	28.0	0.2	27.7	0.7	26.1	0.6	22.5	0.6	18.3	0.7
5	27.1	0.4	27.9	0.2	27.6	0.5	26.0	0.5	22.5	0.5	18.3	0.8
10	26.9	0.3	27.7	0.1	27.5	0.5	26.0	0.5	22.5	0.5	18.3	0.7
15	25.4	0.8	27.2	0.4	26.9	0.5	25.9	0.5	22.5	0.5	18.3	0.7
20	19.3	2.1	20.3	2.3	21.7	2.3	23.3	1.8	22.2	0.4	18.1	0.7
25	16.7	1.3	16.8	1.3	17.1	1.3	17.4	1.5	18.3	2.5	17.9	0.9
30	15.8	1.1	15.8	1.0	16.0	1.1	16.0	1.1	16.1	1.2	16.3	1.1
35	15.5	1.0	15.6	1.1	15.6	1.0	15.7	1.0	15.7	1.0	15.9	1.2
40	15.4	1.0	15.5	1.0	15.5	1.0	15.6	1.0	15.5	1.0	15.6	1.2

Table 20 shows the monthly interannual variation and the corresponding standard deviation of the mean monthly temperature for the period 1969–1975. These data allow a detailed description of the successive sequences of the annual thermal cycle of the lake.

Lake Kinneret is a warm, monomictic lake with only one mixed period in winter. The complete turnover down to 42 m occurs generally in late December or in January. Exceptionally, during very mild winters such as the 1969–1970 winter, the turnover took place in late February. The minimum temperature of the lake water is usually observed in February; therefore, the full turnover period extends, on the average, from January to early March. The temperature at which the turnover takes place governs important chemical processes; for example, it determines the order of magnitude of oxygen storage. The duration of the turnover affects also the biological environment; in particular, the occurrence and intensity of the *Peridinium* bloom is related to the intensity of water movements as described in page 286. It is also during this period that the lake sediments are resuspended by turbulent mixing. This input of mechanical energy results in a substantial release of nutrients and in the recirculation of benthic bacteria.

In March–April, a thermal gradient develops in the upper layers generating a labile thermocline at a depth varying from 8 to 10 m. This thermal gradient, 2 to 3°C difference between the upper and the lower layer, already constitutes a

160

barrier; the oxygen starts to decrease and CO_2 appears on the bottom where the pH decreases.

In May–June, the continuous warming generates a steady stratification; then the difference of temperature between the epilimnion and the hypolimnion may reach 8°C. The thermocline deepens down to 16–18 m and the thermal gradient is about 0.6–0.8°C m^{-1}. This thermal stratification divides the lake into two almost independent water masses: a warm, oxygenated, CO_2-depleted epilimnion with an abundant algal flora and a cold, oxygen-depleted hypolimnion with dissolved H_2S and CO_2.

From July to October, the thermocline slowly deepens from 18 m down to 22–23 m and the thermal gradient then reaches 1.2°C m^{-1}. The chemical characteristics of the two layers described in the above paragraph are accentuated.

In October, November and December, the rate of deepening of the thermocline increases until the complete turnover. An important factor in the fast deepening of the thermocline is the occurrence of the first easterly storms. This process is clearly observed, for example, between November and December 1971, where the 18°C isotherm, corresponding at this period to the thermocline boundary, remained at a depth of about 25–27 m until the end of November, with a sharp deepening from 28 to 35 m between the 5th and 12th of December due to a particularly windy week: westerly storms on the 6th and 7th, southerly on the 8th, and strong easterly on the 10th and 11th (up to 11.6 m s^{-1}).

The non-linear variation of water density with temperature is such that the difference of density corresponding to a difference of 16°C is twice as high in the range 14–30°C (0.003598) (Lake Kinneret) as in the range 4–20°C (0.001770) (temperate lakes). It follows that during the stratification period, Lake Kinneret is much more stable than the lakes of the temperate zone. This is also the case in comparison with the tropical lakes, which reach higher temperatures but with a very small difference between the epilimnion and the hypolimnion.

3. Internal waves

At the time scale of the weekly routine measurements, the thermocline location seems very steady and the deepening trend stable. However, with shorter interval measurements, 10 or 20 minutes, the temperature distribution at the thermocline level shows a very high variation, up to 10°C during a few hours. These variations correspond to typical internal seiches caused by the oscillation of the thermocline as a response to wind stress and were first observed in the Kinneret in 1968.

The use of thermistor chains during the stratified period confirmed the presence of temperature waves. These waves correspond to the warm epilimnic and to the cold hypolimnic water masses successively measured at the level of the thermistors moored at fixed depths in the thermocline zone. The temperature variations obtained from a thermistor chain moored during a characteristic period of seiches, at station G in the northern part of the lake, are displayed in Fig. 59. The thermistors located at depths of 6, 9 and 13 m showed a steady epilimnic temperature, the 15, 17, 19 and 21 m depth thermistors located in the metalimnic

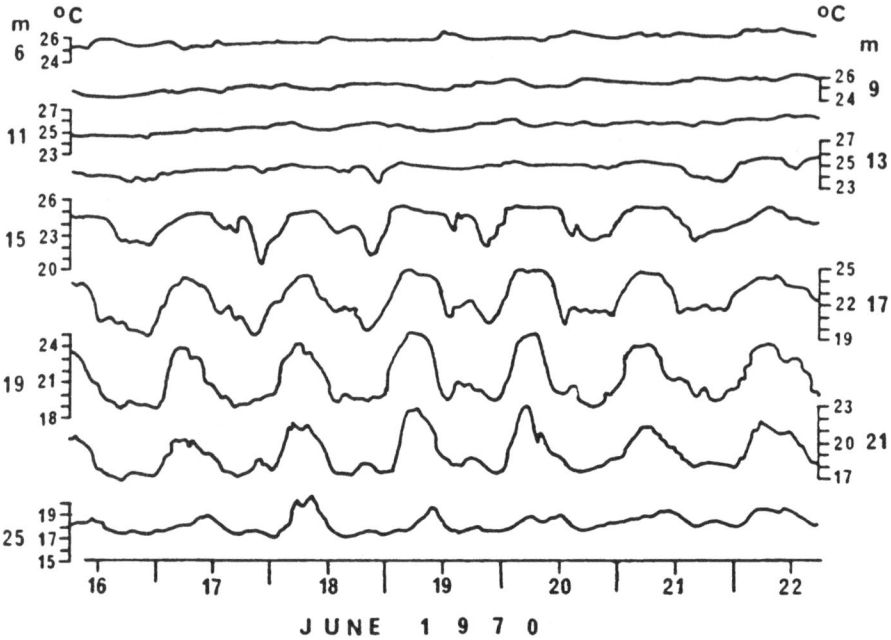

Fig. 59. Temperature variation at station G at the indicated depths. June 16–22, 1970 (from Serruya, S., 1975, Verh. Internat. Verein. Limnol., 19).

zone recorded the maximum temperature variations as daily oscillations, and the 25 m thermistor, moored in the hypolimnion, only recorded very slight temperature variations.

From the early spring to the late autumn, these daily temperature waves are a constant characteristic feature of the Kinneret thermocline. The daily stress of the westerly wind applied on the surface results in a piling up of the epilimnic waters on one side of the lake and in an upwelling of cold hypolimnic waters windward. The concomitant timing of the wind stress, the upwelling of lower cold water at a leeward station (K) and the piling up of superficial warm waters at a windward station (I) are clearly shown in Fig. 60. The displacement of the thermocline has been computed from the temperature data recorded by the thermistor chains for sensors fixed at 2 m distance all through the thermocline zone; it was expressed by the depth variations of the 21°C isotherm which corresponds to the temperature of the metalimnic layer. Fig. 60 displays, from the 13th to the 24th July 1973, the wind recorded at Tabgha, the depth of the 21°C isotherm at stations I, A and K, located on the same NE–SW section, and furthermore, for stations I and K, the depth of the 24 and 18°C isotherms corresponding respectively to the lower epilimnion and to the upper hypolimnion. The peak to peak displacement of the thermocline is larger at station K, up to 10 m, than at station I, up to 5 m, during the days of stronger wind (above 12 m s⁻¹), whereas the displacement is only 3 to 4 m at both stations with winds below 12 m s⁻¹. The oscillations observed at stations I and K are out of phase, and station A seems to be located at the nodal point: very small fluctuations are recorded at station A, about 1 m amplitude, sometimes in phase, sometimes out of phase with either stations I or K. At these latter stations, a marked 24-hour period of oscillation is observed during the days of strong wind, whereas only a 12-hour period is recorded during the days of weaker winds. The existence of a 24-hour and 12-hour period results from the combined effect of an imposed periodicity due to the steady wind frequency of one

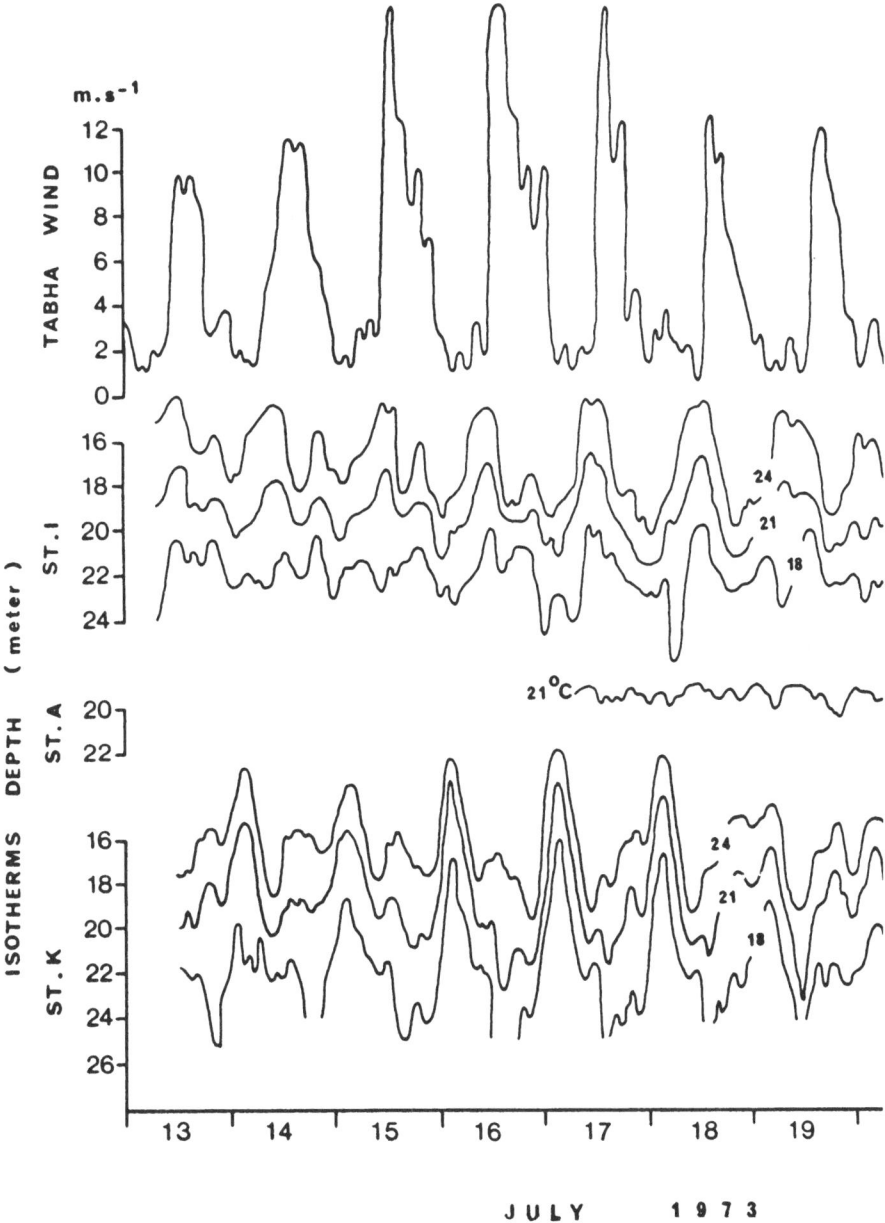

Fig. 60. Wind velocity, depths of isotherms 18–21–24 C at stations K and I and depth of isotherm 21 C at station A.

163

cycle per day and to the resonance caused by the presence of a 12-hour natural free period of oscillation, probably of the third mode.

To explain the differences of amplitude and periodicity of the thermocline displacement between the first and last days of the record, the timing of the wind and not only its relative strength (above or below 12 m s^{-1}) should also be taken into account. During the days of stronger winds, on the 15 and 16 July 1973, the wind began to blow at 1200 hours whereas during the days of weaker wind, on the 22 and 23 July, it started after 1200 hours. At station K, on the 15 and 16 July, between 0200 and 1000 hours, the thermocline deepened by 4–5 m and started to

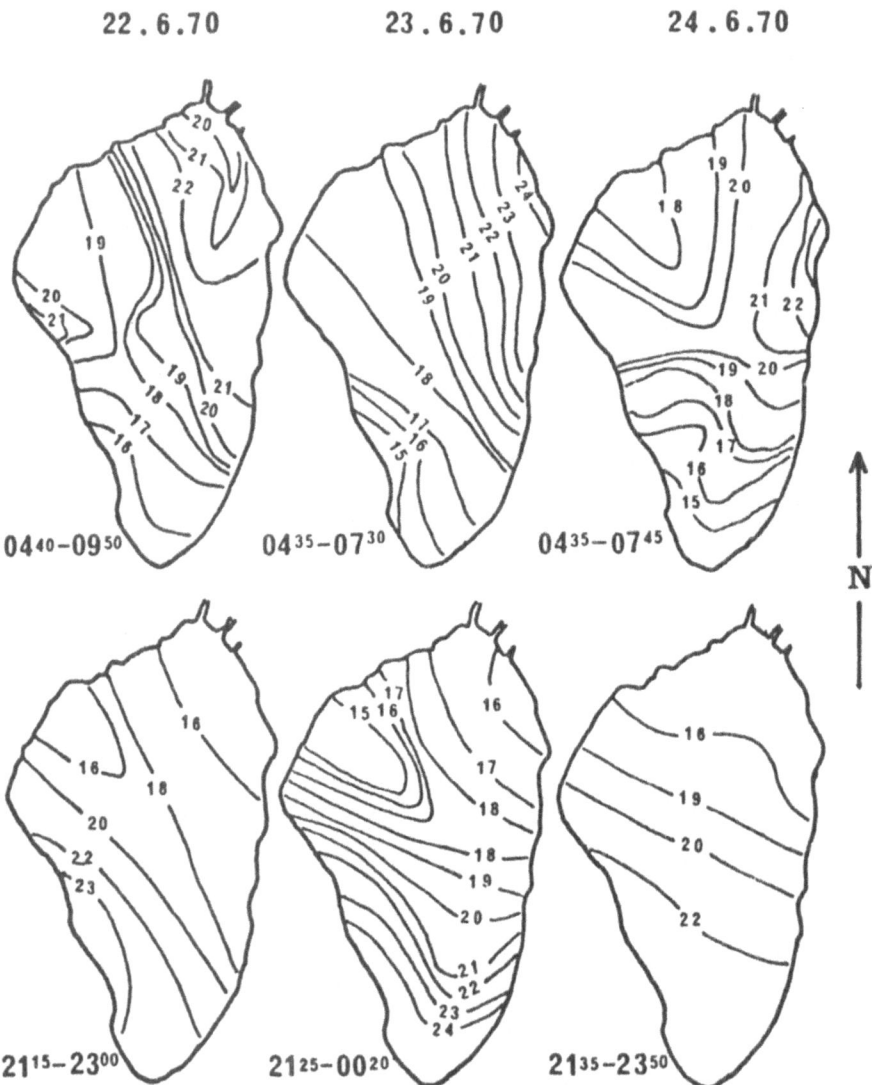

Fig. 61. Isopleth maps of the thermocline (21°C isotherm) measured twice a day – June 22-23-24, 1970. Depth values are in meters (from Serruya, S., 1975, Verh. Internat. Verein. Limnol., 19).

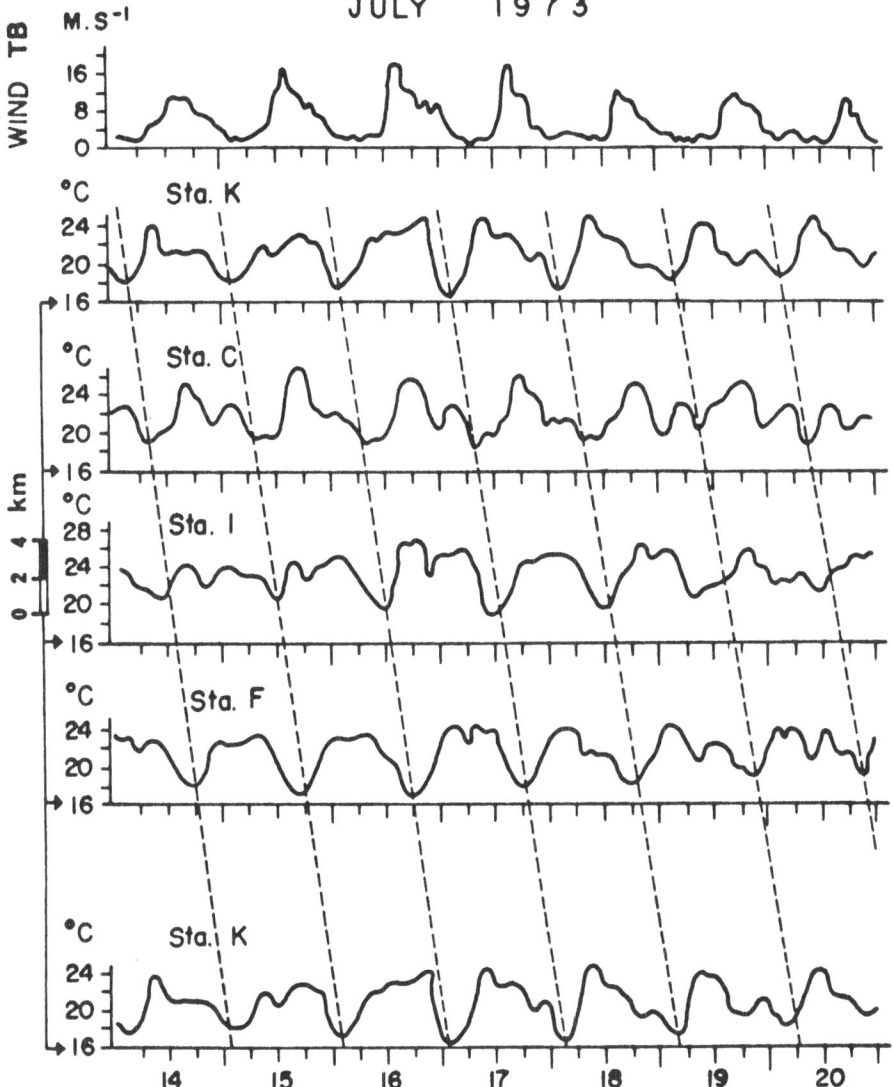

Fig. 62. Rotation of the crest of the thermocline 'wave' at 4 stations arranged in a counterclockwise order around Lake Kinneret.

move about 0.5 m towards the surface when the influence of the oscillation of the warm waters in the upper layer took place. Therefore, the thermocline stopped its upward motion and began to deepen for the second time, with a final displacement and period equal to twice its normal value.

Conversely, during the days of wind lower than 10 m s^{-1}, the 22 and 23 July, the thermocline, after deepening by 2–3 m, again between 0200 and 1000 hours began to move towards the surface and had already reached its higher position when the daily wind started. The wind again lowered the thermocline, but this time from an upper level and therefore with a smaller final displacement.

To follow the thermocline topography and its daily variation all over the lake, a bathythermograph survey was carried out twice a day during three consecutive days, 22, 23 and 24 June 1970. During these three days, the location of the thermocline, identified with the 21°C isotherm, is shown in Fig. 61, where the curves indicate the depth of this isotherm. The same pattern was observed on all three days: a downward tilt of the thermocline occurred at night in the SW area with a corresponding upwelling in the NE part of the lake, whereas in the early morning, the thermocline deepened towards the NE with a corresponding upwelling at the SW.

A westerly wind blowing on a N–S elongated lake should generate a transverse internal seiche oscillating E–W along a N–S axis. In Lake Kinneret, a NE–SW oscillation along a NW–SE axis is observed. This deviation can be accounted for partly by the deflection to the right caused by Coriolis forces, and partly by the constraint imposed by the eastern coastal boundary.

The wind-driven water masses pile up in the evening in the southern part of the lake, and when the wind stress ceases completely at night, a dominant transverse-like internal seiche is set in motion. This wind-induced pattern repeats itself the next day almost at the same hours, and consequently imposes the regularity of the observed 24-hour period.

To obtain the pattern of the thermocline wave moving all over the lake, the temperature was simultaneously recorded at stations F, K, C and I. The temperature variation at the thermocline level from the 14 to the 20 July 1973 is displayed in Fig. 62. Each station has been spaced on the graph at a distance relative to its location in the lake. The crest of the thermocline 'wave' corresponding to the lowest temperature has been traced from station to station and can be regularly followed. The crest wave rotates in an anticlockwise direction from station K to station C, then to station I, and after station F goes back to station K. This rotation takes about 24 hours and the distance covered is about 40 km. The wave celerity is then about 0.5 m s^{-1}, that is, about 2 to 3 times the velocity of the water flow which also circulates in the same anticlockwise direction.

C. Water motions

S. Serruya

To visualize the water movements, we have computed progressive vector diagrams (PVD) (Webster, 1964) which show, approximately, the path of the water flowing at the point recorded.

1. General circulation

Two patterns of circulation have been recorded in Lake Kinneret, depending only on the wind stress.

a. A dominant, relatively steady counterclockwise circulation is observed associated with wind speeds above 3–4 m s^{-1}, no matter the season, depth or wind direction. This occurs during all the westerly wind period, i.e. all the summer and most of the spring and autumn.

At the beginning of June 1973, during the daily westerly wind period, the current observed at the four stations, F, I, C and K, shows a counterclockwise pattern. Fig. 63 displays the currents of the upper layer (about 3 m depth), which

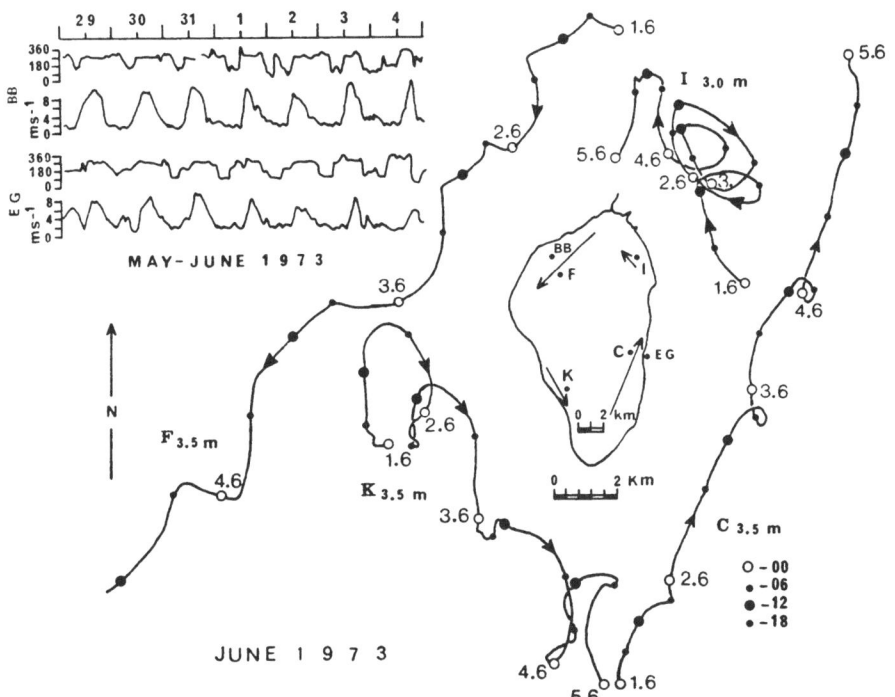

Fig. 63. Progressive vector diagram from data recorded at 3.5 m depth at stations F, K, C, I on June 1–5, 1973. Insert = wind daily direction and speed at Tabgha (BB) and Ein Gev (EG) stations during the same period (from S. Serruya, 1975, Verh. Internat. Verein. Limnol., 19).

167

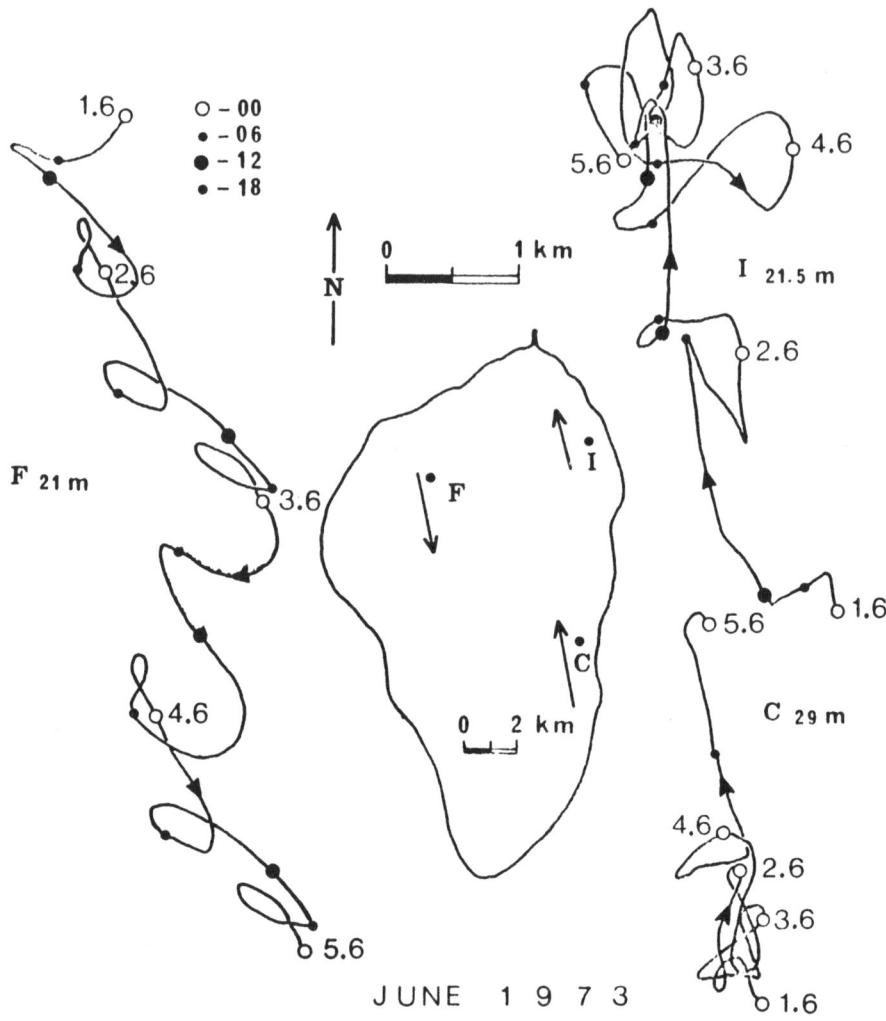

Fig. 64. Progressive vector diagrams computed from data obtained at 21 m depth (stations F and I) and at 29 m (station C) on June 1–5, 1973 (from S. Serruya, 1975, Verh. Internat. Verien. Limnol., 19).

can be described as wind-driven currents, and Fig. 64 displays the lower hypolimnic currents (F and I, 21 m; C, 29 m depth), which are closely associated with the internal seiche oscillations.

Due to the compact form of the lake and to the Coriolis deflection, the easterly storms in winter also give rise to a counterclockwise current.

b. An occasional clockwise circulation is observed with wind speeds below 3 m s⁻¹. This kind of circulation takes place in winter or during particular days in spring and autumn with no wind at all.

Fig. 65 represents the PVD computed from the current measured at stations F, C, K and I in the upper layers (3 m depth) in mid-February 1973, during the

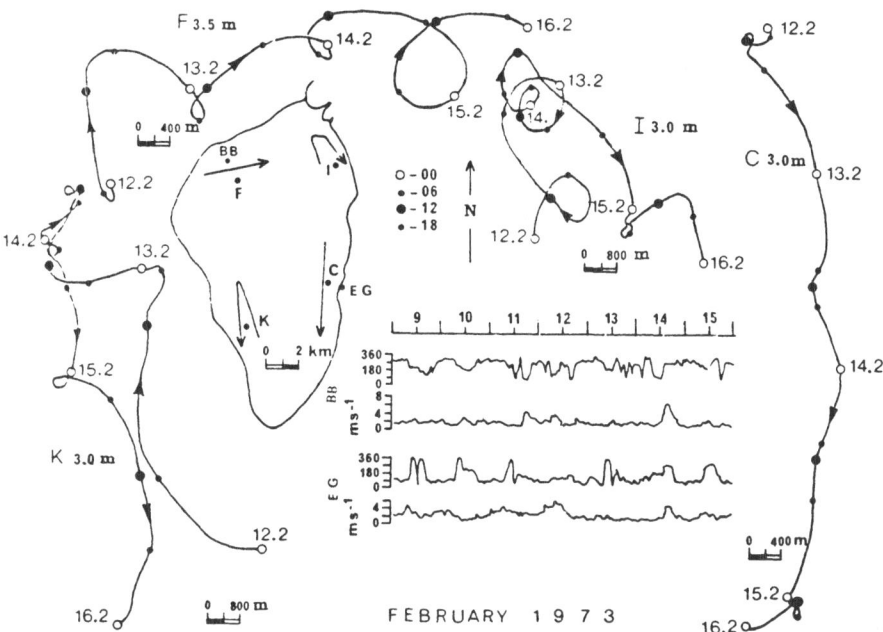

Fig. 65. Progressive vector diagram at stations K, C and I from data recorded at 3 m depth, February 12–16, 1973. Insert = wind daily direction and speed at BB and EG during the same period (from S. Serruya, 1975, same source as Fig. 63).

turnover. Moderate winds were recorded during these days, except on February 14 (7.5 m s⁻¹). The clockwise circulation is observed at stations F, C and partly at I and K.

2. Particular motions

a. Current associated with the internal seiche

To observe the currents during a typical period of internal seiche, the available current meters were moored at various depths at the same station. The hourly current vectors corresponding to different layers at station I on July 1971, 20–21 are shown in Fig. 66, together with the E–W wind component measured at the Tabgha station.

At 3 m, a southerly wind-driven current was recorded in the afternoon simultaneously with the westerly wind, whenever the northerly counterclockwise direction is observed during the other hours of the day. The maximum speed recorded was 20.2 cm s⁻¹ as an hourly mean.

At 10.5 m depth, in the lower epilimnion, the currents are already of opposite direction to those at 3 m, and correspond to the return currents of the upper layer: a high northern current, up to 25.6 cm s⁻¹ during the wind time, and lower southern velocities otherwise.

At 17 m, in the thermocline level, downward about 0600 hours, upward about

169

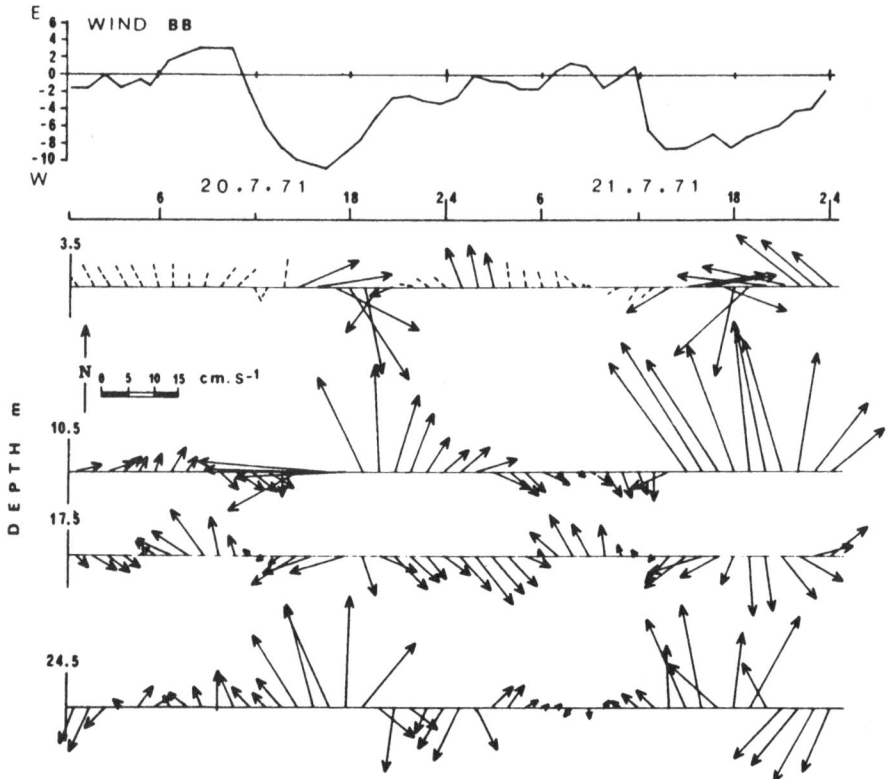

Fig. 66. Hourly E–W wind component at station BB and hourly current vector at station I at four depths; July 20–21 1971 (same source as Fig. 63).

1800 hours, the current direction is opposed to that in the upper layers and in the layer below. The main direction recorded is clockwise, the maximum speed 11.3 cm s^{-1}.

At 24.5 m, on the lake bottom, a reversal of the current is observed; the water masses flow counterclockwise with high velocities, up to 19.6 cm s^{-1}.

b. Inertial-like motions

In many PVD, a circular pattern rotating in a clockwise direction has been observed. This pattern was recorded at different stations throughout the year: in winter, during the turnover, in the upper layer, with a clockwise circulation (Fig. 65, station F), and in summer, during the stratification period, with a counter-clockwise circulation either in the epilimnion (Fig. 63, stations I and K) or in the thermocline level (Fig. 67), or even in the hypolimnion (Fig. 64, station F, 21 m).

This pattern is very similar to the intertial motion generally observed in the oceans, far from the shore constraint, which appears as a clockwise circle in the northern hemisphere with a radius equal to V/2ω Sin φ (V = water speed, ω = 0.73 × 10^4 the angular velocity of the earth's rotation, φ = latitude) and with a

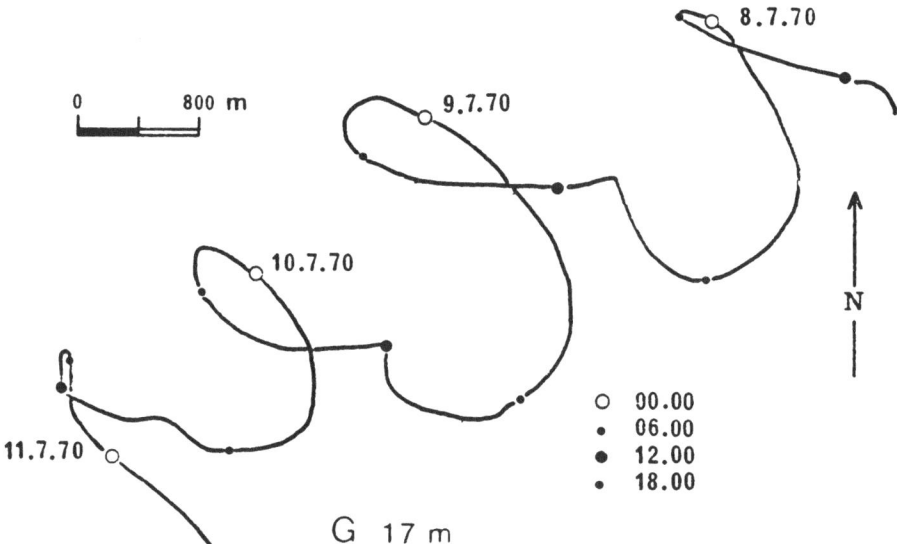

Fig. 67. Progressive vector diagram computed from data obtained at 17 m depth at station G, July 8–11, 1970 (same source as Fig. 63).

period of rotation Tp corresponding to the inertial period Tp $= 12/\mathrm{Sin}\,\phi$.

For the circular-like paths shown in Figs. 63, 64, 65 and 67, the radius of the inertial circle has been computed and compared to the radius of the inertial-like circle observed; both values are of the same order of magnitude.

D. Energy balance and evaporation

G. Stanhill & J. Neumann

The radiation balance at the lake's surface and the thermal regime of its waters are the two major elements determining the lake's energy balance and evaporation loss. The different flux components of the energy balance are considered in this chapter using the following balance equations:

$$Q_* = K\downarrow + L\downarrow - L\uparrow - R = LE + S + A + V + M$$

where: Q_* is the net radiation balance at the water surface
  $K\downarrow$ is the global, short-wave radiation from sun and sky
  $L\downarrow$ is the atmospheric, long-wave radiation from the sky
  $L\uparrow$ is the long-wave radiation emitted by the water surface
  R is the short-wave radiation reflected by the water surface
  LE is the latent heat of evaporation
  S is the change in heat storage in the water
  A is the sensible heat exchange with the air
  V is the net advective heat exchange via the water fluxes
  M is the change in metabolic heat storage

All fluxes are expressed per unit water surface area, positive values representing fluxes to the water surface.

1. Global radiation

In the absence of direct measurements, 25 years of observations of the cloud cover at Degania A were used with the following empirical equation:

$$K\downarrow = K\downarrow\downarrow (0.7985 - 0.0311\ C - 0.001\ C^2)$$

where: $K\downarrow$ is mean global radiation in cal cm^2 day^{-1}
  $K\downarrow\downarrow$ is mean solar radiation at the top of the atmosphere in the same units
  C is mean cloud cover from observations at 08, 14 and 20 hours local time, in tenths of sky covered

The equation was derived from 94 pairs of monthly data obtained prior to 1960 from three stations in unpolluted areas of Israel (Stanhill, 1962). As no difference was found between the relationships at the three stations, despite an altitude difference of more than 700 m, it may be assumed that the relationship will also yield monthly estimates for Lake Kinneret with a mean standard error of 4.4%, as observed at the other three stations.

The annual total insolation calculated, 199.3 Kcal cm^{-2} yr^{-1}, agrees to within 1% of the nearest measured value at Amir, 52 km to the north of the lake in the Upper Jordan Rift (Stanhill, 1970). The calculated mean monthly values at Lake

Kinneret, listed in Table 21, are somewhat higher in the winter and slightly lower in the summer than those measured at Amir.

2. Reflected short-wave radiation, R

The short-wave radiation reflected from the lake has not been systematically measured. In Table 21, the values presented are calculated with the mean monthly albedo values listed by Kondratyev (1972), appropriate to the latitude and mean cloud cover of the lake. Two days of measurement in the center of the lake during the early summer of 1966 gave a mean albedo of 6.0%, in good agreement with the tabulated values for that time of the year.

3. Long-wave emission, L↑

Emission was calculated as $L\uparrow = \varepsilon\sigma T^4$, where T is the water temperature on the absolute scale. The values used are mean monthly values, each based on an average of six dates of measurement, each of which is based on 15 points of measurement selected to sample the lake surface (Stanhill, 1969). σ is the Stefan–Bolzmann constant and ε is the emissivity of the water surface; the value used for ε was 0.97 (U.S. Geological Survey, 1954).

4. Net radiation balance, Q∗

Data from four years of continuous measurement of the radiation balance over the water surface are available from a special station established at Ginnosar as part of the Lake Kinneret Evaporation Project, initiated by Water Planning for Israel Ltd. under the supervision of the late Dr. J. Frenkiel, and now under the supervision of F. Miro. A description of the instrumentation, exposure and calibration procedures can be found in the first report (Miro & Kahanovitz, 1969). The mean monthly values presented in Table 21 are taken from data presented in three progress reports (Miro & Kahanovitz, 1972, 1973, 1974). Fluxes for the individual periods are shown in Fig. 68.

The areal variation in the radiation balance is probably small. Measurements during August 1964 near the shore at Ginnosar and Ein Gev show a 5% difference in the mean daily total (Fig. 69), almost certainly due to the difference in water surface temperature occurring in the summer months (Stanhill, 1969).

5. Long-wave sky radiation, L↓

Values of this flux, calculated as the difference from the radiation balance equation, are listed in Table 21. It may be noted that the values calculated exceed those for the short-wave global radiation received at the water surface.

6. Heat storage change in the water, S

Four series of lake water temperature measurements are available from which this flux may be calculated. The first, presented by Ashbel (1945), is for a 14-month

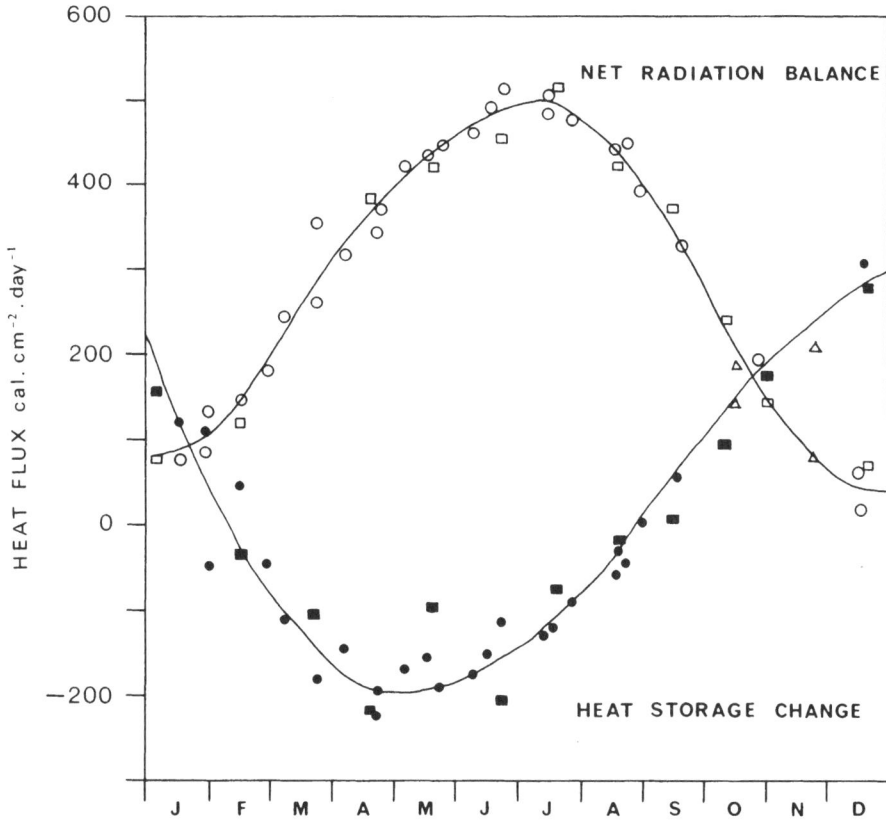

Fig. 68. Measured values of net radiation balance over Lake Kinneret at Ginnosar and heat storage changes in the Lake for corresponding periods. Original data in Miro & Kahanovitz, 1972 and 1973 and Kahanovitz & Karni, 1974.

period between 1943 and 1945 and is published as mean monthly values for 5 depths based on two fixed points of measurement near the shore at Ein Gev and Hamei Tiberias and an unspecified number of points on the lake. The annual heat storage change, i.e. the sum of changes through the calendar year neglecting signs, totalled 67,000 Kcal cm^{-2} yr^{-1}.

Oren (1962) reports the results of a series of measurements taken during 27 months between 1948 and 1951. Twenty sets of measurements were made at four depths along two transects, one from Tiberias to Ein Gev and a second from Ginnosar to El-Kursi. The annual flux totalled 47,433 Kcal cm^{-2} yr^{-1}.

A third series of measurements presented by Stanhill (1969) consists of 72 sets of data collected at ten depths at 15 stations chosen to sample representatively the lake's volume. The fluxes for the two years of measurement, from 1965 to 1967, were 40,000 and 47,240 Kcal cm^{-2} yr^{-1}. The standard deviation of temperature differences were used to calculate the error of the estimates of heat storage change. For the great majority of the periods examined, the coefficient of variation was less than 3%.

175

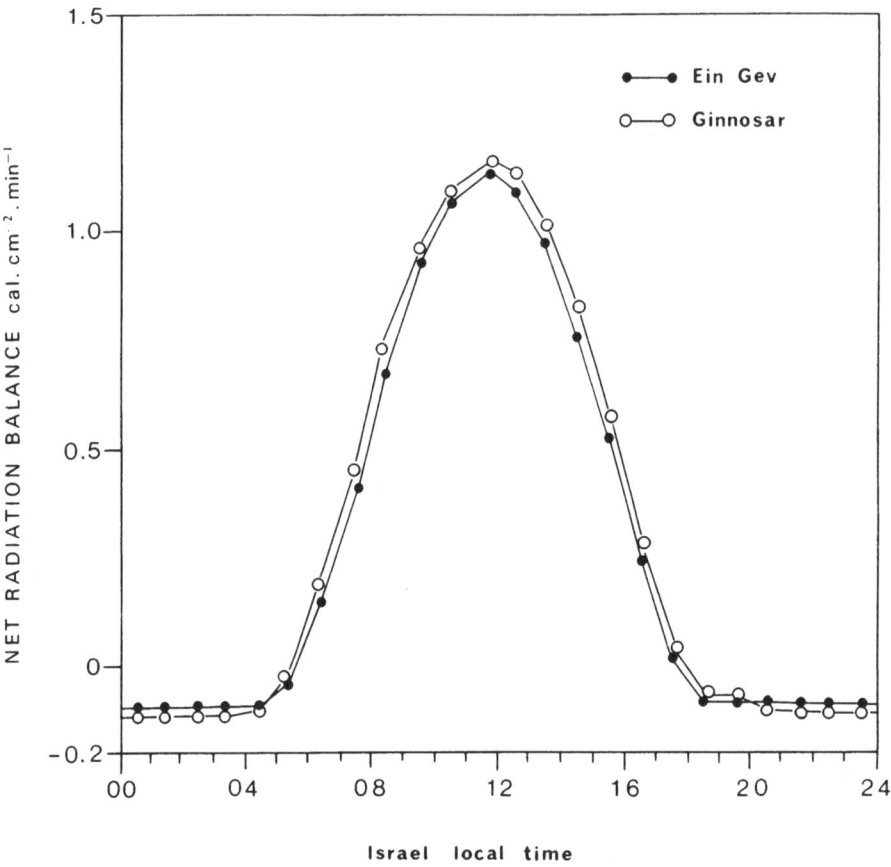

Fig. 69. Measured values of net radiation balance over Lake Kinneret at Ginnosar and Ein Gev, mean monthly values for August 1964.

The heat storage changes shown in Fig. 68 and presented as mean monthly values in Table 21 are based on the 40 sets of measurements taken over the same time periods as the net radiation measurements. During the four-year period, the number of sampling stations varied, starting at 15 and increasing to 45 during the last two years of the period. There were ten depths of measurement. The original data are presented in the same reports referred to for the net radiation data. The annual heat storage fluxes averaged 50,321 Kcal cm^{-2} yr^{-1}.

The degree of annual variation in heat storage can also be seen in Fig. 57, reproduced from Serruya (1975), which shows the isotherms measured in the center of the lake.

7. Advective heat exchange, V

Heat is advected into the lake in the water entering the lake through surface flow, rainfall and subterranean springs, and out of the lake in evaporating water vapor and in the water pumped into the National Water Carrier. Neither the

176

Table 21. Radiant and heat flux densities at Lake Kinneret. Mean monthly values cal cm^{-2} day^{-1}, negative values represent fluxes from the lake surface.

Flux component		J	F	M	A	M	J	J	A	S	O	N	D	Annual total Kcal cm^{-2} yr^{-1}
Global radiation	K↓	320	406	511	622	692	769	757	705	610	482	367	307	199.34
Albedo %		9.6	7.8	6.9	6.0	5.9	5.7	5.7	5.8	6.7	7.0	8.7	10.1	6.8
Reflected short wave	R	−31	−32	−35	−38	−41	−44	−43	−41	−41	−34	−32	−31	−13.49
Emitted long wave	L↑	−806	−801	−812	−843	−880	−921	−937	−950	−942	−915	−885	−840	−319.04
Sky radiation	L↓	596	572	624	613	660	677	717	715	711	675	658	616	240.22
Net radiation balance	Q*	79	145	288	354	431	481	494	429	341	208	108	52	107.03 +251.6
Heat storage change	S	125	−13	−120	−190	−180	−161	−105	−55	48	140	215	292	−251.6
Net heat flux	Q*-S-V-M	200	128	164	160	248	317	386	371	386	345	319	340	105.71
Evaporation, mm day^{-1}		2.9	2.9	2.7	3.9	4.4	5.0	6.8	6.8	6.7	5.3	4.0	4.8	1,716
Latent heat	LE	−171	−171	−159	−230	−260	−295	−401	−401	−336	−313	−236	−283	−101.13
Sensible heat exchange	A	−29	43	−5	70	12	−22	15	30	−50	−32	−83	−57	−4.58

temperature nor the volume of the various terms are precisely known; an estimate can, however, be made based on the estimates of the lake's water balance (Negev & Keller, 1964), measurements of water temperature in the Jordan and the assumption that the water pumped from the lake is at surface temperature. The annual totals of heat advected into the lake are 4.72 Kcal cm^{-2} lake surface in Jordan inflow, 3.05 Kcal cm^{-2} via the subterranean hot springs and 0.41 Kcal cm^{-2} in rainfall. The heat advected from the lake in evaporation was calculated to be 4.20 Kcal cm^{-2} and in the pumped water, 5.00 Kcal cm^{-2}. Thus the net advective heat exchange is 1.00 Kcal cm^{-2} yr^{-1} from the lake.

8. Metabolic heat, M

The solar radiation fixed metabolically by algal photosynthesis (primary production) forms a very small fraction of the incident global radiation, less than 0.2%, but is of great significance for the quality of the lake's water and also provides the food base for the 1,672 tons of fish harvested from the lake each year. Average values of primary production (Lake Kinneret Report, 1974), with the heat of combustion of algae (5.5 Kcal g^{-1}), gave a total annual net metabolic heat fixation of 0.32 Kcal cm^{-2} lake surface, with a maximum flux of 1.4 cal cm^{-2} day^{-1} in April.

9. Latent heat of evaporation, LE

Approximately the same volume of water leaves the lake in water vapour as is pumped from the lake into the National Water Carrier. No doubt because of its significance in the national water balance, the size of this flux has been the subject of numerous investigations using a variety of methods. The results have been summarized in Table 22 in the form of the various estimates of annual evaporation. The average of the six estimates is 175 cm yr^{-1}, equivalent to a latent heat flux of 103.25 Kcal cm^{-2} yr^{-1}. The 7% variation between the estimates includes the various errors of measurement and estimation involved in each method plus the year-to-year variation introduced by considering different periods. Data given in the references numbered 2, 4, 5 and 6 in Table 22 suggest that this year-to-year variation can be considerable, as do the values of year-to-year variation in the size of the heat storage term previously presented.

Estimates of monthly evaporation obtained by the different investigations show much more variation than do the annual totals. The first energy balance estimate, based on Ashbel's early heat storage measurements (Neumann, 1953), showed an autumn maximum and late spring minimum, whereas all other estimates indicate midsummer maxima and early winter minima. Estimates by the combined water balance and mass-transfer method (Stanhill, 1969) differed in that the midsummer maximum (353 mm in July) was almost 50% more than the amount estimated in the other three investigations (nos. 2, 3 and 6 in Table 22). These three estimates, all based on the more recent series of heat storage measurements, show a fair agreement on monthly values.

All three energy balance estimates contain a systematic error by ignoring the unmeasurable but marked diurnal variation in the heat storage term. The most re-

Table 22. Estimates of annual Evaporation from Lake Kinneret

Method of estimation	Annual evaporation, cm	Reference
1. Energy Balance Lakeside climatological data Ashbel's thermal survey data	163	Neumann 1953
2. Energy Balance Lakeside climatological data Oren's thermal survey data	178	Neumann 1961
3. Combined Energy Balance and Mass Transfer Lakeside climatological data Oren's thermal survey data	171	Stanhill 1963
4. Combined Water Balance and Mass Transfer Lakeside climatological data Lake water balance data	184	Stanhill 1969
5. Evaporation Pan Class A evaporation pan and reduction constant, mean of five lakeside stations for three years	167	Bezer 1970
6. Energy Balance Micrometerorological measurements at Ginnosar with concurrent thermal surveys, 4 year mean	187	Miro & Kahanovitz, 1969–1973 Kahanovitz & Karni, 1974

cent estimate (no. 6, Table 22) is the most precise in that micrometeorological measurements made at Ginnosar over the water surface were used to derive hourly values of Bowen's ratio needed to partition the available energy between latent and sensible heat fluxes. In contrast, the earlier estimates used mean monthly climatological values from lakeside stations and thermal survey data for the same purpose.

Unfortunately, the greater precision of the Ginnosar measurements introduces the problem of areal variation in evaporation over the lake, the Ginnosar data representing predominantly windward shore conditions. The marked increase in vapour pressure at the eastern shore, in particular during midday ill the summer months, has already been referred to in a previous chapter. Further evidence for the areal differences in atmospheric evaporation potential can be found in the values of Class A evaporation measured at the different lakeside stations (Bezer, 1970). The annual value measured at Migdal, 3 km south of Ginnosar, was 18% more than that measured at Kafer Aaqeb, almost on the same latitude on the downwind, eastern bank, and 13% more than that measured at Ha'on, on the southern section of the eastern shore.

Clearly, accurate estimates of the seasonal variation in evaporation from the lake require detailed information on a diurnal basis of the areal variation in the factors controlling this flux. In the absence of such information, and because, for the previously stated reasons, evaporation at Ginnosar is believed to be higher than over the lake as a whole, the lower, earlier estimates of evaporation based on

Table 23. A Comparison of the Annual Radiation and Energy balance in three Mid-latitude Lakes

		Mean annual values in cal cm⁻² day⁻¹ Negative values represent fluxes from surface		

Lake		Kinneret	Mead*	Hefner*
Coordinates		33°N, 35°E	36°N, 114°W	35°N, 97°W
Altitude		−210 m	400 m	375 m
Flux components				
Global radiation	K↓	546	506	420
Reflected short wave	R	−37	−37	−26
Emitted long wave	L↑	−874	−842	−781
Sky radiation	L↓	658	672	619
Net radiation balance	Q*	293	299	232
Advective	V	−3	52	4
Latent heat	LE	−277	−344	−222
Sensible heat	A	−13	−5	8
Evaporation mm day⁻¹		4.7	5.8	3.8

* Data from Harbeck *et al.* (1958).

lakeside climatological data (Stanhill, 1963) have been used for the monthly values given in Table 21.

10. Sensible heat exchange with the air, A

There is a small sensible heat flux from the lake to the air totalling 4.58 Kcal cm⁻² yr⁻¹. The monthly values of this flux given in Table 21 should be regarded as an approximate indication only, being calculated by difference and incorporating all the uncertainties in the heat balance. The values suggest that the lake warms the passing air from late summer to early winter, with a marked cooling effect confined to late spring. This pattern is consistent with the previously-noted variation in air temperature around the lakeside, but it is believed that other factors, including the topographical influences discussed elsewhere, and the much greater sensible heat exchange from the land surfaces during the summer months, are probably of greater significance.

11. Conclusions

The size of the major terms in the annual heat balance of Lake Kinneret are by now well established. A comparison of these data with that of other lakes shows that the annual flux of radiant heat and the changes in its heat storage are, per unit area, among the highest recorded (Table 23). The seasonal variation of the radiation balance components and of the changes in heat storage are also satisfactorily established, but the accuracy of the latent and sensible heat fluxes is less accurately known. To obtain this information, a detailed study of the areal and diurnal variation in these heat fluxes is required, information which is also needed to model the circulation of heat, water and momentum in the lake and its surroundings.

180

References

Aberg, B. & W. Rodhe. 1942. Uber die Milieufaktoren in einigen südschwedischen Seen Symb. Bot. Ups. 5:1–256.

Ashbel, D. 1945. The temperature of sweet water lakes in Palestine. Hateva 2:72–74 (in Hebrew).

Berman, T. 1976. Light penetrance in Lake Kinneret. Hydrobiol. 49:1, 41–48.

Bezer, E. 1970. Evaporation from Kinneret-measurements with evaporation pans, October 1965–September 1969. Mekorot Water Co. Ltd., Jordan District, Upper Nazareth (in Hebrew).

Dubinsky, Z. & T. Berman. 1976. Light utilization efficiencies of phytoplankton in Lake Kinneret (Sea of Galilee). Limnol. Oceanogr. 21:226–230.

Duval, W. S., J. Brown & G. H. Geen. 1969. A submersible spectroradiometer and data acquisition system. J. Fish. Res. Bd. Can. 30:313–316.

Harbeck, G. E., M. A. Kohler, G. E. Koberg et al. 1958. Water loss investigations: Lake Mead studies. U.S. Geol. Survey Prof. Paper 298.

Hutchinson, G. E. 1957. A treatise on Limnology. Vol. 1: Geography, physics and chemistry. John Wiley and Sons, New York. 1015 p.

Kondratyev, K. Y. 1972. Radiation processes in the atmosphere. (I. M. O. Lecture 1970). World Meteorological Organisation. No. 309. XXXIV + 220.

Lake Kinneret Report. 1974. Kinneret Limnological Laboratory, Tiberias. June 1975.

Miro, F. & Z. Kahanovitz. 1969. Estimates of evaporation from Kinneret Report 740. Water Planning for Israel Ltd., Tel Aviv (in Hebrew).

Miro, F. & Z. Kahanovitz. 1972. Estimates of evaporation from Kinneret. Prel. Report. HR. 72/061. Water Planning for Israel Ltd., Tel Aviv (in Hebrew).

Miro, F. & Z. Kahanovitz. 1973. Estimates of evaporation from Kinneret. Prel. Report. Water Planning for Israel Ltd., Tel Aviv (in Hebrew).

Miro, F. & Z. Kahanovitz. 1974. Estimates of evaporation from Kinneret. Prel. Report. Water Planning for Israel Ltd., Tel Aviv (in Hebrew).

Negev, M. & P. Keller (eds.). 1964. Water balance of Kinneret. Report 369. Water Planning for Israel Ltd., Tel Aviv (in Hebrew).

Neumann, J. 1953. Energy balance of and evaporation from sweet water lakes of the Jordan Rift. Bull. Res. Counc. Israel. 2:337–357.

Oren, G. H. 1962. Physical and chemical characteristics of Lake Tiberias. Bull. Res. Counc. Israel. 11G:1–33.

Rechnitzer, D. 1967. Penetration of sun radiation in Lake Kinneret. Internal Report for the year 1966–1967.

Rodhe, W. 1965. Standard correlations between pelagic photosynthesis and light. Rev. 1st. Ital. Idrobiol. (suppl.) 18:365–381.

Rodhe, W. 1969. Primary production and its conditions in Lake Kinneret. Internal Report to the Oceanographic & Limnological Research Ltd., Haifa.

Rodhe, W. 1972. Evaluation of primary production parameters in Lake Kinneret (Israel). Verh. int. Verein. Limnol. 18:93–104.

Serruya, S. 1975. Wind, water temperature and motions in Lake Kinneret: general pattern. Verh. int. Verein Theor. Ang. Limn. 19:73–87.

Smith, R. C., J. E. Tyler & C. R. Goldman. 1973. Optical properties and color of Lake Tahoe and Crater Lake. Limnol. Oceanogr. 18:189–199.

Stanhill, G. 1962. Solar radiation in Israel. Bull. Res. Counc. Israel. 11G:34–41.

Stanhill, G. 1963. Evaporation in Israel. Bull. Res. Counc. Israel. 11G:160–172.

Stanhill, G. 1969. The temperature of Lake Tiberias. Israel J. Earth Sci. 18:83–100.

Stanhill, G. 1969. Evaporation from Lake Tiberias: an estimate by the combined water balance–mass transfer approach. Israel J. Earth Sci. 18:101–108.

Stanhill, G. 1970. Measurements of global solar radiation in Israel. Israel J. Earth Sci. 19:91–96.

Talling, J. F. 1965. The phytosynthetic activity of phytoplankton in East African lakes. Int. Rev. Gesamten Hydrobiol. 50:1–32.

Webster, F. 1964. Processing moored current meter data. Woods Hole Oceanogr. Inst. Ref. 64–55, unpublished Manuscript. 35 pp.

Whitney, L. V. 1938. The transmission of solar energy and the scattering produced by suspensoids in lake waters. Trans. Wisc. Acad. Sci. Arts Lett. 31–201–221.

181

III. The chemical environment

A. History

The analysis published by W. A. K. Christie (1913), and performed on a sample taken by Annandale in October 1912 is certainly the first accurate and complete analysis of major ions in Lake Kinneret water (Table 24). In 1950, Oren was given the task of investigating the chemical and physical features of the lake in the framework of a common project of the Israel Hydrological Service, the National Planning Board and the Research Council of Israel. He carried out three surveys in 1950 and 1951. His measurements concerned dissolved oxygen, pH and chloride. In the fifties, Taussig, from Water Planning of Israel, also performed numerous analyses of the lake water. In early 1963, Yashouv investigated the influence of the Jordan water on the northern part of the lake and measured essentially chlorides, oxygen and dissolved nitrogen. From 1963 until 1967, the Water Company, Mekorot, carried out a weekly routine survey which included the major ions and the various forms of nitrogen and phosphorus. From 1968 until the present, the Kinneret Limnological Laboratory continued and developed the weekly routine program.

Table 24. Chemical composition of Lake Kinneret waters at different periods. Results in mg. l^{-1}

Author	Year	Na	K	Ca	Mg	Cl	SO_4	CO_3	SiO_2
Annandale	1912	121	—	49	23	239	16	75	13
Taussig	1959	166	—	52	33	331	55	119	—
Oren	1961	160	8.0	62.2	38.4	345	—	—	—
Oren	1967	161	8.1	57.8	35.8	346.3	—	—	—

Until recently, the sediments of Lake Kinneret were little studied. Nir (1963) determined the granulometry and the calcium carbonate content of the sediments. Serruya (1971) studied the distribution of iron, manganese, phosphorus, carbonates, organic nitrogen and organic carbon of the sediments all over the lake. In 1972, Banin et al. studied the clay fraction of the bottom deposits. In 1973, Serruya reported the first systematic measurements of the sedimentation rates at different stations during a full hydrological cycle and estimated the relative contribution of the allogenic and authigenic origin of the $CaCO_3$ to the sediments. In 1974, Serruya et al. reported the seasonal fluctuations of the chemical composition of the interstitial water; in the same year, Stiller & Magaritz studied the ^{13}C content of the pore water. In 1974, Stiller studied the rates of transport and sedimentation in the lake; in 1976, Stiller, studying the isotopic composition of the sediments, attempted to quantify the amounts of autochtonous and allochtonous carbonates and organic matter.

B. Water chemistry

C. Serruya

1. General

As already mentioned in a previous chapter, the chemical composition of the lake water differs largely from that of the River Jordan, the main inlet of the lake. The combined influence of the sublacustrine springs and of the intense evaporation (1.6 meter per year) gives the lake water its unusual composition and its relatively high concentrations. In Table 25, the ionic composition of Lake Kinneret water is compared to the standard freshwater composition of the world and to the composition of the Jordan and hot springs water.

Table 25. Ionic composition of Lake Kinneret, River Jordan and Hot Springs waters compared to the freshwater world standard. Results in percentage of Milli-equivalents (Serruya C. and Pollingher U., 1971 Mitt. Internat. Verein. Limnol. 19)

	Kinneret	Hot Springs	River Jordan	Freshwater standard composition
Ca	19.1	31.4	63.3	63.5
Mg	26.5	11.1	21.1	17.4
Na	52.5	55.7	14.7	15.7
K	1.4	1.7	0.97	3.4
Cl	67.4	95.8	12.8	10.1
SO_4	7.9	3.2	13.9	16.0
HCO_3	24.6	0.5	73.7	73.9

It is clear from Table 25 that whereas the River Jordan water is typical freshwater, Lake Kinneret water belongs to a completely different type. This original feature which results from the geological history of the area has a deep influence of the fauna and flora of the lake and its surroundings.

2. Factors influencing the composition of the lake water

The considerable difference of composition and concentrations existing between the lake water and its main inlet explains why the winter floods have a considerable dilution effect. This effect is maximum in the northern and western part of the lake. In the unusually rainy 1968/1969 winter, about one-fourth of the lake volume was replaced by river water and the chloride concentration of the lake water dropped from 290 ppm to 240 ppm. The same dilution effect applies to magnesium, sodium and potassium. Conversely, the river waters are more concentrated than the lake water in nitrate, calcium and phosphorus.

Among the internal factors which modify the lake water composition, thermal stratification occupies a major position. It induces successively a rapid exhaustion of the hypolimnic oxygen, the disappearance of nitrate (denitrification) and the decrease of sulfates (sulfate reduction). It also causes the accumulation of ammonia and CO_2.

The strong winds, frequent in the Kinneret area, resuspend large amounts of deposited fine sediments, causing sudden increases of total solids and enhancing the recirculation of dissolved nutrients.

3. Average composition

The chemical record of the Kinneret Limnological Laboratory covers the period 1968–1976. It is based on weekly sampling carried out at a number of stations which has varied from three to seven according to the years. At each station, various depths were sampled depending on the depth of the station and the existence or absence of thermal stratification; for example, the central station A, 40 m deep, was sampled at 11 depths during the stratification period and at 8 depths during the mixed period. The data have been interpolated between the sampling depths and averaged on a monthly basis. In Table 26, we present the data corresponding to three different periods: the mixed period (February), the peak of the bloom (April) and the stratification (September). For a given month, each value represents the average value of monthly averages over the years 1969–1972 inclusive. The range given is the maximum and minimum values of the four monthly averages concerned. For the stratified period, two numbers are given representing the epilimnic and hypolimnic averages. Because the epilimnion is well mixed, epilimnic averages correspond to the actual values found anywhere in this layer. The hypolimnion shows a strong gradient from the thermocline to the bottom which the hypolimnic averages do not reflect.

It should be noted that our weekly sampling has always been carried out from 0500 to 1000 hours. This causes a systematic underestimation of certain parameters such as pH and oxygen. At noon, the combined effect of increased photosynthesis and migration of algae may raise the pH of upper layers to 9.0 and the oxygen concentration to 19 mg l^{-1}.

4. Seasonal and interannual fluctuations

In this section, we present the variation of the main chemical parameters for the period 1969–1974. This presentation is based on monthly averages of the composition of epilimnic and hypolimnic water.

a. Conductivity and chloride (Fig. 70)

In the epilimnion, the conductivity ranges from 1,223 to 984 umho cm^{-1}. A clear seasonal pattern is observed: the minimal values occur in May at the end of the rainy period, the maximal values occur during the dry season with a peak from August until the first flood.

Table 26.   The Chemical Features of Lake Kinneret Water for Three Representative Periods*

Parameter	Mixed (February)	Peak of Bloom (April)	Stratified (September)
Total solids	674 (640–714)	644 (620–669)	642 (628–657)
		676 (650–702)	672 (666–677)
pH	8.1 (7.9–8.3)	8.4 (8.3–8.5)	8.5 (8.5–8.6)
		7.9 (7.7–8.5)	7.6 (7.5–7.6)
O_2	8.5 (6.6–9.3)	8.5 (7.1–9.3)	7.4 (6.9–7.8)
		3.3 (1.7–7.1)	0
CO_2	1.6 (0–3.4)		0
		8.3 (1.2–14.5)	10.6 (9.3–12.3)
H_2S	0	0	0
		0	3.7 (3.0–4.3)
Cl^-	246 (229–265)	234 (224–240)	236 (231–245)
		240 (228–255)	240 (233–250)
Org-N	0.389 (0.310–0.407)	0.620 (0.580–0.650)	0.400 (0.340–0.460)
		0.490 (0.420–0.580)	0.350 (0.270–0.540)
NH_3-N	0.034 (0.014–0.065)	0.024 (0.010–0.038)	0.034 (0.014–0.057)
		0.034 (0.013–0.058)	0.439 (0.388–0.481)
NO_2-N	0.029 (0.005–0.058)	0.008 (0.006–0.009)	0.002 (0.001–0.005)
		0.012 (0.009–0.017)	0.002 (0.001–0.005)
NO_3-N	0.327 (0.179–0.493)	0.277 (0.102–0.612)	0.022 (0.002–0.056)
		0.273 (0.121–0.591)	0
Total P	0.017 (0.011–0.032)	0.022 (0.020–0.024)	0.007 (0.006–0.011)
		0.016 (0.011–0.024)	0.011 (0.006–0.018)
PO_4–P	0.002†		
Dissolved P	0.004†		
Alkalinity $CaCO_3$	117 (113–121)	100 (96–106)	98 (93–106)
Hardness $CaCO_3$	225 (250–259)	230 (216–247)	228 (222–233)
		253 (247–267)	260 (255–267)
Dissolved Fe	0.003 (0.002–0.004)	0.005 (0.003–0.006)	0.006 (0.005–0.010)
		0.004 (0.002–0.006)	0.021 (0.013–0.032)
Ca	53 (51–56)	45 (42–50)	44 (42–46)
		53 (50–57)	55 (52–58)
Na	118 (115–123)	113 (109–118)	114 (113–116)
K	6.2 (6.1–6.3)	5.8 (5.7–6.0)	6.1 (5.8–6.3)
SO_4.	58 (54–60)	58 (57–59)	59 (52–67)
		57 (56–59)	53 (48–61)
Elect. Cond. (umho/cm)	1105 (1051–1179)	1070 (1013–1187)	1054 (1014–1098)
		1099 (1062–1176)	1104 (1067–1167)

* All values (except pH) given in mg/l, range of values in parentheses. Where two sets of figures are given, the upper and lower values refer to epilimnion and hypolimnion, respectively.
† Values for 1968–1969 only.

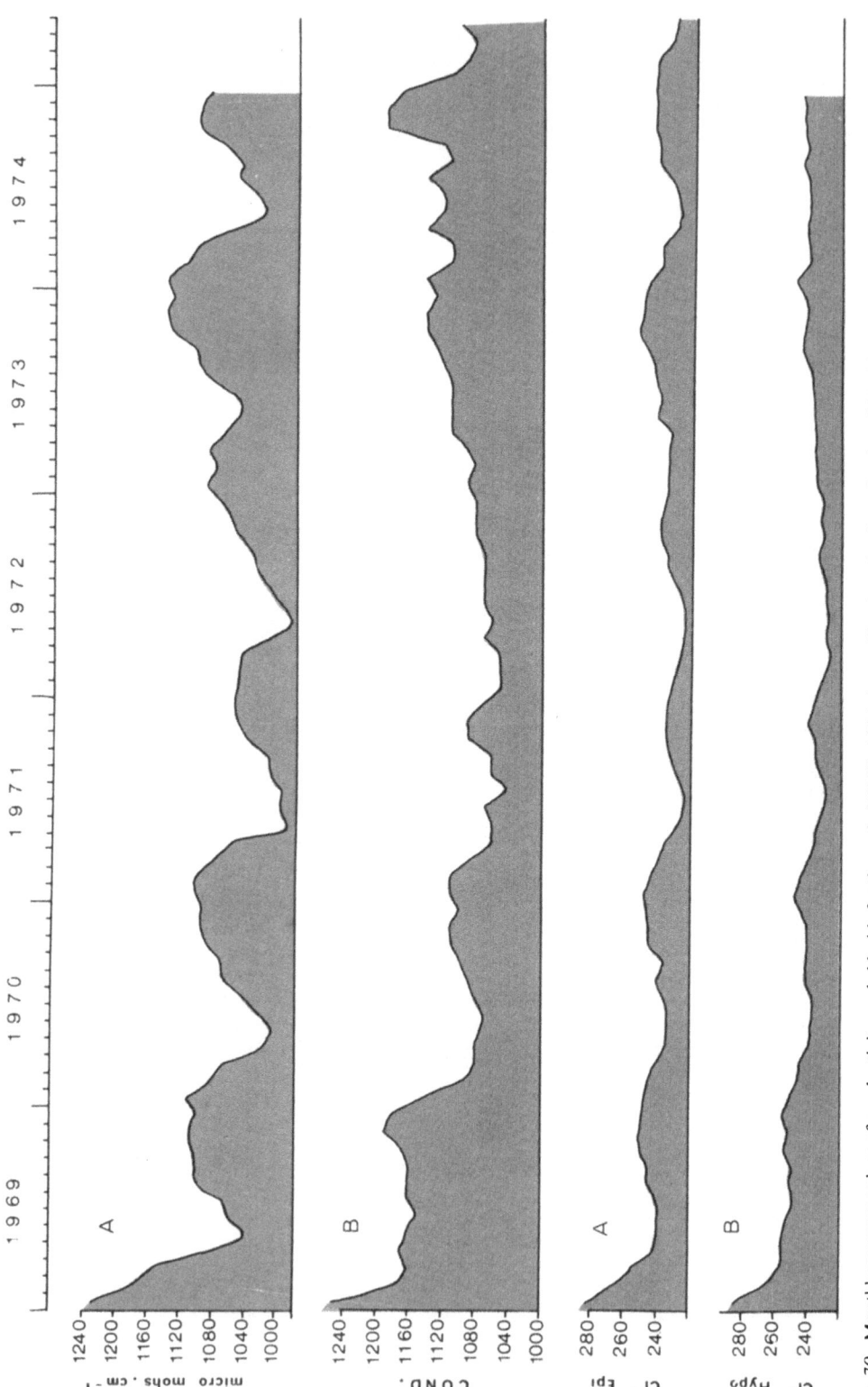

Fig. 70. Monthly average values of conductivity and chloride for the period 1969–1974. A = epilimnion, B = hypolimnion. Chloride in ppm.

In the hypolimnion, the conductivity ranges from 1,250 to 1,040 umho cm^{-1}. The seasonal pattern is less clear than in the epilimnion. This results from the fact that the river water penetrates into the deep layers only in January and February. In March and April, the river water floats on the lake water (see p. 136). In the rainy 1969 winter, for example, the January–February floods caused, in the deep layers, a drop of conductivity from 1,250 down to 1,170 umho cm^{-1}, but the later floods did not affect the deep layers. Curiously, in winter 1970, the relatively late turnover caused a much more pronounced drop of conductivity in the deep layers than in other years.

The interannual fluctuation results from the diversion of the salty springs located on the shores, which was completed in 1965, and from the amount of flood water in winter. The rainy years 1969 and to a lesser extent 1971 are marked by a decreasing trend, whereas dry years such as 1972–1973 are marked by a rapid increase of conductivity values. The chloride concentration varies as the conductivity.

b. pH

In the epilimnion, pH values range from 7.5 up to 8.9 (Fig. 71). A clear seasonal pattern is observed with low pH values during the turnover period and peak values during the late period of the bloom in April, May and June.

In the hypolimnion, the pH ranges from 7.4 to 8.3. The lowest values are observed in summer and are related to the increasing concentrations of CO_2 (Fig. 72). The highest values are found in February–March in the early stages of the bloom. This pattern indicates that the increase of pH caused by the increasing photosynthesis in the upper layer affects the deeper layers as long as active circulation occurs. When the vertical circulation diminishes in March–April, the pH continues to increase in the upper layers but the high pH waters no longer reach the deeper zone. No prominent interannual fluctuations were observed.

c. Ca, CO_2 *and alkalinity*

The waters of Lake Kinneret are rich in calcium (55 ppm during the mixed period). At the rather high pH prevailing in the lake, the bicarbonates are the most common species of inorganic carbon. However, during the mixed period, dissolved CO_2 is present (2 to 3 ppm) in the whole water column during a few weeks. The alkalinity then reaches 120 ppm $CaCO_3$. As the algal bloom develops, the dissolved CO_2 disappears and the alkalinity decreases to approximately 80 ppm $CaCO_3$. The CO_3 then becomes more abundant and at the peak of the bloom, the milky aspect of the water as well as the composition of the tripton in sediment traps indicate an active precipitation of calcium carbonate.

In the hypolimnion, we can observe a rapid build-up of CO_2 which may reach 18 ppm (Fig. 72). The peaks of calcium observed in deep waters in November and December (Fig. 72) are clearly due to redissolution of calcium carbonate by CO_2.

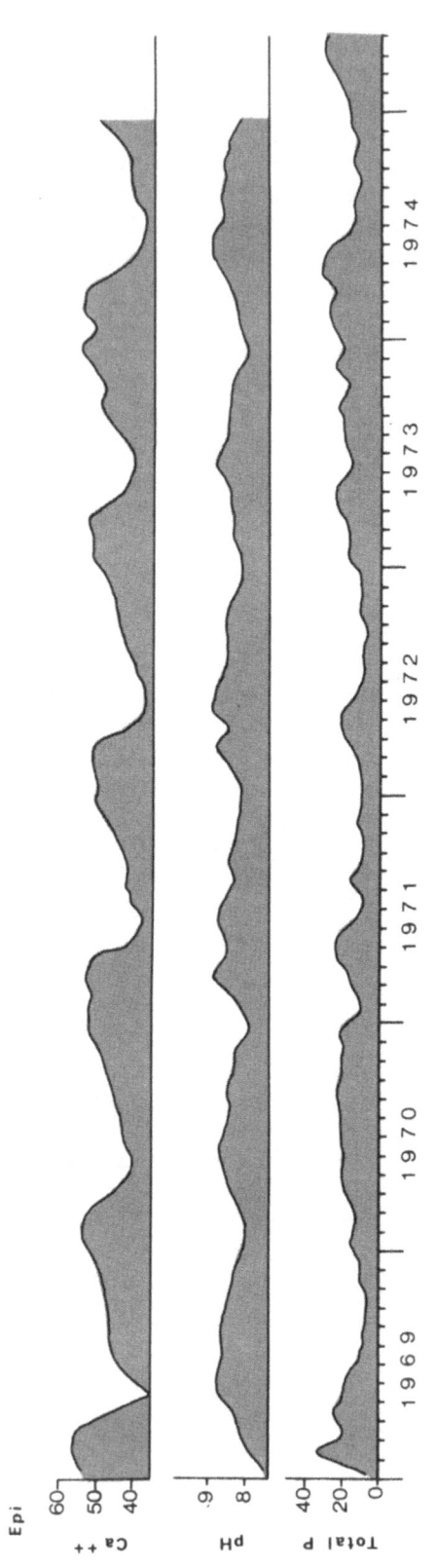

Fig. 71. Monthly averages of Ca^{++}, pH and total P in the epilimnion for the period 1969–1974. Ca^{++} in ppm, total P in ppb.

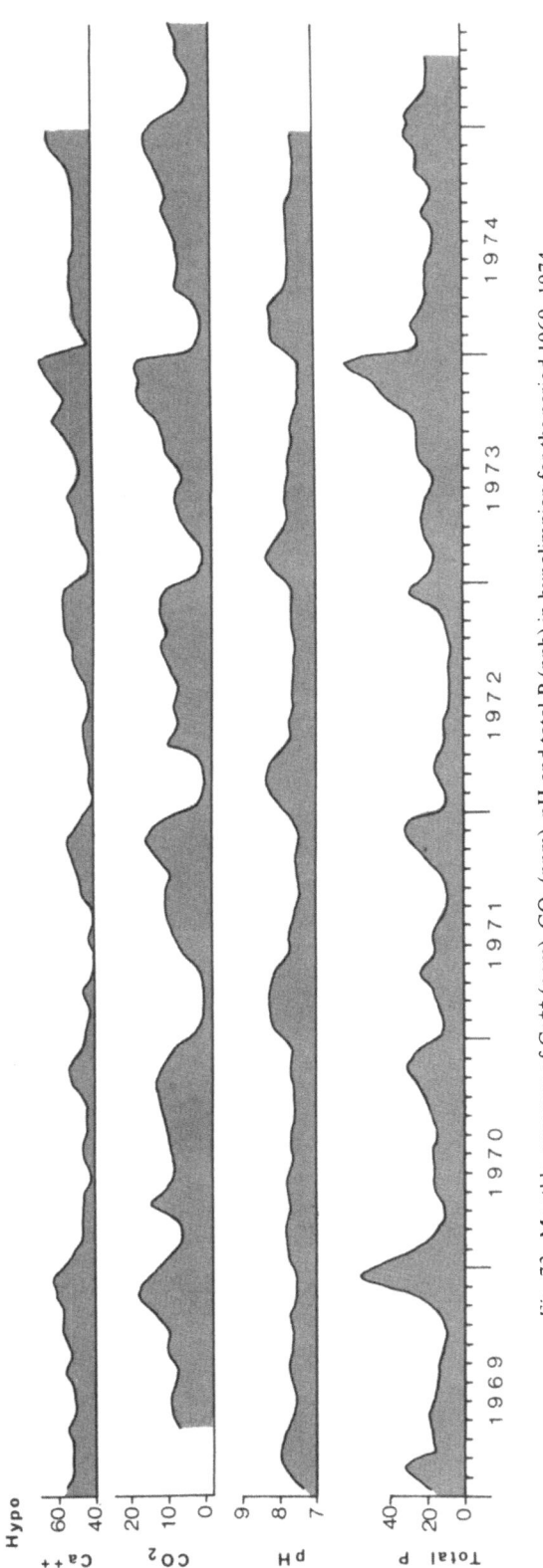

Fig. 72. Monthly averages of Ca^{++} (ppm), CO_2 (ppm), pH and total P (ppb) in hypolimnion for the period 1969–1974.

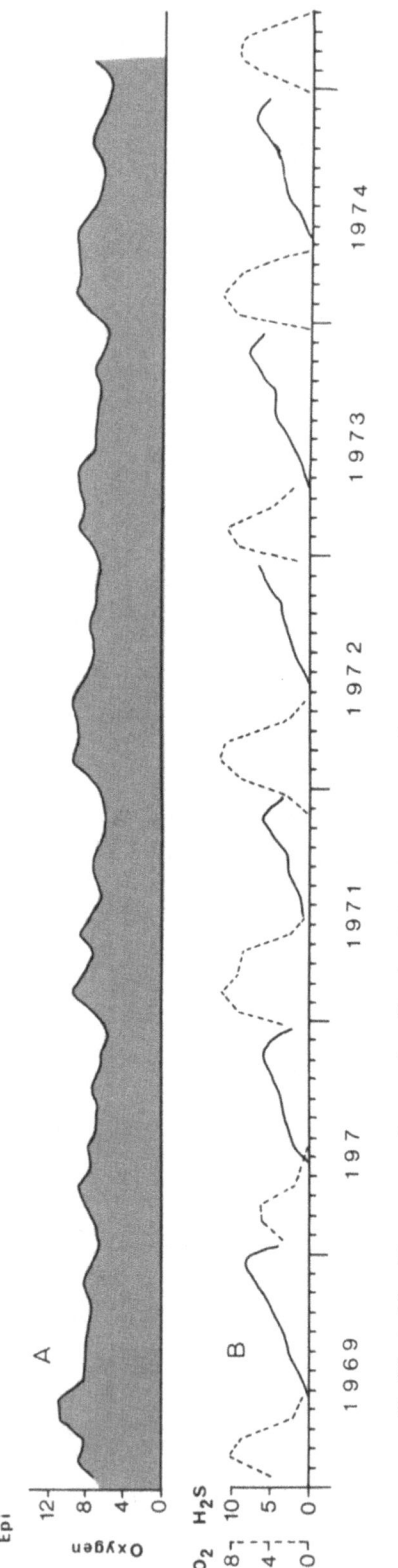

Fig. 73. A. Monthly averages of oxygen in epilimnion. B. Monthly averages of oxygen (dotted line) and sulfides (full line) in hypolimnion. Results in ppm.

d. Oxygen

In the epilimnion, the monthly averaged concentrations of oxygen range from 5.8 to 9.2 ppm. However, at noon, during the peak of the bloom, values of 19 ppm have been measured. The hypolimnion values range from 0 to 9.2 ppm (Fig. 73).

The hypolimnion of the lake is depleted of oxygen from May to December. The vertical currents due to superficial cooling are the driving force which brings oxygen to the deep layers. During this period, the rate of oxygenation is considerable. For example, during a period of thirty days in January 1974, the content of oxygen of the lake increased at a rate of 2 $g.m^{-2} d^{-1}$. Moreover, during the same period, 4,500 tons of sulfides and 1,200 tons of ammonia were oxidized, consuming approximately 3 $g m^{-2} d^{-1}$. Some 5,000 mg $m^{-2} d^{-1}$ or 850 tons per day represent a minimum value of the daily rate of oxygenation during this period. This does not include the oxygen consumed by the fauna and flora of water and by sediments.

The concentration of oxygen at the end of the mixing period depends on the duration of the turnover. The low values of oxygen in the deep layers in 1970 resulted from the fact that the turnover occurred one month later than in other years.

As soon as the intensity of vertical motions decreases in March, the oxygen level begins to decrease in the deep layers, although at this period the lake is not stratified. When the stratification becomes steady in May, the hypolimnion is completely devoid of oxygen. Therefore the estimation of the hypolimnic areal deficit can only be calculated during the period preceding stratification and has to be corrected for diffusion. For the period 28 March–28 April 1970, the hypolimnic areal deficit was found to be 1.7 $g m^{-2} d^{-1}$. Compared with the values given by Hutchinson (1957), the hypolimnic areal deficit of Lake Kinneret is very high and reflects its high winter–spring productivity.

The interannual fluctuations of the levels of oxygen in the epilimnion and hypolimnion are due to specific meteorological events, but no general trend of evolution can be detected during the period presented here.

e. NO_2–NO_3

When the hypolimnic waters are depleted of oxygen in April–May, nitrate is the first oxygenated anion which is utilized as a source of oxygen. It is therefore reviewed here.

The nitrate in Lake Kinneret originates from the washing of the peat area of the Hula plain in the northern part of the watershed and from the nitrification of the ammonia which accumulates in summer below the thermocline. The nitrification process is accompanied by an early peak of nitrites. A second and more modest peak of nitrites accompanies the denitrification process in May. The concentration of nitrate depends then on the timing and intensity of floods, on the amount of hypolimnic ammonia recirculated in winter and on the rate of uptake of algae and denitrifying bacteria. Since denitrification takes place before the complete stratification of the lake, the nitrate concentrations are nearly similar in the epilimnion and hypolimnion, and therefore we have represented only the average con-

Table 27. Monthly N and P balance. Hydrological cycle 1970–1971. Results in metric tons. R = lake reserve; R₂ = lake reserve at end of a given month; R₁ = lake reserve at end of previous month. NI = net income. (from Serruya C., 1975 Verh. Internat. Verein. Limnol. 19)

	Total N		Org. N particulate + soluble		NH₃-N		NO₃–NO₂–N		Total P	
R Sept	2,300		1,760		500		40		54.60	
NI Oct	36	$R_2-(R_1+NI)$ (R_2-R_1)	3	$R_2-(R_1+NI)$	0	$R_2-(R_1+NI)$	39	$R_2-(R_1+NI)$	3.06	$R_2-(R_1+NI)$ (R_2-R_1)
R + NI	2,336	+244	1,763		500		79		57.66	−3.06
R Oct	2,580	(+280)	1,520	−243	1,000	+500	60	+19	54.60	(0)
NI Nov	41		4		0		35		2.72	
R + NI	2,621	+99	1,524		1,000		95		57.32	−4.02
R Nov	2,720	(+140)	1,580	+56	1,110	+110	40	−55	53.30	(−1.3)
NI Dec	93		35		−4		61		8.19	
R + NI	2,813	−33	1,615		1.106		101		61.49	+16.89
R Dec	2,780	(+60)	1,730	+115	910	+196	140	+39	44.60	(−8.7)
NI Jan	69		17		−1		53		2.17	
R + NI	2,849	+391	1,747	+293	909		193		46.77	+5.03
R Jan	3,240	(+460)	2,040	(310)	60	−849	1,140	+947	51.80	(+7.2)
NI Feb	362		119		7		233		51.95	
R + NI	219	+628	2,159	+101	67		1,373		103.75	−39.75
R Feb	3,602	(+990)	2,260	(220)	170	+103	1,800	+427	64.00	(+12.20)
NI Mar	4,230		51		5		162		18.23	

R + NI	4,449	−119	2,311	+449	175	+5	1,962	−572	82.23	+6.97
R Mar	4,330	(+100)	2,760	(500)	180		1,390		89.20	(+25.2)
NI Apr	566		151		7		406		108.90	
R + NI	4,896	−1,876	2,911	−791	187	−117	1,796	−966	198.10	−102.3
R Apr	3,020	(−1,310)	2,120		70		830		95.80	(+6.6)
NI May	125		25		1		97		8.84	
R + NI	3,145	+45	2,145	+215	71	+49	927	−217	104.64	−12.84
R May	3,190	(+170)	2,360		120		710		91.80	(−4.0)
NI Jun	83		19		−1		64		5.32	
R + NI	3,273	−513	2,379	−199	119	+191	774	−504	97.12	−41.32
R Jun	2,760	(−430)	2,180		310		270		55.80	(−36.0)
NI Jul	−2		−18		1		15		2.08	
R + NI	2,758	+172	2,162	+108	311	+179	285	−115	57.88	−14.48
R Jul	2,930	(+170)	2,270		490		170		43.40	(−12.4)
NI Aug	16		1		−2		17		2.51	
R + NI	2,946	−96	2,271	−251	488	+252	187	−97	45.91	−2.11
R Aug	2,850	(−80)	2,020		740		90		43.80	(+0.4)
NI Sept	153		49		1		102		9.87	
R +NI	3,003	−713	2,069	−739	741	+149	192	−122	53.67	−14.67
R Sept	2,290	(−560)	1,330		890		70		39	(−4.8)

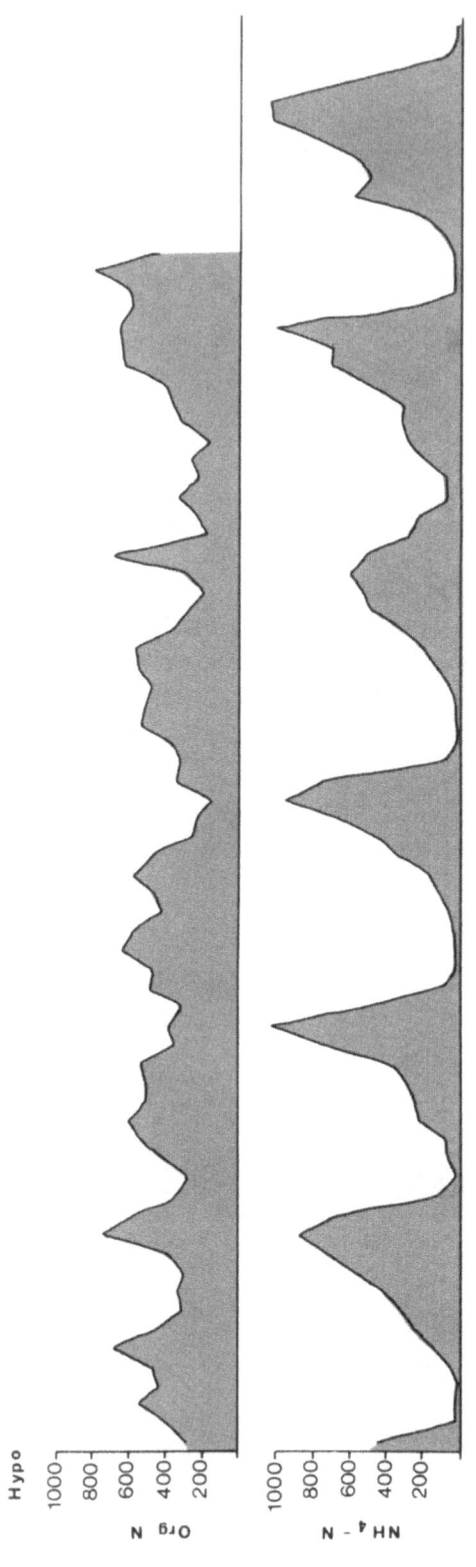

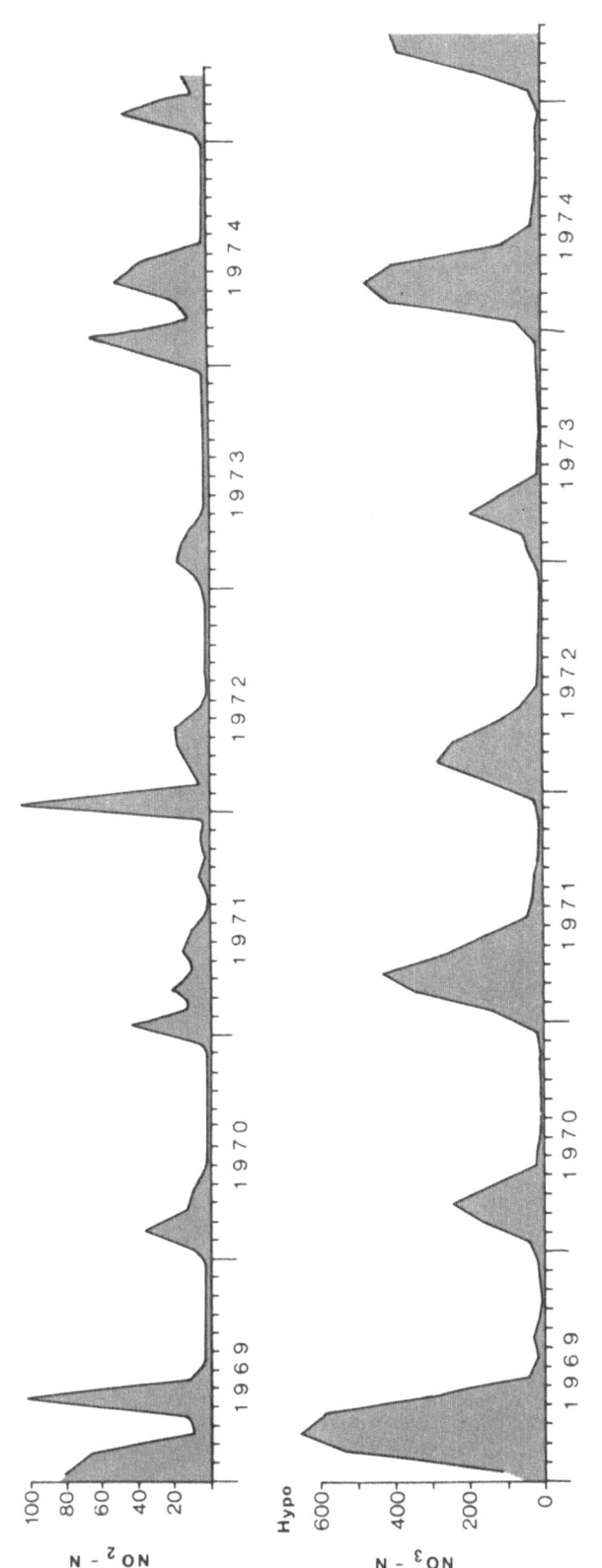

Fig. 74. Monthly averages of organic N, ammonia, nitrites and nitrates in the hypolimnion for the period 1969—1974. Results in ppb.

centrations in the hypolimnion (Fig. 74). The numerous factors regulating the concentrations of nitrate explain the wide range of its fluctuations, from 0 to 650 ppb. The highest level of nitrate was observed in 1969, a rainy year, when one-fourth of the lake water was replaced by the nitrate-rich water of the Jordan River. The lowest level of nitrate was observed in 1973, the driest year of the record. The second low level observed in 1970 was due to the fact that the nitrification and denitrification occurred nearly simultaneously.

Nitrogen balances (Serruya, 1971, 1975) have shown that 60% of the nitrogen yearly entering the lake is denitrified (Table 27). Denitrification is the main process which regulates the nitrogen cycle and prevents the accumulation of this substance in Lake Kinneret. The denitrification often begins in the metalimnion area. The reason for the unusual profiles of Fig. 75 is probably the accumulation of organic detritus which can be seen easily in microscopic observations. In late May, 90% of the lake nitrate has disappeared, and the reduction of sulfates starts in the hypolimnion, now well isolated from the upper layers.

f. SO$_4$–H$_2$S

In the epilimnion, the sulfate ranges from 50 to 65 ppm and in the hypolimnion from 38 to 60 ppm (Fig. 76). The seasonal variations are more strongly marked in

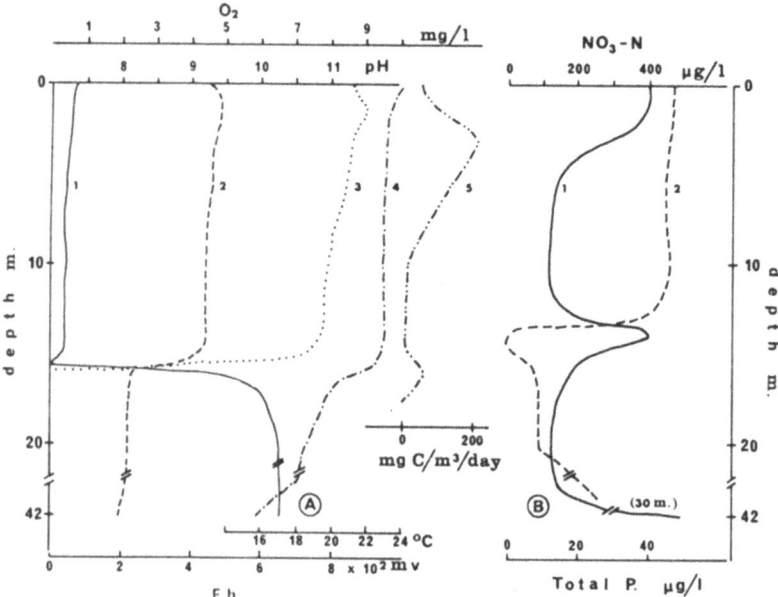

Fig. 75. Variations of some chemical and biological factors at the central station, 42 m deep. A. 1-Eh potential (the results are stated as differences with the maximum values of the profile and not as absolute potentials), 2-pH, 3-Oxygen, 4-Temperature, 5-Primary productivity (profiles measured in June 1970). B. 1-Total P, 2-Nitrate-Nitrogen (profiles measured in May 1969) (from Serruya C., 1972, Hydrobiologia, 40).

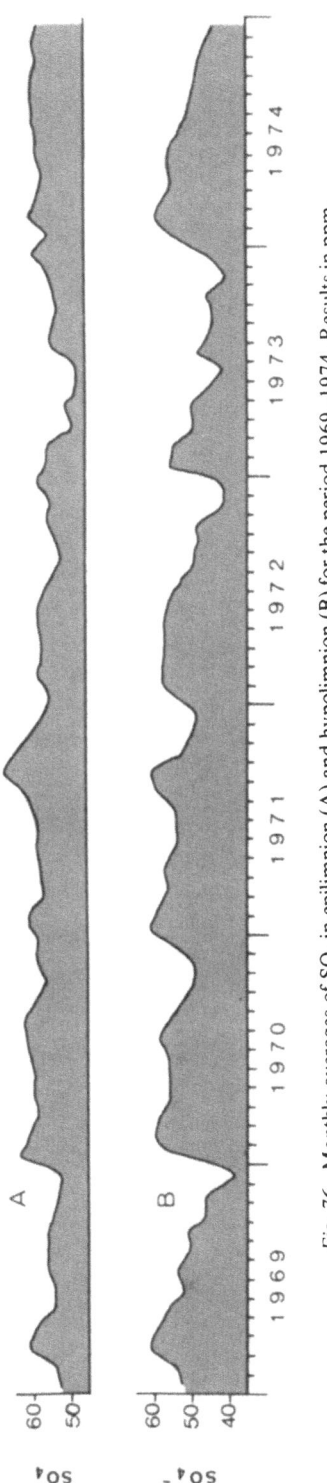

Fig. 76. Monthly averages of SO_4 in epilimnion (A) and hypolimnion (B) for the period 1969–1974. Results in ppm.

the hypolimnion than in the epilimnion. The maximal values are observed during the turnover and decrease during the summer by 20 ppm SO_4, which corresponds to a drop of 7 ppm sulfur. The drop of sulfate–sulfur is stoichiometric to the simultaneous increase of sulfide–sulfur.

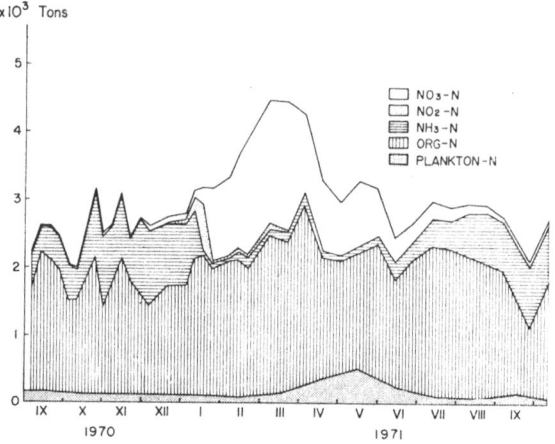

Fig. 77. Fluctuation of the amounts of the different forms of nitrogen in the lake for the hydrological cycle 1970–1971 (from C. Serruya, 1975, Oïkos, 26).

g. *Nitrogen compounds*

In Fig. 77, we have represented the fluctuations of the different forms of nitrogen in Lake Kinneret during one limnological cycle. The organic nitrogen is the main component and represents from 55 to 73% of the total nitrogen (Serruya *et al.* 1975). The transformation of ammonia into nitrate is clearly shown in Fig. 74. This process is discussed in more detail in the section 'Bacteria'. An amount of 2,500–3,000 tons of total nitrogen is the 'base level' nitrogen value. In winter, another 1,000 tons are contributed by the floods; in May, the denitrification process brings back the nitrogen amount to its base level values.

The nitrogen locked in living material represents only a very small portion of the total organic nitrogen (TON), as shown in Table 28. The dissolved organic

Table 28. Distribution of the different forms of nitrogen in lake water. Results in tons (%) (Serruya C. *et al.* 1975, Oikos, 26)

	Ton	Plankton N	Non-living organic N
Nov 1970	1,900 (100)	120 (6.3)	1,780 (93.7)
Jan 1971	2,100 (100)	100 (4.8)	2,000 (95.20)
Mar 1971	2,500 (100)	150 (6.0)	2,350 (94.0)
May 1971	2,200 (100)	520 (23.6)	1,680 (76.4)

Table 29. Fluctuations of DON in comparison with total organic nitrogen at 3 and 30 m depths in 1973. Results in ppb (%) (Serruya C. *et al*. 1975, Oikos, 26).

	Jan	Feb	Mar	Apr	May	June
3 m	89 (64)	169 (51)	321 (57)	425 (77)	363 (70)	294 (51)
30 m	80 (50)	100 (62)	265 (83)	212 (76)	134 (67)	49 (30)
	Jul	Ang	Sep	Oct	Nov	Dec
3 m	276 (67)	321 (75)	332 (69)	468 (82)	545 (84)	527 (83)
30 m	40 (15)	157 (58)	147 (33)	343 (46)	—	513 (65)

nitrogen (DON) fraction represents a large part of the total organic nitrogen as shown in Table 29.

Very little is known concerning the nature of the dissolved organic nitrogen. Berman (1974) measured the concentrations of urea from April 1972 to April 1973. His findings indicate that although the occurrence of urea is rather sporadic, a general seasonal pattern can be observed. In April–May, when the algal bloom is maximum, low amounts (10–20 ppb) of urea-N are measured. In summer, higher concentrations are observed in the upper layers (0–10 m), rich in phytoplankton, and in the hypolimnion. The annual maximum of urea concentration has been measured in August 1972 (150 ppb). These data show that most of the year, urea-N represents no more than 10% of DON. Conversely, the maximum value measured in summer in the hypolimnion represents approximately 80% of DON at this depth.

The dissolved organic nitrogen is rapidly transformed into ammonia as shown in Fig. 78.

h. Phosphorus

The monthly averaged concentration of total phosphorus ranges from 6 to 34 ppb in the epilimnion. The minimal values are usually observed in summer (Fig. 71). In the hypolimnion, the total P ranges from 6 to 60 ppb. The seasonal pattern is very clear: the maximal concentrations are observed in November–December. In this period, the volume of the hypolimnion is considerably reduced; the elements released from the bottom sediments are dispersed in a smaller volume of water generating a sudden and rapid increase of concentrations. In particular, the concentration of CO_2 in the hypolimnion increases by about 70% during November–December (Fig. 72). It is very likely that there is a causal relationship between this increase of the bicarbonate reserve and the simultaneous increase of phosphorus in the deep layers of the lake.

In contrast to the results obtained for nitrogen, a large fraction of the total phosphorus of the lake is included in the planktonic biomass, as shown in Fig. 79.

The high pH and calcium concentration of the lake water are favorable to the precipitation of phosphate into apatites. Therefore, the concentrations of dissolved inorganic phosphorus in the epilimnion rarely exceed 5 ppb and generally remain below 1 ppb. It is then obvious that the lake waters are, chemically speaking,

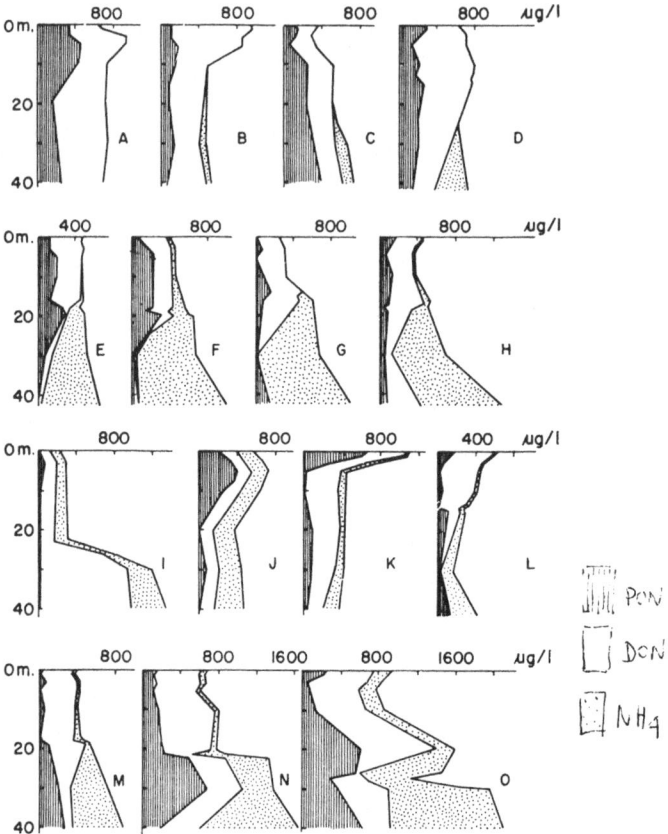

Fig. 78. Vertical distribution of PON, DON and ammonia at station A. (From C. Serruya *et al.*, 1975, Oikos, 26.)

A = 9 Apr 1972	F = 22 Aug 1972	K = 24 Apr 1973
B = 30 Apr 1972	G = 19 Sep 1972	L = 13 May 1973
C = 14 May 1972	H = 15 Oct 1972	M = 12 Aug 1973
D = 4 Jun 1972	I = 4 Dec 1972	N = 21 Oct 1973
E = 13 Aug 1972	J = 18 Feb 1973	O = 7 Dec 1973

extremely limited in phosphorus. We shall see in the section on biology that certain planktic algae have found ways to overcome this chemical limitation.

In spite of the anoxic conditions prevailing in the hypolimnion during eight months in summer and fall, no massive release of phosphorus has been observed from the sediments (Fig. 72) as described by Mortimer (1941, 1942) in Windermere. The problem has been investigated by Serruya (1974). It appears that in the sediments of Lake Kinneret, only a small fraction of the phosphorus is bound as iron-phosphate. Most of the phosphorus is bound as calcium phosphate and organic phosphorus. Consequently, the phosphorus dynamics in the Kinneret sediments are governed by pH variations more than by redox fluctuations. It is then easy to understand why we do not observe any significant hypolimnic increase of phosphorus during the anoxic period and why the phosphorus concentrations respond very rapidly to any decrease of pH. It follows that the lowering of the thermocline does not increase the phosphorus concentration and does not

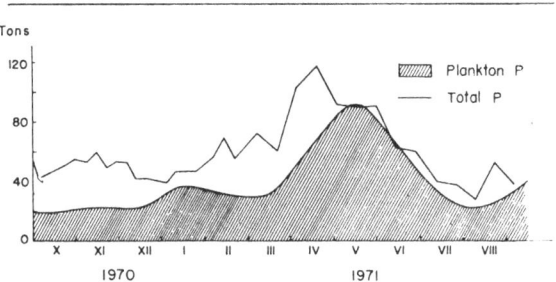

Fig. 79. Fluctuation of the amounts of total P and plankton P in the lake for the hydrological cycle 1970–1971 (from C. Serruya, 1975, Oïkos, 26).

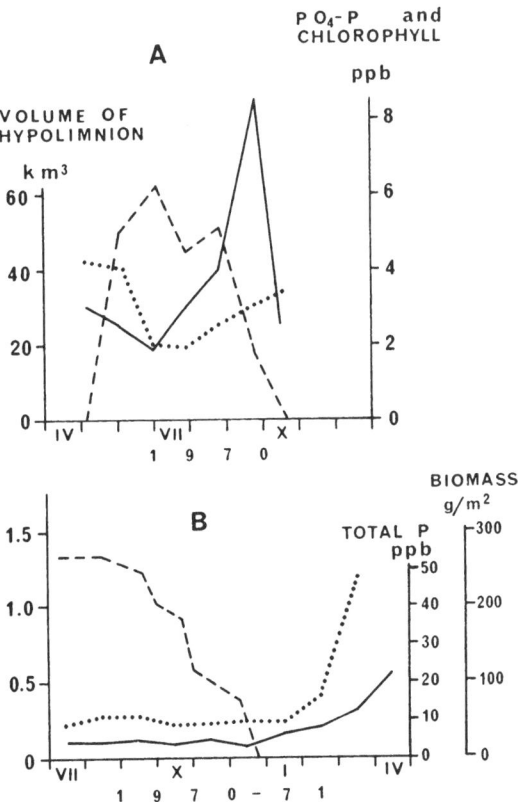

Fig. 80. The influence of distratification on phosphorus and biomass concentration in the water of Lake Erie (A) and Lake Kinneret (B). Dashed line = volume of hypolimnion; dotted line = phosphorus concentration (as PO_4–P for Lake Erie and total P for Lake Kinneret); full line = chlorophyll for Lake Erie and phytoplankton weight biomass for Lake Kinneret. Data on Lake Erie are from Burns & Ross(1972). (from Serruya C. 1975, Verh, Internat. Verein. Limnol., 19).

cause any immediate algal development. In Fig. 80, where the destratification process and concomitant biomass of Lake Erie (Burns & Ross, 1972) and of Lake Kinneret are shown together, we see that the algal biomass of Lake Kinneret remains unaffected by the introduction of hypolimnic water. In contrast, the destratification in Lake Erie generates an immediate and simultaneous rise of phosphorus and chlorophyll.

i. Fe *and* Mn

The dissolved forms of iron and manganese have also been measured on a routine basis. The vertical profiles indicate a clear metalimnic peak of dissolved iron in summer. In the hypolimnion, the release of dissolved iron is limited by the formation of iron sulfides which are found abundantly as black spherules in the sediments. They reach a particularly large size when the iron sulfide is precipitated within the body of dead organisms. The dissolved manganese reaches high concentrations in the hypolimnion probably because the field of solubility of manganese extends to much lower values of redox potential than that of iron.

C. Sediments

C. Serruya

1. General description

The upper layer of the bottom sediments of Lake Kinneret is composed of a black, jelly-like, low density material of a thickness of 2 to 5 cm. The underlying sediments are grey and compact. An outstanding feature of these sediments is the complete absence of varves. The very well-marked seasonal events (floods restricted to four months of the year, authigenic formation of $CaCO_3$ in March–April, organic sedimentation in June–July) are very favourable to the formation of varves. We interpret their absence as evidence of physical mixing of the upper layer during the winter turnover. This interpretation is strengthened by the rapid increase, at this period, of sulfates in the interstitial waters of this layer, since the rate of this increase cannot be accounted for by diffusion.

2. Granulometry and gross composition

Three main features characterize the sediments of Lake Kinneret: the high percentage of fine-grained material, the abundance of calcium carbonate and the dominance of montmorillonite in the clay fraction. It is worth noting the contrast between the coarse gravel and boulders of the littoral and supralittoral area and the fine material of the deeper zones. Depending on locations and water levels, the boulder zone attains 2 to 5 m depth. Between the boulder zone and a depth of 10–12 m, sandy mud dominates but the percentage of the coarse sand decreases rapidly with depth. Below 12 m, the fraction smaller than 20μ represents 90 to 95% of the sediments (Table 30).

Whereas the suspended matter of the River Jordan contains a maximum of 25% (dry weight) of calcium carbonate, the Kinneret sediments (with the exception of the shallow sands) reach concentrations of 40 to 55%. This difference emphasizes the importance of the formation of authigenic calcium carbonate in Lake Kinneret. As shown in Table 30, most of the calcium carbonate belongs to the silt fraction.

Montmorillonite constitutes approximately 60% of the total silicates which account for the very high specific surface of the Kinneret sediments.

3. Distribution of iron, manganese, phosphorus, nitrogen, organic carbon and calcium carbonate in the surface sediments

A special sampling survey was carried out in order to study the distribution of detrital and redox-sensitive elements and of authigenic material (Serruya, 1971).

The detrital elements (phosphorus, iron, manganese) show a clear distribution pattern (Fig. 81): high concentrations in a N–S tongue-shaped area located south of the Jordan River mouth, intermediate concentrations in the western area of the

Table 30. Main Granulometric, Physical and Mineralogical Features of the Sediments (Data of Banin et al., 1972).

Stations	Granolometry			CaCO₃			
	Sand 20	Silt 2–20	Clay 0.2	Total CaCO₃ % of total sediment	Granulometric·fractionation: % of total sediment		
	%	%	%	sediment	sand	silt	clay
G	7.4	47.0	45.7	37.6	3.5	32.7	1.6
F	4.5	54.4	41.1	45.1	1.6	33.7	9.8
A	9.5	42.9	47.6	56.0	7.1	34.3	14.6
C	7.8	59.1	33.1	54.6	4.6	44.0	6.0

Stations	Specific surface		Silicate composition			
	Total sediment m²/g	Carbonate free m²/g	Montmo-rillonite	Kaolinite	Palygor-skite	Quartz
			% of total silicates			
G	244	394	62	32	3	3
F	217	433	57	37	2	4
A	207	454	74	13	10	3
C	138	433	70	16	10	4

lake and minimal concentrations in the central and eastern areas. This distribution is primarily caused by the sedimentation pattern of the suspended matter brought by the river. As we have seen (p. 136), in early winter, the Jordan waters can be traced as a freshwater front progressing in a N–S tongue-shaped area. It is clear that the N–S tongue of sediments rich in detrital elements results from the deposition of the suspended matter of the river associated with the early floods.

The river water and suspended load coming with late floods remain in the upper layers of the lake and are deflected towards the western coast by the general circulation. This dispersion over large areas accounts for the intermediate concentration of detrital elements in the sediments of the western side. Conversely, the hydrodynamic pattern prevailing in this lake leaves the central and eastern regions outside of the depositional area of the River Jordan suspended matter, which explains the low values of the detrital elements in the sediments of this area.

Iron and manganese are of detrital origin but their post-depositional behaviour is governed mostly by redox potential. In particular, under anoxic conditions, the iron and manganese of the sediments become soluble and they are released into the overlying water. The central and eastern regions of the lake receive relatively little iron and manganese-rich detrital material, as we have seen in the previous paragraph. These regions are also the deepest areas of the lake where the anoxic conditions develop early and last longer. The post-depositional release of these elements is then an additional factor explaining the low content of these elements in the central and eastern sediments.

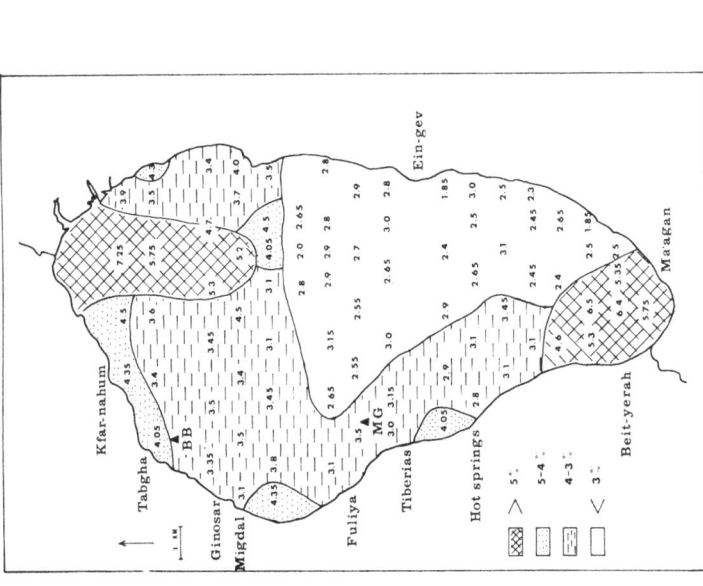

Fig. 81.3. Distribution of phosphorus in surface sediments. Results stated as percentages of dry weight (same source).

Fig. 81.2. Distribution of manganese in surface sediments. Results stated as μg/g dry wet (same source).

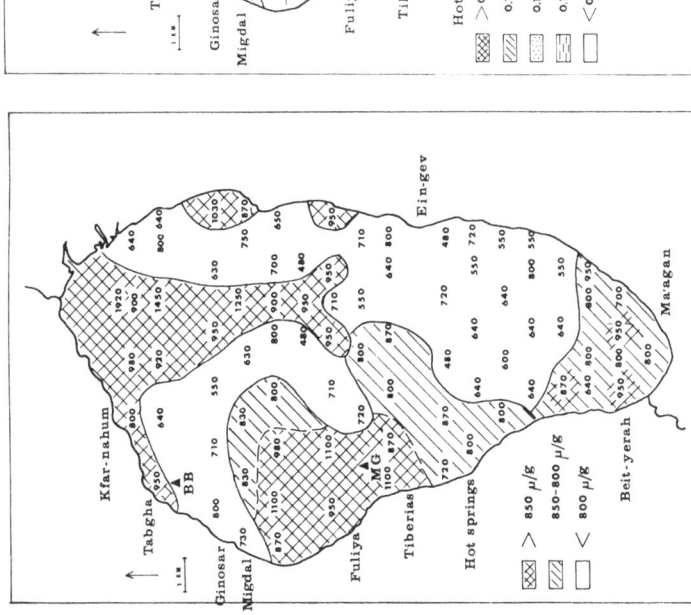

Fig. 81.1. Distribution of iron in surface sediments. Results stated as percentages of dry weight (C. Serruya, 1971, Limnol. Ocean., 16.3).

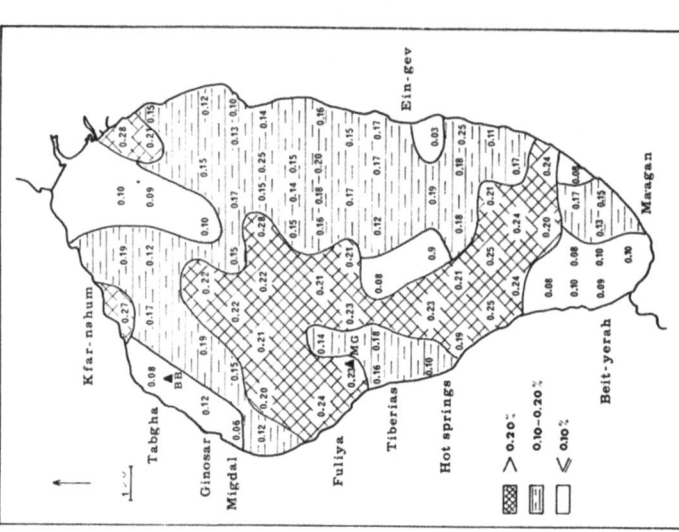

Fig. 81.6. Distribution of nitrogen in surface sediments. Results stated as percentages of dry weight (same source).

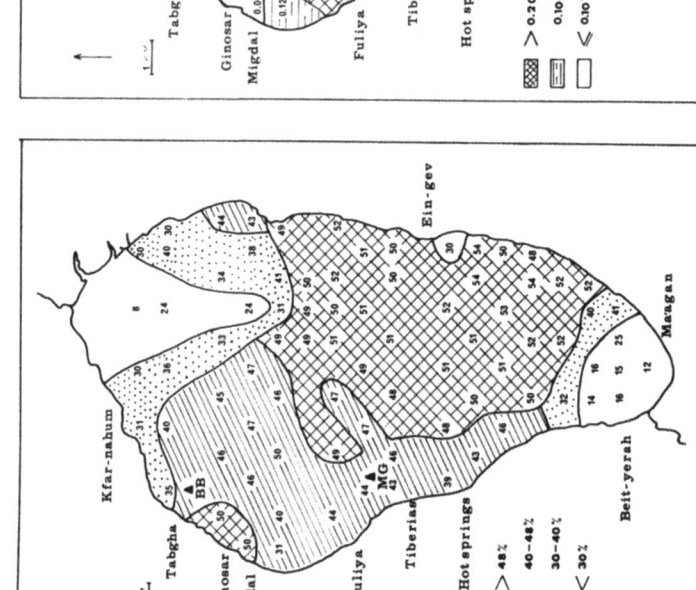

Fig. 81.5 Distribution of carbonates in surface sediments. Results stated as percentages of CaCO₃ (same source).

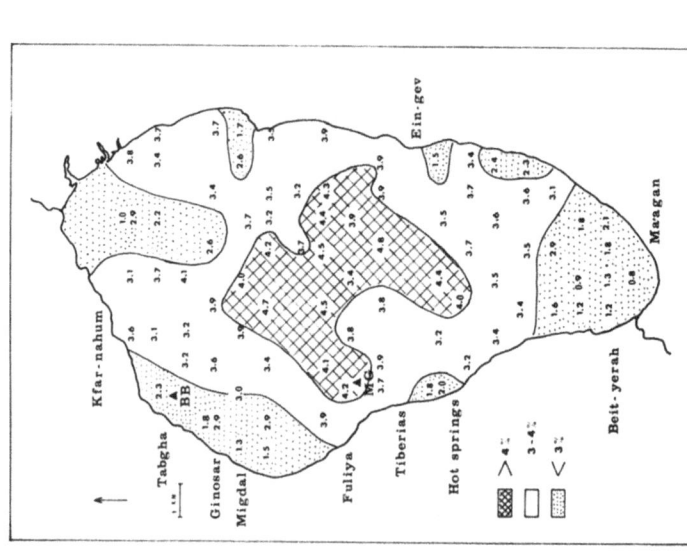

Fig. 81.4. Distribution of reducing capacity in surface sediments. Results stated as percentages of organic carbon (same source).

The distribution of non-detrital elements shows an inverse pattern of distribution with higher values in the central-eastern sediments and low values in the N–S tongue area. The concentrations of organic carbon in the sediments range from 1.0% in the northern tongue area to 4.8% in the central sediments. The concentrations of organic nitrogen amount to 0.06 and 0.25 in these respective areas. The most frequent C:N atomic ratio is approximately 30:1, instead of 7:1 generally found in algae and organic matter. These unusual ratios are due to the rapid transformations of organic nitrogen in the sediments (ammonification, nitrification) which finally lead to denitrification.

4. Sedimentation rates

The sedimentation rates have been measured with sediment traps placed one meter above the bottom (Serruya, 1973). Their contents were collected, weighed and analyzed.

a. Spatial variation of the annual rate of sedimentation

From October 1972 to October 1973, eight stations were equipped with sediment traps which were placed and withdrawn simultaneously at all stations (Fig. 82). The annual amount of tripton collected at the various stations is shown in Table 31.

During the considered period, the sedimentation rate varied from 640 to 2,158 g m^{-2} y^{-1}. By ascribing to each station a representative area, the total amount of tripton sedimented was estimated to be 198,000 tons (dry weight basis) or 1,214 g m^{-2} y^{-1}. The sedimentation rate is maximum at the northern station G, located about 3 km south of the Jordan mouth, but is also high in the other two northern stations, F and B. These stations also have the highest amount of silicates which are of purely detrital origin. The minimum rate is observed in the central zone of the lake. There are two reasons for this: this area is the deepest part of the lake, and on their way down many soluble elements may be dissolved or particles may drift away; however the second reason is probably the most important one. The active counterclockwise circulation of the lake (see p. 167) generates a centrifugal effect which leaves the center of the lake outside the area of main circulation.

The relatively high values measured at station L are due both to local erosion of the cliffs of the southeastern shore of the lake and to active resuspension caused by resedimentation in the shallow waters around this station, which is only 9 meters deep.

b. Seasonal distribution

In all the deep stations, there is a considerable difference between winter and summer rates of sedimentation. The winter–spring rates range from 8 to 12 g m^{-2} d^{-1} whereas the summer–fall values seldom exceed 1 to 2 g m^{-2} d^{-1}. It follows that approximately 85% of the material sedimented during the whole hydrological cycle reaches the bottom from January to April. At the deep

209

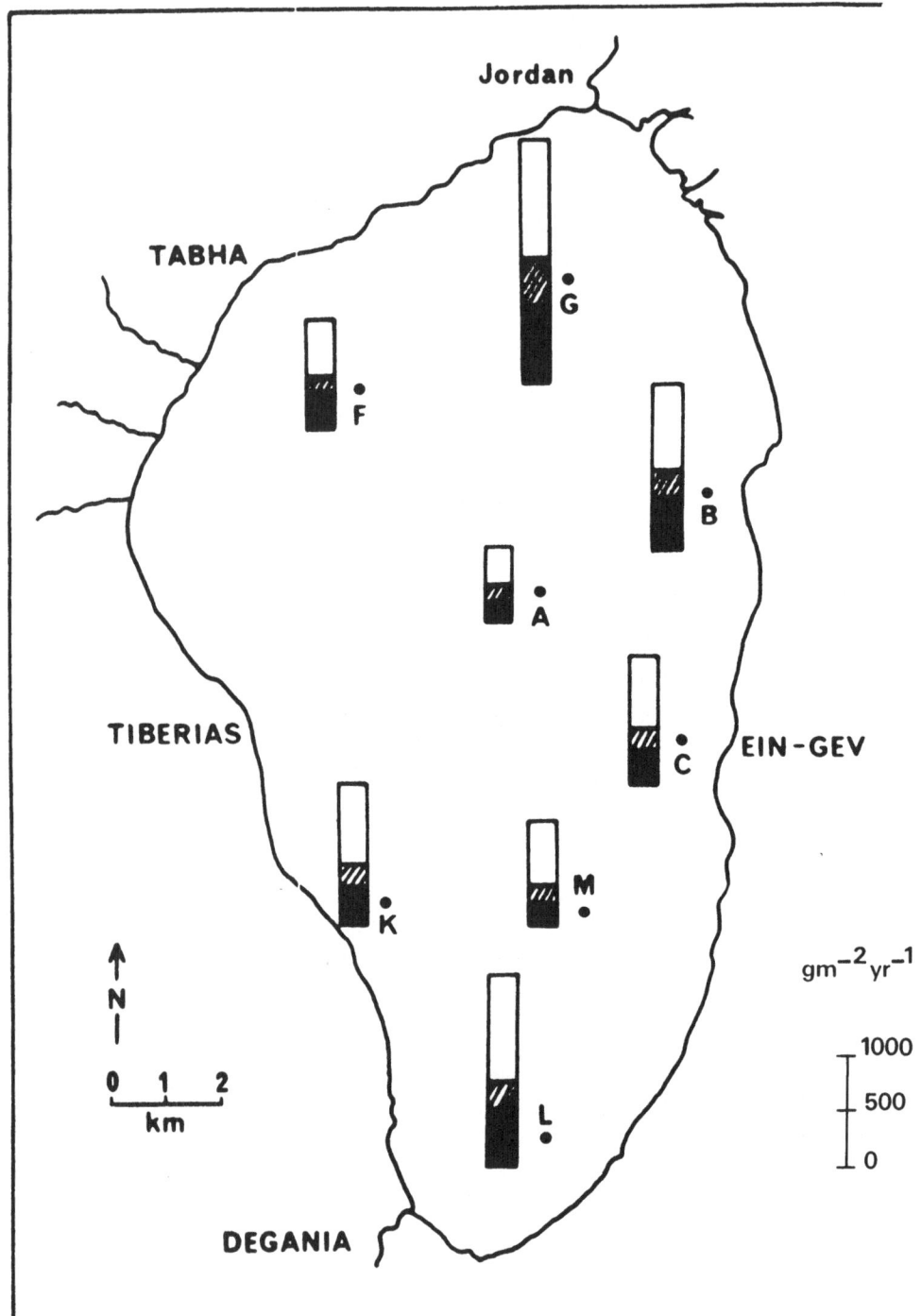

Fig. 82. Annual sedimentation rate at different stations. Results in g m⁻² y⁻¹. Black area = silicates, striped area = organic matter, white area = calcium carbonates (from Serruya C. 1977, Intern. Symp. Amsterdam 1976).

Table 31. Annual amounts of total tripton and its main components at 8 stations during the period October 1972–September 1973. Results in g m^{-2} y^{-1}. Percentages of different components in the tripton given in brackets (same source as fig. 82).

Stations	Total tripton	CaCO$_3$	Organic matter	Silicates	Organic carbon	Organic nitrogen	Total phosphorus
G	2,158	1,044 (48.4)	324 (15.0)	790 (36.6)	93.5 (4.3)	8.5 (0.39)	4.8 (0.22)
F	1,951	933 (47.8)	365 (18.7)	635 (32.5)			
B	1,428	734 (51.4)	225 (15.8)	469 (32.8)	65.0 (4.6)	5.8 (0.41)	3.0 (0.21)
C	1,191	706 (59.3)	160 (13.4)	325 (27.8)	49.6 (4.2)	4.8 (0.40)	2.1 (0.18)
A	640	340 (53.1)	123 (19.2)	177 (27.7)	44.2 (6.9)	3.7 (0.58)	1.4 (0.22)
M	909	546 (60.0)	131 (14.5)	232 (25.5)	45.4 (5.0)	3.9 (0.43)	1.2 (0.13)
L	1,615	873 (54.0)	261 (16.2)	481 (29.8)	80.0 (4.9)	7.2 (0.45)	5.7 (0.35)
K	1,254	735 (58.6)	173 (13.8)	346 (27.6)	57.4 (4.6)	5.1 (0.41)	3.4 (0.27)

stations, the winter–spring maximum is caused by the input of allochtonous matter of the floods and by the autochtonous precipitation of CaCO$_3$. In shallow stations, such as station L, where the process of resuspension is active, this Gauss-type distribution of tripton is not observed, since then the rate of sedimentation is related to wind-stress which in our area shows little seasonal variation.

c. Interannual variation

In Fig. 83, we can compare the amount of material deposited at station F during two consecutive but very different hydrological cycles. As far as precipitations are concerned, winter 1971–72 belongs to the average category while winter 1972–1973 can be classified as dry. Consequently, the cumulated water yield of the River Jordan for the period November–April was 50% lower in 1972–1973 than during the previous year. The total amount of tripton collected at station F during the November–April period was 1,680 g m^{-2} in 1971–1972 and 808 g m^{-2}

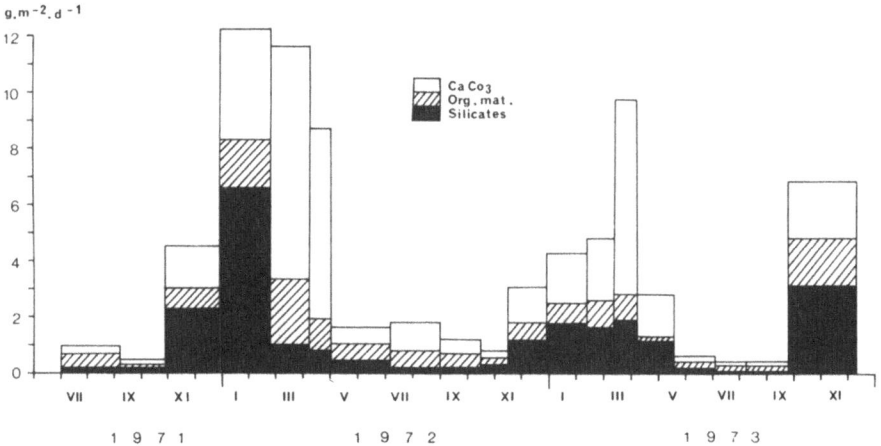

Fig. 83. Sedimentation rate at station F for the period July 1971–November 1973. Results in g m^{-2} d^{-1}. (same source as *Fig. 82*).

211

in 1972–1973. These data indicate that, at least in the northern part of the lake, the river material contributes an important fraction of the tripton.

d. Chemical composition

The chemical composition of the tripton is shown in Table 31. The tripton is very rich in $CaCO_3$ (47–60%). These values are very close to the concentration of $CaCO_3$ in *in situ* sediments; this indicates that the inorganic carbon accumulating in the interstitial water does not originate mainly from redissolution of calcium carbonate. A similar conclusion has been reached by Stiller & Magaritz (1974) who showed that the enrichment in ^{13}C of the dissolved carbonates in interstitial waters over the dissolved carbonates of the lake water cannot be explained by dissolution of $CaCO_3$; it is more probably due to the anaerobic decomposition of organic matter in the sediments.

An attempt has been made to evaluate separately the contribution of $CaCO_3$ from the watershed and from the lake to the total tripton based on the fact that the watershed is the sole source of silicate (Serruya, 1973). These calculations show that only 31% of the $CaCO_3$ deposited annually comes from the watershed and 69% is authigenic $CaCO_3$ produced in the lake mostly during the bloom period. These findings indicate that an amount of authigenic $CaCO_3$ ranging from 70,000 to 100,000 tons (corresponding to 412 to 588 g m^{-2} y^{-1}) precipitates annually. Very similar results were found by Stiller (1976), who investigated the stable isotopic composition of the carbonates and organic matter of the watershed and the tripton.

The organic carbon content of the tripton ranges from 4.2 to 6.9% of the tripton dry weight, while the level of this element in the superficial *in situ* sediments is only 3.5%, indicating that a rapid decomposition of a certain fraction of the organic carbon occurs after deposition.

Comparing the rate of sedimentation of organic carbon to the average rate of photosynthetic assimilation of carbon, we see that an average amount of 600 g C m^{-2} of carbon is annually fixed by photosynthetic organisms while only 59 gC m^{-2} reach the lake floor. In other words, no more than 10% of the carbon fixed by photosynthesis is sedimented. In fact, it is much less because the tripton carbon also includes detrital carbon brought with floods.

The main post-deposit transformations affect organic nitrogen. The N content of the tripton ranges from 0.40 to 0.58% in comparison with 0.19% in the *in situ* sediments. This difference gives the order of magnitude of the successive processes of the tripton ranges from 0.40 to 0.58% in comparison with 0.19% in the *in situ* sediment–water interface (see p. 317).

In contrast, the phosphorus content of the tripton ranges from 0.13 to 0.22% in comparison with 0.11 to 0.20% in the *in situ* sediments, showing a greater stability of this element. The higher values found in the tripton of the southern stations K and L (0.27 and 0.35%) are an additional indication of resuspension of local sediments which, in this part of the lake, have a high P content (Fig. 81).

5. Chemistry of interstitial waters

The chemistry of interstitial waters has been studied during twenty months on the basis of bi-weekly sampling (Serruya *et al.*, 1974) in the upper layer of the sediments.

The behaviour of dissolved iron is of particular interest. The highest levels of dissolved iron were observed from December to March, i.e. during the turnover

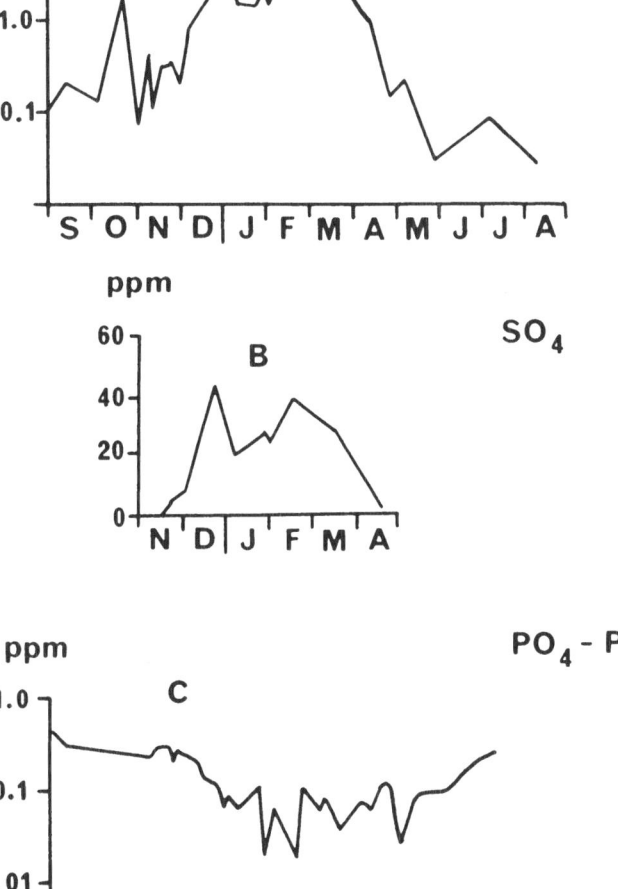

Fig. 84. Fluctuation of dissolved iron (A), SO_4 (B) and dissolved PO_4–P (C) in the interstitial water of superficial sediments. Years 1971–1972 (from C. Serruya *et al.*, 1973, Limnol. Ocean., 19, 3).

period. An opposite picture would have been obtained if iron oxides had been formed as a result of the presence of oxygen on the mud–water interface during this period. In fact, in summer, the low redox potential values maintain the sediments in the field of stability of the iron sulfide. When the interface is oxygenated, the system passes into the field of stability of siderite (iron carbonate). The considerable oxygen consumption of the sediments never allows the system to reach the field of stability of oxide. Since the iron has a maximum solubility in the field of siderite (Stumm & Morgan, 1970), the turnover is accompanied by a steep increase of interstitial dissolved iron. The rapid decrease of dissolved iron observed in May is accompanied by a similar decrease of the interstitial sulfate (Fig. 84), indicating that the dissolved iron is precipitated with the newly-formed sulfides. The fluctuations of the dissolved interstitial phosphorus show an accumulation of this element in summer and a one order of magnitude decrease in winter which may indicate that, in contrast to many other lakes, the Kinneret sediments lose phosphorus in winter, probably by mixing of the upper layers of sediments.

References

Banin, A., A. Singer & M. Gal. 1972. Amounts, composition and physico-chemical properties of clays in Lake Kinneret sediments. Tech. Rep. Fac. Agri. The Hebrew Univ. Jerusalem. 78 p. (in Hebrew).

Berman, T. 1974. Urea in the waters of Lake Kinneret (sea of Galilee). Limnol. Oceanogr. 19(6): 977–980.

Burns, N. M. & C. Ross. 1972. Oxygen-nutrient relationships within the central basin of Lake Erie. In: Project Hypo., C.C.I.W. (Burlington, Ont. Can.) EPA (Region IV, U.S.A.).

Christie, W. A. K. 1913. The composition of the water of Lake Tiberias. J. Asiatic Soc. of Bengal. IX(1): 25–29.

Hutchinson, G. E. 1957. A treatise of limnology. Vol. 1: John Wiley & Sons Inc. N.Y. 1015 p.

Mortimer, C. H. 1941. The exchange of dissolved substances between mud and water in lakes, 1 and 2. J. Ecol. 29: 280–329.

Mortimer, C. H. 1942. The exchange of dissolved substances between mud and water in lakes 1 and 2. J. Ecol. 30: 147–201.

Nir, Y. 1963. Preliminary Report on the Bottom Sediments of Lake Tiberias. The Geological Survey of Israel (mimeographed Report).

Oren, O. H. 1957. Physical and chemical characteristics of Lake Tiberias. Water Planning for Israel Ltd., No. 1445/58.

Serruya, C. 1971. A tentative nitrogen and phosphorus budget of Lake Kinneret (1968–1971). Internal Report. Kinneret Limnological Laboratory. Tiberias, Israel.

Serruya, C. 1971. Lake Kinneret, the nutrient chemistry of the sediments. Limnol, Oceanogr. 16: 510–521.

Serruya, C. 1973. Sediments: Sedimentation Rates. In: Lake Kinneret data record. N.C.R.D. 13–73. T. Berman (ed.).

Serruya, S. 1974. The mixing patterns of the Jordan River in Lake Kinneret. Linmol. Oceanogr. 19(2): 175–181.

Serruya, C., M. Edelstein & U. Pollingher. 1974. Lake Kinneret sediments: nutrient composition of the pore water and mud water exchanges. Limnol. Oceanogr. 19: 489–508.

Serruya, C. 1975. Nitrogen and phosphorus balances and load–biomass relationship in Lake Kinneret (Israel). Verh. Internat. Verein. Limnol. 19: 1357–1369.

Serruya, C., U. Pollingher and M. Gophen 1975. N and P distribution in lake Kinneret (Israel) with emphasis on dissolved organic nitrogen. Oïkos 26: 1–8.

Serruya C., 1977. Rates of sedimentation and resuspension in lake Kinneret. In: Intern. Symp. on Interactions between sediments and freshwater. Amsterdam 1976.

Stiller, M. 1974. Rates of transport and sedimentation in Lake Kinneret. Ph.D. thesis Weizmann Institute of Science. Rehovot. 240 p. (in Hebrew).

Stiller, M. 1976. Origin of sedimentation components in Lake Kinneret, traced by their isotopic composition. Int. Sympos. Interaction between sediments and freshwater. Amsterdam, 1976 (in press).

Stiller, M. & M. Magaritz. 1974. Carbon-13 enriched carbonate in interstitial waters of Lake Kinneret sediments. Limnol. Oceanogr. 19 : 849–853.

Stumm, W. & J. Morgan. 1970. Aquatic chemistry. Wiley.

Yashouv, A. 1963. Chemical and biochemical characteristics of the northern part of Lake Kinneret. Water Planning for Israel Ltd., Mekoroth Water Co., Internal Report.

Part three

The Planktic community

I. Phytoplankton

A. The algal population

1   The algae of the River Jordan

U. Pollingher

The algae of the River Jordan water were studied in order to estimate the amount of living organic material which is conveyed by the river into Lake Kinneret.

The water samples were collected at the Pardes Huri station and the algae were determined and counted using Utermohl's sedimentation method. The species which were observed during the period of the survey are included in List 1.

The monthly average of algal biomass ranges from 0.5 to 10 g m^{-3}. Minimal values are observed during the flood period (winter–spring); in late summer, when the river yield is at its lowest, the algal biomass reaches its maximal values (Fig. 85).

The dominant algae are coenobial forms or unicellular species of large size. The Cyanophyta, represented by *Microcystis* and *Anabaena* species, is the principal contributor to the algal biomass (Fig. 86). Diatoms form the second most important group with *Melosira*, *Cyclotella*, *Synedra* species and *Rhoicosphaenia curvata* (Kutz) Grun. The Euglenophyta are not abundant, but the large size of the *Euglena* and *Phacus* cells gives this group the third place in terms of biomass.

The Chlorophyta are represented by numerous taxa; however, *Pediastrum* and *Scenedesmus* are the only species which have a quantitative importance, and their contribution to the total biomass is moderate.

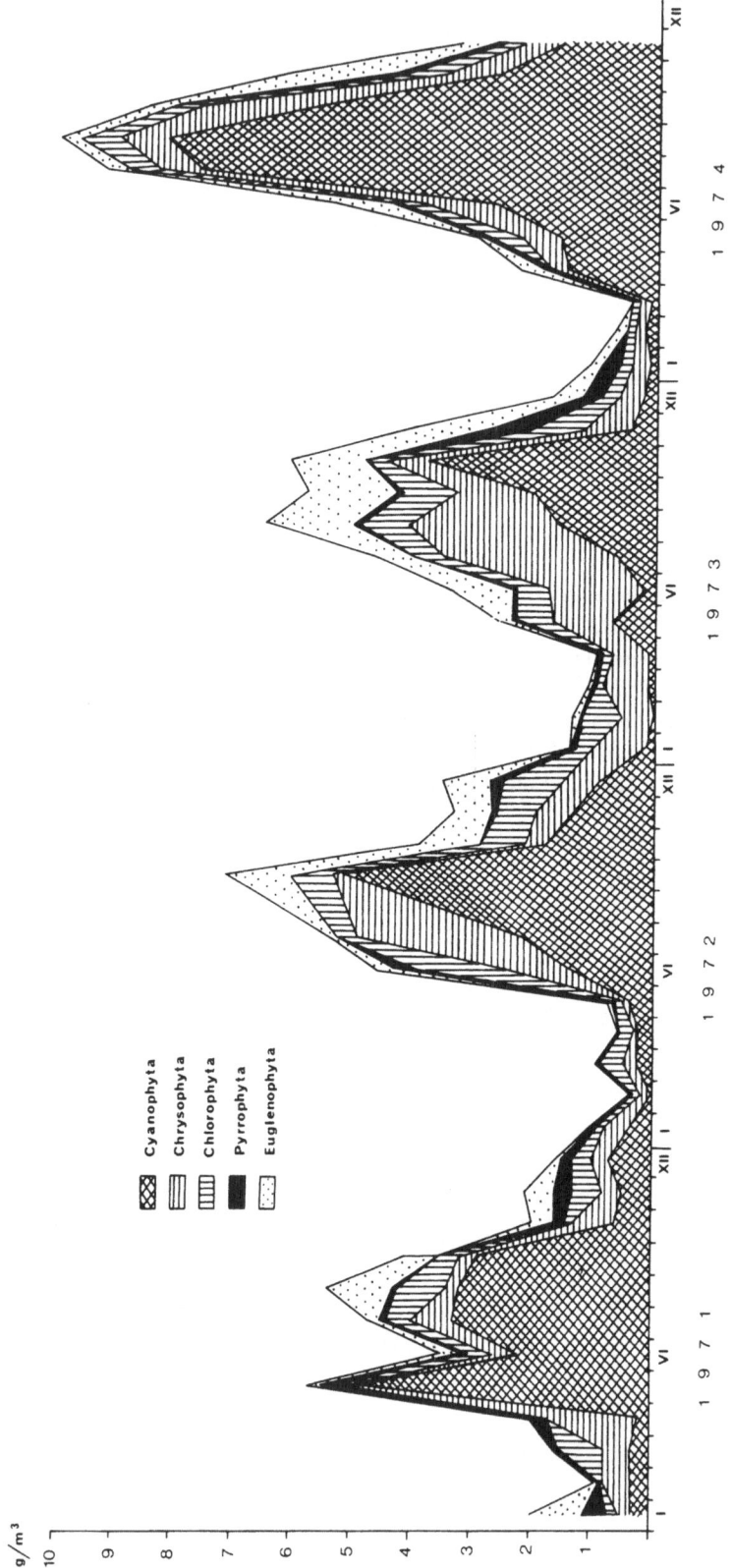

Fig. 85. Fluctuation of the biomass of the main groups of algae in the River Jordan, at Pardes Huri station. Results in g m⁻³.

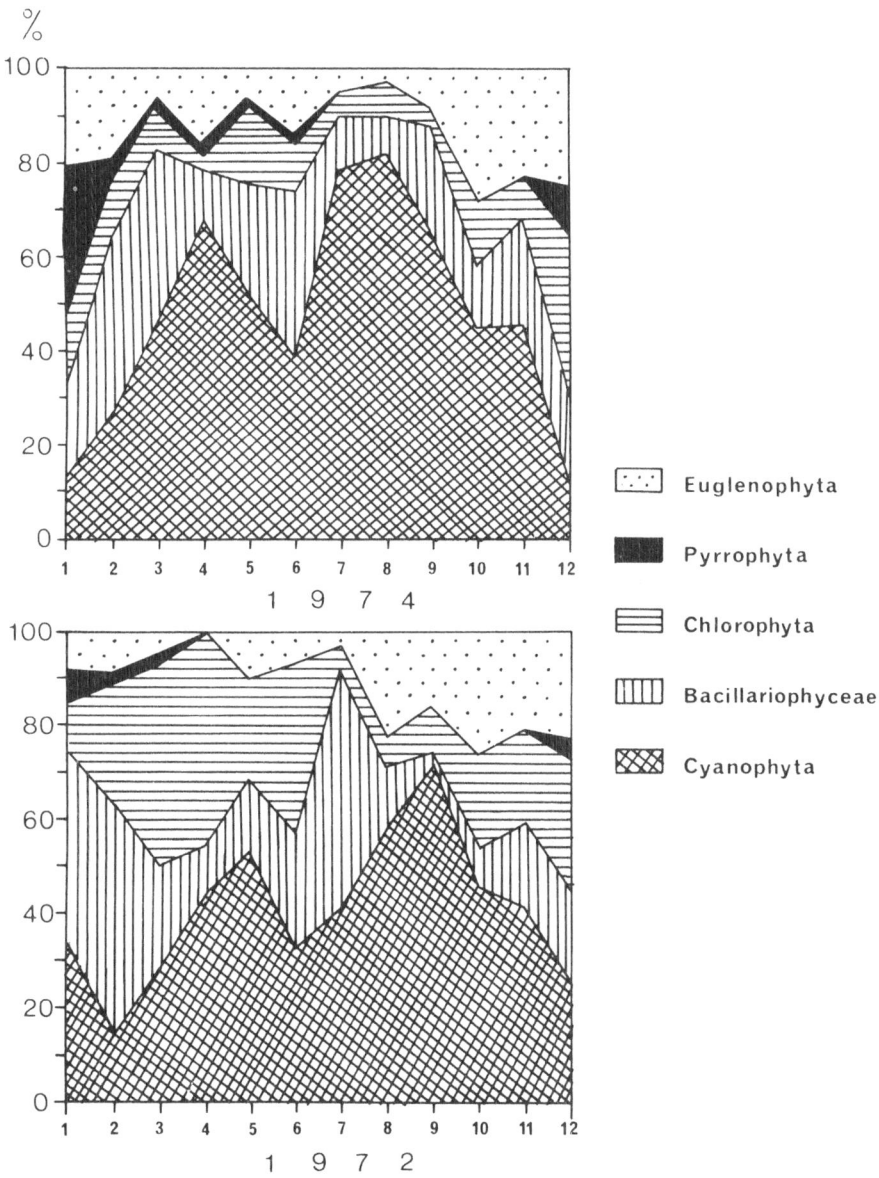

%

Fig. 86. Biomass contribution of each group of algae to the total algal biomass of the River Jordan.

The algae from the Jordan River

(Water Sampled at Huri Bridge)

CYANOPHYTA

Chroococcales

Microcystis aeruginosa (Kutzing)
Microcystis wesenbergii Kom
Microcystis botrys Teiling
Microcystis pulverea var. *incerta* (Lemm.) Crow.
Microcystis marginata (Meneghini) Kutzing
Aphanocapsa delicatissima W. et G. S. West
Chroococcus minutus (Kütz) Naeg.
Merismopedia tenuissima Lemm.
Coelosphaerium dubium Grunow

Hormogonales

Oscillatoria agardhii var. *isothrix* Skuja
Oscillatoria sp.
Spirulina platensis (Nordstedt) Geitler
Aphanizomenon gracile Lemm
Anabaena planctonica Brunnthalev
Anabaena spiroides Klebahn
Anabaena spiroides var. *crassa* Lemm.

BACILLARIOPHYTA

Centrales

Melosira granulata (Ehr) Ralfs
Melosira granulata var. *angustissima* Müll
Melosira granulata var. *muzzanensis* Meister
Melosira italica sbsp. *subarctica* O. Müll
Melosira varians C.A. Ag.
Melosira arenaria Moore
Cyclotella meneghiniana Kütz
Cyclotella stelligera Cl. u. Grun.
Cyclotella kützingiana Thwaites
Stephanodiscus sp.
Attheya zachariasi J. Brun

Pennales

Diatoma vulgare Bory
Synedra ulna (Nitzsch) Ehr.
Synedra ulna var. *oxyrhynchus* (Kütz) van Heurck
Synedra affinis Kütz
Synedra acus Kütz
Cocconeis placentula var. *euglypta* (Ehr.) Cleve
Rhoicosphaenia curvata (Kütz) Grun.

Gyrosigma acuminatum (Kütz) Rabh.
Gyrosigma attenuatum (Kütz) Rabh.
Stauroneis anceps Ehr.
Navicula cryptocephala Kütz
Navicula rhynchocephala Kütz
Navicula spp.
Pinnularia sp.
Amphora sp.
Cymbella ventricosa Kütz
Cymbella sp.
Gomphonema acuminatum Ehr.
Epithemia argus Kütz
Bacillaria paradoxa Gmelin
Nitzschia sublinearis Hust
Nitzschia sigmoidea (Ehr) W. Smith
Nitzschia vermicularis (Kütz) Grun.
Nitzschia acicularis W. Smith
Cymatopleura solea (Bréb.) W. Smith
Cymatopleura elliptica (Bréb) W. Smith
Surirella biseriata Brébisson
Surirella robusta Ehr.
Surirella robusta var. *splendida* (Ehr) v Heurck
Surirella capronii Brébisson
Surirella elegans Ehr.

CHRYSOPHYTA

Chrysomonadales

Mallomonas sp.

Dinobryon sp.

226

CRYPTOPHYTA

Cryptomonas spp.

Rhodomonas minuta Skuja n.sp.

DINOPHYTA

Glenodinium sp. (colourless)

EUGLENOPHYTA

Euglena acus Ehrb.
Euglena tripteris (Duj.) Klebs
Euglena spirogyra Ehrb.
Euglena elastica Presc.
Phacus caudatus Hübn.
Phacus longicauda (E.) Duj.

Phacus longicauda var. insecta Koczw.
Phacus acuminatus Stokes
Phacus helicoides Pochm.
Trachelomonas volvocina Ehrenb.
Trachelomonas hispida (Perty) Stein em. Defl.

CHLOROPHYTA

Volvocales

Gonium pectorale Muller
Pandorina morum (Muller) Bory
Eudorina elegans Ehrbg
Chlamydomonas sp.

Tetrasporales

Gloecoccus shroeterii (Chod) Lemm.

Chlorococcales

Characium limneticum Lemm.
Pediastrum duplex Meyen
Pediastrum simplex (Meyen) Lemm
Pediastrum clathratum (Schroter) Lemm
Pediastrum sturmii Reinsch.
Chlorella sp.
Micractinum pusillum Fresen.
Chodatella citriformis Snow
Franceia ovalis (Franie) Lemm
Oocystis solitaria Wittrock
Oocystis submarina Lagerh.
Nephrocytium agardhianum Naegeli
Kirchneriella lunaris (Kirchn.) Moeb.
Kirchneriella obesa (W. West) Schmidle
Kirchneriella contorta (Schmidle) Bohlin
Kirchneriella subsolitaria G. S. West
Tetraëdron minimum (Naeg.) Hansg.
Tetraëdron trigonum (Naeg.) Hansg.
Tetraëdron trigonum var. gracile Reinsch fa.
Tetraëdron regulare Kützing
Tetraëdron caudatum (Corda) Hansg. var.
  incisum Lager

Polyedriopsis spinulosa Schmidle
Treubaria triappendiculata Bernard
Tetrastrum staurogeniaeforme (Schröder)
  Lemm.
Scenedesmus hulensis Rayss
Scenedesmus acuminatus (Lagerh.) Chod.
Scenedesmus bijuga (Turp.) Lagerheim
Scenedesmus ecornis var. disciformis f.
  spinosus Hortob et Németh
Scenedesmus quadricauda (Turp) Bréb.
Scenedesmus quadricauda var. longispina
  (Chod.)
Scenedesmus javanensis Chod.
Scenedesmus opoliensis P. Richt
Scenedesmus carinatus (Lemm.) Chod.
Actinastrum hantzschii Lagerh.
Dictyosphaerium pulchellum Wood
Dictyosphaerium ehrenbergianum Naeg.
Crucigenia rectangularis (A.Br.) Gay
Crucigenia tetrapedia (Kirchn) W. et G. S. West
Coelastrum microporum Naeg.
Coelastrum proboscideum Bohlin
Coelastrum reticulatum (Dang.) Senn.
Coelastrum sphaericum Naeg.
Coelastrum scabrum v. torbolense Kirchn.
Selenastrum bibraianum Reinsch
Selenastrum minutum (Naegeli) Collins
Selenastrum gracile Reinsch
Ankistrodesmus falcatus (Corda) Ralfs.
Ankistrodesmus falcatus var. acicularis (A.
  Braun) G.S. West
Ankistrodesmus falcatus var. duplex (Kütz.)
  G. S. West
Ankistrodesmus falcatus var. mirabilis W. et
  G. S. West

Ankistrodesmus convolutus Corda
Ankistrodesmus nannoselene Skuja
Ankistrodesmus setigerus (Schröder) G. S. West
Ankistrodesmus nitzschioides G. S. West
Elakatothrix gelatinosa Wille

Desmidiales

Closterium acerosum (Schrank) Ehrnb.
Closterium moniliferum (Bory) Ehrnb.
Closterium polystictum Nygaard
Closterium acutum v. *variabile* (Lemm.) Krieger
Closterium gracile Bréb.
Cosmarium spp.
Staurastrum volans W. et W.
Staurastrum sp.

PRASINOPHYTA

Platymonas cordiformis (Carter) Korschikow

2 The phytoplankton of Lake Kinneret

U. Pollingher

The first list of algae of Lake Kinneret was published in 1873 by Petit & Brun and concerned mainly benthic and littoral species (List 2).

List No. 2

List of diatoms of Lake Tiberias

by M. Petit

PLACOCHROMATICEAE

Cocconeis Pediculus, EH.
— *Placentula*, EH.
Achnanthidium delicatulum, KTZ.
Gomphonema capitatum, EH.
— *affine*, KTZ.
Amphora ovalis, KTZ.
Epithemia Sorex, KTZ.
— *Argus*, KTZ.
— *Musculus*, KTZ.
— *Zebra*, EH.
— *gibba*, KTZ.
Cymbella, turgida, GREG.
Cocconema lanceolatum, EH.
Navicula alpestris, GRUN.
— *firma*, KTZ, var. *Scoliopleuroides*, P.-P.
— *limosa*, GRUN.
— *menisculus*, SCHUM.
— *Reinhardti*, GRUN. (var. *Vernalis*, DONK.)
— *Scutelloïdes*, W.S.M.

Navicula Trochus, EH. (var. *Schumanniana*),
  GRUN., Synop., VAN HEURCK.
— *viridis*, EH.
Stauroneis dilatata, W.SM.
Pleurosigma attenuatum, W.SM.
Hantzschia (Nitzschia) amphioxys (W.SM.).
  GRUN.
Tryblionella angustata, W.SM.
Surirella biseriata, BREB.
— *elegans*, AD. SCH.
— *striatula*, TURP.
Campylodiscus Echeneis, EH. (*C. cribrosus*
  W.SM.).
— *spiralis*, W.SM.
Cymatopleura Solea, W.SM.
— *elliptica*, BREB.
— *Bruni*, NOV.SP.
Synedra danica, KTZ. (*S. Ulna* var.).
Staurosira construens, EH. (var. *binodis*).
— *mutabilis*, W.SM.

COCCOCHROMATICEAE

Fragilaria brevistriata, GRUN. (*Synopsis* Van
  Heurck)
Terpsinoe musica, EH. (found by M.
  Schlumberger).

Gallionella granulata, FRITCH.
— *arenaria*, MOORE.

Desmids from the Mud of Lake Tiberias

M. Brun

Scenodesmus quadricauda, RALFS.
Desmidium aptogenum, BREB.
Polyedrum trigonum, NAEG.
Pediastrum simplex MEYEN

In April–May 1893, Barrois organized an expedition to Lake Kinneret, and the planktic material that he collected was divided between several biologists who produced detailed lists of zooplankton species. Unfortunately, this team did not include a phycologist.

However, Barrois noticed the extreme abundance of planktic algae in the lake water, as he wrote in his description of the lake 'C'est surtout à l'extrême surabondance de ces algues qu'il faut attribuer dans le cas présent le faible coefficient de la limite de visibilité'. This observation indicates that Lake Kinneret was already a productive water body at the end of the last century.

Although Barrois did not determine the planktic flora of the lake, he mentioned *Ceratium hirundinella* among the Protozoans, and noted that it is a good food source for *Asplanchna*. This was the first time that dinoflagellates were mentioned in Lake Kinneret. We have to take into account that *Ceratium hirundinella* was described for the first time by Müller in 1773, and later on redescribed by Schrand (1802), Perty (1849), Berg (1882) and Limhof (1884). *Peridinium* was described by Ehrenberg in 1839, but appeared in two other publications only: Stein (1883) and Schiling (1891).

In October–November 1912, another expedition was organized by Annandale. The results concerning planktic forms referred only to zooplankton; discussing the abundance of Mollusca, Annandale wrote 'the latter live to a large extent on the minute algae that are extremely abundant . . .'.

It was not until 1944 that, Rayss, in 'Materiaux pour la flore algologique de la Palestine I. Les Cyanophyceae' and later in 1951 in 'II. Les algues de eaux continentales', described thirty species of algae from Lake Kinneret (four species of Cyanophyta, three species of diatoms, three species of dinoflagellates, including *Peridinium westii*, and twenty species of Chlorophyta).

In the same year (1951), Komarovsky described the water bloom of *Peridinium westii* Lemm and *Fragilaria crotonensis* Kitton (January–February, 1948) in Lake Kinneret. The same author published in 1959 a list of eighty-three species of planktic algae from the lake (List 3).

In 1961, Yashouv gave a list of twenty-eight species of algae; among them, only two species are different from the above-mentioned lists.

In 1965, Kimor & Pollingher published a booklet describing one hundred and fifty species collected not by net, but by centrifugation of water samples; this method allowed the identification of some nannoplanktic species but at the same time destroyed the minute flagellates.

List No. 3

The phytoplankton of Lake Kinneret

B. Kimor

CYANOPHYCEAE

CHROOCOCCALES

Family CHROOCOCCACEAE

Aphanocapsa pulchra (Kütz.) Rabenh.
Aphanocapsa endophytica G. M. Smith
Aphanothece nidulans Richter
Chroococcus limneticus Lemmerm.
Chroococcus minutus (Kütz.) Naegeli
Chroococcus turgidus (Kütz.) Naegeli
Coelosphaerium dubium Grunow

Holopedia dieteli (Richter) Migula
Merismopedia glauca (Ehrbg.) Naegeli
Merismopedia minima Beck
Microcystis aeruginosa (Kütz.) Elenk.
Microcystis firma (Bréb. and Lenorm.) Rabenh.
Microcystis flos aquae (Wittr.) Kirchn.
Microcystis ichthyoblabe Kütz.

HORMOGONALES

Family OSCILLATORIACEAE

Lyngbia cryptovaginata Schkorbatow
Lyngbia limnetica Lemmerm.
Oscillatoria constricta (Szafer) Geitler

Oscillatoria sp.
Spirulina platensis (Nordst.) Geitler

Family NOSTOCACEAE

Anabaenopsis circularis G.S. West) V. Miller

DIATOMEAE

Family COSCINODISCACEAE

Cyclotella catenata Brun.
Melosira granulata (Ehrbg.) Ralfs

Melosira granulata v. *angustissima* Müller

Family FRAGILARIACEAE

Fragilaria crotonensis Kitton
Synedra affinis Kütz. v. *fasciculata* (Kütz.)
   Grunow

Synedra ulna (Nitzsch) Ehrbg.

Family ACHNANTHACEAE

Cocconeis sp.

231

Family NAVICULACEAE

Amphora ovalis Kütz.
Cymbella affinis Kütz.
Cymbella cymbiformis (Ag.) Kütz.
Gomphonema sp.

Gyrosigma acuminatum (Kütz.) Rabenh.
Navicula sp.
Stauroneis anceps Ehrbg.

Family SURIRELLACEAE

Cymatopleura solea (Bréb.) W. Smith
Surirella ovalis Bréb.

Surirella robusta Ehrbg.

CHLOROPHYCEAE

PROTOCOCCALES

Family CHARACIACEAE

Characium debaryanum (Reinsch) De Toni

Characium hookeri (Reinsch) Hansg.

Family HYDRODICTYACEAE

Pediastrum boryanum (Turpin) Menegh.
Pediastrum simplex (Meyen) Lemm.
Pediastrum duplex Meyen f.t.
Pediastrum duplex v. *clathratum* (A.Br.) Lag.

Pediastrum duplex v. *coronatum* Racib.
Pediastrum ovatum (Ehrbg.) A. Braun
Pediastrum tetras (Ehrbg.) Ralfs

Family OÖCYSTACEAE

Ankistrodesmus falcatus (Corda) Ralfs f.t.
Ankistrodesmus falcatus v. *duplex* (Kg.) G. S. West
  G. S. West
Ankistrodesmus longissimus (Lemmerm.) Wille
Kirchneriella lunaris (Kirchner) Moebius

Lagerheimia longiseta Lemmerm.
Oöcystis solitaria Witr.
Oöcystis sp.
Tetraëdron muticum (A.Br.) Hansg.

Family DICTYOSPHAERIACEAE

Dictyosphaerium ehrenbergianum Naegeli

Dictyosphaerium pulchellum Wood

Family SCENEDESMACEAE

Crucigenia tetrapedia (Kirchn.) W. & G. S. West
  West
Scenedesmus acuminatus (Lagerh.) Chod.
Scenedesmus bijugatus a. *seriatus* Chod.

Scenedesmus bijugatus β. *alternans* (Reinsch) Har
  Hansg.
Scenedesmus obliquus (Turp.) Kg.
Scenedesmus quadricauda (Turp.) Bréb.

Family COELASTRACEAE

Coelastrum microporum Naegeli
Coelastrum proboscideum Bohlin

Coelastrum reticulatum (Dangeard) Senn

Family MICRACTINIACEAE

Golenkinia radiata Chod.

VOLVOCALES

Family VOLVOCACEAE

Eudorina elegans Ehrbg.

TETRASPORALES

Family PALMELLACEAE

Gloeocystis botryoides (Kütz.) Naegeli   *Sphaerocystis schroeteri* Chod.

ULOTRICHALES

Family MICROSPORACEAE

Microspora stagnorum (Kütz.) Lag.

HETEROKONTAE

Family CHLOROBOTRYDACEAE

Chlorobotrys regularis Bohlin

Family BOTRYOCOCCACEAE

Botryococcus braunii Kutz.

CONJUGATAE

Family DESMIDIACEAE

Spondylosium moniliforme Lund   *Staurastrum muticum* Bréb.
Staurastrum gracile Ralfs

PERIDINEAE

Family KYRTODINIACEAE

Glenodinium cinctum Ehrbg.

Family KROSSODINIACEAE

Ceratium hirundinella O. F. Müller   *Peridinium bipes* Stein
Diplopsalis acuta (Apstein) Entz-fil.   *Peridinium westii* Lemmerm.
Gonyaulax apiculata (Penard) Entz-fil.

233

EUGLENINAE

Family EUGLENACEAE

Euglena acus Ehrbg.

CRYPTOMONADINAE

Family CRYPTOMONADACEAE

Cryptomonas erosa Stein

Since 1968, we have used Uthermohl's sedimentation method which permitted us to find and identify all minute species existing in the water. As a result, we can give here, for the first time, the complete list of algae found and described in the water of Lake Kinneret. Following every species are the initials of the author who described it in the lake for the first time (List 4).

Not all algae which appear in our long list are common in the lake. We therefore present a second list of the more common species (List 5).

The Chlorophyta are principally represented by typically eutrophic forms such as species of *Pediastrum* and *Scenedesmus*. Other species are characteristic of waters rich in organic matter and with a neutral or alkaline reaction: species belonging to the genera *Scenedesmus, Oocystis, Tetraëdron, Golenkinia* and *Chlorella*.

Some of the common forms of green algae present in Lake Kinneret have been described as typically alkalophilic by Van Oye & Gillard (1949–1950), such as *Coelastrum microporum* Naeg., *Ankistrodesmus falcatus* Corda, Ralfs, *Actinastrum hantzschii* Lagerh. and most of the species of *Crucigenia*. The Volvocales, *Gonium, Pandorina* and *Eudorina* were found especially near the littoral.

The desmids recorded in the lake plankton (*Closterium, Cosmarium* and *Staurastrum*) are considered as alkalophilic, although the typical forms of this group prefer water with low calcium content and with an acid reaction. It is worthwhile to underline the presence of a large number of a new morpha of *Cosmarium laeve* Rbh.

Among the Cyanophyta, the Chroococcales are far more common in the plankton than the Hormogonales. The Bacillariophyta are represented by typical euplanktic forms such as *Melosira granulata* (Ehr) Ralfs, *Melosira granulata* var. *angustissima* Mull, *Cyclotella meneghiniana* Kutz, *Cyclotella stelligera* Cl. u. Grun, *Synedra* and littoral forms such as species of *Nitzschia* and *Navicula*. The Pyrrhophyta species are typical cosmopolitan, widely distributed in the inland waters of all continents, but the dominant species *Peridinium cinctum* fa. *westii* (Lemm) Lef. was never described.

The Euglenophyta, represented by species of *Euglena* and *Phacus*, are more common in the littoral zone but never in significant numbers *Trachelomonas* sp. are found in the metalimnic layer, in warm lakes.

Among the flagellates, we found some species which are described as typical, cold, stenotherm forms such as *Erkenia subaequiciliata* Skuja (Chrysophyta) and *Rhodomonas minuta* var. *nannoplanctica* Skuja (Cryptophyta). These species are known only from temperate lakes in Europe and North America, but have never been reported in warm lakes.

In Lake Kinneret, *Erkenia* appears in large numbers in fall, reaching its maximum during the winter and spring months when the temperature ranges from 15 to 20°C. By May, *Erkenia* begins to decline; during the summer it is found only in the metalimnion in very small amounts.

According to Ruttner (1952), Skuja (1948) and Bourrelly (1957), the Chrysophyta are absent or very rare in warm and tropical zones, and when represented, it is only by *Dinobryon* and *Malomonas* spp. Species of *Dinobryon*, *Synura* and *Malomonas* are very rarely found in Lake Kinneret, only occasionally in winter in the littoral.

Rhodomonas minuta var. *nannoplantica* (Skuja) appears in a relatively large amount in November, reaches its maximum in January–February, then decreases suddenly and nearly disappears during the summer months.

The Cryptophyta are represented in the lake also by other species: *Cryptomonas erosa* Ehrenbg, *Cryptomonas reflexa* (Marsson) Skuja and *Rhodomonas lacustris* Pascher et Ruttner. These species accompany the water bloom of *Peridinium* and are abundant during the winter and spring months.

Ruttner (1952), in his work on tropical lakes during the Sunda Expedition, insisted on the fact that the genus *Rhodomonas* is completely absent and species of *Cryptomonas* are scarce.

The presence of species such as *Erkenia subaequiciliata* Skuja and *Rhodomonas minuta* Skuja in Lake Kinneret demonstrates that cold stenotherm forms with a short life cycle can thrive during the short period of 'winter' in warm lakes, like the polytherm species which thrive in the warm period of temperate lakes.

The phytoplankton of Lake Kinneret have the following characteristics: (a) a great variety of species; (b) a bloom of *Peridinium*; (c) the presence of a large number of a morpha of *Cosmarium laeve*; (d) the presence of temperate and cold stenotherm forms during the short winter.

The algal population of Lake Kinneret, dominated by Pyrrhophyta and Chlorophyta, is very different from the Cyanophyta-Bacillariophyta assemblage of African warm lakes. Lake Kinneret and Lake Victoria, for example, have in common the presence of a great variety of species, the presence of a large number of dominant species (twenty-four in Lake Victoria and thirty in Lake Kinneret), and the presence of species with important seasonal fluctuations, such as *Peridinium* and *Cosmarium* in Lake Kinneret and *Anabaena flos-aquae* in Lake Victoria. A relatively constant population of some green algae is found in Lake Kinneret, and green algae and *Microcystis wesenbergii* in Lake Victoria.

The algae of Lake Kinneret can be divided into two groups. One group is composed of species characteristic of warm water; they are also found in African lakes. The second group harbours species such as *Peridinium cinctum*, *Erkenia* and *Rhodomonas*, which belong to the freshwater flora of the temperate zone in Europe.

List No. 4

Algae found in the plankton of Lake Kinneret

U. Pollingher

CYANOPHYTA

Chroococcales

Aphanocapsa delicatissima W. et G. S. West
Pol
Aphanocapsa elachista W. et G. S. West Pol
Aphanocapsa pulchra (Kütz.) Rbh. R
Aphanothece castagnei (Breb.) Rbh. Pol
Aphanothece nidulans P. Richter. K
Chroococcus limneticus Lemm. K
Chroococcus minutus (Kütz.) Naeg. K
Chroococcus turgidus (Kütz.) Naeg. K
Coelosphaerium dubium Grunow K
Gomphosphaeria aponina var. *delicatula*
Virieux Pol
Merismopedia glauca (Ehrnb.) Naeg. K
Merismopedia minima G. Beck K
Merismopedia punctata Meyen Pol
Merismopedia tenuissima Lemm. Pol
Holopedia dieteli = *Micrococis geminata*
(Richter) Migula K
Microcystis aeruginosa Kutzing R
Microcystis botrys Teiling
Microcystis wesenbergii Kom. Pol.
Microcystis viridis (A. Braun) Lemm. Pol
Microcystis marginata (Meneghini) Kutzing Pol
Microcystis pulverea var. *incerta* (Lemm.)
Crow. Pol
Radiocystis geminata Skuja Pol

Hormogonales

Lyngbya limnetica Lemm. K
Oscillatoria agardhii Gomm. Pol
Oscillatoria agardhii var. *isothrix* Skuja Pol
Oscillatoria amoena Gom. Pol
Oscillatoria mougeotti Kutzing. Pol
Phormidium mucicola Hub.-Pest. et Naumann
Pol
Spirulina platensis (Nordstedt) Geitler K
Spirulina albida var. *tenuior* Pol
Pseudanabena catenata Lauterb. Pol
Anabaena planctonica Brunnthaler Pol
Anabaena spiroides Klebahn Y
Anabaena spiroides var. *crassa* Lemm Pol
Anabaenopsis circularis (G. S. West) V. Miller K
Aphanizomenon flos-aquae (L) Ralfs. Pol

Pelonematales

Acroonema angustum (Koppe) Skuja Pol
Acroonema lentum Skuja Pol

BACILLARIOPHYTA

Centrales

Melosira arenaria Moore P
Melorira granulata (Ehr) Ralfs P
Melosira granulata var. *angustissima* Mull K
Melosira italica subsp. *arctica* O. Mull Pol
Cyclotella kuzinginiana Thwaites Pol
Cyclotella meneghiniana Kütz Pol
Cyclotella stelligera Cl. U. Grun. Pol
Stephanodiscus hantzschii Grun. Pol
Attheya zachariasi J. Brun. Pol.

Pennales

Fragilaria crotonensis Kitton K
Fragilaria construens var. *subsalina* Hystedt
Pol
Synedra ulna (Nitzch.) Ehr K
Synedra ulna var. *ramesi* (Heribaud et
Peragallo) Hust. Pol
Synedra ulna var. *oxyrhynchus fa. mediocontracta* Hust. Pol
Synedra ulna var. *acus* Kütz. Pol

236

Synedra ulna var. affinis Kütz K
Synedra ulna var. fasciculata (Kütz) Grun. Pol
Cocconeis pediculus Ehr. P
Cocconeis placentula (Ehr.) P
Rhoicosphaenia curvata (Kütz.) Grun. Pol
Navicula cryptocephala Kütz. Pol
Navicula ryncocephala Kütz. Pol
Gyrosigma attenuatum (Kütz.) Rabh. P
Gyrosigma acuminatum (Kütz) Rabh. K
Gyrosigma spencerii var. nodifera Grun. Pol
Pleurosigma angulatum var. strigosum Pol
Stauroneis anceps Ehr. K
Gomphonema parvulum Kütz.
Gomphonema ventricosum Y
Amphora ovalis Kütz P
Cymbella affinis Kütz K
Cymbella cymbiformis (Agard? Kütz) v. Heurck K
Cymbella helvetica Kütz. Pol
Epithemia muelleri Fricke Pol

Epithemia zebra (Ehr) Kütz. P
Nitzschia acicularis W. Smith Pol
Nitzschia holsatica Hust. Pol
Nitzschia hungarica Grun. Pol
Nitzschia sigmoidea (Ehr.) W. Smith Pol
Nitzschia stagnorum Rabh. Pol
Bacillaria paradoxa Gmelin Pol
Cymatopleura elliptica (Breb.) W. Smith P
Cymatopleura solea (Breb.) W. Smith P
Cymatopleura solea var. apiculata (W. Smith)
    Ralfs Pol
Surirella capronii Brebisson Pol
Surirella ovalis Brebisson K
Surirella robusta Ehr. K
Surirella robusta var. splendida (Ehr) V.
    Heurck Pol
Surirella tenera var. nervosa Mayer Pol
Surirella spiralis Kütz. Pol

CHRYSOPHYTA

Erkenia subaequiciliata Skuja Pol
Monas mediovacuolata Skuja Pol
Malomonas sp. Pol

Synura sp. Pol
Dinobryon sp. Pol

DINOPHYTA (PYRRHOPHYTA)

Gymnodiniales

Gymnodinium lacustre Schiler Pol

Peridiniales

Glenodinium gymnodinium Penard Pol
Glenodinium oculatum Stein R
Glenodinium sp. (Colourless) Pol
Sphaerodinium cinctum Wol. Pol
Peridinium cinctum (Muller) Ehrb. R
Peridinium cinctum fa. westii (Lemm) Lef. K
Peridinium palatinum Lautb. Pol

Peridinium pseudolaeve Lef. Pol
Peridinium bipes Stein R
Peridinium inconspicuum Lemm. Pol
Peridinium cunnigtonii (Lemm.) Lemm. Pol
Diplopsalis acuta Entz K
Gonyaulax apiculata (Penard) Entz Fil. K
Ceratium hirundinella fa. gracile (Bachm.) Bar.
    Pol
Ceratium hirundinella fa. scotticum (Bachm.)
    Pol
Ceratium hirundinella fa. robustum (Amberg.
    Bachm.) Pol

CRYPTOPHYTA

Cryptomonadales .

Rhodomonas lacustris Pascher et Ruttner Pol
Rhodomonas minuta Skuja Pol
Rhodomonas minuta var. nannoplanctica Skuja
    Pol

Cryptomonas erosa Ehrenbg. K
Cryptomonas ovata Ehrenbg. Pol
Cryptomonas reflexa (Marsson) Skuja Pol
Cyatomonas truncata (Fres) Fisch. Pol
Chroomonas acuta Utermohl Pol

EUGLENOPHYTA

EUGLENOPHYCEAE

Euglena acus Ehrb. K
Euglena oxyuris Schmarda Pol
Euglena texta (Duj.) Hubner Pol
Euglena tripteris (Duj.) Klebs Pol
Euglena spathirhyncha Skuja Pol
Phacus caudatus Hubn Pol

Phacus longicauda E. (Duj.) Pol
Phacus longicauda var. *insecta* Koczw. Pol
Trachelomonas hispida var. *punctata* Lemm.
  Pol
Trachelomonas volvocina Ehrenb. Pol
Rhizaspis sp. Skuja Pol

CHLOROPHYTA

CHLOROPHYCEAE

Volvocales

Collodyction triciliatum Carter Pol
Aulacomonas submarina Skuja Pol
Chlamydomonas sp. Pol
Gonium pectorale Muller R
Pandorina morum (Muller) Bory Pol
Eudorina elegans Ehrbg. K

Tetrasporales

Asterococcus superbus (Cienk.) Scherffel Pol
Gloeocystis ampla (Kütz) Rabenhorst Pol
Gloeococcus schroeterii (Chod) Lemm. R

Chlorococcales

Golenkinia radiata Chodat K
Chlorococcum humicola (Naeg.) Rab. Pol
Acanthosphaera zachariasi Lemm. Pol
Characium debarianum (Reinsch) de Toni K
Characium hookeri (Reinsch) Hansg. K
Pediastrum sturmii Reinsch. R
Pediastrum simplex (Meyen) Lemm. B
Pediastrum clathratum (Schroter) Lemm. Pol
Pediastrum clathratum var. *microporum* Lemm.
  Pol
Pediastrum clathratum var. *duodenarium*
  (Bailey) Lemm. Pol
Pediastrum integrum Naegeli Pol
Pediastrum duplex Meyen R
Pediastrum duplex var. *clathratum* Al. Brun R
Pediastrum duplex var. *rotundatum* Lucks Pol
Pediastrum duplex var. *reticulatum* Lagerheim
  Pol
Pediastrum tetras (Ehr.) Ralfs K
Pediastrum tetras var. *excisum* Rabenhorst Pol
Pediastrum boryanum (Turp.) Menegh. R

Pediastrum boryanum var. *longicorne* Reinsch
  Pol
Coelastrum sphaericum Naeg. R
Coelastrum microporum Naeg. K
Coelastrum reticulatum (Dang.) Senn. R
Coelastrum scabrum var. torbolense Kirchn. Pol
Coelastrum cambricum var. *intermedium* (Bohl)
  G. S. West Pol
Botryococcus braunii Kuetzing R
Chlorella pirenoidosa Chick Pol
Chlorella vulgaris Beyerinck Pol
Dictyosphaerium elegans Bachman Pol
Dictyosphaerium ehrenbergianum Naeg. K
Dictyosphaerium pulchellum Wood K
Oocystis elliptica var. *minor* West et West Pol
Oocystis solitaria Wittrock K
Oocystis lacustris Chodat Pol
Oocystis novae-semliae Wille R
Oocystis submarina Lagerh. Pol
Nephrocytium agardhianum Naegeli Pol
Nephrocytium lunatum W. West Pol
Nephrocytium limneticum (G. M. Smith) Skuja
  Pol
Chodatella ciliata (Lag.) Lemm. Pol
Chodatella longiseta Lemm. K
Chodatella citriformis Snow Pol
Lagerheimia genevensis Chodat Pol
Franceia ovalis (France) Lemm. Pol
Ankistrodesmus falcatus (Corda) Ralfs R
Ankistrodesmus falcatus var. *acicularis* (A.
  Braun) G. S. West Pol
Ankistrodesmus falcatus var. *duplex* (Kütz) G.
  S. West K
Ankistrodemus falcatus var. *mirabilis* G. S.
  West Pol
Ankistrodesmus falcatus var. *spirilliformis* G. S.
  West Pol

238

Ankistrodesmus nannoselene Skuja Pol
Ankistrodesmus setigerus (Schroed.) G. S. West
Pol
Selenastrum bibraianum Reinsch Pol
Selenastrum gracile Reinsch Pol
Selenastrum minutum (Naegeli) Collins R
Selenastrum capricornutum Printz Pol
Kirchneriella lunaris (Kichn.) Moeb. Pol
Kirchneriella obesa (W. West) Schmidle Pol
Nephrochlamys subsolitaria (G. S. W) Korch.
Pol
Tetraëdron minimum (Naeg.) Hansg. Pol
Tetraëdron caudatum var. *incisum* Lagerh. Pol
Tetraëdron trigonum (Naeg.) Hansg. B
Tetraëdron trigonum var. *minor* Reinsch Pol
Tetraëdron quadratum fa. *minor obtusum*
Reinsch. Pol
Tetraëdron regulare fa. *minor* Reinsch. Pol
Treubaria triapendiculata Bernard Pol
Scenedesmus acuminatus (Lagerh.) Chod. R
Scenedesmus acuminatus var. *minor* G. M.
Smith Pol
Scenedesmus acutiformis Schroder Pol
Scenedesmus arcuatus Lemm. Pol
Scenedesmus armatus Chod. R
Scenedesmus bijuga (Turp.) Lagerheim Pol
Scenedesmus bijuga var. *seriatus* Volk R
Scenedesmus bijuga var. *alternans* Pascher K
Scenedesmus denticulatus Lagerh. Pol
Scenedesmus ecornis (Ralfs) Chod. Pol
Scenedesmus bicellularis Chod. Pol

Scenedesmus acutus (Meyen) Chod. Pol
Scenedesmus acutus f. *alternans* Hortob. Pol
Scenedesmus opoliensis P. Richt. Pol
Scenedesmus perforatus Lemm. Pol
Scenedesmus spinosus Chod. Pol
Scenedesmus quadricauda (Turp.) Breb. B
Scenedesmus quadricauda var. *longispina*
(Chod) Pol
Scenedesmus quadricauda var. *maximus* W. et
G. S. West Pol
Scenedesmus quadricauda var. *parvus* G. M.
Smith Pol
Scenedesmus houlensis Rayss Pol
Actinastrum hantzschii Lagerh. Pol
Crucigenia quadrata Morren Pol
Crucigenia triangularis Chodat Pol
Crucigenia tetrapedia (Kirchn.) W. et G. S.
West K
Crucigenia emarginata (W. et G. S. West)
Schmidle Pol
Crucigenia minima (Fitschen) Brunnth R
Crucigenia rectangularis (A. Br.) Gay Pol
Tetrastrum pulloideum Teiling Pol
Tetrastrum staurogeniforme (Schroder) Lemm.
Pol
Tetrastrum heteracanthum fa. *elegans* (Playf.)
Ahlstr. et Tiff. Pol
Micractinium pusillum Fresen Pol
Planctococcus sphaerocystiformis Korschik. Pol
Coccomyxa coccoides Rodhe and Skuja Pol
Elakatothrix gelatinosa Wille Pol

ZYGNEMAPHYCEAE

Desmidiales

Closterium aciculare T. West Pol
Closterium aciculare var. *subpronum* W. et G.
S. West Pol
Closterium acutum var. *variabile* (Lemm)
Krieger Pol
Closterium polystictum Nygaard Pol
Closterium pronum Breb. Pol
Closterium acerosum (Schrank) E. Pol

Closterium gracile Breb. Pol
Cosmarium laeve Rbh. Pol
Cosmarium sphagnicolum West et West Pol
Cosmarium granatum Breb. Pol
Staurastrum longipes (Nordst) Teiling Pol
Staurastrum paradoxum var. *parvum* W. West
Pol
Staurastrum muticum Breb. R
Staurastrum gracile Ralfs R
Staurastrum volans W. et W. Pol

PRASINOPHYTA

Platymonas cordiformis (Carter) Korschikow Pol

BACTERIOPHYTA

Chlorobacteriales

Chlorobium limicola Nadson Pol
Planctomyces bekefii Gimesi Pol

239

Bar =	Barrois 1894		R =	Rayss 1941, 1951
B =	Brun 1883		Y =	Yashuv 1961
K =	Kimor 1950		Pol =	Pollingher
P =	Petit 1883			

The most common algae in the plankton of Lake Kinneret

CYANOPHYTA

Chroococcales

Chroococcus limneticus Lemm.
Chroococcus minutus (Kütz) Naeg.
Chroococcus turgidus (Kütz) Naeg.
Microcystis aeruginosa Kutzing
Microcystis bothrys Teiling
Microcystis wesenbergii

Hormogonales

Anabaena planctonica Brunnthaler
Anabaena spiroides Klebahn

Pelonematales

Acroonema angustum Skuja
Acroonema lentum Skuja

BACILLARIOPHYTA

Centrales

Melosira granulata (Ehr) Ralfs
Cyclotella meneghiniana Kütz
Cyclotella stelligera Cl. U. Grun

Stephanodiscus hantzschii Grun

Pennales

Synedra spp.
Navicula spp.

CHRYSOPHYTA

Erkenia subaequiciliata Skuja

DINOPHYTA (PYRRHOPHYTA)

Glenodinium gymnodinium Penard
Peridinium cinctum fa. *westii* (Lemm.) Lef
Peridinium inconspicuum Lemm

Peridinium cunnigtonii (Lemm.) Lemm
Ceratium hirundinella (Bachm.)

CRYPTOPHYTA

Rhodomonas minuta Skuja

Cryptomonas spp.

CHLOROPHYTA

Volvocales

Collodictyon triciliatum Carter

Chroococcales

Pediastrum sturmii Reinsch
Pediastrum simplex (Meyen) Lemm.
Pediastrum clathratum (Schroter) Lemm.
Pediastrum duplex Meyen

Pediastrum boryanum (Turp.) Menegh.
Pediastrum tetras (Ehr.) Ralfs
Coelastrum microporum Naeg.
Coelastrum reticulatum (Dang.) Senn.
Botryococcus braunii Kuetzing
Chlorella vulgaris Beyerinck
Dictyosphaerium elegans Buchman
Oocystis lacustris Chodat
Oocystis submarina Lagerh.
Oocystis solitaria Wittrock

241

Chodatella citriformis Snow
Ankistrodesmus falcatus (Corda) Ralf
Tetraëdron minimum (Naeg.) Hansg.
Tetraëdron trigonum var. minor Reinsch.
Scenedesmus armatus Chod.

Scenedesmus ecornis (Ralfs) Chod
Scenedesmus quadricauda (Turp) Breb.
Crucigenia rectangularis (A.Br.) Gay
Elakatothrix gelatinosa Wille

ZYGNEMAPHYCEAE

Closterium aciculare var. subpronum W. et
    G. S. West

Cosmarium laeve Rbh

BACTERIOPHYTA

Chlorobium limicola Nadson

3 Annual pattern of algal succession

U. Pollingher

The following is based on data obtained from the routine weekly sampling which has been carried out from 1964 until the present and on available data from various sources for the period prior to 1964.

The prominent feature of the algal community in Lake Kinneret is the unique bloom of the dinoflagellate *Peridinium cinctum* fa. *westii* which dominates from February until May–June. Its decline is followed by the rise of Chlorophyta and Cyanophyta which form the bulk of algal biomass in summer and fall. Although Barrois did not identify the abundant algae that he noticed in the water in April 1893, the peak period of the present *Peridinium* bloom, it is likely that what he observed was a bloom of dinoflagellates. His mentioning *Ceratium hirundinella* confirms this hypothesis. Barrois' observation indicates that the great abundance of algae in spring is far from being a new feature of Lake Kinneret. This information is of considerable importance in evaluating correctly the present trophic status of the lake.

This general seasonal pattern of succession suffered a few exceptions: in 1948, Kimor observed a winter bloom of *Fragilaria crotonensis*; in 1964 and to a lesser extent in 1971, a bloom of *Microcystis* spp. delayed the development of *Peridinium*. It is worth noting that the biomass values of winter 1964 were one order of magnitude lower than in a '*Peridinium* year'.

During the period of continuous record, besides the previously mentioned exceptions, the succession of algae followed the general pattern: winter – *Peridinium* bloom and summer – Chlorophyta and Cyanophyta; however, certain interesting modifications were observed from 1973 onwards. Therefore, both periods will be described separately.

a. The period prior to 1973 (Fig. 87)

With the exception of 1970, very regular sequences were observed. The period January–February was characterized by a peak of diatoms, mainly *Melosira* spp., and green algae, mainly *Pediastrum* spp., *Coelastrum* spp. and *Closterium aciculare* var. *subpronum*. We note that these algae developed during the mixed period; it is certainly significant that the heavy filaments of *Melosira* spp. appeared in the water only during the time of maximum turbulence of the lake water. This peak declined with the end of the turbulent period. Then, the more quiescent conditions favored the development of *Peridinium* (see Section B, where the development and characteristics of the *Peridinium* bloom are described). In May–June, the *Peridinium* declined and the biomass of the three other groups increased. Among the Chrysophyta, *Cyclotella* and *Stephanodiscus* spp. appeared in large amounts. The Cyanophyta were dominated by *Microcystis* spp. In the Chlorophyta group, *Oocystis*, *Tetraëdron*, *Scenedesmus*, *Pediastrum* and

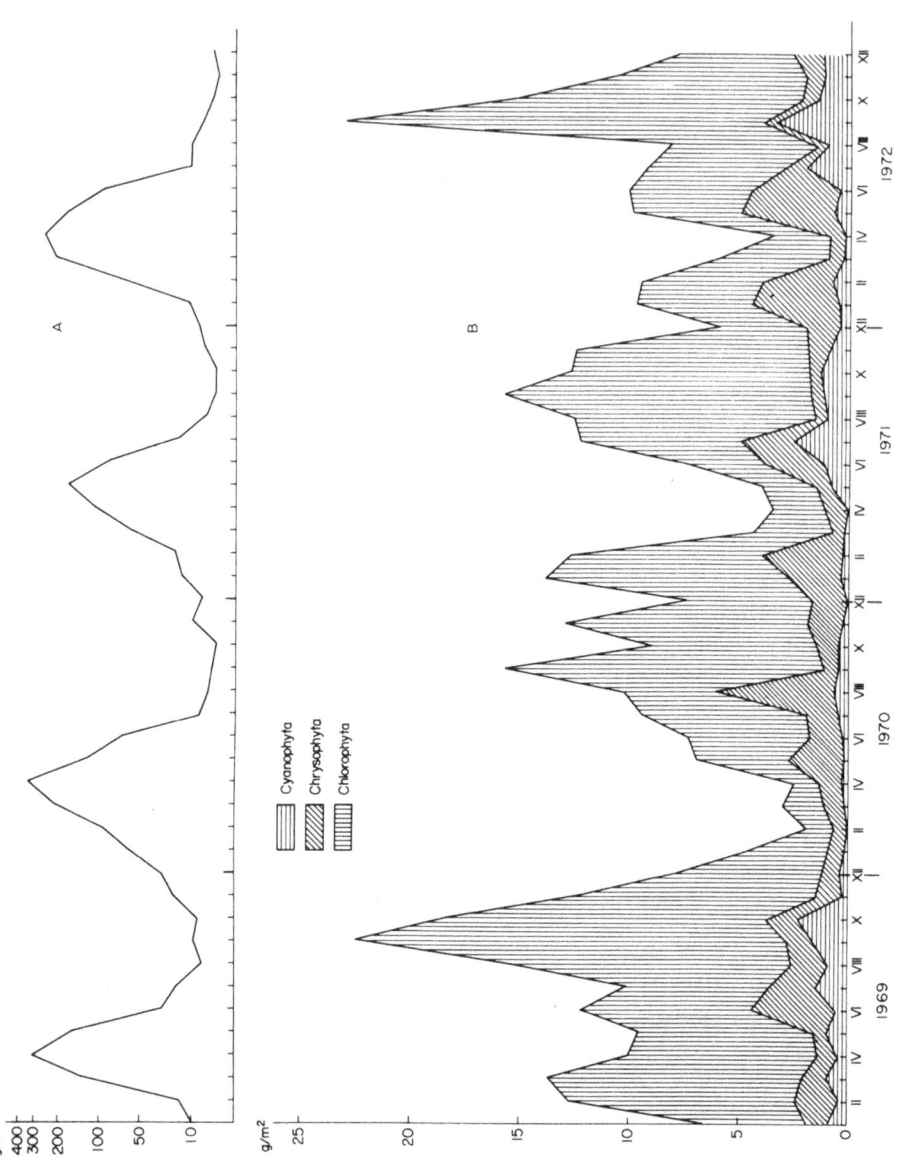

Coelastrum spp. were the dominant genera. In autumn, the large number of *Cosmarium* generated a second and modest biomass peak.

b. The period 1973–1975 (Fig. 88)

The winter peaks of diatoms and Chlorophyta appeared according to the same pattern as in the previous period, but their quantitative importance increased. Moreover, these peaks were accompanied by a noticeable increase of the

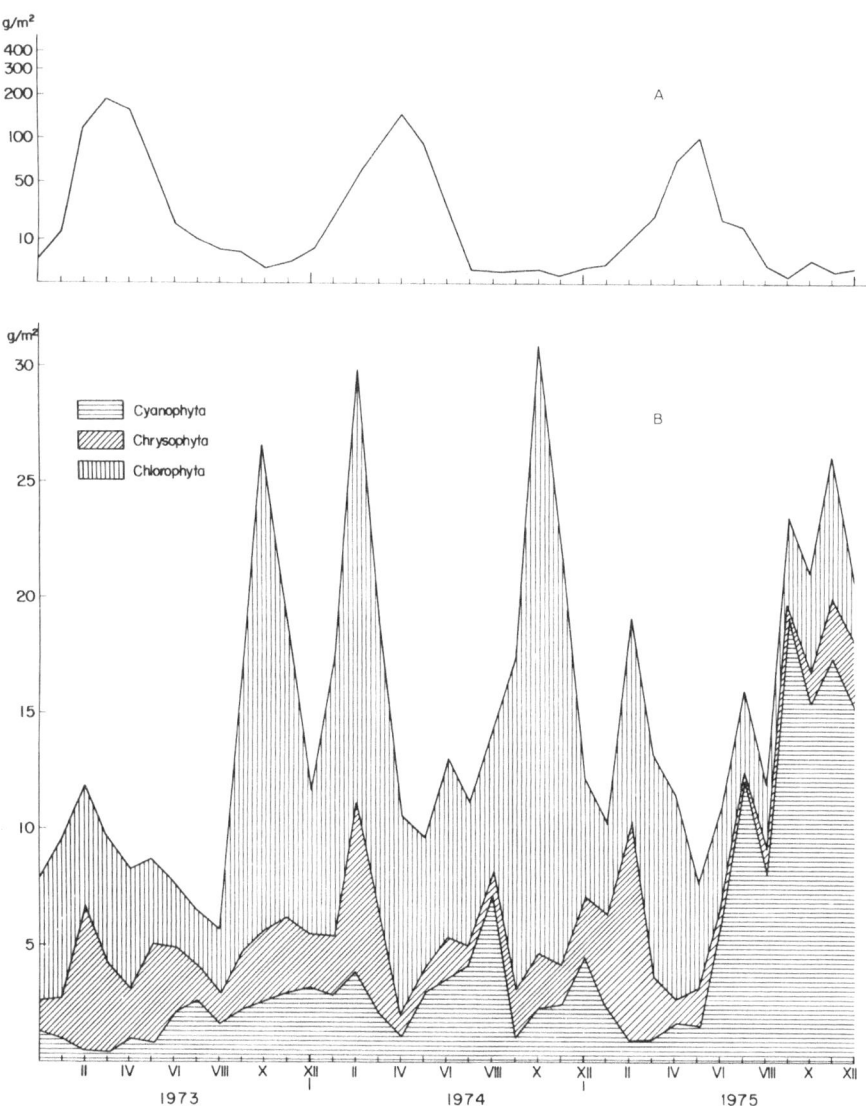

Fig. 88. Fluctuation of the different algal groups from 1973 to 1975. A = Pyrrhophyta (semilogarithmic scale), B = other groups. (modified from Pollingher U. and Berman T., 1977, Oikos 29).

245

Cyanophyta *Microcystis*. The *Peridinium* bloom was in general delayed, and its maximum biomass value declined gradually from 1973 to 1975. In summer, we observed a concomitant increase of Chlorophyta and Cyanophyta biomass in 1973 and 1974. In 1975, the blue–green *Chroococcus* spp. constituted almost the whole algal biomass and Chlorophyta remained at very low levels. One should not interpret these data as a trend towards a dominance of Cyanophyta, since the unusual 1975 bloom of *Chroococcus* did not occur in 1976.

4   Biomass and chlorophyll

U. Pollingher & T. Berman

a. Range of values of total biomass and contribution of main groups

During the period 1969–1975, the extreme values of fresh weight (fw) biomass were 12 g m^{-2} in December 1972 and 341 g m^{-2} in April 1970, with corresponding values of chlorophyll of 53 and 1,000 mg m^{-2}.

The annual averages reported in Table 32 show clearly the extreme dominance of the Pyrrhophyta which contribute up to 88.5% of the total biomass. The modifications which have occurred since summer 1973 diminished the biomass values of the Pyrrhophyta by 50% in 1974 and 65% in 1975; consequently, the contribution of this group to the total biomass dropped from 88.5% in 1970 to 59% in 1975. In contrast, the Cyanophyta, which represented no more than 1% of the total biomass, contributed 20.5% in 1975.

Table 32.   Annual averages of phytoplankton biomass (g m^{-2}) and chlorophyll (mg m^{-2}) in the trophogenic layer. Percentage contribution of every algal group to total algal biomass given in brackets.

Year	Total algal biomass	Chloro-phyll mg m^{-2}	Pyrrho-phyta g m^{-2}	Chloro-phyta g m^{-2}	Cyano-phyta g m^{-2}	Chryso-phyta g m^{-2}	Bacterio-phyta g m^{-2}
1969	87	—	70 (80.0)	10 (11.0)	1.0 (1.2)	2.0 (2.3)	4.0 (5.0)
1970	87	287	77 (88.5)	6 (7.0)	0.3 —	2.0 (2.3)	2.0 (2.3)
1971	54	217	43 (79.6)	7 (13.7)	1.0 (2.0)	1.3 (2.2)	1.2 (2.2)
1972	80	307	69 (86.2)	7 (8.8)	1.0 (1.2)	2.0 (2.5)	1.0 (1.2)
1973	63	319	51 (80.9)	17 (11.0)	2.0 (3.2)	3.0 (5.0)	0.3 —
1974	49	331	30 (61.0)	13 (27.0)	3.0 (6.0)	2.0 (4.0)	1.0 (2.0)
1975	39	201	23 (59.0)	6 (15.4)	8.0 (20.5)	2.0 (5.1)	0.2 —

The concentration of chlorophyll did not vary with the fw biomass. The highest values of fw biomass were accompanied by relatively low levels of chlorophyll; this is due to the dominance of *Peridinium*, which is known to have a low specific content of chlorophyll. The increase of chlorophyll which occurred simultaneously with the decrease of total biomass is accounted for by the replacement of *Peridinium* by species having a higher specific content of chlorophyll.

b. Carbon:Chlorophyll ratios

The carbon to chlorophyll ratios found in the Kinneret reflect the seasonal changes of phytoplankton composition. Low values of 17–34:1 (for the trophogenic zone) and 9:1 (at best depth) were observed before and after the bloom; higher values of 72–134:1 (trophogenic layer) and 186:1 (at best depth) were found during the *Peridinium* season. For both marine and freshwater algae, this ratio usually varies from about 30 to 90, in progression from eutrophic to oligotrophic waters. High carbon:chlorophyll ratios seem to be typical of nutrient-depleted cells which grow slowly (Eppley, 1972), and ratios of 150:1 have been reported for nitrogen-limited chemostat cultures of marine algae (Thomas & Dodson, 1972). The high values measured in Lake Kinneret during the bloom are certainly due to some extent to the heavy polyglucan theca of the *Peridinium* (Nevo & Sharon, 1969).

During the years 1972 through 1975, the C:chlorophyll ratio decreased, reflecting the qualitative changes which had occurred in the phytoplankton.

c. Monthly variations of biomass

In Fig. 89, we have represented the monthly variation of the fw biomass and the chlorophyll concentration for the year 1972, a typical *Peridinium* year, and for the three following years. We note that the March–April peak decreased gradually from 250 g m^{-2} in 1972 to 100 g m^{-2} in 1975. The autumn peak which

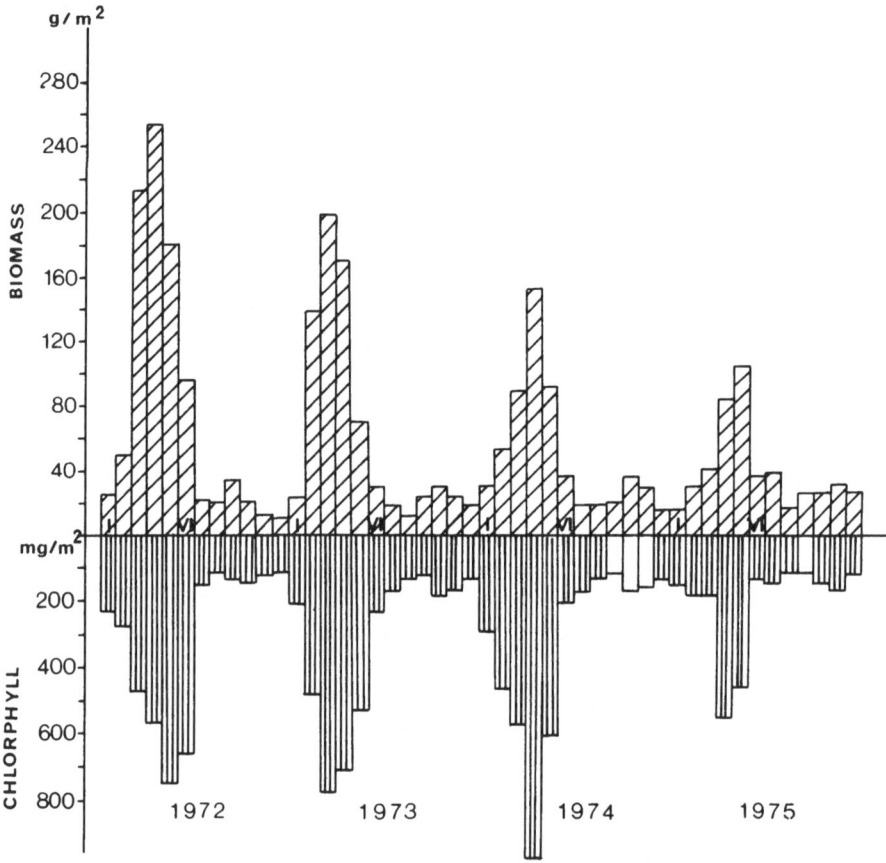

Fig. 89. Monthly averaged values of algal biomass (upper graph) and chlorophyll (lower graph). The white columns in lower graph correspond to calculated values of chlorophyll. (from Pollingher U. and Berman T., 1977, Oikos, 29).

generally occurred in September until 1973 shifted to October in the following years and showed a slight increase.

The peak of chlorophyll in April 1974 was remarkable and was mainly due to the fact that the *Peridinium* cells were of much smaller size and had an unusually high chlorophyll content.

The contribution of each group to the monthly averages of fw biomass is represented in Fig. 90.

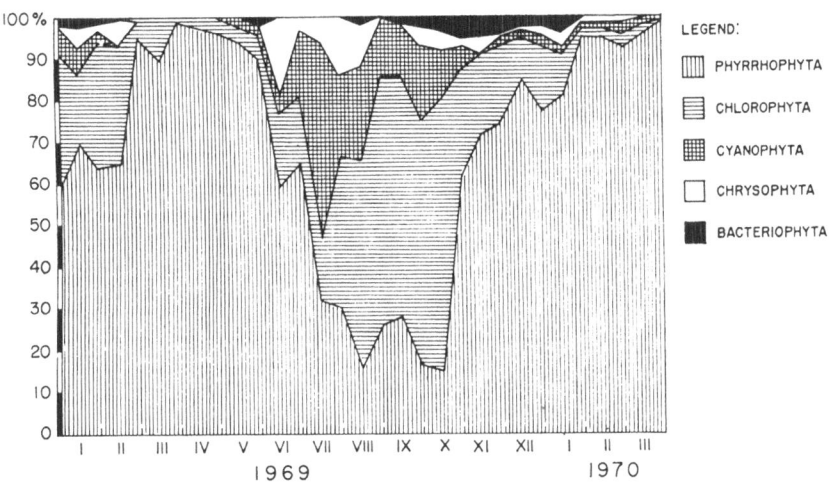

Fig. 90. Percentage distribution of the phytoplankton groups in the epilimnion. Results stated in percentages of total phytobiomass. (From Serruya, C. & Pollingher, U., 1971, Mitt. Internat. Verein. Limnol., 19.)

d. Horizontal distribution

The algae in Lake Kinneret are distributed rather homogeneously in winter, summer and fall (Table 33). Conversely, the spring values may vary from station to station: this is mainly due to the patchiness of *Peridinium*, discussed in Section B. (page 292).

Table 33. Spatial distribution of phytoplankton in 1970. Numbers represent average annual biomass at four stations in g m^{-2} wet weight

Stations →	G	A	C	D
Winter	44.6	50.3	24.7	40.1
Spring	325.1	195.4	135.5	112.6
Summer	64.6	47.0	42.7	40.9
Fall	17.6	26.4	22.1	23.6

Station G generally has higher biomass values than other stations. Among other reasons, this is caused by the proximity of the Jordan River and its rich input of algae and nutrients.

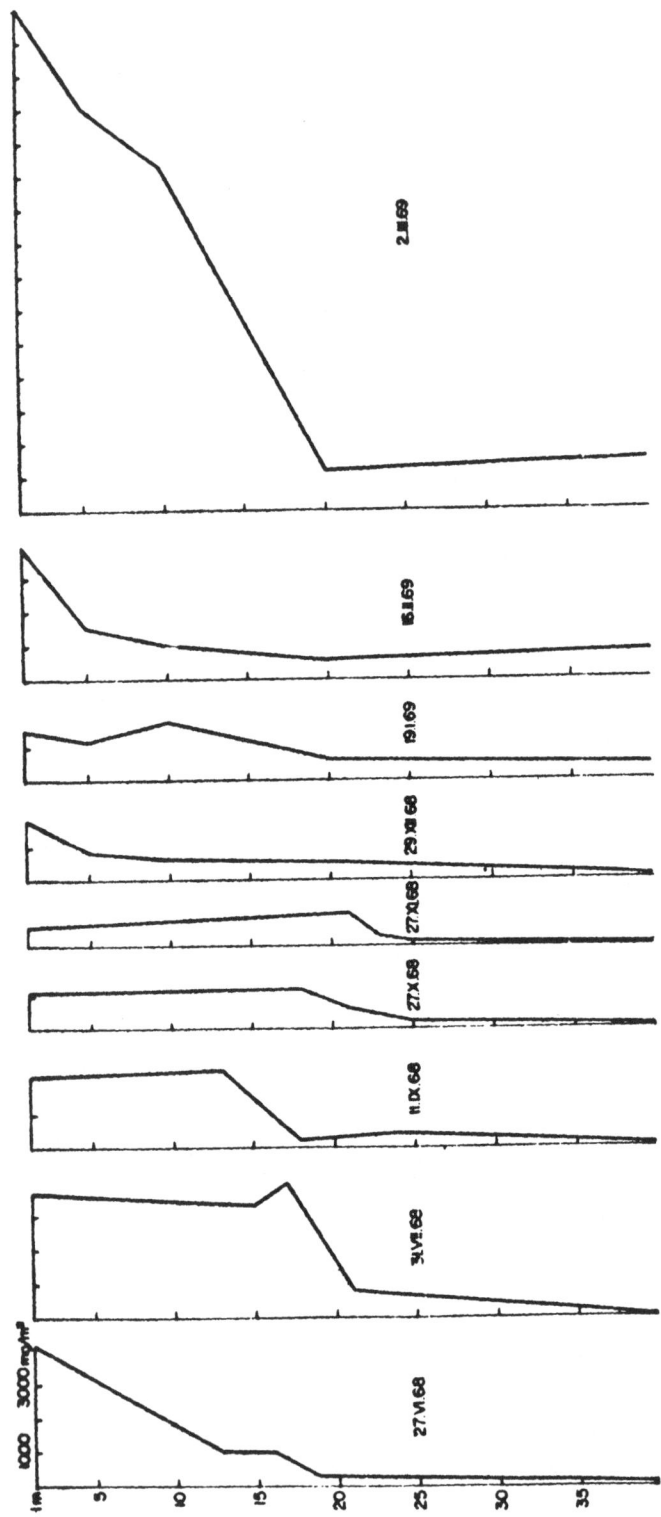

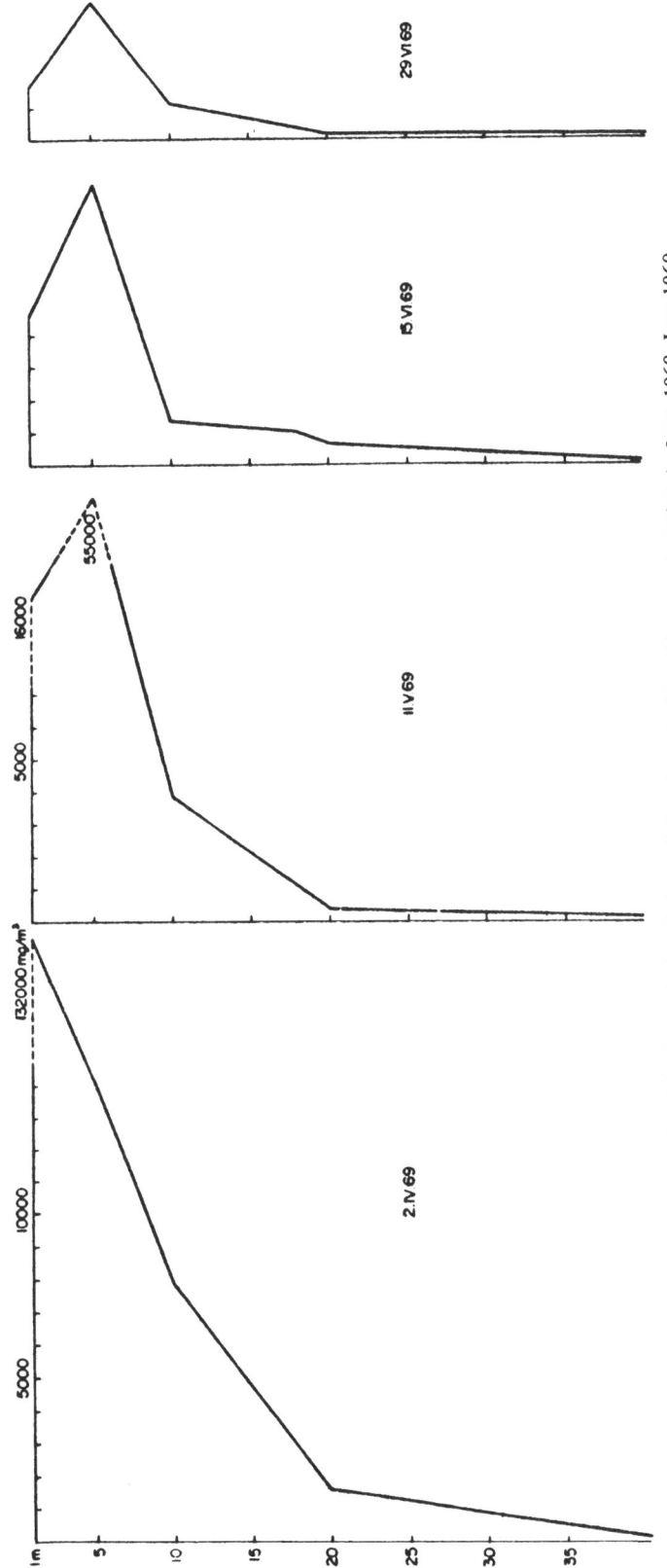

Fig. 91. Seasonal and bathymetrical variations of the phytoplankton biomass at station A. June 1968–June 1969.

Berman & Elias (1973) measured chlorophyll during one full year at seven stations in the lake, and estimated the total amount of chlorophyll in the lake by attributing a certain volume of water to each station on the basis of morphological and hydrodynamic data. This technique allowed the estimation of an "average lake profile" to which were compared the profiles obtained in each particular station (Table 34).

Table 34. Comparison between average concentration of chlorophyll at each station and average lake profile. Concentrations at each station are stated as percentage of the lake profile value. Standard deviation given in brackets

Station	I 12 months (Dec. 1971– Nov. 1972)	II Bloom period (Mar.–June 1972)	III Non-bloom
A	91 (18)	75 (17)	101 (13)
B	100 (33)	110 (39)	99 (34)
C	126 (67)	161 (122)	112 (17)
D	96 (72)	110 (127)	90 (38)
E	104 (62)	74 (53)	96 (35)
F	93 (29)	104 (35)	88 (26)
G	105 (36)	119 (18)	102 (41)

e. Vertical distribution (Fig. 91)

The bathymetric distribution of the algae is closely related to the formation of the thermocline. In December–January, during homothermy, the algae are homogeneously distributed throughout the water column. The stratification of the phytoplankton begins with the appearance of *Peridinium* in February, preceding the thermal stratification of the lake which occurs in April–May. In February, March, April and May, the majority of algal cells are found between 0 and 5 meters when sampled during the morning hours. However, large differences in cell numbers result from the active vertical migrations of *Peridinium* and also from dramatic variations in horizontal patchiness (Berman & Rodhe, 1971).

From June, at the end of the *Peridinium* bloom, until December, the algae are fairly evenly distributed throughout the epilimnion. Maxima are usually found between 0–5 m and a smaller peak is occasionally noted in the metalimnic layer. Dense populations of photosynthetic sulfur bacteria, copepods and algae are found in this region (Pollingher, 1972; Serruya, 1972). During the stratification, the hypolimnion is anoxic and devoid of algae.

5. Primary productivity and photosynthetic efficiency of light utilization by phytoplankton

T. Berman

Yashuv & Alhunis (1961) made the first measurements of primary production in Lake Kinneret in 1956–1957, at a small, shallow bay station near Degania at the southern tip of the lake. A more extensive study was made during the years 1965 and 1966 by Hepher & Langer (1970) who determined the rates of photosynthesis and respiration and also chlorophyll concentrations at several lake stations. Rodhe (1972) introduced the ^{14}C method of Steemann–Nielsen in 1968 and initiated a series of weekly measurements, beginning in 1969. This work was continued, with some modifications in methodology, first by C. Serruya, and from 1971 by Berman & Pollingher, who have summarized their results in several papers (Berman & Pollingher, 1974; Pollingher & Berman, 1975, 1977). Most of their results were derived from a central lake station (A) which proved to be quite representative, in terms of chlorophyll and primary productivity, for the pelagic waters of Lake Kinneret (Berman & Elias, 1973; Pollingher & Berman, 1975).

a. Range of photosynthetic carbon fixation rates

During the period 1969 to 1972, the highest assimilation rates were 1,460 mg C m^{-3} d^{-1} (April 1970) and 1,187 mg C m^{-3} d^{-1} (May 1972). These values corresponded to extreme peaks of *Peridinium* with concentrations of chlorophyll of 151 and 148 mg m^{-3} respectively. Although there was a marked decrease in the biomass of phytoplankton from 1973 through 1976, the rates of primary production and chlorophyll concentrations did not fall correspondingly. Pollingher & Berman (1977) attributed this to the partial replacement of large, relatively slowly metabolizing *Peridinium* cells by more active smaller algae with higher intracellular chlorophyll levels. The greatest rates of photosynthetic carbon fixation measured during the period 1973 to 1975 were 2,589 mg C m^{-3} d^{-1} (April 1973) and 2,587 mg C m^{-3} d^{-1} (April 1974) when chlorophyll concentrations were 230 and 274 mg m^{-3} respectively.

The areal photosynthetic rate, based on integrated values for six to eight depths in the trophogenic layer and averaged from 1969 through 1975, was 1,739 mg C m^{-2} d^{-1}. Monthly averages for this period ranged from 968 mg C m^{-2} d^{-1} (December) to 2,419 mg C m^{-2} d^{-1} (April). Monthly averages of chlorophyll concentrations in the trophogenic layer varied from 108 mg m^{-2} (September) to 628 mg m^{-2} (April). The annual and semi-annual averages for areal primary productivity from 1969 to 1975 are shown in Table 35. These figures imply an average annual carbon fixation of 635 g C m^{-2} yr^{-1} by phytoplankton in Lake Kinneret, which is similar to that calculated by Lewis (1974) for Lake Lanao and Lake Victoria.

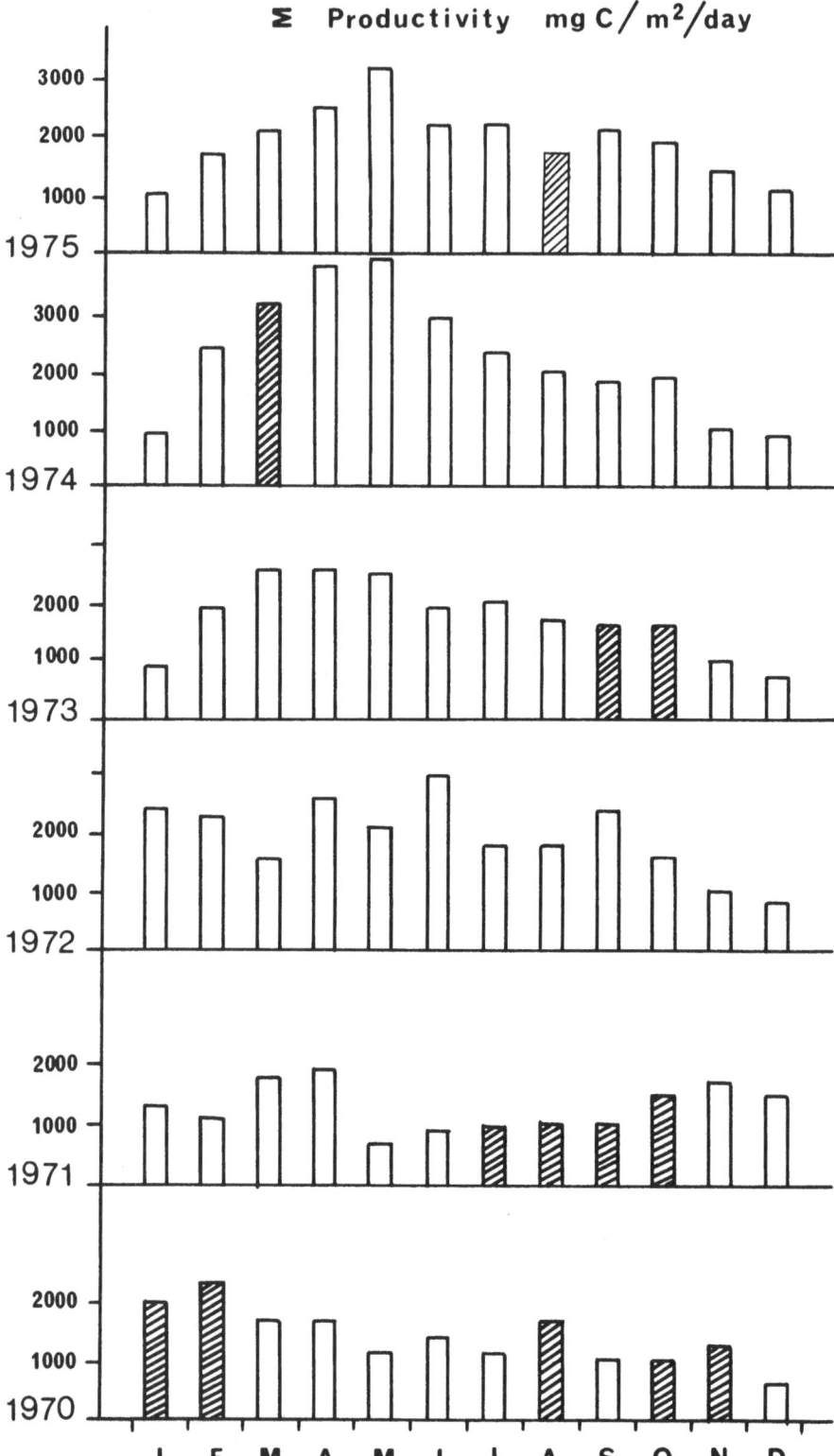

Fig. 92. Seasonal variations in primary productivity, averaged monthly values for the period 1970–1975 (from Pollingher U. & Berman T., 1977, Oikos, 29).

Table 35. Semi-annual and annual averages of primary productivity. Results in mg C m^{-2} d^{-1}

	1969	1970	1971	1972	1973	1974	1975
Jan–Jun	2036*	1503	1344	2307	2123	2900	2386
Jul–Dec	797	937†	837‡	1612	1459	1722	1798
Annual	1292	1260	1175	1960	1791	2311	2091

Missing months: * Jan–Feb
                 † Aug–Oct–Nov
                 ‡ Sept–Oct

b. Seasonal variations of primary production and chlorophyll

Typical yearly patterns of primary production are given in Fig. 92. From January to April–May, there is a progressive increase from about 900 to 2,500–4,000 mg C m^{-2} d^{-1}, at the peak of the *Peridinium* bloom. However, at this time, assimilation numbers, a measure of photosynthetic efficiency, are at their minimum (about 0.4 mg C mg Chlorophyll^{-1} h^{-1} measured at the optimum depth).

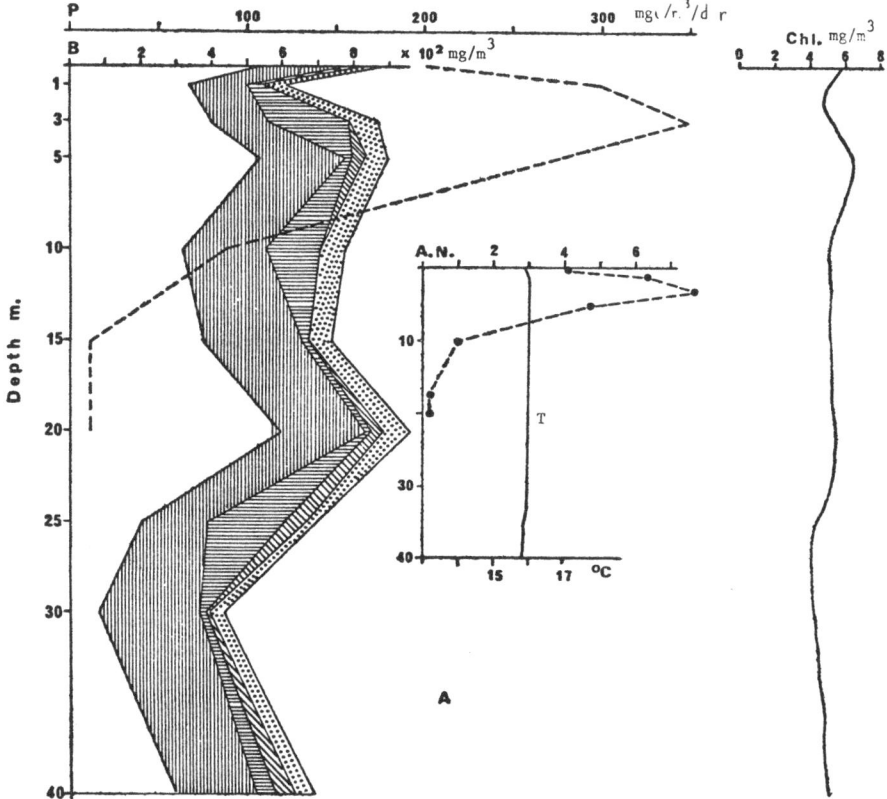

Fig. 93. Profiles of phytoplankton biomass, primary productivity and chlorophyll. A.N. = assimilation numbers (mg C/m^3/hr : mg chlorophyll/m^3). 5 January 1972. (From Berman, T. & Pollingher, U., 1974, Limnol. Oceanogr., 19.)

From June to August, primary production decreases to 800–1,200 mg C m^{-2} d^{-1}. This level is maintained until the following January. Assimilation numbers during the fall reach their maximum (about 5 mg C mg Chlorophyll^{-1} h^{-1}).

Typical seasonal vertical profiles of primary production and chlorophyll are shown in Figs. 93 to 96. In January (Fig. 93), significant photosynthetic activity is found down to 15–20 m and the metabolic state of the phytoplankton is characterized by relatively high assimilation numbers. In March (Fig. 94), when *Peridinium* constitutes 98% of the algal biomass, high carbon fixation rates are restricted by self-shading to a shallow layer (1 to 3 m) with little photosynthesis occurring below 6 m. However, it should be emphasized that the picture presented in Fig. 94 is somewhat artificial and reflects only the vertical distribution of algal biomass in mid-morning. Vertical migration of the dinoflagellates will modify this situation throughout the day (Berman & Rodhe, 1971), and therefore primary productivity values at the bloom season may be somewhat underestimated.

In August (Fig. 95), the depth of maximum photosynthesis is located from 2 to

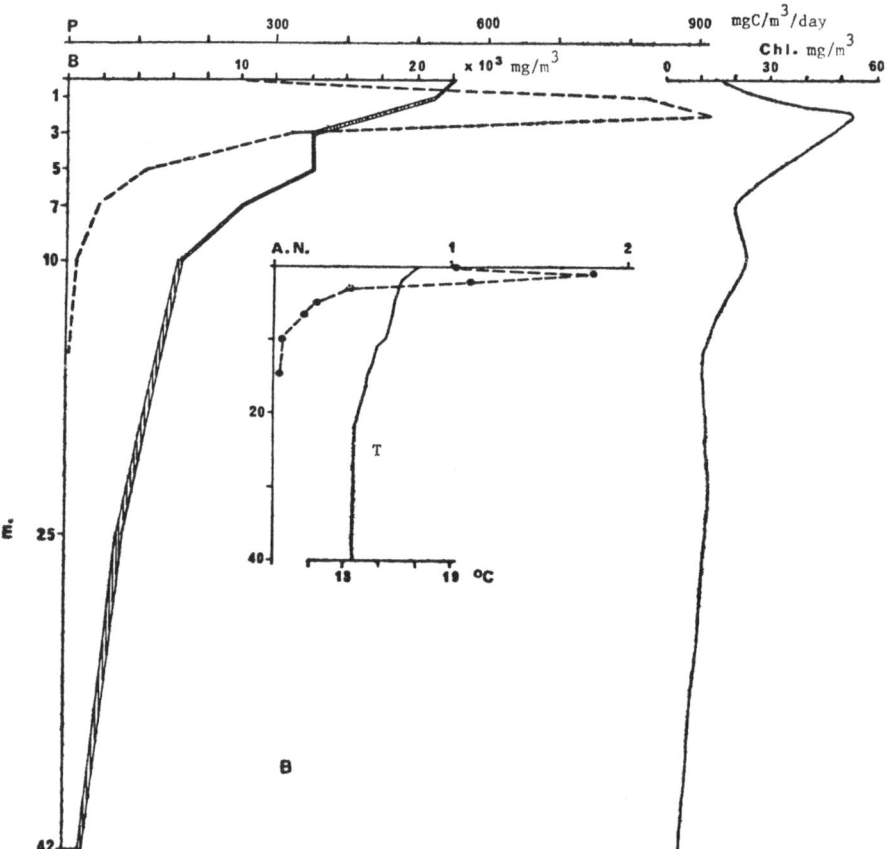

Fig. 94. Profiles of phytoplankton biomass, primary productivity and chlorophyll. A.N. = assimilation numbers (mg C/m³/hr : mg chlorophyll/m³). 23 March 1972. (Same source as Fig. 93.)

256

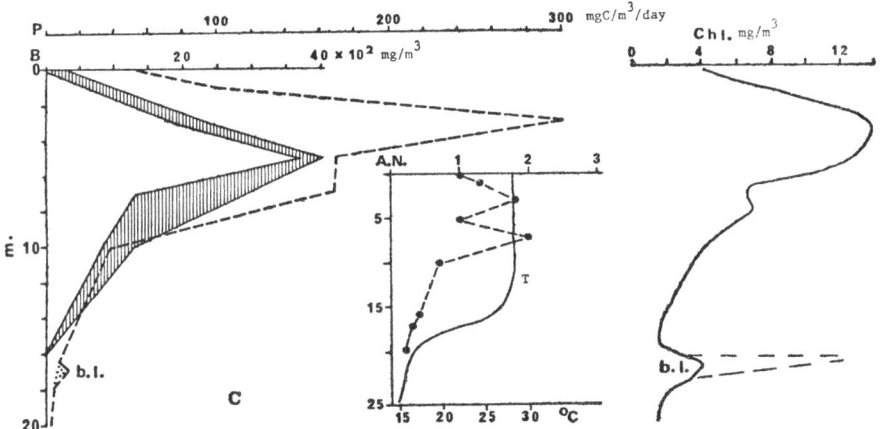

Fig. 95. Profiles of phytoplankton biomass, primary production and chlorophyll. A.N. = assimilation number (mg C/m³/hr : mg chlorophyll/m³). b.l. = bacterial layer. 15 August 1972. (Same source as Fig. 93.)

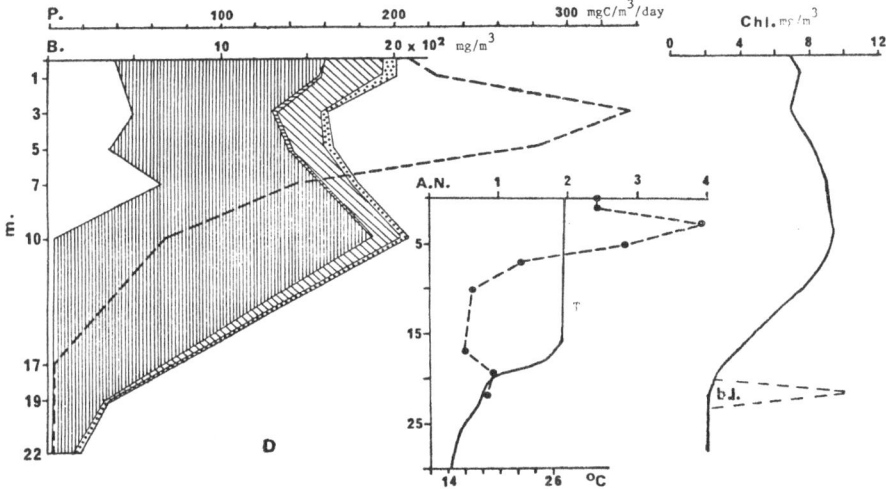

Fig. 96. Profiles of phytoplankton biomass, primary production and chlorophyll. A.N. = assimilation number (mg C/m³/hr : mg chlorophyll/m³). b.l. = bacterial layer. 12 September 1972. (Same source as Fig. 94.)

6 m. At the metalimnion, there is a small peak of light-stimulated carbon fixation due to a layer of photosynthetic green sulphur bacteria, *Chlorobium phaeobacteroides*. In autumn (Fig. 96), Chlorophyta generally dominate the phytoplankton biomass, and some increase in both primary productivity and assimilation number is observed.

c. Release of organic matter by photosynthesizing algae in Lake Kinneret

Berman (1976) found a direct relationship between the rates of photosynthetic carbon fixation and the release of dissolved organic carbon (DOC) by

257

phytoplankton. Rates of DOC release varied from 6.8 mg C m^{-2} d^{-1} to 211 mg C m^{-2} d^{-1}. The percentage of extracellularly released carbon to fixed carbon (PER), based on integrated measurements for the euphotic zone, averaged 3.7% and was lowest (mean 1.9%) during the exponential phase of *Peridinium* growth. It appears, therefore, that this flux of carbon is not a major pathway for heterotrophic nutrition of the lake microflora.

d. Efficiency of light utilization by phytoplankton

Dubinsky & Berman, using an ISCO Spectroradiometer specially adapted for underwater, determined the energy and spectral characteristics of downwelling light at the central station A during the period July 1973 to June 1974. Concomitant measurements of photosynthesis and chlorophyll permitted these workers to estimate the ability of the phytoplankton to use photosynthetically available radiation (Ph. A.R.) to assimilate carbon (Dubinsky & Berman, 1976). The overall efficiency of light utilization by phytoplankton for the integrated water column (calculated as the percentage of daily areal photosynthetic carbon fixation to total incident subsurface light) ranged from 0.34% (August 1973) to 4.01% (April 1974) when chlorophyll concentrations in the euphotic layer were 50 and 580 mg m^{-2} respectively. These values are comparable to those reported by Talling *et al.* (1973) for highly productive Ethiopian lakes (0.51 to 3.34%). Photosynthetic efficiencies, when measured at discrete points in the water column, increased with depth through the euphotic zone. Maximum photosynthetic efficiencies (carbon fixation to the Ph.A.R. attenuated in a water layer of 1 m thickness) were 10 to 12%; when expressed as quantum efficiencies (the percentage of carbon fixation to Ph.A.R. absorbed by algal chlorophyll only), these values were 25 to 30%, which is close to the maximum quantum efficiencies for laboratory cultures reported by Kok (1960).

Seasonal changes of light spectral quality were found by Dubinsky & Berman (in preparation). The ratios of quanta per joule (Q : W) varied from 2.57 × 10^{18} to 2.96 × 10^{18}, reflecting spectral changes which were attributed mainly to seasonal variations in pigment composition of the phytoplankton. The specific attenuation coefficients for chlorophyll and water were calculated as 0.0067 ln units mg^{-1} m^{-1} and 0.49 ln units mg^{-1} m^{-1} respectively.

B. *Peridinium cinctum* fa. *westii* (Lemm.) Lef.

1 Morphology and systematics

U. Pollingher

Peridinium cinctum fa. *westii* (Lemm.) Lef. is a unicellular organism with a sub-sphaerical form. Its size in Lake Kinneret varies from 42 to 66 μ (longitudinal diameter) and from 36 to 60 μ (transversal diameter).

The body of *Peridinium* is surrounded by a theca which is made of a definite number of polygonal and reticulated plates of unequal sizes and which consists mainly of a non-cellulosic glucan (Nevo & Sharon, 1969).

The theca belongs to the Cleistoperidinoid type (without an apical pore). A transversal furrow, cingulum, divides the theca into two parts: epitheca (or epicone or epivalva) and hypotheca (or hypocone or hypovalva). A longitudinal furrow or sulcus (Plate 16) is located in the central area of the hypotheca.

The epitheca is composed of 14 plates with plate formula 4′3a7″ and the hypotheca consists of 7 plates with formula 5″′2″″; the cingulum includes 5 cingular plates and the sulcus 1 to 4 plates (described by Bourrelly, 1968, 1970 and Boltovskoy, 1975). The sulcal plates are thinner than those of the rest of the theca and have much less marked ridges (Dodge & Crawford, 1970).

Between the thecal plates, there are intercalary bands which are narrow in young organisms; in aging cells they are wider and have a striated appearance. These spaces allow for a certain expansion and contraction of the cytoplasm. The mechanism of plate construction is unknown. The surface features of *Peridinium cinctum* have been studied by scanning electron microscope by Dodge who obtained further information about the sulcal plates and flagella pores (Dodge, 1971).

Peridinium has two flagella, a longitudinal one located in the sulcus and a transversal one located in the cingulum (plates 17, 18, 19, 20, 21) (Berman T. & I. L. Roth). The longitudinal flagellum imparts a strong forward impulse in movement and the transversal flagellum imparts a rotational motion to the cell. *Peridinium* has a brown or sometimes a slightly greenish colour; it has a single central nucleus of large size with a great number of chromosomes in a condensed state during all its life.

The fine structure of *Peridinium cinctum* fa. *westii* from Lake Kinneret was studied by Messer & Ben-Shaul (1969, 1970, 1971). The thecal structure of *Peridinium cinctum* from India was studied earlier by Venkataraman & Mehta (1961), the structure of *Peridinium cinctum* cultivated at Birkbeck College was studied by Dodge & Crawford (1970) and Dodge (1971), and the structure of *Peridinium cinctum* from different water bodies from Argentine was studied by Boltovskoy (1975).

The basic structure of the theca consists of a single outer membrane (plasma membrane) surrounding the cell, beneath which lies a single layer of flattened vesicles containing the wall plates. The plates are 0.6–1.2 μ thick and each plate is perforated by one or two round or oval porelike openings, 0.2–0.3 μ in diameter (Messer & Ben-Shaul, 1969). The pores allow the discharge of trichocysts (Venkataraman & Mehta, 1961). Subthecal microtubules are present below the theca.

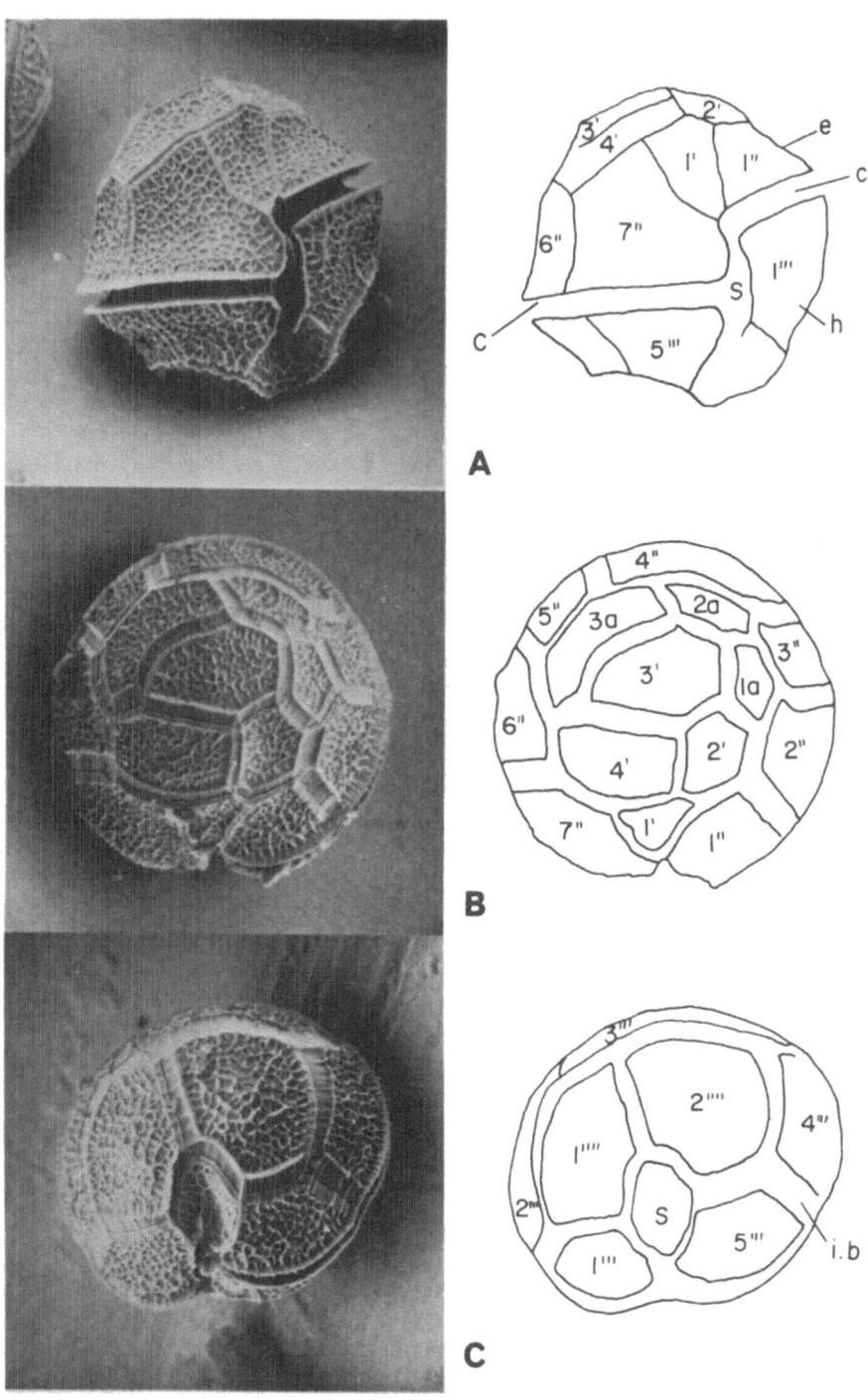

Plate 16. *Peridinium cinctum* fa. *westii*. A. Ventral view. Numbers refer to thecal plates (see text). e = epitheca, h = hypotheca, c =cingulum, s = sulcus, i.b. = intercalar band. B. Apical view of the epitheca. C. Antiapical view of the hypotheca. (Courtesy of Dr. F. E. Round).

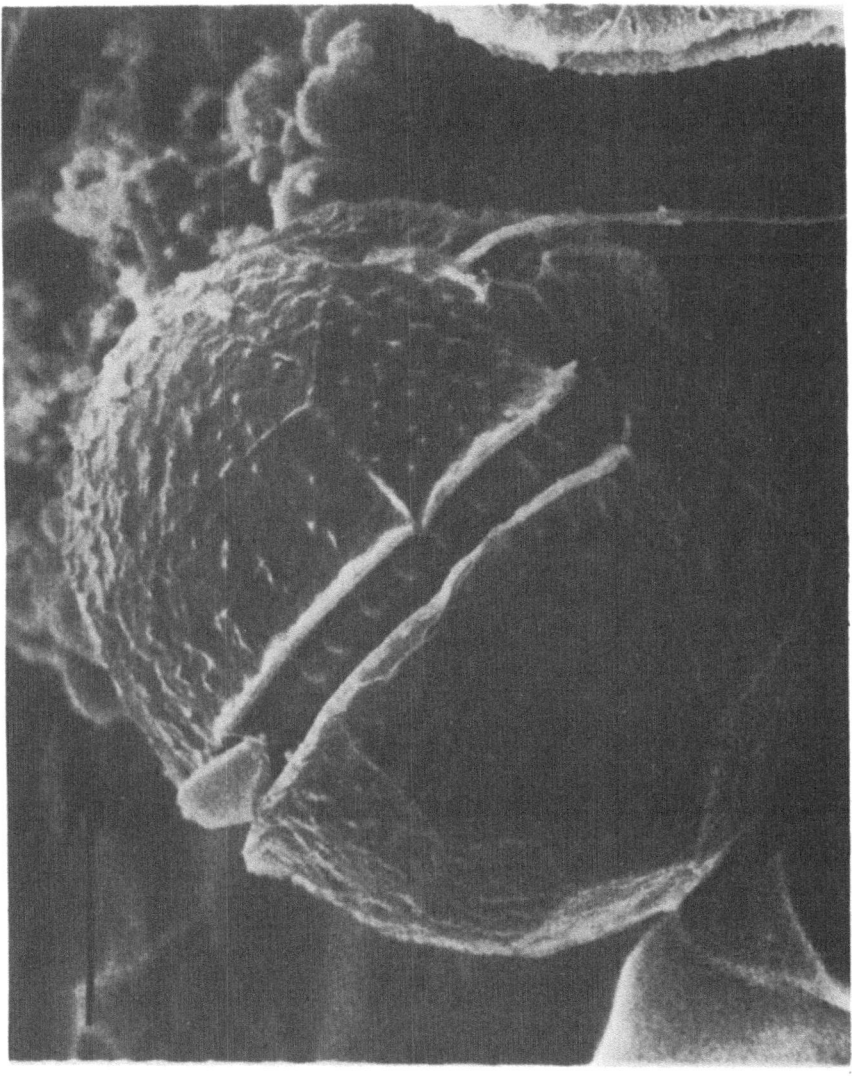

Plate 17. General view of *Peridinium* cell showing outer membrane, transverse and longitudinal flagella. (Live sample fixed with Perducz fixative, 1% Os Oy and Saturated HgCl$_2$ (6:1). Note difference in appearance to specimens in which the outer membrane is removed in preparation). Scalar bar 10 μm. By courtesy of Drs. T. Berman and I. L. Roth.

The cytoplasm is surrounded by three membrane layers. The chloroplasts, in large number and of varying size, are usually situated at the cell periphery. Each chloroplast contains numerous lamellae composed of three thylakoids, which run the whole length of the organelle. The nucleus contains a central nucleolus and numerous coil-like chromosomes. The nuclear membrane is often continuous with the surrounding cytoplasmic vesicles. A region of well-developed Golgi vesicles is situated near the nucleus. A number of densely-stained bodies were found in Golgi regions. Their relation to lysosomes is suggested by Messer & Ben-Shaul (1969);

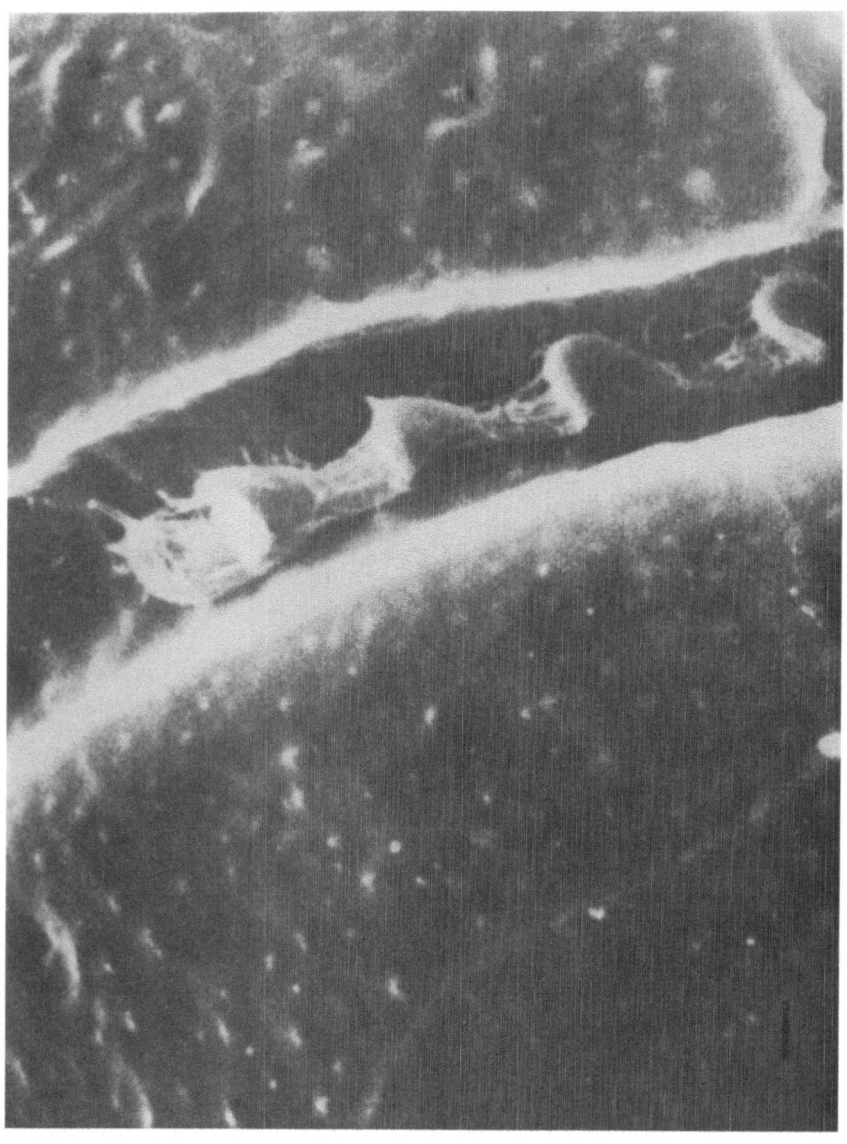

Plate 18. View of transverse flagellum tip. Scalar bar 1 μm. By courtesy of Drs. T. Berman and I. L. Roth.

Dodge (1971) does not exclude the possibility that these bodies are in fact small phagotrophic vesicles, because they look similar to such structures which have been noted in *Ceratium hirundinella* and other dinoflagellates (Dodge & Crawford, 1970).

Ribosomes are present within the chloroplasts. Pyrenoid-like structures are observed both embedded within chloroplasts and as protrusions from chloroplasts. Starch-like storage grains lie around a central dense granular pyrenoid. Oil droplets are frequently found. Eyespot granules are present in an organelle which resembles a chloroplast, and are located in the cell periphery.

264

About 10–12 eyespot granules were observed. They are spherical or hexagonal and vary in staining intensity. Numerous trichocysts are present in the peripheral region of the cell at or near the cell surface. The detailed structure of the trichocyst fibrils was studied by Messer & Ben-Shaul (1971). The same authors did not describe the fine structure of *Peridinium*'s flagella. The study of the complicated structure of dinoflagellate flagella was made by Leadbetter & Dodge (1967) and Dodge (1971).

Studies of aging cells of *Peridinium cinctum* fa. *westii* (Messer & Ben-Shaul, 1970) have shown the presence of degenerated chloroplasts and of large amounts

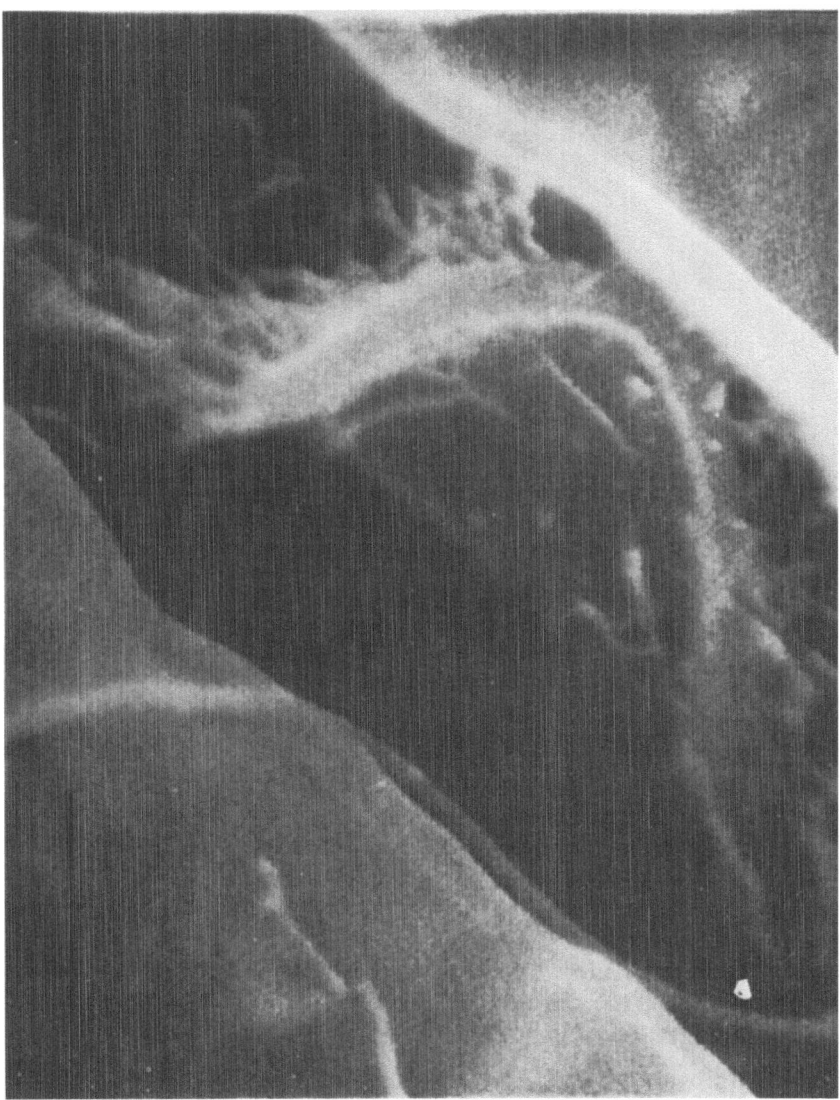

Plate 19. Detail of portion of transverse flagellum. Scalar bar 1 μm. By courtesy of Drs. T. Berman and I. L. Roth.

265

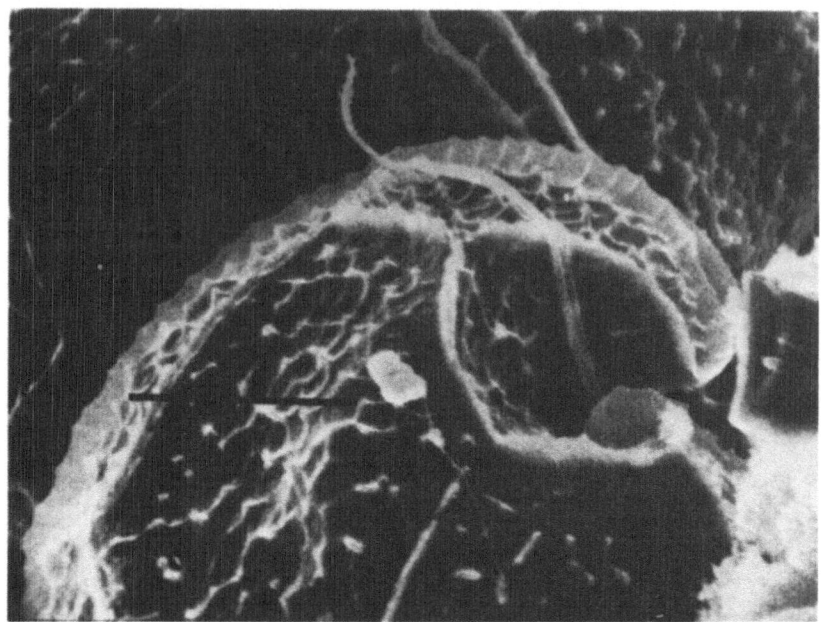

Plate 20. Longitudinal flagellum. Note apparent double stranded structure. Scalar bar 10 μm. By courtesy of Drs. T. Berman and I. L. Roth.

of lipid material and collagen-like rods. The collagen-like rods are certainly trichocysts that have discharged internally. Similar changes, although to a lesser degree, are found in cells taken from cultures during the late stationary phase. Furthermore, the typical cell membranes are replaced by an electron-dense, thick, rigid envelope.

The chromosomes of Dinophyceae have attracted considerable attention because of their lack of typical histones. Although their chromatin does contain chromosomal proteins, these organisms have an apparently primitive form of nuclear organization and division (Dodge, 1965). In *Peridinium balticum*, both mesocaryotic and eucaryotic nuclei have been described coexisting in the same cell (Thomas *et al.*, 1973). These characteristics indicate the rather primitive evolutionary status of the dinoflagellates. Rae (1970) described the dinoflagellate nucleus as a possible contemporary manifestation of the primeval eucaryotic nucleus. No detailed studies of the nuclear organization of *Peridinium westii* have been made. However, Gressel *et al.* (1975), using polyacrilamate gel electrophoresis, have recently shown that although the lighter component of rRNA had an apparent molecular weight identical to that of higher plants, the heavy fraction of rRNA had an apparent molecular weight of 1.23×10^{-6}, intermediate between that of procaryotes (typically $1.05 - 1.1 \times 10^{-6}$) and eucaryotes. Thus, if these molecular weights are regarded as a measure of evolution without considering all other factors, one might consider *Peridinium* as an early eucaryote of lower evolutionary status than plants and fungi. A similar apparent molecular weight was reported for the heavy rRNA subunit of red algae (Howland & Ranus,

266

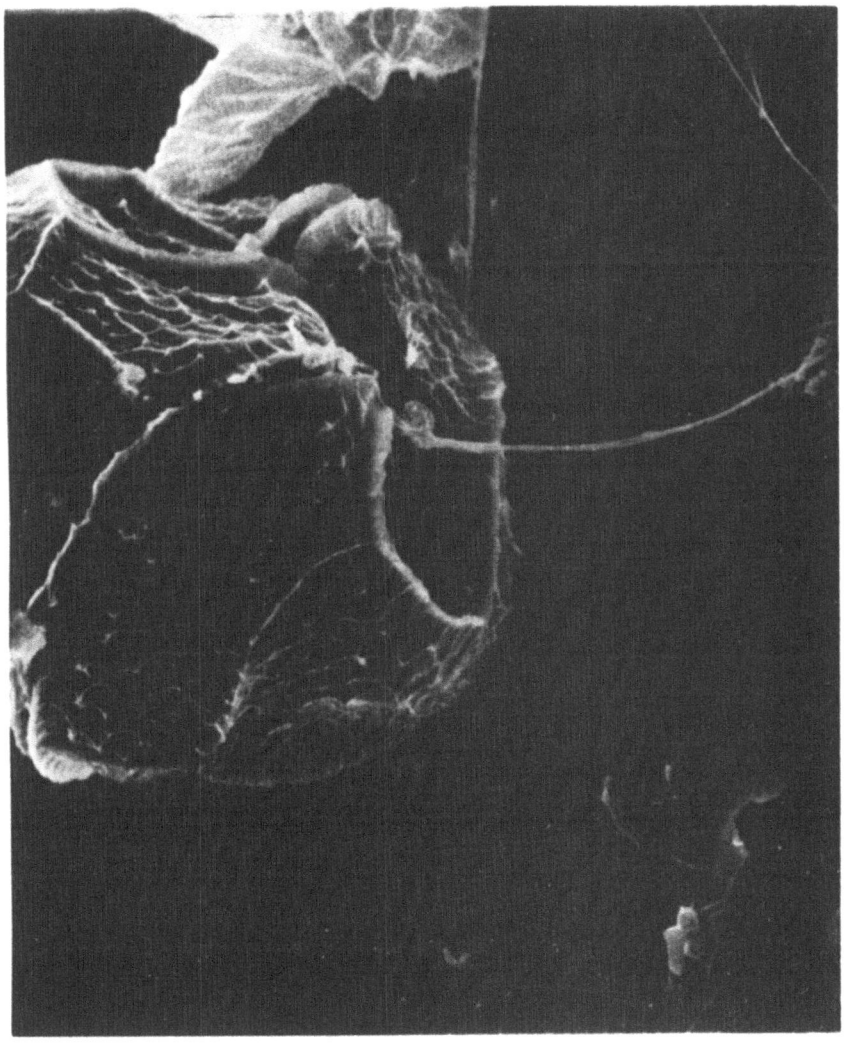

Plate 21. Longitudinal flagellum with basal lobed structures. Scalar bar 10 μm. By courtesy of Drs. T. Berman and I. L. Roth.

1971). Possibly this form of RNA is an archaic intermediate in the evolutionary pathway from procaryotic to eucaryotic rRNA.

2 General biochemical features

T. Berman

a. Gross chemical composition

As can be seen in Table 36, carbon comprises an unusually high fraction of the total dry weight of the *Peridinium* cell. Much of this carbon undoubtedly resides in the theca, which has been characterized by Nevo & Sharon (1968) as a non-cellulosic glucan. About 95% of the skeletal cell wall was found to be a polymer of D-glucose linked by $\beta - (1-4)$ and β linkages. It is worth noting, in order to give a correct idea of the amount of carbon synthesized by *Peridinium*, that upon each cell division, the theca of the mother-cell is thrown off.

Table 36. Some characteristics of the *Peridinium* cell

Cell volume	70,000–125,000 μ^3
Cell wet weight	0.075 μg
Cell dry weight	0.03 μg (0.02 to 0.05 μg)
Chlorophyll content	0.3% of wet weight
Carbon content	50% of dry weight
Phosphorus content	0.1–0.26% of dry weight
Nitrogen content	3% of dry weight

Depending upon growth conditions, the intracellular ratios of carbon, nitrogen and phosphorus can vary (Serruya & Berman, 1975). For an average carbon content of 12 ng per cell, the intracellular N ranges from 0.46 to 0.86 ng and intracellular P varies from 0.018 to 0.048 ng. The usual values of 106 : 15 : 1 of the atomic C : N : P ratios generally found in plankton (Redfield, 1958) have never been observed in Lake Kinneret during the bloom period. Serruya *et al.* (1974) found ratios of 235 : 18 : 1 in the early stages of the bloom and 645 : 25 : 1 at late bloom. Levels of alkaline phosphatases vary inversely with the availability of external phosphorus (Berman, 1969, 1970), and probably also with the intracellular levels of phosphorus (Wynne, 1977). In the lake, during the bloom, a range of 1 to 6 · 10^{-6} μM p-nitrophenol released per minute per milliliter of water was observed. The presence of extracellular and intracellular phosphatases implies a potential for the utilization of organic phosphorus compounds as phosphorus sources. Monoalgal cultures of *Peridinium* will indeed grow on compounds such as glucose 6 phosphate, glycerophosphate and adenosine triphosphate as their only phosphorus source.

Although detailed investigations of changes with aging in the composition of major chemical components of the *Peridinium* cell have not been made, both light and electron microscope studies have noted morphological changes, including the formation of vesicles which may contain lipid-like materials. In the lake, an increase in the amounts of dissolved organic matter released by photosynthesizing

269

Peridinium cells has been noted towards the end of the bloom season from about 1% extracellular release of carbon in January–April to 3% in May–June (Berman, 1976).

b. Pigments

Intracellular chlorophyll concentrations depend upon growth conditions. In growing cultures of *Peridinium*, average levels of chlorophyll were found to constitute 0.3% of the wet weight. In the lake, during the peak of the *Peridinium* bloom in April 1970, the concentration of chlorophyll *a* within the cells decreased to a minimum of 0.1% of wet weight. In absolute terms, *Peridinium* cells in culture contain about 225×10^{-6} μg chlorophyll *a* per cell. For *Ceratium hirundinella*, Talling (1969) reported a value of 250×10^{-6} μg per cell.

In *Peridinium*, rather high levels of chlorophyll *c* were also found, possibly reaching 25 to 30% of chlorophyll *a*. Carbon to chlorophyll ratios of 48:1 have been measured in exponentially growing cultures.

In addition to chlorophyll, carotenoid pigments, especially peridinin, are also present in amounts and proportions which vary with the maturity of the cells. Recently, peridinin, which is the principal carotenoid pigment of the Dinophyceae, has been chemically characterized. This pigment has been isolated both from freshwater and marine species, including the endozooic symbionts (Zooxanthellae) of marine corals, clams and sea anemones. The compound is orange-red, with maximal absorbance at 445 and 485 nm (in hexane) and has a molecular formula of $C_{39}H_{50}O_7$. The C_{37} skeleton is unusual; in most isoprenoid carotenoids, a C_{40} type skeleton occurs. It would be of considerable interest if some physiological or ecologically useful role could be attributed to the peridinin, which appears to increase in relation to the other pigments with cell aging (Dubinsky & Polna 1976).

In 1973, the lake water pigments were extracted and separated by thin layer chromatography and their composition followed during the bloom period. During all the stages of the bloom, peridinin represented over half of the total carotenoid content. The ratio of carotenoid to chlorophyll pigments varied from 2:1 at the bloom peak to 3.5:1 at late bloom. The decline of the bloom was accompanied by a decrease of all the pigments except peridinin, which continued to increase. An interesting hypothesis considers that the two red bodies of *Peridinium* cysts are composed of peridinin; if this is the case, the increase of this pigment in senescent cells may correspond to a preencystment stage.

It is noteworthy that the absorption region of peridinin lies between that of the chlorophyll absorption peaks. Possibly this enables *Peridinium* to trap and utilize light energies at maximum penetration wave-lengths in the Kinneret (450–650 nm), and thus the abundance of such a pigment might represent a considerable ecological advantage.

3 Life cycle

U. Pollingher

a. The cysts

The life cycle of *Peridinium* includes a resting form, the cyst, characterized by a smaller size than the vegetative cell, the absence of a theca, the absence of motility and the presence of two red bodies (plate 22).

The cyst is denser than water and therefore is generally found only in sediments. This explains why cysts are observed in littoral water samples only after strong autumn and early winter storms. It is surprising that, in the thousands of samples we have examined, we never observed the stage of excystation itself (Rahat, 1968). However, we found, from November until February, swimming *Peridinium* cells still having two red bodies. These forms were found only in littoral water samples, from 1700 to 2000 hours and never during the night. A similar timing was observed in the formation of cysts. In contrast with other lakes, we never observed a massive encystation at the end of the bloom. Cysts are formed during the whole bloom period.

b. The division process

Whatever their origin, the swimming cells divide by binary fission. The growing mass of the cell exerts an internal pressure on the theca, which breaks open at the cingulum. In *Peridinium cinctum*, the division process begins within the theca. A constriction appears and develops in the middle of the protoplast, and this leads to the formation of two globular equal daughter cells, both naked. Then each daughter cell immediately secretes a new theca. We note that in this respect, *Peridinium cinctum* differs from most of the armoured dinoflagellates where only one-half of the theca has to be rebuilt.

c. The division rate

The division of *Peridinium cinctum* takes place only at night between 0100 and 0700 hours, with a peak between 0200 and 0400 hours (Table 37). We note that the timing of the peak does not vary from month to month.

Although division occurs in complete darkness, the dividing cells are found only from the surface to 7 m. Measurements of oxygen carried out at intervals of 1 m during the night in April showed that the concentration dropped suddenly from 10 ppm at 8 m to 3 ppm at 10 m. It is likely that the low concentration of oxygen below 10 m at this point prevents the oxidative processes associated with the division.

The complete cycle was observed in 1973–1974 (Table 37). A division rate of 10% was observed from 15 November 1973 until 20 February 1974. Then it increased and remained between 30 and 40% until late March, and dropped again

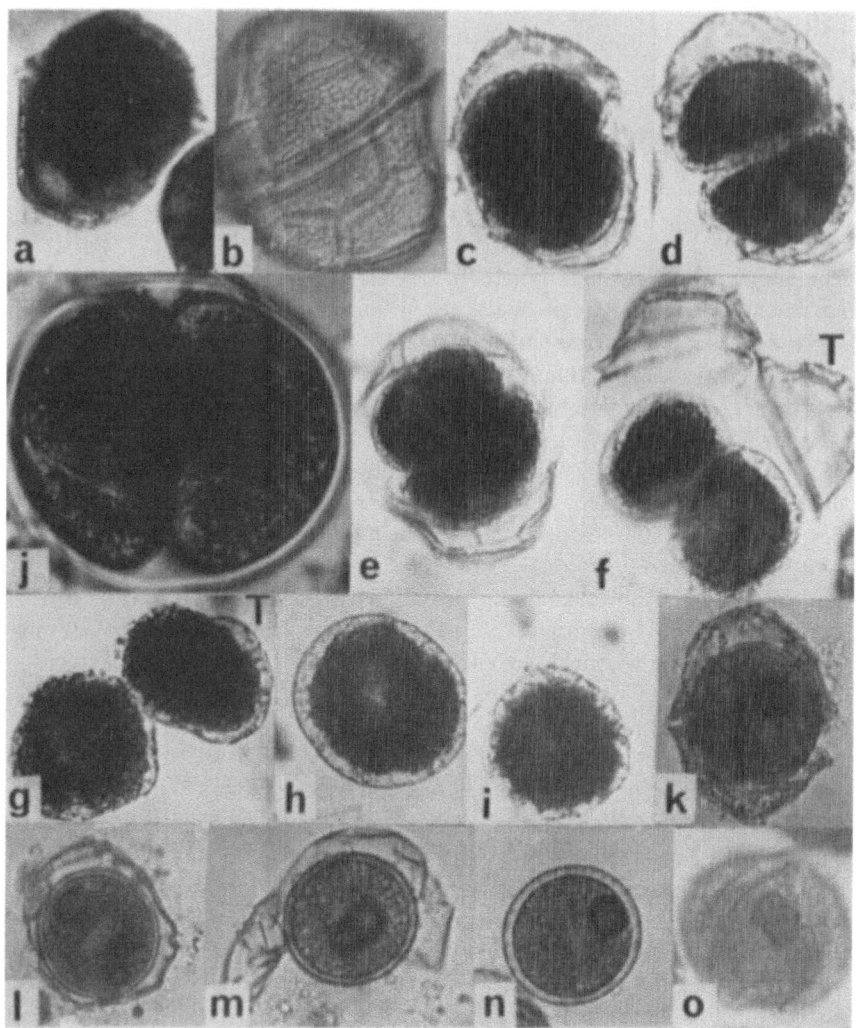

Plate 22. Life history of *Peridinium cinctum* fa. *westii.* (fixed material unless noted) a. Vegetative cell. b. Empty theca. c-f. Successive stages of cell division. g-i. Daughter cells (T = theca). j. same as d, live material (greater magnification) showing complete development of daughter cells. k-n. Successive stages of cyst formation. O. Young cell emerged from cyst with one red body, Scale = 30 μm from Pollingher U. and Serruya C., 1976, Journal of Phycology 12).

Table 37. The division rate during the bloom periods (1973 and 1974) at different hours (from Pollingher and Serruya, 1976, J. Phycol., 12).

Date	24	01	02	03	04	05	06	07
1973								
29 Mar				10.5	21.2			
30 Mar				8.0	13.7			
19 Apr				1.2	2.07	2.5	0	
24 May					6.1			
25 May					1.7	3.3		
28 Jun			0	0	0	5.5	1.9	
15 Nov		1.2	10.0	10.0	10.0	2.4	1.4	
5 Dec			5.1	12.7	9.06	3.6	0.6	
13 Dec		4.0	10.0	9.2	3.8	1.4	0.6	
27 Dec		1.3	7.0	7.0	9.3	1.6		
1974								
3 Jan		1.4	2.7	4.5	3.7	1.9		
10 Jan			2.2	6.0	8.0	1.8		
1 Feb			0	3.2	0	0		
6 Feb			7.7	10.7	5.3	4.3		
21 Feb	1.2	2.9	8.1	10.0	3.5	2.2		
28 Feb		27.0	45.0	36.9	37.3	26.0	8.0	2.3
5 Mar	9.3	19.7	23.5	30.6	28.5	19.0	13.2	
14 Mar		0	3.0	8.3	37.0	32.4	7.2	1.8
26 Mar	0	0	6.1	33.6	41.0	32.8	19.1	
4 Apr	0	0	0	5.2	5.3	11.0	5.0	
16 Apr		0		13.2	10.0	15.0	3.8	
2 May	5.3	3.4	5.0	10.2	10.5	6.8	0	
4 Jun	0	0	0	0	15.0	22.5	6.4	

to 10% until the end of May. In summer, the division rate was not higher than 2.5% and similar values were obtained during three consecutive summers (Pollingher & Serruya, 1976).

The variability of the division rate in space was checked by sampling simultaneously at the central station A (40 m depth) and in a shallow station H (10 m) on the western shore of the lake. The division rate was similar at both stations. The variability of division rate from day to day was checked by measurements carried out during two consecutive days. The results are shown in Table 38.

We have no data concerning the duration of the division process: if this process

Table 38. The division rate during two consecutive days (1973) (same source as table 37)

Date	Hour	Division rate	Hour	Division rate
29 Mar	03	10.5	04	21.2
30 Mar	03	8.0	04	13.7
24 May	04	6.1		
25 May	04	1.7		

extends over several hours, it is possible that the same division is counted several times during the night. The peak observed between 0200 and 0400 hours would then represent the cumulative effect of the growing number of cells in division during the night period. Therefore, we considered the maximum percentage of cells in division found during the night as the closest to the actual daily division rate.

4 Growth characteristics

W. Rodhe

Margalef (1960) classified the dinoflagellates among 'the species which are difficult to culture'. *Peridinium cinctum* fa. *westii*, found in Lake Kinneret, is no exception to the rule. The history of experimental research on *Peridinium* is so far, first of all, a history of the efforts to culture it in a defined and bacteria-free medium.

a. The growth medium

The first successful cultures of *Peridinium cinctum* fa. *westii* were achieved by Eren (1969). Whereas motile cells from the free water of Lake Kinneret failed to grow *in vitro*, Eren got resting cysts to hatch and to produce offspring in a semi-synthetic medium. He based its composition on the culture solution of Carefoot (1968) which he supplemented with vitamins and trace metals and diluted with Kinneret water. Other workers diluted the Carefoot–Eren medium with various proportions of lake water (e.g. 1 + 4 or 1 + 2) or lake water and distilled water (1 + 1 + 1). The addition of natural water was always found to be absolutely necessary for continuous growth of *Peridinium*.

Work at the Kinneret Limnological Laboratory demonstrated after a while the fact that motile *Peridinium* cells, taken fresh from the lake, are able to continue their growth in cultures provided they get sufficient time for adaptation to gradually changing conditions. To that end, the phototactic response and active migration of the cells can be used. At first, a mixture of *Peridinium* and other phototactic organisms is achieved in lake water, and then the dominance of *Peridinium* in the semi-synthetic medium can be secured. Eventually, individual cells are picked up from the 'raw' culture, washed repeatedly and isolated to form unialgal strains (clones). In that way, Rodhe (unpubl.) obtained several strains originating from cells of different shape and size. He made the interesting observation that their descendents, after a series of subcultures, assumed a common appearance. Four years later, in 1974, Pollingher examined some of them and found their morphological features to be in accord with those of the typical Kinneret *Peridinium*.

The methods used to attain bacteria-free (axenic) algal cultures include antibiotics and ultraviolet light, but less harmful to the algae are repeated washings. It would be a hopeless task, however, to wash away every bacterium from the large and ornamented shell of *Peridinium*. Therefore, Miss Anna Panders, laboratory assistant of Prof. Rodhe, trained herself to pick up newborn cells at the very moment of their release from the mother cell. By means of that technique, she succeeded, in 1971–1972, in preparing five axenic *Peridinium* clones. Later on, Gressel *et al.* (1975) removed the bacteria from some of Berman's multi-algal (non-clonic) cultures by treating them with the antibiotics gentamycin and mycostatin.

Although Eren found some stimulation with additions of peat or soil extract, a clear-cut effect of organic supplements has not been confirmed. Carefoot (1968) reported enhanced growth in the light with additions of glucose, fructose, galactose, malate, malonate and pyruvate. In the dark, succinate permitted growth above survival level. On the basis of these findings, Carefoot suggested a heterotrophic mode of nutrition for his strain of *Peridinium westii*. However, in an autoradiographic study of natural Kinneret populations, Pollingher & Berman (1976) were never able to show any uptake by *Peridinium* of glucose, amino acids, acetic acid or glycolic acid either in short (3 hours) or in long (24 hours) incubation in light or dark experiments.

The undefined Carefoot–Eren medium, though useful for culturing and other purposes, is unsuitable for quantitative determinations of the chemical requirements of *Peridinium*. In 1972, Rodhe started a research project at the Institute of Limnology, Uppsala (Sweden), to elucidate the environmental requirements and growth conditions of *Peridinium* in axenic cultures. K. Lindström took on the task of replacing the composite lake water component of the medium by the active growth factor(s). All kinds of organic substances were rapidly ruled out, because the 'activity' of natural water (from Lake Erken, east of Uppsala) remained unaffected by ignition of its residue. To screen the metals contained in lake water was a much more tedious job, but in 1974, Lindström found one single element, namely selenium (Se), to be the growth factor he sought. It cannot yet be excluded that other trace elements, if added to the culture solution in the form of impurities in the chemicals used, may likewise be required for the growth of *Peridinium*, but selenium alone proved to be a substitute for the entire lake-water portion in the Carefoot–Eren medium. Moreover, in the presence of water from Lake Erken, a minute addition of Se (0.13 μg per litre) trebled the final yield in the semi-synthetic medium, from the ordinary level of 8–10 to about 30 thousand cells per ml (Fig. 97). This indicates that selenium, in addition to being an indispensable constituent of *Peridinium*, may play the part of a limiting factor for its production in nature.

An 'ideal' synthetic medium has to provide all essential but no more constituents in optimal balance for maximal rates of exponential growth. The concentrations should not exceed the natural ranges more than necessary to maintain unlimited growth for a period of 2–3 weeks. If prolonged growth and maximal final yields are desired, a suitable multiple of the basic recipe can be used to raise the carrying capacity of the medium.

According to these guiding principles, Lindström is on the way to settling a completely balanced medium for *Peridinium*, based on results from separate growth experiments for every single constituent. So far (March 1976), his medium Nr. 9 is the most advanced step towards that goal (Table 39). It deviates from the Carefoot–Eren medium in several respects, above all in the absence of lake water. Omitted also are the vitamins B$_{12}$, biotin and thiamine. *Peridinium* cells living in the Kinneret may be vitamin 'addicts' and unable to accept a vitamin-free medium suddenly, but Lindström proved their ability, potential at least, to multiply as well on an entirely mineral basis. Another striking feature of his medium Nr. 9 is its low conductivity, four times smaller than that of Kinneret water where the contents of major ions are higher than *Peridinium* demands. As regards pH, however, Lindström Nr. 9 is in agreement with the lake; there, the levels during the period

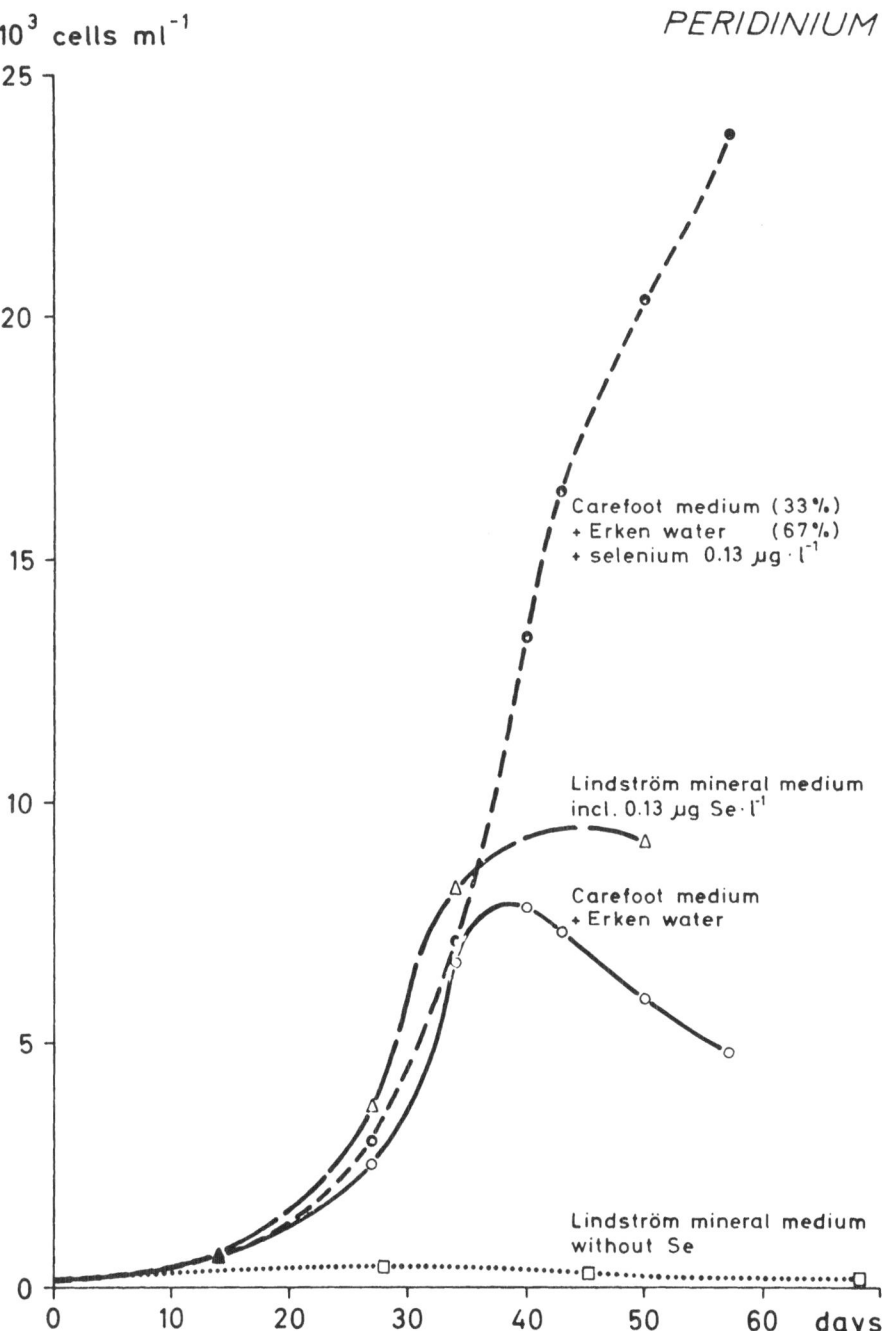

10^3 cells ml^{-1}

PERIDINIUM

Carefoot medium (33%)
+ Erken water (67%)
+ selenium 0.13 μg·l^{-1}

Lindström mineral medium
incl. 0.13 μg Se·l^{-1}

Carefoot medium
+ Erken water

Lindström mineral medium
without Se

days

Fig. 97. Growth in semi-synthetic Carefoot–Eren medium and mineral Lindström medium No. 4, with and without selenium. Dense cultures (Lindström, unpubl.).

Table 39. Lindström medium Nr. 9 for *Peridinium cinctum* fa. *westii*

Compound	mg ml^{-1}	ml A in 1,000 ml B	mg l^{-1}	mmol l^{-1}	Element	mg l^{-1}
	A. Stock solutions			**B. Culture medium**		
NaHCO$_3$	13.44	10.0	134.4	1.6	C	19.22
NaHCO$_3$				1.6	Na	36.80
NaNO$_3$	2.55	10.0	25.5	3.0·10^{-1}	Na	6.90[43.7]
NaNO$_3$				3.0·10^{-1}	N	4.20
MgSO$_4$·7 H$_2$O	6.16	10.0	61.6	2.5·10^{-1}	Mg	6.08
MgSO$_4$·7 H$_2$O				2.5·10^{-1}	S	8.02
CaCl$_2$·2 H$_2$O	2.21	10.0	22.1	1.5·10^{-1}	Ca	6.01
CaCl$_2$·2 H$_2$O				1.5·10^{-1}	Cl	10.64
K$_2$HPO$_4$	0.348	10.0	3.48	2.0·10^{-2}	P	0.62
K$_2$HPO$_4$				2.0·10^{-2}	K	1.56
Na$_2$SiO$_3$·9 H$_2$O	0.114	10.0	1.14	4.0·10^{-3}	Si	0.11
Metal mix:						
EDTA	0.37		0.37	1.0·10^{-3}		
ZnCl$_2$	0.068		0.068	5.0·10^{-4}	Zn	0.0327
FeCl$_3$·6 H$_2$—	0.108		0.108	4.0·10^{-4}	Fe	0.0223
Na$_2$MoO$_4$·2 H$_2$O	0.048	1.0	0.048	2.0·10^{-4}	Mo	0.0192
MnCl$_2$·4 H$_2$O	0.010		0.010	5.0·10^{-5}	Mn	0.0028
CoCl$_2$·6 H$_2$O	0.0048		0.0048	2.0·10^{-5}	Co	0.0012
NaSeO$_3$·5 H$_2$O	0.00526	0.1	0.000526	2.0·10^{-6}	Se	0.00016

Conductivity: 255 μS cm^{-1}
Alkalinity: 1.6 meq l^{-1}
pH: 8.3

of *Peridinium* blooms increase from pH 8.0 to 9.0 in the morning and 9.5 to 10.0 for midday values.

b. Influence of pH

Dubinsky (1975) demonstrated that short-term photosynthetic carbon fixation in cultures is strongly influenced by pH (his Fig. 3). It is probable that the pH acts mainly through the CO_2, HCO_3^-, $CO_3^=$ system. The pattern shown in Dubinsky's figure indicates that dissolved CO_2 is preferentially utilized by *Peridinium*, in contrast to bicarbonate. These results may have relevance to the growth pattern of *Peridinium* in nature, where a pH-imposed carbon limitation might occur during the bloom.

c. Influence of light and temperature

The effects of light and temperature on growth and photosynthesis are intimately interdependent.
    In *Peridinium* cultures at three different temperatures between 15 and 25°C and

light/darkness periods of 12/12 hours, Dubinsky (1975) found the yields to increase with temperature at the two highest but not at the lowest intensity of light (his Fig. 4). In continuous light, the yield level was raised, with the exception of the highest combination of light and temperature which induced a significant reduction. There, the cell division appeared to require a daily period of darkness. This is in accordance with the fact that *Peridinium* in the lake divides most frequently between 0300 and 0400 hours (Pollingher & Serruya, 1976). In another experiment of Dubinsky, *Peridinium* grew well at temperatures as high as 29°C for six weeks.

Lindström (unpubl.) measured the apparent, or net, photosynthesis and the respiration by means of a recording oxygen electrode. Dense cultures of axenic *Peridinium*, kept in temperature-controlled incubators between 5 and 33°C, were exposed to fluorescent light for periods of 10–15 minutes at intensities that were increased, step-by-step, from 500 to 6,000 lux. Gross photosynthesis was obtained by summation of the oxygen released in light (net photosynthesis) and the corresponding rate of oxygen consumption in darkness (respiration). The results are presented in Fig. 98. The assimilatory and respiratory processes both start at about 4°C, reach a maximum level or peak in the upper part of the temperature range, and then decline more or less suddenly to zero at 33°C. Respiration is more accelerated than gross photosynthesis by increasing temperature, up to the peak at 29–30°C. Hence the temperature optima of net photosynthesis (which delivers the energy for growth) are lower than those of gross photosynthesis, in particular

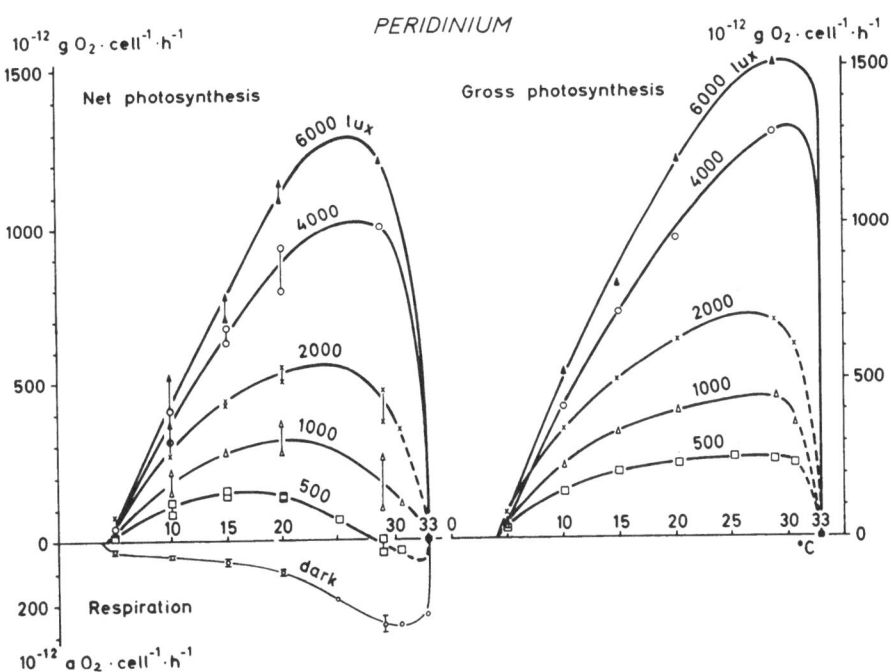

Fig. 98. Measured rates of net photosynthesis and respiration, calculated rates of gross photosynthesis, at various combinations of light and temperature. Dense cultures, oxygen electrode (Lindström unpubl.).

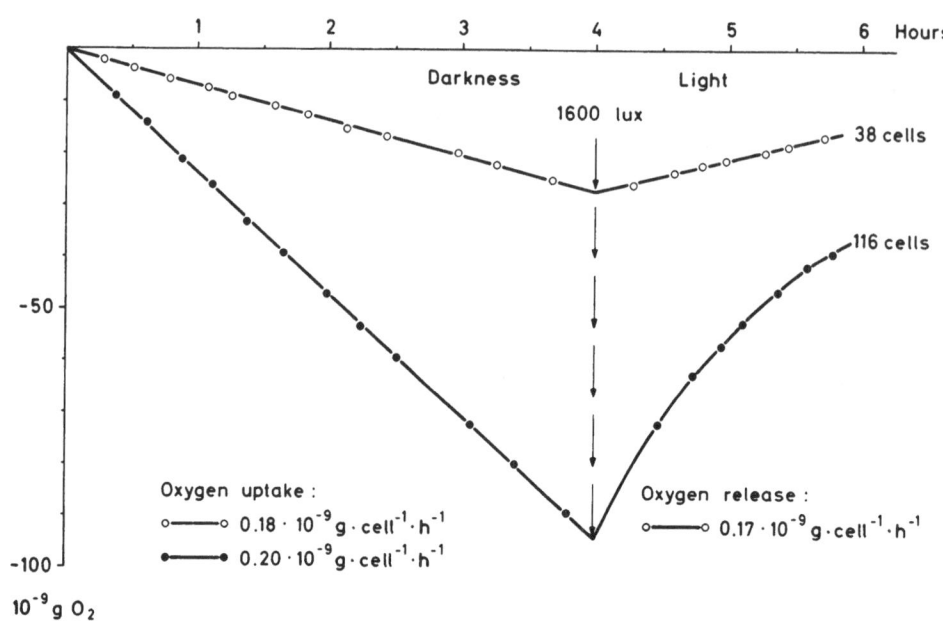

Fig. 99. Uptake and release of oxygen at 23°C. Cartesian diver (Brammer unpubl.).

at low light intensities where they are between 15 and 25°C. This combined effect at low temperature/light levels may contribute to produce the first *Peridinium* bloom in the beginning of the year. On the other hand, in occasional hot spells during summer, the surface-water temperature can get close to its lethal limit as a single factor, but that happens only after the main season of *Peridinium* blooms.

E. Brammer, another member of the Uppsala team on *Peridinium*, adopted the so-called Cartesian-diver technique to measure uptake and release of oxygen by small numbers of cells. The method was originally based on the vertical displacements of a freely floating capillary (a 'diver') caused by changes in its interior gas volume. As few as 10 cells of *Peridinium*, swimming in 2 μl of the enclosed medium, are sufficient to permit precise measurements of respiration and photosynthesis in short-time experiments. As shown in Fig. 99, drawn from results obtained at 23°C, linear progression is proof of reliability whereas a non-linear curve indicates some kind of deficiency (here the beginning of a lack of CO_2 caused by the number of cells). Brammer studied the effect of temperature on respiration and found the numerical relationship to be largely influenced by the preceding conditions, the 'history' of each culture. For example (Fig. 100), cells adapted to 15°C had a comparatively low oxygen consumption throughout the range of higher temperatures and a maximum uptake of 165 picogram O_2/cell · hour at 33°C (where they died shortly afterwards). In contrast, cells adapted to 23°C responded along a curve of similar shape but at higher levels of respiration, culminating with 400 picogram O_2/cell · hour at 31°C.

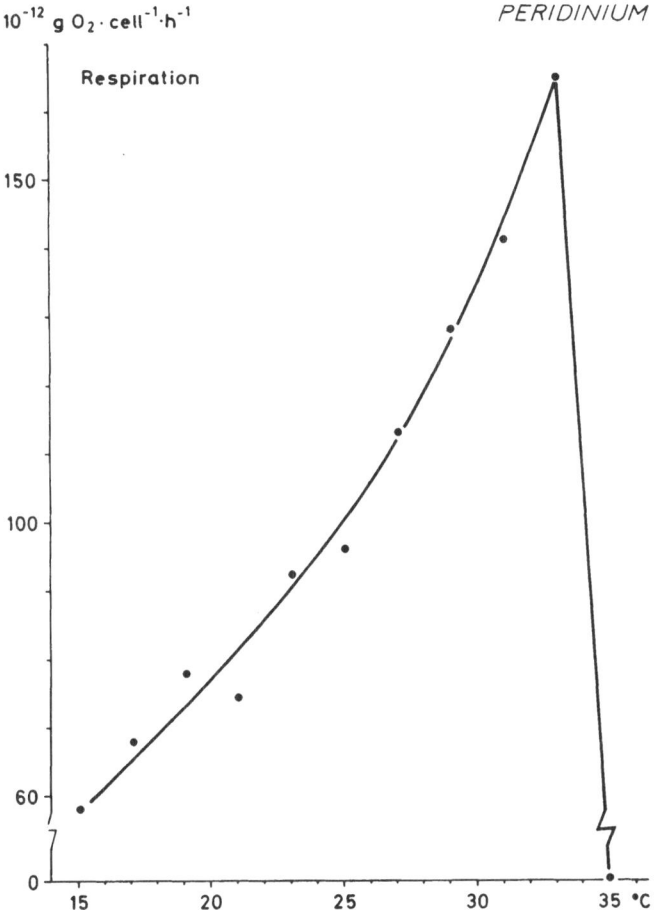

10^{-12} g $O_2 \cdot$ cell$^{-1} \cdot$h^{-1} *PERIDINIUM*

Fig. 100. Oxygen uptake at various temperatures, from 15°C to 35°C. Cartesian diver with 494 cells which had been cultured at 21°C but were kept at 15°C for 12 hours before the measurements (Brammer unpubl.).

The impact of pre-exposure conditions on the actual photosynthetic response has been studied by Rodhe and his student M. Remberger (unpubl.). They found the rate of oxygen release at a given intensity to be higher in increasing than in decreasing sequences of light, and that was true for *Peridinium* taken directly from the lake as well as from axenic cultures (Fig. 101). (Harris & Lott (1973) obtained the same result from plankton algae in Lake Ontario.) Such a pattern may explain, partially at least, the well-known fact that phytoplankton production is higher before than after midday, even if the insolation is symmetrical all the day. There are several reasons, in terms of algal metabolism, for that difference. At the end of the night period respiration has restored the external CO_2 supply, and within the cell, the photosynthetic pathways are free for the transformation of energy. Conversely, after previous photosynthesis, depletion of CO_2 and jamming accumulation of metabolic products tend to slow down the photosynthetic rate.

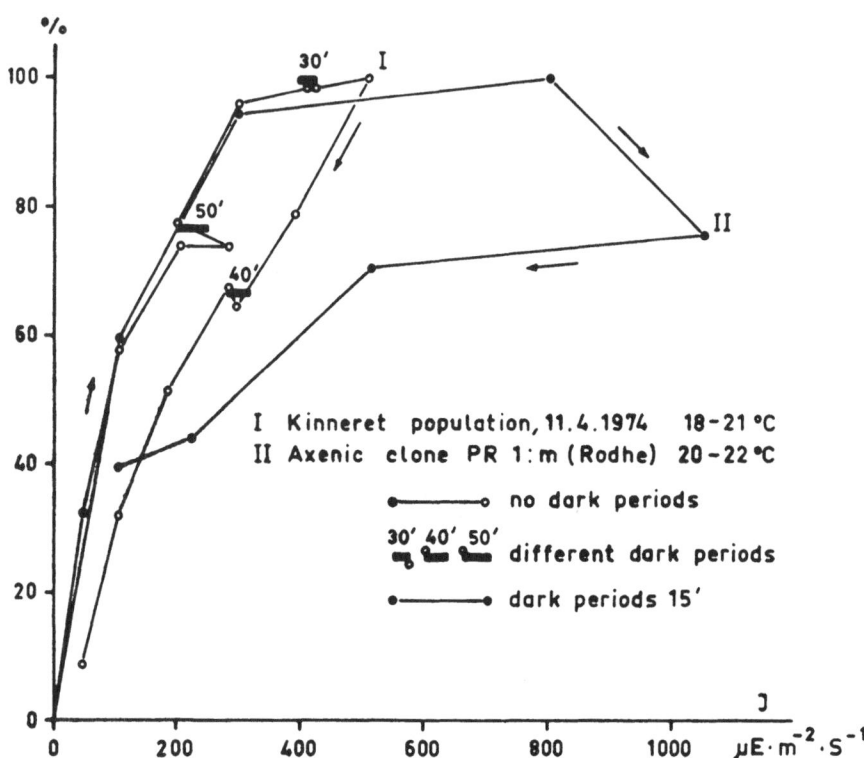

Fig. 101. Gross photosynthesis in increasing and decreasing light, with various breaks of darkness. Indigenous and cultured cells, for density see attached table. Oxygen electrode (Rodhe & Remberger unpubl.).

Thus, as seen from series I in Fig. 101, the saturation level (74%) obtained in successively increasing light was lifted (to 100%) after the insertion of 50 minutes darkness. Regular dark breaks of 15 minutes, alternating with light exposures of equal length (and conveniently used for respiration measurements), allowed photosynthesis to approach its full potential at rising intensities (Fig. 102). After exposures to optimal light, and then, at decreasing intensities, however, photosynthesis would need much longer periods of darkness to recover its inherent capacity.

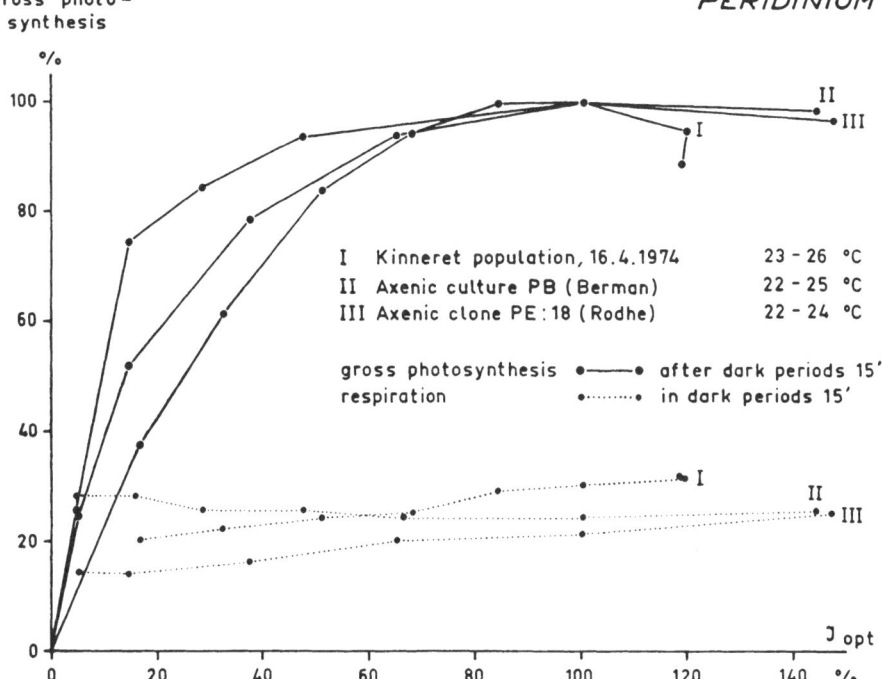

Fig. 102. Gross photosynthesis in increasing light, with regular dark breaks. Indigenous and cultured cells, for density see attached table. Oxygen electrode (Rodhe & Remberger unpubl.).

Table with basic data for Fig. 101 and Fig. 102.

	Cells ml^{-1}	Max. gross photosynth. 10^{-12} g O_2 cell^{-1} h^{-1}	Opt. light intens. μE† m^{-2}s^{-1}
Fig. 101: I	9,200	615	510
101: II	13,900	1,465	800
Fig. 102: I	13,900	1,005	595
102: II	5,700	1,632	1,060
102: III	9,500	1,527	1,060

† microEinstein: lux = approx 1:60

5 Growth pattern in the lake

U. Pollingher & C. Serruya

The development of *Peridinium* in Lake Kinneret is similar to the algal growth in cultures: after an exponential phase of growth (November–late March), the *Peridinium* population remains stable, forming 99% of the total phytoplankton biomass until late April, and then declines.

In Fig. 103, we have represented the increase of number of cells (as monthly averaged values) of the natural population as a function of time for the different years of the record. The straight lines observed are represented by the general expression:

$$N = N_0 \, e^{kt}$$

where N is the number of cells obtained in comparison with the initial number N_0 after a period of time, t, and k is the constant of increase. The values of k for the six curves of Fig. 103 as well as the doubling time (Table 40) were calculated.

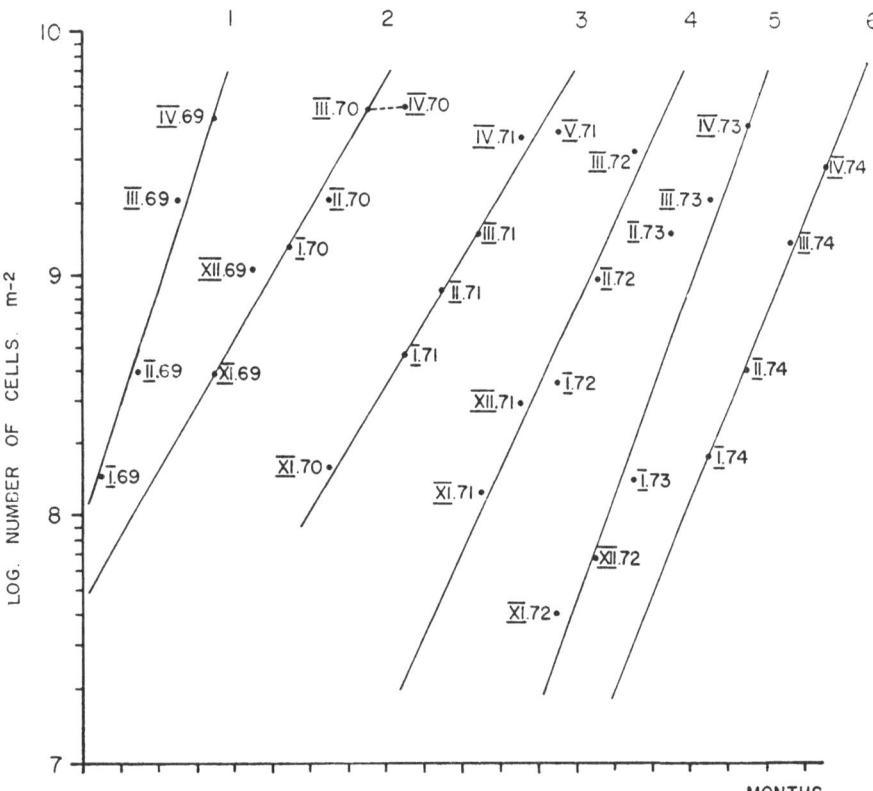

Fig. 103. Monthly increase of *Peridinium cinctum* populations *in situ* during six consecutive years. (From Pollingher, U. & Serruya, C., 1976, J. Phycol., 12.)

Table 40. Constant of increase k (in ln day units) and mean doubling times (in days) (same source as fig. 103)

	1969	1970	1971	1972	1973	1974
k	0.040	0.017	0.023	0.026	0.035	0.030
Doubling time	17.3	40.8	30.1	26.6	19.8	22.6

The values of k are approximately one order of magnitude lower than those found by Nordli (1957) and Kain & Fogg (1960) in cultures of dinoflagellates. In our cultures of *Peridinium cinctum*, from an inoculum sampled in the lake and grown in dark/light conditions at 18–20°C, the rate of increase was found to be 0.063 (in ln day units) and the doubling time 11 days.

In their model concerning the mechanism of development of the dinoflagellates producing red tides, Wyatt & Horwood (1973) came to the conclusion that water stability generated by thermal stratification was a prerequisite of the onset of the bloom. In Lake Kinneret, the *Peridinium* has never been observed during the stratification period. As shown in Fig. 104, it starts developing during the mixed period, but the maximum rate of division is generally observed in March.

Considerable effort has been devoted to the understanding of the factors governing the bloom. We note that there are two crucial stages which have to be elucidated: the excystation process and the sudden increase of the division rate.

As mentioned in the section on life cycle, it is surprising that we never observed the stage of excystation. A possible explanation is that, at least in Lake Kinneret, there might be no excystation. In such a case, the cyst with its red bodies would, triggered by as yet unknown factors, acquire a theca and two flagella and become converted to a vegetative cell. After the first division, the red bodies are still apparent in the daughter cells. They are not visible after the next division. The factors which induce the secretion of theca and formation of flagella are not known.

The spring increase of temperature has been invoked as a possible factor inducing the increase of division rate. In 1974, Pollingher measured the in situ division rate of the *Peridinium* on a weekly basis. The daily percentages of cells in division jumped from 5% to 37% during the last week of February, when the whole water column was still at a constant temperature of 14°C.

In contrast, it seems that the wind pattern and the related water movements have a definite regulatory effect on the behaviour of the *Peridinium* population. The wind pattern has been described in detail (S. Serruya, 1975). Between the winter easterly storms and the summer westerly winds, there is a nearly windless period, the timing and duration of which vary slightly from year to year. Fig. 105 shows that the period of maximum division rate of *Peridinium* cells in 1974 corresponded exactly to this intermediary 'low wind' period. In January and February 1974, the stress of the wind reached 2 dynes cm^{-2} daily average, which corresponds to a wind velocity of 8 m s^{-1} daily average (1.47 dynes cm^{-2} running average), and the division rate of *Peridinium* cells never exceeded 10%. In early March, the wind stress dropped below 0.4 dynes cm^{-2}, which corresponds to a wind velocity of 3.5 m s^{-1}, and remained below this value for 27 days. As soon as

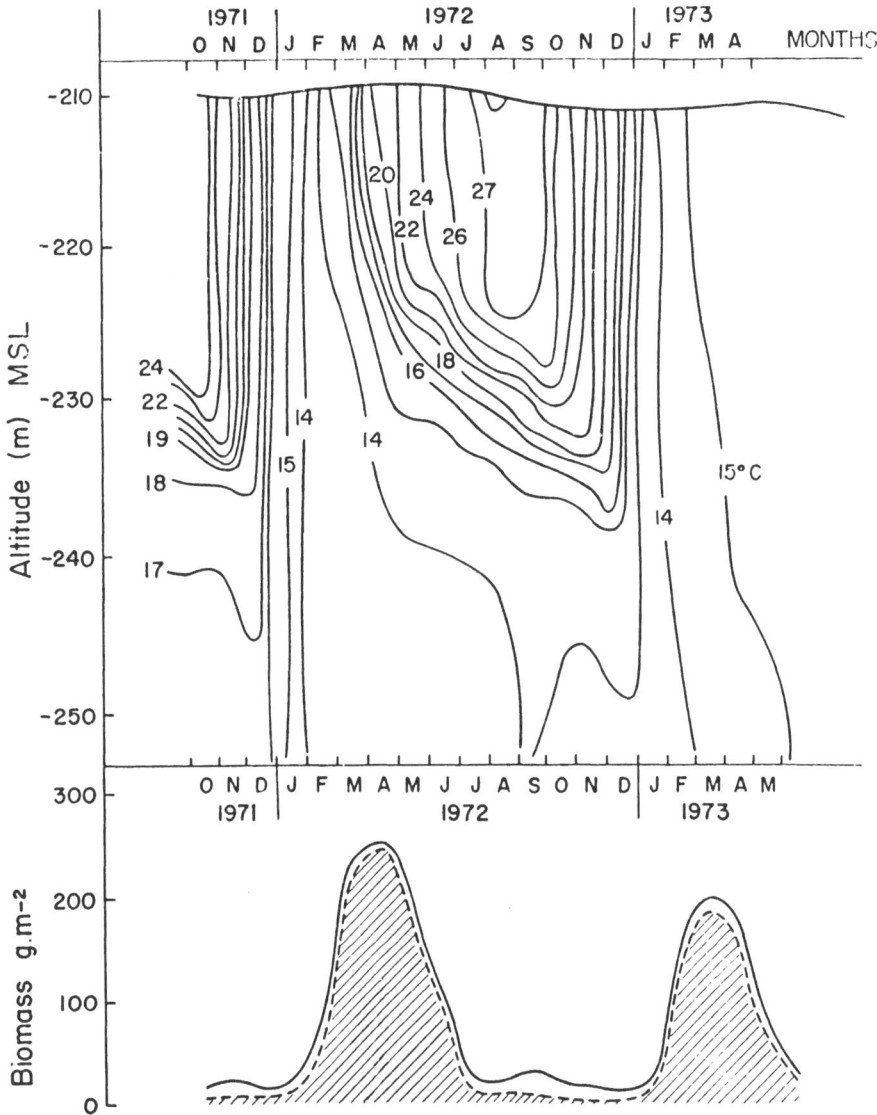

Fig. 104. Isothermal curves (A) and monthly variations (B) of total plankton biomass (———) and dinoflagellate biomass (– – –). The measured water level was used as a basis for the isothermal curves. (Same source as Fig. 103.)

the wind stress diminished, the division rate rose to 32% and remained between 30 to 40% during the whole quiescent period. The *Peridinium* biomass increased significantly during this period. In early April, with the onset of westerly winds, the cell division rate dropped back to its winter values. However, the biomass continued to increase, peaked in mid-April and afterwards decreased rapidly. In 1973, the windless period occurred earlier than usual – from 1–17 February. The division rate was not measured then, but the start of the *Peridinium* bloom coincided

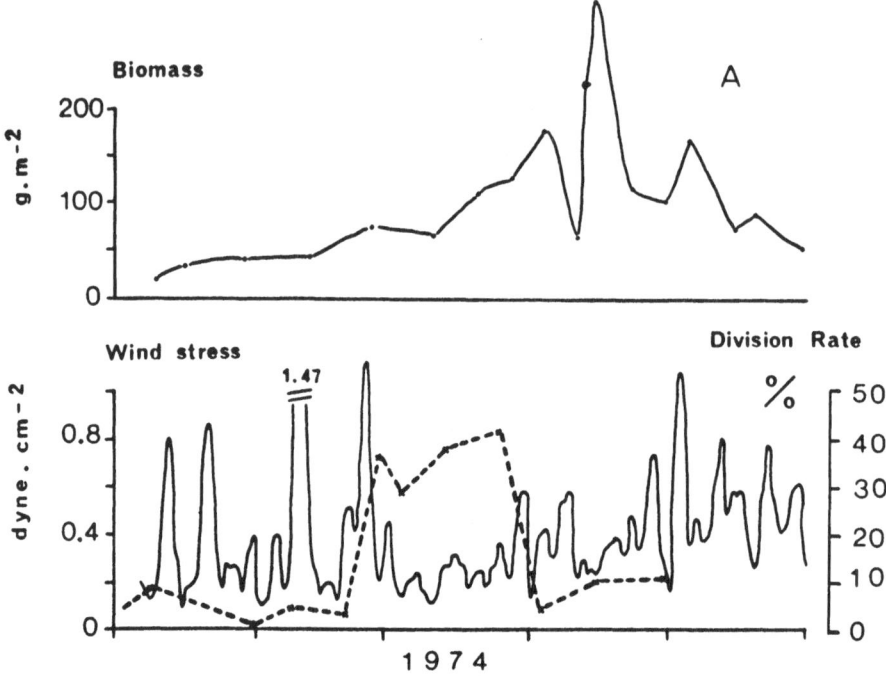

Fig. 105. Upper graph: *Peridinium* biomass in g m⁻² wet weight. Lower graph: Wind stress in dyne cm⁻² (running average and daily division rate of *Peridinium* cells in % of total *Peridinium* population. (From Serruya C. *et al.*, 1978, Verh. Internat. Verein. Limnol., 20).

with this period, that is, one full month earlier than in other years. In 1969–1970, the bloom started in November and reached the highest biomass recorded during our 1968–1976 record. The monthly average wind stress of November and December 1969 was 0.12 dynes cm⁻², corresponding to a wind velocity of 2 m s⁻¹, that is, 50% lower than the long-term average for those months (0.24 dynes cm⁻² or 2.7 m s⁻¹). Moreover, the high number of consecutive calm days (up to 13 days) occurring in this year was not repeated in other years when at most, 3 consecutive calm days are recorded.

These data show that the burst of the division rate generating the bloom occurs only during quiescent weather. Complementary information can be obtained from the vertical distribution of algal cells (Fig. 106), and the simultaneous profile of temperature and oxygen (Fig. 106). During the period which preceded and followed the recording of the profile of 9 January 1974, the wind velocity was permanently high (above 3.5 m s⁻¹). The homogeneous thermal profile indicates a complete turnover. The concentrations of oxygen observed in surface and bottom layers were higher than those found at intermediate depths. This suggests a circulation pattern of Langmuir type with convergence and divergence lines along which mass transfer occurs. Similar conditions of temperature and oxygen prevailed on 27 February, but the wind velocity was lower than 2 m s⁻¹. Conversely, on 27 March, at the end of the windless period, the incipient thermal stratification was accompanied by a marked decrease of oxygen in the bottom water. These physico–chemical data clearly indicate that in January and

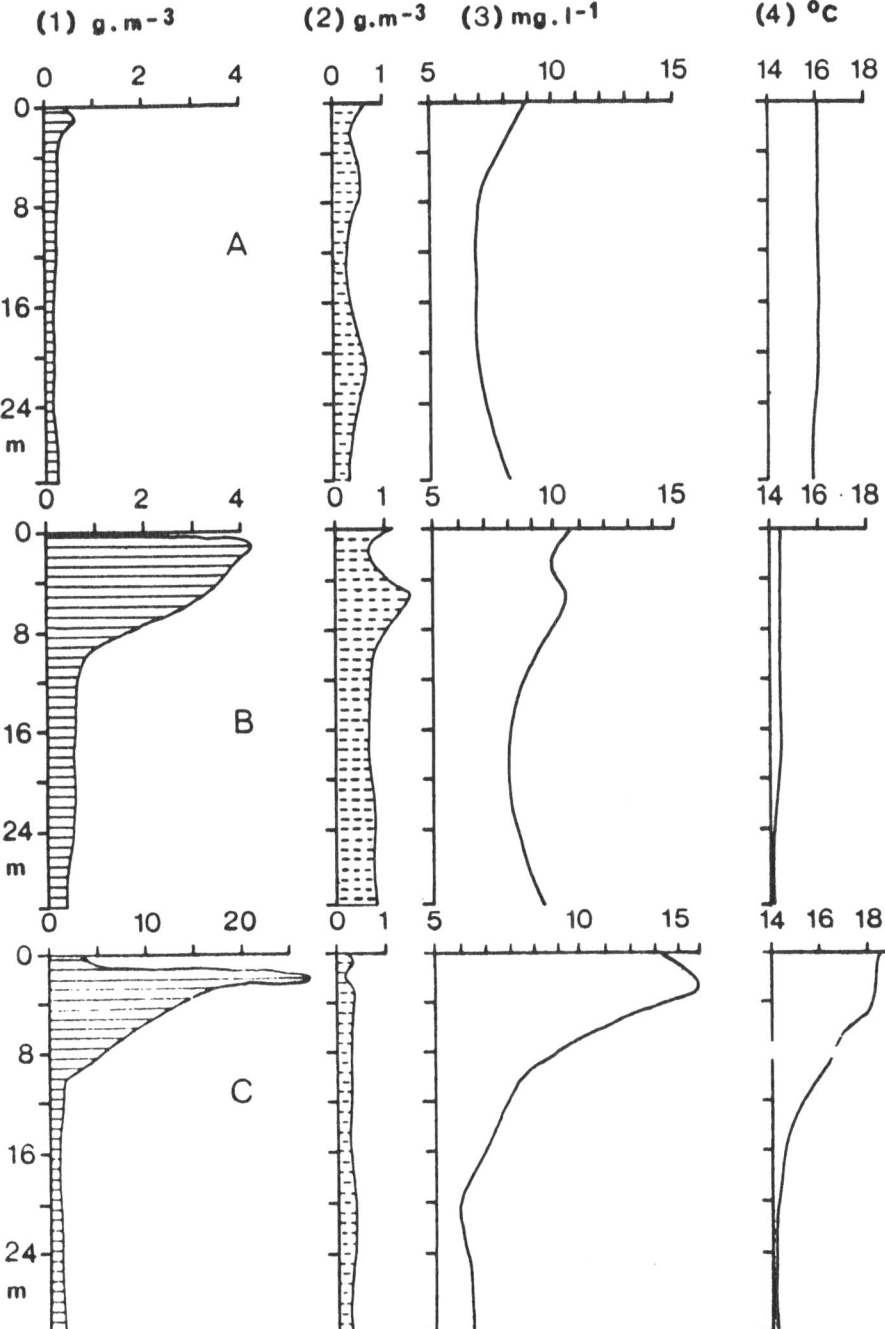

Fig. 106. Vertical distribution of motile cells of dinoflagellate g m^{-3} fresh weight biomass (1) and of non-motile algae g m^{-3} fresh weight biomass (2), oxygen concentration mg l^{-1} (3) and temperature, °C (4). 9 January 1974 (A), 27 February 1974 (B) and 27 March 1974 (C). (from Serruya C. *et al.,* 1978, Verh. Internat. Verein. Limnol., 20).

February, the mass transfer was active enough to maintain high oxygen values at the mud–water interface but this was no longer the case in late March. Figure 106 represents the behaviour of the motile Pyrrhophyta cells and of the non-motile cells of Chlorophyta, Bacillariophyta and Cyanophyta. On 9 January (A), both types of cells, (1) and (2), were nearly evenly distributed at all depths; in particular, in a 30 m profile, the 0–10 m Pyrrhophyta biomass represented no more than 44% of the biomass of the whole profile. On 27 February (B), although the water circulation still reached the bottom, the intensity of vertical water movements was smaller as a result of the reduced wind velocity. It was still enough to maintain a nearly homogeneous concentration of the non-motile cells (2) at all depths, but it was already too low to disperse the motile cells (1); at this time, 74% of the total biomass of Pyrrhophyta became concentrated in the upper 10 m. This phenomenon was accentuated during March, and on 27 March (C), 80% of the motile cells were in the upper 10 m.

These data indicate that the motile cells of Pyrrhophyta are homogeneously distributed with depth only when the vertical water movements are at their maximum intensity. This occurs only in January–February as a result of the influence of the easterly storms on a homothermal water body, which at that time offers a minimum resistance to mixing. A very slight decrease in the vertical displacement of water allows the motile cells of Pyrrhophyta to remain in the trophogenic layer and this selectively enhances the photosynthesis of this group. This considerable advantage conferred on motile cells, enabling them to occupy the photogenic zone exclusively, leads to the 'population explosion' resulting from the sudden increase of division rate as shown in Fig. 105.

6 Spatial distribution of *Peridinium*

T. Berman

a. Vertical distribution

Peridinium cells are capable of independent swimming motions by virtue of their two flagella (see plate 17, p. 263).

In January, viable *Peridinium* cells are more-or-less homogeneously distributed throughout the water column. During the months February–March, the organisms are located close to the surface; subsequently, during the morning hours, they are found mainly concentrated at depths from 0 to 3 meters. As surface light intensity increases, we note a deepening of the major peaks of dinoflagellates (April through June). The phototrophic response of the *Peridinium* cells leads to stratification in February, much earlier than thermal stratification of the water. A secondary peak of organisms appears during the months May, June and July, just above the thermocline.

In the lake, a very characteristic diurnal depth dispersion has been noted (Berman & Rodhe, 1971). Generally, the cells tend to be dispersed fairly homogeneously throughout the water column at night, towards morning they tend to rise, and during the early morning hours a population peak is often found either at or close to the surface (1 to 2 m). Later, the cells swim downwards – by late afternoon two peaks are often observed at 3–4 m and 8–10 m or a single lower peak is found (Fig. 107). No evident physiological differences have been noted between organisms in the upper or lower peaks (Rodhe, personal communication).

We assume that these migrations permit the *Peridinium* to optimize their underwater light climate (Talling, 1971) and to exploit effectively the nutrients in the

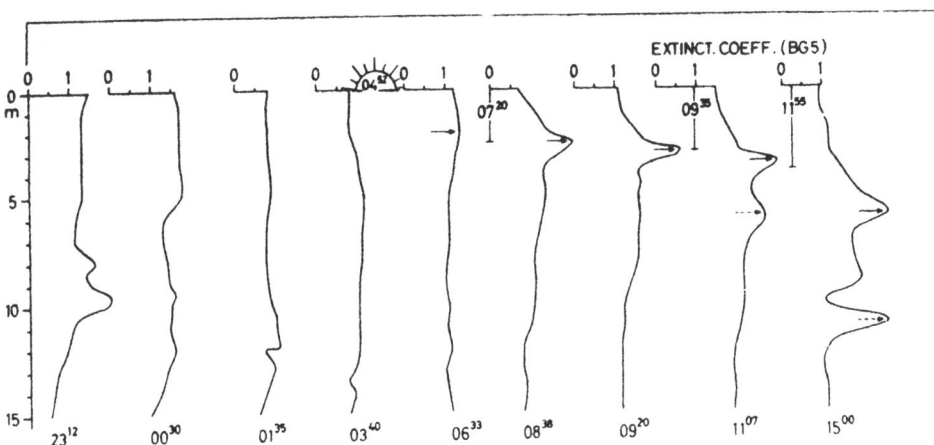

Fig. 107. Vertical distribution of *Peridinium* at different hours of the day. Berman T. & Rodhe W., 1971, Mitt. Internat. Verein. Limnol. 19).

water column. In a recent paper, however, Sibley *et al.* (1974) suggested that the swimming motions of *Peridinium* in Clear Lake were at least partly population-dependent. The vertical swimming ratios of *Peridinium* ranged from 0.7 to 1.2 m hr^{-1} and were rather similar to the rates observed for marine dinoflagellates.

In the laboratory, the motion of *Peridinium* is generally phototrophic – and indeed this property was first utilized to obtain monoalgal cultures by Eren, Rodhe, Berman & Messer. However, in culture flasks, complex swimming motions can often be observed, and it seems probable that chemotactic effects are also important in this phenomenon.

b. Horizontal distribution

Routine sampling at different lake stations indicates an overall pattern of horizontal distribution which has been consistently observed each year.

Deep open waters, as represented by station A, generally have fewer organisms in the trophogenic layer than shallower locations. This layer extends to about 30 m prior to thermal stratification and is limited thereafter to the epilimnion (i.e. to about 15–20 m depth). The northern stations, represented by G, H and F, generally show higher numbers of organisms at earlier dates than other locations. At the peak and late phase of the bloom, all shallow stations (F, G, H, K and L) tend to have greater numbers of organisms than station A in the epilimnetic waters (see p. 252).

Towards the end of the bloom, when the regular pattern of westerly winds during the afternoon has begun (Serruya, 1975), fairly sharply defined strips of *Peridinium* can be seen on the surface close to the western shore.

Berman & Rodhe (1971) described the formation, in April, of a 'strip' formed at right angles to the wind direction about 1 or 2 km off the western shore, possibly by upwelling water bringing the *Peridinium* 'coffee' layer to the surface. These strips were interpreted later as the boundaries of Langmuir cells (Serruya *et al.*, 1978).

During the period when *Peridinium* is located at the surface, the strong winds, common in this lake, tend to concentrate the algal population in distinct patches accumulating near the shoreline. In these patches, the concentration of cells reaches 4,000 to 5,000 cells per ml, that is, one order of magnitude higher than outside the patch.

References

Annandale, N. 1916. A report on the biology of Lake Tiberias. The distribution and origin of the fauna of the Jordan River system, with special reference to that of Lake Tiberias. J. Asiat. Soc. Bengal. 11:437–476.
Barrois, Th. 1894. Contribution á l'etude de quelques lacs de Syrie, Rev. biol. Nord, France, VI: 224–314.
Berg, R. S. 1882. Quoted in Schiller, J. 1937.
Berman, T. 1969. Phosphatase release of inorganic phosphorus in Lake Kinneret. Nature 224:1231–1232.

Berman, T. 1970. Phosphatase and phosphorus availability in Lake Kinneret. Limnol. Oceanogr. 15:663–674.

Berman, T. 1976. Release of dissolved organic matter by photosynthesizing algae in Lake Kinneret, Israel. Freshwat. Biol. 16: 13–18

Berman, T. & I. L. Roth (submitted) Scanning electron microscopy of Peridinium cinctum fa. westii (Lemm.) Lef. J.: Gen. Microbiology.

Berman, T. & Y. Elias. 1973. Lake Kinneret: synoptic studies of chlorophyll concentrations. Proc. 4th Israel Ecol. Soc. Conf. F : 19–27.

Berman, T. & U. Pollingher. 1974. Annual and seasonal variations of phytoplankton chlorophyll and photosynthesis in Lake Kinneret. Limnol. Oceanogr. 19:31–55.

Berman, T. & W. Rodhe. 1971. Distribution and migration of Peridinium in Lake Kinneret. Mitt. Int. Ver. Limnol. 19:266–276.

Boltovskoy, A. 1975. Estructura y estereoultraestructura tecal de dinoflagelatos. II Peridinium cinctum (Müller) Ehrenberg. Physis 34:73–84.

Bourrelly, P. 1957. Reprint 1971. Recherches sur les Chrysophycees. Bibliotheca Phycologica. Bd. 12. Verlag von Cramer, New York. 409 pp.

Bourrelly, P. 1968. Notes sur les Peridiniens d'eau donce. Protistologica 4:5–14.

Bourrelly, P. 1970. Les algues d'eau douce. Tome III. Les Algues bleues et rouges, les Eugleniens, Peridiniens et Cryptomonadines. Ed. Boubée, Paris. 512 pp.

Brun, M. 1883. Desmidiees des vases du lac de Tiberiade. Arch. Mus. Hist. Nat. Lyon. 3.

Carter, J. 1849. Quoted in Schiller, J. 1937.

Carter, J. 1871. Quoted in Schiller, J. 1937.

Carefoot, R. 1968. Culture and heterotrophy of the freshwater dinoflagellate Peridinium cinctum f. ovoplanum Lindeman. J. Phyc. 4:129–131.

Dodge, J. D. 1965. Chromosome structure in the dinoflagellates and the problem of the mesocaryotic cell. Progress in Protozool. 2:p. 264.

Dodge, J. D. 1971. Fine structure of Pyrrophyta. Bot. Rev. 37:481–508.

Dodge, J. D. & R. Crawford. 1970. A survey of thecal fine structure in Dinophyceae. Bot. J. Lim. Soc. 63:53–67.

Dubinsky, Z. 1976. The effect of light and other environmental conditions on the growth of Peridinium cinctum fa. westii in Lake Kinneret and in laboratory cultures Ph.D. dissertation, Bar-Ilan University of Tel Aviv.

Dubinsky, Z. & T. Berman. 1976. Light utilization efficiencies of phytoplankton in Lake Kinneret (Sea of Galilee). Limnol. Oceanogr. 21:226–230.

Dubinsky, Z. & Polna, M. 1976. Pigment composition during a Peridinium bloom in Lake Kinneret (Israel). Hydrobiologia 51:239–243.

Ehrenberg, C. G. 1839. Quoted in Schiller, J. 1937.

Eppley, R. W. 1972. Temperature and phytoplankton growth in the sea. Fish. Bull. 70:1063–1085.

Eren, Y. 1969. Studies on development cycle of Peridinium cinctum f. westii. Verh. Internat. Verein. Limnol. 17:1013–1016.

Gophen, M. 1973. Zooplankton in Lake Kinneret. In: Data record. Isr. Nat. Counc. Res. Dev. Public. 13–73, Jerusalem (Berman, T.).

Gressel, J., T. Berman & N. Cohen. 1975. Dinoflagellate Ribosomal RNA – an evolutionary relict? J. Mol. Evolution 5:307–313.

Harris, G. P. & J. N. A. Lott, 1973. Net phytoplankton production studies. The relationship between light intensity and photosynthetic rates. Journ. Fish. Res. Bd. Canada.

Hepher, B. & J. Langer. 1970. On the primary production of phytoplankton in Lake Kinneret (Tiberias). Bull. Sea. Fish. Res. Stn. Haifa. 55:21–62.

Howland, G. P. & J. Ramus. 1971. Analysis of blue-green and red algal ribosomal-RNAs by gel Electrophoresis. Arch. Mikrobiol. 76:292–298.

Kain, J. M. & G. E. Fogg. 1960. Studies on the growth of marine phytoplankton. III Prorocentrum micans Ehrenberg. J. Mar. Biol. Assoc. U.K. 39:33–50.

Kimor, B. & U. Pollingher. 1965. The plankton algae of Lake Tiberias. Sea Fish Res. Stn. Haifa. Bull. Cont. 7, Ser. A:1–76.

Kok, B. 1960. Efficiency of photosynthesis, p. 563–633. In: W. Ruhland (Ed.) Handbuch der Pflanzenphysiologie 5:part 1, Springer.

Komarovsky, B. 1951. Some characteristic water blooms in Lake Tiberias and fish ponds in the Jordan Valley. Int. Ass. Theor. Appl. Limnol. 14:219–223.

Komarovsky, B. 1959. The plankton of Lake Tiberias. Bull. Sea Fish. Res. Stn Haifa. 25:1–94.

Leadbetter, B. & J. D. Dodge. 1967. An electron microscope study of nuclear and cell division in a dinoflagellate. Arch. Mikrobiol. 57:239–254.

Lewis, W. M. Jr. 1974. Primary production in the plankton community of a tropical lake. Ecol. Monogr. 44:377–409.

Limhof, M. 1884. Quoted in Schiller, J. 1937.

Margalef, R. 1960. Temporal succession and spatial heterogeneity in phytoplankton 323–349. In A. Buzzati-Traverso (Ed.) Perspectives in marine biology. University of California Press, Berkeley.

Messer, G. & Y. Ben Shaul. 1969. Fine structure of *Peridinium westii*, a freshwater dinoflagellate. J. Protozool. 16(2):272–000.

Messer, G. & Y. Ben Shaul. 1970. Ultra structure of aging cells of *Peridinium westii* (Abstr.) 7th Congr. Int. Microsc. Electron. Grenoble, pp. 411–412.

Messer, G. & Y. Ben Shaul. 1971. Fine structure of trichocyst fibrils of the dinoflagellate *Peridinium westii*. J. Ultrastructure Res. 37:94–104.

Muller, O. F. 1773. Quoted in Schiller, J. 1937.

Nevo, Z. & N. Sharon. 1969. The cell wall of *Peridinium westii*, a non cellulosic glucan. Biochim. Biophys. Acta. 173:161–175.

Nordli, E. 1957. Experimental studies on the ecology of Ceratia. Oïkos 8:201–265.

Perty, M. 1849. Quoted in Schiller, J. 1937.

Petit, M. 1883. Liste des Diatomees du lac de Tiberiade. Arch. Mus. Hist. Nat. Lyon. 3.

Pollingher, U. 1972. The Protista of the metalimnic layer of Lake Kinneret, Israel. J. Protozool. 19:(suppl.) 55.

Pollingher, U. & T. Berman. 1975. Temporal and spatial patterns of dinoflagellate blooms in Lake Kinneret, Israel (1969–1974) Verh. int. Ver. Limnol. 19:1370–1382.

Pollingher, U. & T. Berman. 1977. Quantitative and qualitative changes in the phytoplankton of Lake Kinneret, Israel (1972–1975). Oïkos. vol. 29, 3:418–428.

Pollingher, U. & C. Serruya. 1976. Phased division of *Peridinium cinctum* fa. *westii* and the development of the blooms in Lake Kinneret. J. Phycol. 11:155–162.

Rae, P. M. M. 1970. The nature and process of Ribosomal ribonucleic acid in a dinoflagellate. J. Cell. Biol. 46:106–113.

Rayss, T. 1944. Materiaux pour la flore algologique de la Palestine; les Cyanophycées. Palest. J. Bot. Jerusalem 3:94–113.

Rayss, T. 1951. Materiaux pour la floore algologique de la Palestine. II Les algues des eaux continentales. Palest. J. Bot. Jerusalem 5:51–95.

Reffield, A. C. 1958. The biological control of chemical factors in the environment. Am. Sci. 46:205–222.

Rodhe, W. 1972. Evaluation of primary production parameters in Lake Kinneret (Israel). Verh. int. Ver. Limnol.

Ruttner, F. 1952. Planktonstudien der Deutschen Limnologischen Sund-Expedition. Arch. Hydrobiol. (Suppl.) 21(1/2):1–274.

Schiling, A. J. 1891. Quoted in Schiller, J. 1937.

Schiller, J. 1937. Dinoflagellate (Peridineae) 2. Teil in Dr. L. Rabenhorst's Kryptogamen-Flora von Deutschland, Österreich und der Schweiz. Zehnter Band Flagellatae. Leipzig Akademische Verlagsgesellschaft M.B.H. 590 pp.

Schrand, K. 1802. Quoted in Schiller, J. 1937.

Serruya, C. 1972. Metalimnic layer in Lake Kinneret (Israel). Hydrobiol. 40:355–359.

Serruya, C., M. Edelstein, U. Pollingher & S. Serruya. 1974. Lake Kinneret sediments: nutrient composition of the pore water and mud water exchanges. Limnol. Oceanogr. 19:489–508.

Serruya, C. & T. Berman. 1975. Phosphorus, nitrogen and the growth of algae in Lake Kinneret. J. Phycol. 11:155–162.

Serruya, C., Serruya, S. & Pollingher U. 1978. Wind, phosphorus release and division rate of Peridinium in lake Kinneret. Verh. int. Ver-Limnol. 20:(in press).

Serruya, C., Pollingher, U. & Cavari, B. Z., Gophen, M., Landau, R. & Serruya, S. 1978. Lake Kinneret: management options. Symp. on lake metabolism and management, Uppsala 1977.

Serruya, S. 1975. Wind, water temperature and motions in Lake Kinneret – general pattern. Verh. Internat. Verein-Limnol 19:73–85.

Sibley, T. H., P. L. Herrgesell & A. W. Knight. 1974. Density dependent vertical migration in the freshwater dinoflagellate *Peridinium penardii* (Lemm) Lemm. fa. *californicum Javorn.* J. Phycol. 10:475–476.

Skuja, H. 1948. Takonomie des Phytoplanktons einiger Seen in Uppland, Schweden. Symb. Bot. Ups. IX:399 pp.

Stein, F. 1883. Quoted in Schiller, J. 1937.

Talling, J. F. 1969. Comparative problems of phytoplankton production and photosynthetic productivity in a tropical and temperate lake. In: Primary productivity in aquatic environments. C. R. Goldman (ed.) Univ. California Press, 1969.

Talling, J. F. 1971. The underwater light climate as a controlling factor in the production ecology of freshwater phytoplankton. Mitt. Intern. Verein. Limnol. 19:214–243.

Talling, J. F., R. B. Wood, M. V. Prosser & R. W. Baxter. 1973. The upper limit of photosynthetic productivity by phytoplankton; evidence from Ethiopian lakes. Freshwater Biol. 3:53–76.

Thomas, W. H. & A. N. Dobson. 1972. On nitrogen deficiency in tropical Pacific oceanic phytoplankton. II. Photosynthetic and cellular characteristics of a chemostat grown diatom. Limnol. Oceanogr. 17:515–523.

Thomas, R. N., E. R. Cox & K. A. Steidinger. 1973. *Peridinium balticum* (Levander) Lemmerman, an unusual dinoflagellate with a mesocaryotic and an eucaryotic nucleus. J. Phycol. 9:91–98.

Oye van & A. Gillard. 1949–1950. Contribution à la connaissance de la distribution géographique de quelques chlorophycées en Belgique. Hydrobiol. 2:322–334.

Venkataraman, G. S. & S. C. Mehta. 1961. The thecal structure of *Peridinium cinctum* Lloydia 23:115–117.

Wyatt, T. & J. Horwood. 1973. Model which generates red tides. Nature 244:238–249.

Yashuv, A. & M. Alhunis. 1961. The dynamics of biological processes in Lake Tiberias. Bull. Res. Counc. Isr. 10B:11–35.

II. Zooplankton

M. Gophen

A. History

The work of Richard (1890) is among the first reports concerning the zooplankton of Lake Kinneret. This author worked on the material collected by Barrois. The observation by Barrois that *Cyclops leuckarti* is most abundant at 1.5 m depth in the early evening is the first indication of the vertical migration of this species in Lake Kinneret. Gurney (1913), who described the zooplankton collected by Annandale in 1912, mentioned *Daphnia lumholtzi* which was not found by Barrois. Bodenheimer (1935) reported four species of Cyclopoida, among them *Cyclops varicans*. This species was mentioned by Barrois not as a lake species but as a dweller of a 'marecage situé à l'Est de l'embouchure Nord du Jourdain dans le lac de Tiberiade'. It is obvious that this species is only a 'visitor' brought by the Jordan River floods in winter and does not belong to the actual lake fauna.

In 1948, the Sea Fisheries Research Station of Haifa initiated a biological investigation of Lake Kinneret. Dr. H. Lissner, the director of the Station, took regular net plankton samples at two stations located on the western side of the lake from November 1948 to October 1949. This material was examined by Komarovsky, who continued the sampling at the same stations in 1950. This author examined also the plankton collected at discrete depths with a water sampler by O. H. Oren, the hydrographer of the Station. The paper of Komarovsky (1959) supplies the first quantitative data concerning the seasonal fluctuations of total plankton in the lake and a detailed list of 22 species of zooplankton.

From May 1956 to July 1957, Yashuv and Alhunis took plankton samples at a shallow station in the southwestern part of the lake. In 1961, these authors published quantitative information concerning the seasonal variations of total plankton. They also showed the relative contribution of phytoplankton and zooplankton to the total plankton biomass at different periods of the year. Their list of zooplankton includes 35 species.

In 1968, the Kinneret Limnological Laboratory initiated an interdisciplinary survey of the lake, and Gophen started a detailed and quantitative investigation of the zooplankton based on sampling at seven stations. The samples were taken at discrete depths with a water sampler and filtered. This sampling technique, combined with periodic measurements of the biometric parameters of the different animals, permitted an accurate determination of the zoobiomass (Gophen, 1972, 1973).

B. The zoo-plankton population

1. The zooplankton of Lake Kinneret

List No. 6 records the forms of the planktic fauna found by Gophen during his continuous survey over the period 1969–1976. It must be emphasized that this list refers only to lacustrine species of the offshore deeper than 10 m and does not include littoral species and fauna carried by the rivers.

We note that the last record of *Daphnia lumholtzi* in Lake Kinneret appears in the list of zooplankton published by Komarovsky in 1959; this list, however, is based on samples taken during the period 1948–1950. In spite of the extensive sampling (thousands of samples taken all over the lake) carried out since 1969, this species was never found. *Daphnia lumholtzi* probably disappeared from the lake in the late fifties or early sixties.

With regard to the species found for the first time in Lake Kinneret by Gophen, it is difficult to decide whether these species recently appeared in the lake or whether they existed before but were not found by previous workers because of different sampling techniques.

List No. 6

Zooplankton of Lake Kinneret

Copepoda
Cyclopidae
Mesocyclops leuckarti (Claus)
Thermocyclops dybowskii (Lande)
Acanthocyclops languidoides (Lilljeborg)
Microcyclops minutus (Claus)
Eucyclops serrulatus (Cyclops agilis) (Koch, Sars)
Calanoidae
Eudiaptomus gracilis (Sars)

Cladocera
Sididae
Diaphanosoma brachyurum (Lieven)
Bosminidae
Bosmina longirostris (O. F. Müller)
Bosmina longirostris var. *cornuta* (Jurine)
Daphnidae
Ceriodaphnia reticulata (Jurine)
Ceriodaphnia rigaudi (Richard)
Moina rectirostris (Leydig)
*Daphnia magna (Baird)
Chydoridae

Alona affinis (Leydig)

Rotifera
Brachionidae
Keratella (*Anuraea*) *cochlearis* (Gosse)
Keratella (*Anuraea*) *valga tropica* (Ehrenberg)
Brachionus budapestinensis (Daday)
*Brachionus calyciflorus (Pallas)
Brachionus angularis (Gosse)
Platyas patulus (Harring)
Asplanchnidae
Asplanchna brightwelli (Gosse)
Asplanchna priodonta (Gosse)
Asplanchna sieboldi (Leydig)
Synchaetidae
*Synchaeta oblonga (Ehrenberg)
Synchaeta pectinata (Ehrenberg)
Polyarthra remata (Skorikov)
Polyarthra vulgaris (Carlin)
Testudinellidae
Filinia longiseta (Ehrenberg)
Gastropidae

Ascomorpha saltans (Bartsch)
Trichocercidae
Trichocerca stylata (Gosse)
Hexarthridae

Pedalia fennica (Ehrenberg)
*Collotheca sp. (Harring)
Kellicotia longispina (Kellicot)

* Recorded for the first time in Lake Kinneret by Gophen.

2. The zooplankton of the River Jordan

List No. 7 gives the zooplankton species found by Gophen (1972) in a one-year survey (1971–1972), based on weekly samples taken in the River Jordan at Almagor Bridge.

The planktic fauna of the Jordan River is dominated by rotifers, common in the channels of the river and in the fishponds. These animals are brought in large amounts to the lake during the flood period. Therefore, the peak of rotifers occurs in the lake in the winter rainy period (December–May). Most of these species are found in the lake according to the pulsatic pattern of the floods and are generally not found in summer.

List No. 7

Zooplankton of the River Jordan

Copepoda
Cyclopidae
Mesocyclops leuckarti (Claus)
Microcyclops minutus (Claus)

Cladocera
Bosminidae
Bosmina longirostris (O.F. Müller)
Daphnidae
Ceriodaphnia reticulata (Jurine)
Chydoridae
Alona affinis (Leydig)

Rotifera
Brachionidae
Brachionus angularis (Gosse)
Brachionus calyciflorus (Pallas)
Brachionus caudatus (Barrois and Daday)
Keratella cochlearis (Gosse)
Keratella valga tropica (Ehrenberg)

Keratella quadrata (O. F. Müller)
Asplanchnidae
Asplanchna brightwelli (Gosse)
Asplanchna priodonta (Gosse)
Synchaetidae
Synchaeta oblonga (Ehrenberg)
Synchaeta pectinata (Ehrenberg)
Polyarthra spp. (Ehrenberg)
Testudinellidae
Filinia longiseta (Ehrenberg)
Gastropidae
Ascomorpha saltans (Bartsch)
Trichocercidae
Trichocerca stylata (Gosse)
Hexarthridae
Pedalia fennica (Ehrenberg)
Collotheca sp. (Harring)
Kellicotia sp. (Kellicot)

Nematoda spp.

3. Biomass of zooplankton in the lake

a. Contribution of main groups and species (Fig. 108)

The biomass of zooplankton in Lake Kinneret ranges from 24 to 56 g m^{-2} wet

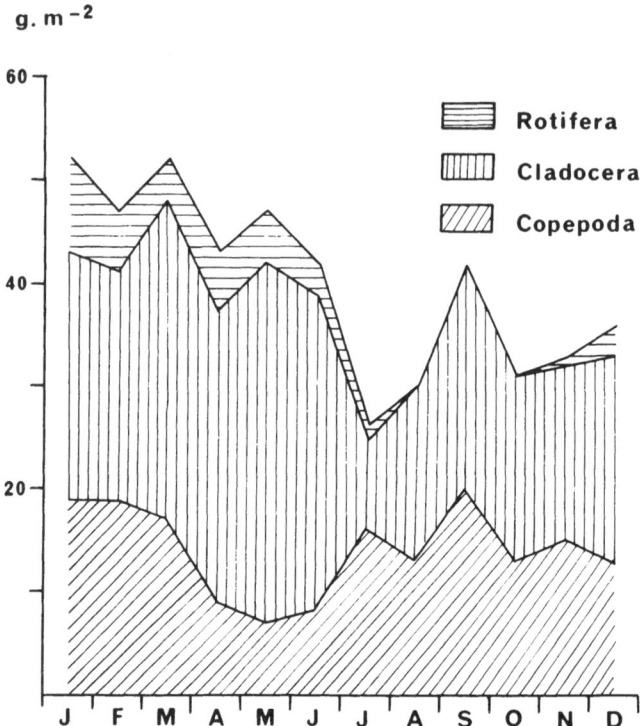

g. m^{-2}

Rotifera

Cladocera

Copepoda

Fig. 108. Seasonal variations of zooplankton biomass. Period 1969–1975.

weight. The average contribution of the different groups, based on the results of seven years, is as follows: Cladocera 58%, Copepoda 35% and Rotifera 7%.

Among the Cladocera, the main contributors to the biomass are *Ceriodaphnia reticulata* (38% of the total biomass of Cladocera), *Diaphanosoma brachyurum* (30%), *Bosmina longirostris* (21%) and *Ceriodaphnia rigaudi* (6%)

Mesocyclops leuckarti represents 90% of the biomass of copepods, whereas *Thermocyclops dybowskii* contributes approximately 9%. This emphasizes the minor role played by all the other copepods.

b. Seasonal fluctuations

In Table 41, the monthly averages of biomass for the main groups of zooplankton are presented. These averages are based on biweekly sampling during the seven-year period, 1969–1975.

The total biomass is greatest during the first six months of the year (42–55 g m^{-2}) and decreases in summer and fall to values ranging from 25 to 36 g m^{-2}, excluding the September value due to the unusually high biomass of Copepoda and Cladocera observed in September 1970.

With the exception of a minimum from April to July, the biomass of copepods varies very little and remains around an average of 18 g m^{-2}. Conversely, the

Table 41. Monthly average of biomass of zooplankton in Lake Kinneret based on the data obtained at stations A, C, D, F, G over the period 1969–1975. Results in g m^{-2} wet weight (standard deviation)

Months	Copepoda	Cladocera	Rotifera	Total
I	19 (5)	24 (15)	9 (7)	55 (14)
II	19 '7)	22 (7)	6 (6)	46 (12)
III	17 (9)	31 (11)	4 (3)	51 (19)
IV	9 (1)	28 (15)	6 (4)	43 (12)
V	7 (3)	35 (13)	5 (3)	46 (13)
VI	8 (3)	31 (13)	3 (2)	42 (12)
VII	16 '7)	9 (7)	1 (1)	25 (6)
VIII	13 (3)	17 (9)	0	30 (10)
IX	20 (14)	22 (18)	0	43 (31)
X	13 (3)	18 (4)	0	31 (6)
XI	15 (5)	17 (8)	1 (1)	32 (13)
XII	13 (3)	20 (6)	3 (2)	36 (9)
Average	14 (3)	23 (5)	3 (2)	40 (6)
% of total	35	58	7	100

biomass of Cladocera is greatest in winter (22–35 g m^{-2}) and drops in summer (9–23 g m^{-2}). The Rotifera disappear in summer.

c. Vertical distribution

During the stratified period, the organisms are restricted to the epilimnic and metalimnic layers. Two distinct peaks can be observed (Fig. 109): an upper peak between 5 and 7 m depth and a lower peak in the metalimnic layer which is located at 14 m in May, reaches 20 m in September and gradually deepens later in the year. A minimum of biomass was noted at about 10 m. A large number of organisms was found in the metalimnic layer, the copepods being the dominant group. The large number of Rotifera found in the lake in 1969 was due to the high floods of winter 1968–1969 which brought to the lake an unusually high number of these animals. Such a population of Rotifera is very untypical of the more "normal years" in Lake Kinneret.

It seems that the concentration of organisms in the metalimnic layer is caused by the accumulation, at the end of the *Peridinium* period, of large amounts of organic detritus and the consequent proliferation of bacteria in this layer. Experiments have shown that bacteria are excellent food for the nauplii of *Mesocyclops leuckarti*, and that cladocerans such as *Ceriodaphnia reticulata* feed selectively on bacteria (Gophen *et al.*, 1974).

During the homothermal period, zooplankton organisms are found at all depths but do not necessarily have a uniform distribution. In the upper layers, the number of organisms is very variable, depending on the weather and the hour of sampling. The organisms generally concentrate in the upper layers in early morning and migrate downwards on bright days; if the sky is cloudy, they remain near the surface throughout the day period. Frequently, high numbers of organisms (100–200 per liter) are found at a depth of 40 m.

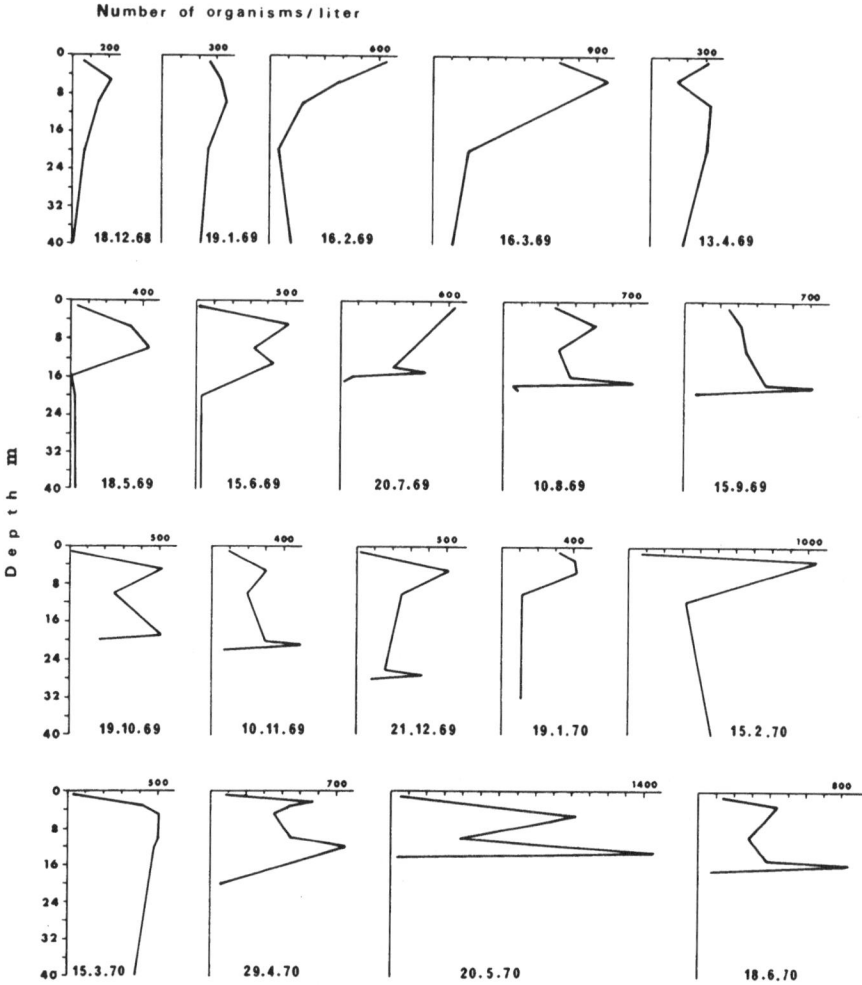

Fig. 109. Seasonal fluctuations of the vertical distribution of zooplankton. Number of organisms per liter. (From Gophen, M., 1972, Isr. J. Zool., 21.)

C. *Mesocyclops leuckarti* (Claus)

1. General data

The average biometric data of the different stages of *Mesocyclops leuckarti* are shown below:

Stages	Volume 10⁶ μ^3	Dry weight μg
Egg	0.30	—
Small nauplius (stages 1–3)	0.58	—
Large nauplius (stages 4–5)	1.21	—
Copepodite I	2.95	—
Copepodite II	3.88	—
Copepodite III	5.86	—
Copepodite IV ♂	7.59	1.0
Copepodite V ♂	8.05	—
Copepodite V ♀	12.55	1.6
Adult ♂	11.30	1.2
Adult ♀	24.00	2.4

In summer, the adult females are approximately 30% smaller than in winter.

2. Feeding

The gut analysis of the different stages of *Mesocyclops leuckarti* and the results obtained from laboratory cultures indicate that nauplius and copepodite I to III feed on bacteria, small algae and small protozoans, while the latter stages are mostly carnivorous and feed on Cladocera. *Ceriodaphnia* and *Diaphanosoma* represent 70 to 90% of the preyed material, but *Bosmina* is not preyed at all. The remaining 10 to 30% of the diet is composed of Chironomid larvae, Copepoda (cannibalism), detritus and algae.

3. Metabolic rates

Gophen (1976), working on cultures of adult females fed with *Artemia* nauplii, investigated the influence of temperature on the metabolic rates.

The results presented in Table 42 indicate that the rate of food intake increases rapidly with temperature and then remains steady; from 15 to 22°C, the rate of food intake increased by 360% but from 22 to 27°C it increased only by 20%. A similar pattern is observed for the excretion of ammonia. In contrast, the rate of oxygen consumption increased more rapidly between 22 and 27°C than between 15 and 22°C.

Table 42. Food intake (wet weight), NH_3–N excretion and oxygen consumption in mg g^{-1} wet weight of body content per day in adult females of *Mesocyclops leuckarti* at three temperatures. Standard deviations are given in parenthesis: ten organisms were tested. (From Gophen M. 1976 Oecologia, 25)

Temp °C	Food intake mg $g^{-1} d^{-1}$	NH_3–N excretion mg $g^{-1} d^{-1}$	Oxygen consumption mg $g^{-1} d^{-1}$
15	282 ($\pm$51)	0.5 ($\pm$0.3)	17.2 ($\pm$3.3)
22	1,299 ($\pm$254)	1.2 ($\pm$0.3)	30.6 ($\pm$4.0)
27	1,525 ($\pm$203)	1.3 ($\pm$0.2)	53.6 ($\pm$4.8)

4. Development rates

Gophen (1976) also studied the influence of temperature on the development rates of the different stages and on the production of eggs by adult females.

The results presented in Table 43 show that the development rates are considerably influenced by temperature; at low temperatures, males develop more rapidly than females. Their lifespan is shorter than that of females, especially at high temperatures. It is worth noting that the parameters related to the reproductive cycle are more influenced in the 15–22°C range than in the range 22–27°C, in particular, from 15 to 22°C the egg production increases by 230%.

Table 43. Duration time of egg development, instars, life span of adults in days and breeding data of females (number of eggs laid, percentage of hatching, laying period and number of clutches per female) at three temperatures. Standard deviations are given in parenthesis (n = 40). (Same source as Table 42)

	15°C		22°C		27°C	
	Male	Female	Male	Female	Male	Female
Egg	6.2		1.4		0.8	
Nauplius	25 ($\pm$8.0)	28 ($\pm$10.0)	15 ($\pm$2.0)	16 ($\pm$2.0)	9. ($\pm$2.0)	9 ($\pm$2.0)
Copepodite	27 ($\pm$2.0)	39 ($\pm$0.4)	10 ($\pm$0.4)	15 ($\pm$0.5)	10 ($\pm$0.3)	10 ($\pm$0.3)
Adult	76 ($\pm$31.0)	149 ($\pm$21.0)	62 ($\pm$21.0)	62 ($\pm$19.0)	8 ($\pm$4.0)	42 ($\pm$19.0)
Total number of eggs	27		100		93	
Hatching %	30		67		86	
Laying period (days)	17		18		10	
Number of clutches	2		4		4	

The accelerating effect of increasing temperature is not uniform on all the physiological functions of *Mesocyclops*. Obviously, the increase of temperature affects the pattern of utilization of energy: at high temperature, the increment of energy resulting from the higher food intake is channeled preferentially towards reproduction. The overall balance of energy is characterized by a smaller flow of

306

energy to maintenance. This generates a shorter longevity and smaller body size of the adults. This might, of course, lead to a decline of the population. However, these two factors are compensated for by the considerable increase of the rate of development of larval stages and egg production of the adult females. This allows *Mesocyclops leuckarti* to maintain abundant populations in summer, when, as we shall see, the losses of nauplius instars are very high.

5. Seasonal fluctuations of the various stages of *Mesocyclops leuckarti* in Lake Kinneret

The population of *Mesocyclops leuckarti* in Lake Kinneret has a very unusual age composition. The average ratio of adult: copepodite: nauplius has been found to be 1:1:2 in a north Russian lake (Meshkova, 1953), 1:1:4 in summer and 1:2:1 in winter in Lago Maggiore (Ravera, 1954), and 1:5:3 in Lake George (Burgis, 1971). In Lake Kinneret, the unusual value of this ratio (1:2:9) emphasizes the large number of nauplius in comparison with the other stages of the population. In Fig. 110, we see that, from January to September, nauplius appear in very large numbers, especially from July to September, but they decrease by 80% in the last three months of the year. The copepodites and adults are at their lowest from April to June and increase by 70% in summer; in contrast to the nauplii, their number remains high until the end of the year. These variations explain why the ratio corresponding to the last three months of the year (1:3:4) is very different from the annual average.

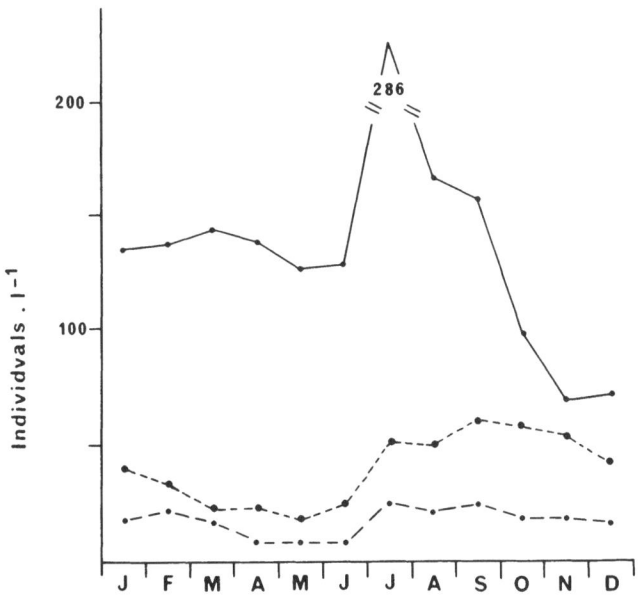

Fig. 110. Seasonal distribution of *Mesocyclops leuckarti.* Number of individuals per liter ——— Nauplius, •– – – –• Copepodite, •— —• Adult. Averages for the period 1969–1975.

6. Productivity and P:B ratio

The productivity of *Mesocyclops* has been calculated on the basis of the development times of all stages at the temperatures corresponding to those prevailing in the lake and on the basis of the biometric data (Fig. 111). The summer values of productivity are approximately twice as high as the winter values.

The daily P:B has an annual average value of 0.10, with summer values ranging from 0.11 to 0.17 and winter values ranging from 0.05 to 0.08.

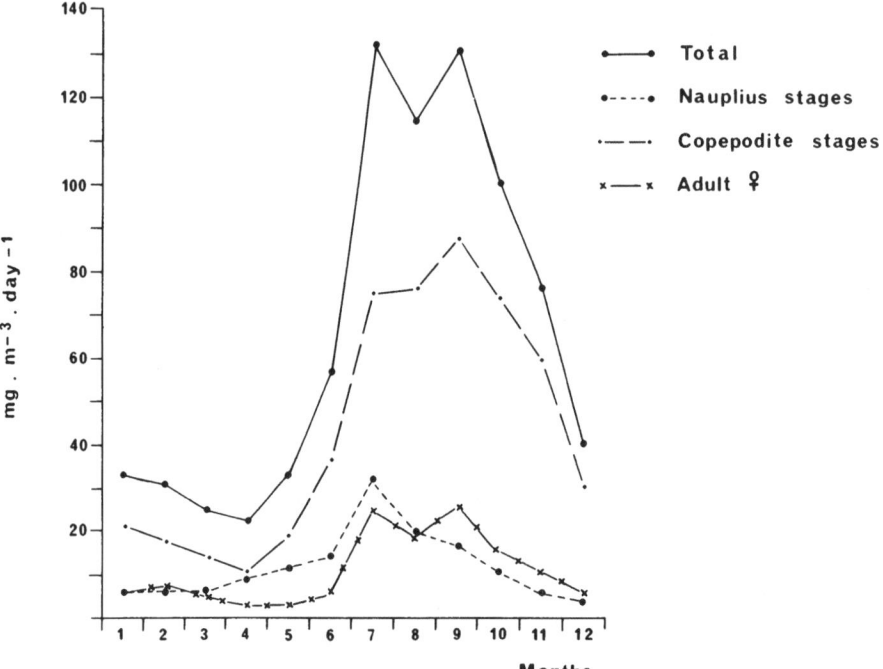

Fig. 111. Productivity of *Mesocyclops leuckarti* in Lake Kinneret. Monthly averages for the period 1969–1975.

D. *Ceriodaphnia reticulata* (Jurine)

1. General data

In Lake Kinneret, *Ceriodaphnia reticulata* has a volume ranging from 9.4 to 30.2 $\times$ 10^6 μ^3 with corresponding dry weight of 0.8 and 2.1 μg. It is abundant in the lake during a short period (March–June), with a peak in May when the temperature of the epilimnion is 21–22°C.

The main species of fish feeding on zooplankton, *Mirogrex terraesanctae terraesanctae* has a negative index of electivity for *Ceriodaphnia* spp. It therefore appears that temperature is probably one of the main factors which determines the seasonal distribution of *Ceriodaphnia reticulata* in Lake Kinneret.

2. Feeding

Gophen *et al.* (1974), by using a technique of differential radioactive isotopic labelling, demonstrated that *Ceriodaphnia reticulata*, when provided with algae and bacteria, shows a preferential assimilation of bacteria. In this animal, which feeds by filtration of large amounts of water, discriminative grazing is unlikely. The greater assimilation of bacteria in comparison with algae is probably related to the relative ease with which the bacterial cells are lysed by *Ceriodaphnia* in contrast to algae with their cell walls of cellulose.

3. Metabolic rates and body size

The influence of temperature on the rates of food intake, excretion of ammonia and oxygen consumption has been investigated (Gophen, 1976). The results are shown in Tables 44 and 45.

The rates of food intake and ammonia excretion increased by 142% and 90% respectively when the temperature was elevated from 15 to 22°C. A lower increase was observed with a temperature increase from 22 to 27°C. Conversely, the respiration rate increased by 6% between 15 and 22°C and by 154% for the temperature rise from 22 to 27°C.

Table 44. Food intake (wet weight), NH_3–N excretion and oxygen consumption in mg g^{-1} wet weight of body content per day for *Ceriodaphnia reticulata* at three temperatures. Standard deviations are given in parenthesis (n = 10). (From Gophen M. 1976, Freshwater Biology, 6)

Temp °C	Food intake mg g^{-1} d^{-1}	NH_3–N excretion mg g^{-1} d^{-1}	Oxygen consumption mg g^{-1} d^{-6}
15	397 ($\pm$36)	1.0 ($\pm$0.1)	16.3 ($\pm$3.4)
22	960 ($\pm$125)	1.9 ($\pm$0.2)	17.3 ($\pm$4.1)
27	1,026 ($\pm$174)	2.4 ($\pm$0.3)	44.0 ($\pm$6.2)

Table 45. Atomic O/N ratios for *Ceriodaphnia reticulata* at three temperatures (15, 22 and 27°C), in mg-atoms/g wet weight of body content day. (Same source as Table 44)

Temp (°C)	Respiration mg-atoms/$g_{(ww)}$ day (O)	NH₃-N excretion mg-atoms/$g_{(ww)}$ day (N)	O/N
15	1.02	0.07	14.57
22	1.08	0.14	7.71
27	2.75	0.17	16.18

In March, the average dry weight of an adult animal was found to be 4.1 μg ($\pm$1.1) but only 2.1 μg ($\pm$1.1) in August.

One would expect the increase of respiration with temperature to be accompanied by an increase of food intake. However as shown in Table 44, the rates of food intake and respiration are not directly correlated. In the filter feeder animals, the beating of the thoracic appendages have two functions, supply of oxygen and supply of food. With increasing temperature, the rate of movements of the appendages increases in order to supply more oxygen, but these movements also cause a quicker filling of the gut. However, it seems that the digestion and defecation processes are not fast enough and the tension on the gut wall lowers the number of particles passing through the mouth of the gut. The increase of temperature causes an increase of respiration and metabolic demand which cannot be satisfied because the digestion and defecation processes are much less influenced by temperature than most of the other physiological functions. The organisms rapidly suffer from a disequilibrium between the requirements in energy and the capability of the digestive tract to supply it. This disequilibrium results in a reduction of body size and egg production: in the period January–April, 20% of the females carry eggs compared to 1% from May to September.

The optimum temperature which permits an equilibriated metabolism is around 22°C. This explains the timing of the *'Ceriodaphnia period'* and the disappearance of this animal in summer when the temperature of the epilimnion is above 28°C.

References

Bodenheimer, F. S. 1935. Animal life in Palestine. 506 p.

Burgis, M. J. 1971. The ecology and production of Copepodes, particularly *Thermocyclops hyalinus*, in the tropical Lake George, Uganda. Freshwat. Biol. 1:169–192.

Gophen, M. 1972. Zooplankton distribution in Lake Kinneret (Israel), 1969–1970. Israel J. Zool. 21:17–27.

Gophen, M. 1973. Zooplankton. In: Lake Kinneret Data Record. N.R.C.D. 13–73.

Gophen, M. 1976. Temperature effect on lifespan, metabolism and development time of *Mesocyclops leuckarti* (Claus). Oecologia 25:271–277.

Gophen, M. 1976. Temperature dependence of food intake, ammonia excretion and respiration of *Ceriodaphnia reticulata* (Jurine) (Lake Kinneret, Israel). Freshwat. Biol. 6:451–455.

Gophen, M., B. Z. Cavari & T. Berman. 1974. Zooplankton feeding on differentially labelled algae and bacteria. Nature 247:393–394.

Gurney, R. 1913. Entomostraca from Lake of Tiberias. J. As. Soc. Bengal. IX : 231–232.

Komarovsky, B. 1959. The plankton of Lake Tiberias. Bull. Res. Counc. Israel 8B (2):65–69 and Bull. Sea Fish. Res. Stn. Haifa (25).

Meshkova, T. M. 1953. Zooplankton of Lake Sevan. Tr. Sevansk. Hidrobiol. st. 13.

Ravera, O. 1954. La Struttura demografica dei Copepodi del lago Maggiore. Mem. 1st Ital. Idrobiol. 8 : 109–150.

Richard, J. 1890. Copepodes recueillis par. Mr. le Dr. Barrois en Egypt, en Syrie et en Palestine. Revue biol. Nord. France 5 : 400–405, 433–443, 458–475.

Yashuv, A. & Alhunis, M. 1961. The dynamics of biological processes in Lake Tiberias. Bull. Res. Counc. Israel 10B, 1–2 : 12–35.

III. Bacteria

B. Z. Cavari

The ecology of bacteria in natural waters is a recent field of study. After the pioneer work of Waksman, ZoBell and Wood in marine environment and of Kuznetsov in lakes, the study of the role of bacteria in chemical transformations in seas and lakes became a popular subject.

With the exception of the counting of pathogenic bacteria in shallow water by Mekorot and the Ministry of Health, there is no published work concerning bacteria in Lake Kinneret. The following is then the first study of bacteria involved in the nitrogen cycle in the lake.

The various bacterial phases of the nitrogen cycle in natural waters, mainly denitrification and nitrogen fixation, have been investigated by Wilson (1958), Dugdale & Dugdale (1962), Jannasch (1960), Stewart (1964), and others.

In Lake Kinneret, the nitrogen budget (Serruya, 1971, 1975) indicated for the first time the probable existence of denitrification as well as its quantitative importance (see p. 198). This finding led to the initiation of a detailed microbiological study of the bacteria involved in denitrification, nitrification and nitrogen fixation.

A. Bacteria of the nitrogen cycle

1. Denitrification

Among the factors affecting the rate of disappearance of nitrates added to lake water in the range of temperatures measured in Lake Kinneret (15–30°C), the concentration of nitrates plays a dominant role. Within the range of nitrate concentrations actually measured in Lake Kinneret, we can see in Fig. 112 that the rate of denitrification varies by one full order of magnitude. The linear increase of the rate of denitrification with the nitrate concentration provides the lake with a remarkable mechanism of regulation of the nitrogen reserve. This explains the considerable differences found in the nitrogen budgets concerning the absolute amounts of nitrogen denitrified from year to year. In 1968–1969 and 1969–1970, the maximum concentrations of nitrate – N in the lake were respectively 1,000 and 220 ppb: the amounts of nitrogen which were denitrified were respectively 3,000 tons and 850 tons.

The process of denitrification depends on the availability of the organic compounds present in the lake water. The algal bloom constitutes the main source of organic matter which, after the bloom dies off, becomes partly available to heterotrophic organisms. In this respect, the amount of nitrate which can undergo denitrification is related to the biomass of algae.

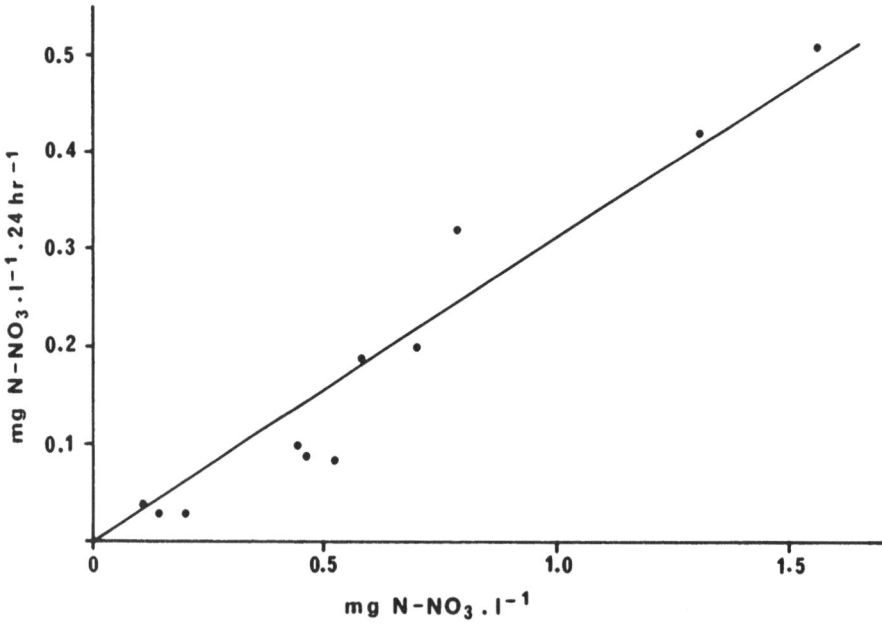

Fig. 112. Rates of denitrification as a function of added nitrate.

In the fight against eutrophication, much effort is devoted to the reduction of the biomass of algae. The amount of nitrates entering the lake should be decreased simultaneously, since the reduction of algal biomass diminishes the denitrification potential of the water.

The chemical balances indicate that most of the nitrogen losses due to denitrification occur almost every year in March–April, when there is still oxygen in the whole water column. At this period, concentrations of 4 ppm oxygen can be found in the deep layers. It follows that the massive *in situ* denitrification of nitrates occurs in the presence of low levels of oxygen. This phenomenon can be explained by the formation of anaerobic microenvironments in the deep layers where the mixing processes are reduced, in the period just preceding thermal stratification. This 'aerobic denitrification' could be reproduced experimentally by incubating pure cultures of *Pseudomonas aeruginosa* isolated from the lake under aerated conditions in the presence of nitrate and an energy source such as glucose. Conversely, in a parallel experiment, conducted under anaerobic conditions, denitrification took place without any addition of glucose. This finding emphasizes the importance of an available carbon source during the period of denitrification. The shortage of such a source due to the shorter and lesser bloom of *Peridinium* in 1975 might explain why the denitrification process was considerably delayed in 1975 compared to previous years.

2. Nitrification

Through nitrification, the labile ammonia is converted into nitrate, the most oxidized and stable form of nitrogen. Nitrification is a bacteria-mediated process; it is a two-stage reaction involving two groups of organisms: the Nitrosomonas family, which transforms ammonia into nitrite, and the Nitrobacter family, which converts nitrite into nitrate. Both types of organisms obtain their energy from the oxidation of ammonia and nitrite and their cellular carbon from the reduction of CO_2. A variety of heterotrophic bacteria and fungi are known to be capable of nitrification. In Lake Kinneret, the latter organisms play no significant part in the nitrification process. This was demonstrated by the total inhibition of nitrification obtained by the addition of N-serve, a specific inhibitor of autotrophic nitrification.

In January and February, ammonia is distributed homogeneously at all depths. In March, we generally observe a drop in the concentration of ammonia accompanied by an equivalent increase of nitrate. The decline in ammonia concentration is more pronounced in the deeper than in the upper layers. This suggests that the algae, and particularly *Peridinium*, abundant in this period, compete successfully with bacteria in the uptake of ammonia or repress the general activity of bacteria. From this we hypothesized that nitrification is probably of minor importance in the euphotic zone during the bloom period.

The fluctuations of the 'ability of the water to nitrify' (nitrification potential = NP) with depth and time have been measured as follows: water sampled at different depths and enriched in ammonia was incubated at 30°C in a dark shaker. The concentrations of ammonia, nitrite and nitrate were measured daily. The time necessary to oxidize a given amount of ammonia is inversely proportional to the

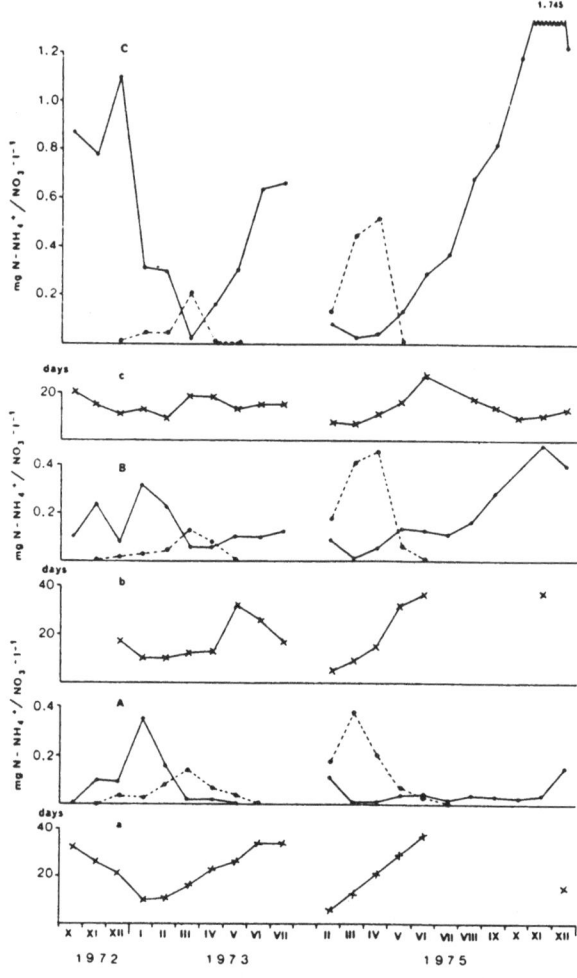

Fig. 113. Ammonia and nitrate concentration at 5 m (A), 25 m (or thermocline) (B) and 40 m (C). Nitrification potential (number of days for complete ammonia oxidation) at 5 m (a), 25 m (or thermocline) (b) and 40 m (c). •———• Ammonia, •–––––• Nitrate. (From Cavari, B.Z., 1977, Oïkos, 28.)

nitrification potential.

The results obtained are summarized in Fig. 113. From June to November, the nitrification potential was extremely low with the exception of the 40 m depth samples. From January to April, the potential was high in the whole water column; in April, during the maximum of the *Peridinium* bloom, the potential decreased in the upper layer. The sequence and timing of the previously-described events show that, in April, most of the lake ammonia was either taken up by algae or converted by nitrifiers into nitrate which was rapidly denitrified by the denitrifying bacteria.

We can also see in Fig. 113 that the nitrification potential at 40 m depth remained high throughout the year, even when, in summer, there was no oxygen at this depth. The sediments, tested according to the same procedure, also showed high values of nitrification potential. The constantly high values of NP at 40 m and in the sediments indicate that nitrifying bacteria can remain alive for several months in anaerobic sediments and resume their activity when oxygen is supplied.

3. Nitrogen fixation

The ability of certain organisms to fix the gaseous dissolved nitrogen permits the development of these organisms even when no other source of mineral dissolved nitrogen is available. Different organisms are able to fix atmospheric nitrogen, among them a number of blue-green algae, photosynthetic bacteria, various anaerobic and aerobic bacteria and many facultative bacteria.

In Lake Kinneret, few species of N-fixing, blue-green algae are found in the planktic community; in the southern part of the lake, benthic colonies of Nostoc have been observed on shallow sediments. Moreover, every year a bloom of photosynthetic bacteria is observed in the metalimnion. The presence of microorganisms capable of N-fixation does not necessarily imply the actual occurrence of fixation. Therefore, nitrogen fixation was determined in lake water by using the acetylene-ethylene technique and the results were negative. Negative results were also obtained in layer of the photosynthetic bacteria. Laboratory experiments have shown that the absence of nitrogen fixation in this layer is due to the limiting number of bacteria and the limiting light intensities.

In the sediment, a slight fixation activity was found at different stations. The average rate observed was about 6 ng N_2 fixed per 1 gr of dry sediment.

B. Photosynthetic sulfur bacteria

Photosynthetic sulfur bacteria. *Chlorobium phaeobacteroides*, have been isolated from the anoxic, sulfide-containing waters of Lake Kinneret. Dense populations (10^7 cell ml^{-1}) of this organism develop in the metalimnic layer of the lake and form a red-brownish layer from July to September almost every year. Laboratory experiments with ^{3}H-labelled *Chlorobium* have shown that these bacteria can be an important food source for zooplankton. The optimal light intensity for the growth of *Chlorobium* has been found to be 250–400 foot-candles; the generation time is then eight hours. In the lake, the light intensity which reaches the bacterial layer is approximately two orders of magnitude less than that of optimal growth. The maximum specific photosynthetic rate of these bacteria was about 20 μg C mg protein^{-1} hr^{-1}. In the bacterial layer, the rate did not exceed 0.25 μg C mg protein^{-1} hr^{-1}.

Glucose and acetate increased the growth rate when added to the medium. The rapid development of the *Chlorobium* in the lake despite very low light intensities found in this area may be explained by myxotrophic growth, which gives a distinct ecological advantage to these organisms.

Double labelling experiments with NaH^{14}CO$_3$ and either glucose or acetate labelled with ^{3}H have shown that addition of organic compounds represses, to some extent, the photosynthetic inorganic carbon uptake. It was also found that, at low light intensity, the concentration of organic compounds in the medium has a greater influence on growth than at high light intensity. The results of this experiment may explain the absence of the *Chlorobium* bloom in 1975. The total algal biomass produced during the bloom of this year was about one third of the usual values. It we assume that the algal bloom is the source of available organic matter for the photosynthetic bacteria, it is probable that the organic matter was limiting to the sulfur bacteria.

References

Dugdale, V. A. & R. C. Dugdale. 1962. Nitrogen metabolism in lakes II. Role of nitrogen fixation in Sanctuary Lake, Pennsylvania. Limnol. Oceanogr. 7:170–177.

Jannasch, H. W. 1960. Denitrification as influenced by photosynthetic oxygen production. J. gen. Microbiol. 23:55–63.

Serruya, C. 1971. A Tentative Nitrogen and Phosphorus Budget of Lake Kinneret (1968–1971). Internal Report, Kinneret Limnological Laboratory.

Serruya, C. 1975. Nitrogen and phosphorus balances and load-biomass relationship in Lake Kinneret (Israel). Verh. Internat. Verein. Limnol. 19:1,357–1,369.

Stewart, W. D. P. 1964. Nitrogen fixation by Myxophyceae from marine environments. J. gen. Microbiol. 36:415–422.

Wilson, P. W. 1958. Asymbiotic nitrogen fixation. p. 9–47. In W. Ruhland, ed, Handbuch der Pflanzenphysiologie, VIII. Springer, Berlin.

Part four

The benthic community

History

The benthic fauna of Lake Kinneret was studied in 1894 by Barrois and in 1913 by Annandale. These authors published a detailed and rather complete list of littoral benthic species which was reviewed by Bodenheimer in 1935.

In 1963, with the construction of the National Water Carrier, the lake became the main freshwater reservoir of the country. Because of concern about the future quality of the lake water, Mekorot Water Company sponsored an extensive biological study of the lake. This program included a detailed investigation of the benthic flora and fauna of the lake which was carried out by a team of scientists from the Hebrew University of Jerusalem and the University of Tel-Aviv. Several groups of animals were studied by this team: Ostracoda (Lerner-Seggev, 1968), Mollusca (Tchernov, 1975a, b), Copepoda (Por, 1967), Chironomidae (Kugler, 1966, 1968, 1973), and others. Endemic animals of particular interest were studied in detail (Tsurnamal, 1968, 1977). Systematic sampling led to a qualitative (Gitay, 1968: Por, 1968) and quantitative study (Por & Eitan, 1970) of the most important benthic groups. Parallel studies concerning benthic algae were carried out by Dor (1974).

This team supplied the first, and so far the only, quantitative information concerning the biomass of benthic fauna and its distribution.

I. The benthic algae

F. E. Round

The benthic algae of Lake Kinneret and surrounding springs were investigated by Dor (1967, 1970, 1971, 1974) and by Rahat & Dor (1968). These authors found that the blue-green algae are represented by numerous species but they do not contribute much to the total benthic phytobiomass. Conversely, the diatom flora is less diversified but shows greater biological activity.

A new survey of the benthic algae has been initiated in 1976 by the present author, and the following is a preliminary report of the results so far obtained.

The phytobenthos of Lake Kinneret is richly developed on all available sites in the littoral, extending down to the depth to which photosynthetically available light penetrates. The individual habitats are varied but all are dominated by growth of diatoms, often resulting in a deep brown, mucilaginous fur-like growth over the substrata. Although absolute values are not yet available, these diatoms probably contribute 90% or more of the photosynthesis within the benthos. The number of species and forms of diatoms involved in this lake flora is well over 200. Amongst and beneath the diatom growths, filamentous and coccoid blue-green algae occur, and at some sites filamentous green algae grow out from the substratum and are visible as macroscopic growths. Photosynthetic flagellate algae (Euglenoids, Crytomonads) occur in some habitats, and rather rarely occasional desmid cells may be found, but only a few species in this latter group can tolerate the chemical conditions of Lake Kinneret. Surprisingly perhaps for such a water body, there do not seem to be any Charophyta present. The algae of the phytobenthos form a rich source of food for the benthic fauna, and in some sites at some times, grazing severely depletes the phytobenthos. The green filamentous algae have been observed parasitized by Phycomycetous fungi, but the diatoms seem unaffected.[*] The diatom growths contribute large quantities of siliceous fragments to the littoral sediments. The algal flora as a whole must supply a considerable amount of detritus to the littoral zone and probably also to the deeper water, since resuspension and drift will carry the material out into the lake. None of these aspects are quantified as yet, but while the photosynthetic activity of the phytobenthos may not provide an appreciable contribution to the overall production of such a large water body, its effect and importance in the littoral girdle must be very considerable.

Two quite different habitats and communities are involved: one free-living and dominated by motile species living on the fine sediments, and the other dominated by non-motile, attached species living on the solid substrata. The former comprises a single community growing on the surface and penetrating a few millimeters into the interstitial spaces but being positively phototactic and constantly moving up to the surface after disturbance. This epipelic community is

---

[*] However samples taken in April 1977 and kept in the laboratory became infected by a Phycomycetous fungus and *Amphora ovalis* was severely parasitized.

developed on sediments all around the lake, but the most extensive areas of sediment occur in the northern delta region, being patchy elsewhere. Between 0–2 m depth the biomass is considerable, but at 3 m and below only a small number of cells occur, presumably due to lack of light. Even between 0–3 m there is a steady drop in numbers, e.g. from around 6×10^4 cells/cm^2 at 0.5 m to 2×10^2 cells/cm^2 at 3 m in one estimation. The attached growths do not form a single entity but three more-or-less distinct communities: the epiphyton growing on the aquatic Angiosperms (no other macrophytes occur in Lake Kinneret), the epilithon on the boulders/stones and the epipsammon on the rare patches of sand. Beds of Angiosperms are abundant along certain shores, but underwater meadows are not common and most of the Angiosperms grow up into the water column; rosette-forming species do not seem to be present. The substratum for epiphytic growth is therefore exposed throughout the water column, extending down to 2–3 m depth, though more usually it is less than 1 m. Rocky shores are the most common type in Lake Kinneret, especially along the western and parts of the eastern shore. The rock is often coated by a growth of several millimeters of diatoms, below which lies a growth of Cyanophyta adjacent to the rock surface. It appears that the newly exposed rock is colonised by both diatoms and Cyanophyta but that the growth rate of the diatoms is so great that the Cyanophyta become smothered by the outwardly growing diatom colonies, many of which are stalked and also produce copious mucilage. Some diatom species grow directly on the rock surface and other diatom genera grow within the mucilage, but a rich metaphyton is not developed here as it is in acidic waters in Europe. A few species may, however, form a very sparse metaphyton amongst the epiphyte/epilithophyte assemblages.

The separate habitats will now be considered in more detail.

A. Epipelon

At most sites around Lake Kinneret, the dominant epipelic alga is *Amphora ovalis*, which occurs in several different forms. Motility or some other means of maintaining a surface position is necessary for epipelic algae, and while *Amphora* is motile, it is rather slow compared to other epipelic algae. Along the northern shore small species of the genus *Navicula* are abundant, and at many sites the large genera *Cymatopleura*, *Surirella* and *Gyrosigma* are conspicuous. The two former show a wide range of form, and the spirally twisted variants seem to be common in the lake. The chemistry of the water excludes many epipelic genera, e.g. *Pinnularia* was found at only one station and *Frustulia*, *Diploneis* and *Stenopterobia* are absent. *Caloneis latiuscula*, *Caloneis bacillum*, *Anomoeoneis sphaerophora*, *Neidium dubium*, *Navicula tuscula*, *Navicula halophila*, *Navicula radiosa*, *Navicula pygmaea*, *Navicula hungarica*, *Navicula cincta*, *Navicula pupula* (and several varieties), *Navicula gastrum*, *Navicula scutelloides*, *Nitzschia palea*, *Nitzschia frustulum*, *Nitzschia tryblionella* and *Nitzschia sigmoidea* are all represented. The final list, especially of species of *Navicula* and *Nitzschia*, will be very long and of great intrinsic interest, since it will confirm habitat requirements of these diatoms which are important indicators of environmental conditions. There are numerous peculiarities of distribution of these diatoms around the lake, e.g. *Nitzschia obtusa* occurs on sediments just south of Ein Gev, *Gyrosigma* species are commoner in the region around Tsemah and *Pinnularia* was recorded only at Ginnosar.

In the Ein Gev region the non-motile filamentous diatom *Fragilaria* is common on the sediments. This raises the problem of how such a diatom with no obvious means of regaining the surface after burial manages to colonise this habitat. Although there is no experimental evidence, it seems likely that its twisted filamentous morphology slows its sinking so that after disturbance it floats down to the surface, while the other particulate matter (especially the inorganic sand grains) settle much faster.

There are what appear to be relicts of the previous, more saline-tolerant flora of the lake still in the epipelic community. These are *Nitzschia tryblionella*, *Anomoeoneis sphaerophora*, *Navicula pygmaea*, *Bacillaria paradoxa* and possibly some other species. They are not excessively common in the lake but sufficiently abundant to give a slight, brackish water stamp to the flora.

B. Epipsammon

The fine sand found in a few sites, and that placed to make beaches, has a sparse attached flora mainly of *Amphora ovalis* v. *pediculus*, *Cocconeis disculus* and several small *Achnanthes* and *Navicula* species. On the sandy sediments a sparse epipelic flora also occurs in which chains of *Fragilaria* species and *Navicula tuscula* are common, and the sand must be washed carefully to remove this unattached flora.

C. Epiphyton

The epiphyton forms thick light-brown coloured, mucilaginous masses on the host plants due to the copious growth of stalked species of *Gomphonema* and *Cymbella*. *Cymbella affinis* and *Gomphonema* seem to be dominant. These tend to coat the host surface such that no other space remains for colonization. The stalks of these genera are host to other short-stalked *Achnanthes* species. The space between the branching stalks is filled with a more fluid mucilage and in this lives a range of non-motile species, e.g. *Fragilaria* colonies and *Synedra*, together with some motile species such as *Mastogloia smithii* and *Navicula radiosa* v. *tenella*. On some hosts the surface supports a rich growth of *Cocconeis placentula*, and these tend to dominate over the stalked forms – this occurs on *Myriophyllum* and *Potamogeton* sampled south of Ein Gev. Different host plants have different dominants, although probably almost all the epiphytic species can be found on every host. The unattached species tend to concentrate in the outer mucilage, while closer to the stem the stalks of the diatoms make up the bulk of the cover. Occasional plants of the green alga *Oedogonium* are also found attached and growing out through the mucilage sheath.

To what extent the non-attached forms comprise a metaphyton remains to be discovered. If they can be relatively easily washed out of the epiphytic coating, then they must be considered as a separate assemblage of species merely occupying a specialised spatial niche. Certainly the small *Cosmarium* species, which is regularly found here, is a metaphytonic species, since it could not maintain itself on a plant stem without the protection of a well-developed epiphyton.

D. Epilithon

Stone fragments vary from large boulders down to gravel-sized pieces. The latter has a very sparse flora and many of the cells found are casuals from other sites. The 2–10 cm sized stones are often richly coated with diatoms as are the boulders. Small *Cocconeis placentula* and *Epithemia* spp. tend to attach directly to the stone, and among these the Cyanophyte *Calothrix* forms pustules. Stalked *Gomphonema* is also present in quantity though in the outer layers the *Gomphonema* cells are motile. At some sites the large *Cymbella lanceolata* grows on stalks, but the most abundant *Cymbella* species is a form with similarities to *Cymbella prostrata*. *Cymbella affinis* also occurs but is less frequently dominant. *Synedra* species are common in the outer layers of mucilage. As with the epiphyton, the problem arises as to how much of the biomass is really metaphyton, but this is complicated by the greater degree of contamination of the flora in such a sedimentary site. Quantitative studies in progress show, in fact, that the contaminants, while always present, never contribute greatly to the total biomass. One interesting species almost always present is a minute, normally planktic *Cyclotella* species, and less frequently an equally small *Stephanodiscus*. These are so constantly present that I believe that they live trapped in the mucilage; maybe it is in this benthic habitat that they maintain small populations which at times seed the plankton.

A major problem which has been discussed briefly by some workers is the question of discretness of the epilithic and epiphytic assemblages. There is no doubt that some species occur in both habitats, but there are some differential species, e.g. *Mastogloia* in the epiphyton and *Epithemia* in the epilithon – at least in Lake Kinneret; elsewhere *Epithemia* is certainly epiphytic. The species of *Gomphonema* and *Cymbella* also show distinct preferences, e.g. *Cymbella affinis* is more abundant in the epiphyton and *Cymbella* cf. *prostrata* in the epilithon. A final analysis will reveal many subtleties, but this must await the detailed quantitative study being undertaken.

References

Dor, I. 1967. Algues des sources thermales de Tiberiade. Sea Fish. Res. Sta. Bull. 48:3–29.
Dor, I. 1970. Production rate of the periphyton in Lake Tiberias as measured by the glass-slide method. Isr. J. Bot. 19:1–15.
Dor, I. 1971. Benthic algae of Lake Tiberias and surrounding springs. Unpublished Thesis. Hebrew University.
Dor, I. 1974. Considerations about the composition of benthic algal flora in Lake Kinneret. Hydrobiologia. 44:255–264.
Rahat, M. & Dor, I. 1968. The hidden flora of a lake. Hydrobiol. 31:186–192.

II. The benthic fauna

A. Protozoa

U. Pollingher

No systematic work has been done on the Protozoa. The following data come from a preliminary survey carried out in the littoral zone.

In the muddy areas of the littoral, various Rhizopoda were found. Some of them belong to the Testacea:

Arcella vulgaris
Centropyxis sp.
Cypoderia sp.
Difflugia sp.

Among the Actinopoda, the following genera were found:

Actinosphaerium sp.
Heterophrys sp.
Actinophrys sp.

The following genera are more commonly found in the biotope accompanying the aquatic plants of the littoral, and are especially abundant in periods of low levels:

Stentor sp.
Paramecium sp.
Strombidium sp.
Vorticella sp.
Lacrymaria sp.

The planktonic Tintinnida are abundant in the littoral zone during the winter and their empty lorica are common in the mud samples.

B. Porifera

C. Serruya

In May 1890, Barrois took a series of bottom samples between 3 and 8 m in the southern part of the lake. In his samples, located between 5 and 8 m and consisting of pebbles and shell fragments, Barrois discovered the first sponge known in Lake Kinneret. This sponge turned out to be an endemic genus and species, *Potamolepis barroisi* Topsent 1892. The unsuccessful efforts of Topsent to find gemmules in the specimens of the new species strengthened the opinion of previous authors that the four genera, *Lubomirskia*, *Lessepsia*, *Uruguaya* and *Potamolepis* do not produce gemmules and should be classified apart.

In October 1912, Annandale carried out a detailed survey of the Porifera of the lake and reported five species, four of which are endemic.

Ephydatia fluviatilis syriaca Topsent was found at shallow depths on the lower surface of stones near Tiberias, Migdal and Tabgha. This soft sponge forms crusts a few millimeters thick and three to four centimeters wide. It is generally of a bright green colour. In contrast to the statement of Topsent ascribing spined as well as smooth macroscleres to the variety *syriaca*, Annandale found only smooth spicules and concluded that this species seems to be intermediate between the variety *syriaca* and the forma typica. This sponge produces gemmules and belongs to the Spongillidae, subfamily Spongillinae. It is very widespread, especially in European water bodies.

The four endemic species were ascribed by Annandale to the subfamily Potamolepidinae, characterized by the absence of microscleres, by their hardness and by the general absence of gemmules.

Cortispongilla barroisi (Topsent) was sampled by Annandale on small pebbles in the channel of the Jordan at the exit of the lake. He identified it with *Potamolepis barroisi* Topsent 1892 by comparing his specimens with one of Topsent's samples. He noted, however, fewer oscula in his samples than in those collected by Barrois and underlined the presence of a skeletal cortex formed by 'the massing together of separate fibers'. He was unable to find any trace of gemmules.

Annandale found only one specimen of *Nudospongilla reversa* sp. nov., at very shallow depth in the southern part of the lake, on the lower surface of a stone. Annandale described it as a hard but friable sponge with short and smooth spicules. No gemmules were found in this species either.

Nudospongilla mappa sp. nov. was found in many places on the lake shores but only in shallow water and on large-size stones. It is a hard, film-forming sponge bright green in sunlight. Annandale noted a transverse reticulation of the skeleton on the surface of the sponge. No gemmules were found.

Nudospongilla aster sp. nov. was also found in the southern part of the lake between 4 and 6 m depth. It is reported to be a hard but friable sponge forming a crust of small size on stones and shells. The spicules resemble those of *Nudospongilla reversa* and no gemmules were found.

Studying that spatial distribution of these various species, Annandale underlined the fact that, in Lake Kinneret where waves and storms are frequent, *Ephydatia* is more widespread than other species in spite of its soft texture. The production of gemmules provides an efficient mechanism of the survival of the species. This author believes that the hardness of the endemic species is an advantage in running water, and this explains why their distribution area is restricted to the Jordan channel.

Annandale divided the four species of Potamolepidinae into two genera: *Cortispongilla* and *Nudospongilla*. The former genus was composed of one species (*Cortispongilla barroisi*) that he compared to *Pachydictyum globosum* Weltner from Celebes. The genus *Nudospongilla* included three species and was compared to the genus *Metschnikowia* Grimm, typical of the Caspian Sea. Annandale expressed the opinion that the sponge fauna of Lake Kinneret showed more affinity with eastern tropical Asia than with Europe and Africa. He also noted the resemblance between *Cortispongilla* and species of *Veluspa* (*Lubomirskia*) from Lake Baikal but he related this to convergence and not to genetic relationship.

Racek (1974) studied the spicular remains of freshwater sponges in a 54 m core sampled in 1963 in Lake Hula in the Upper Jordan Valley, and previously analyzed from mineralogical and chemical points of view by Cowgill (1973a, b). These sediments supplied four gemmule-producing species of the family Spongillidae. *Ephydatia syriaca* Topsent was the only species previously known in Lake Hula, while *Spongilla lacustris* (Linnaeus), *Eunapius fragilis* (Leidy) and *Trochospongilla horrida* Weltner were new records.

The morphology of the megascleres and gemmoscleres of *Ephydatia syriaca* Topsent, found by Racek in the Hula core, identifies this species with those described by Topsent (1910) and Annandale (1913) as *Ephydatia fluviatilis*. However, Racek considered that its spicular characteristics and in particular the large number of spines of the megascleres justify the inclusion of this species in the *Ephydatia ramsayi* group which is known in Indo-West Pacific countries and South America. The association of *Ephydatia syriaca* to the *Ephydatia ramsayi* group makes the Jordan Valley the northwestern limit of extension of this spongillid, characteristic of the southern hemisphere.

The other three spongillid species belong to forms of world-wide distribution.

Besides these four spongillid species, Racek discovered, between 49.4 and 22.9 m depth, megascleres of non-spongillid sponges. These scleres are represented by two different types; completely smooth, stout, slightly curved megascleres belong to the most common type whereas smaller and straight scleres are rarer. Racek attributed these remains to 'thalassoid sponges' which he found identical to the 'previously ill-known and unnecessarily separated' genera described by Annandale. The detailed study of Racek indicates that these 'thalassoid sponges' have two different types of skeleton spicules: large slightly curved megascleres of the main skeleton structure and smaller straight megascleres of the periphery. Moveover, the recent discovery of other closely-related sponges, such as *Malawispongilla echinoides* Brien, showed that they 'display different modes of growth depending on age and depth of occurrence'. The distinction made by Annandale between shallow forms ascribed to the genus *Nudospongilla* and deeper forms ascribed to the genus *Cortispongilla* was then due to a lack of information concerning the considerable morphological variability which may occur in these sponges. Racek then concludes that *'Cortispongilla barroisi* and the three Galilean species of *Nudospongilla*, i.e. *Nudospongilla reversa, Nudospongilla mappa* and *Nudospongilla aster*, are biological identities', and proposes to relegate the *Nudospongilla* to a synonym of *Cortispongilla*.

Arndt (1937) and Stankovic (1960) found common features between *Cortispongilla barroisi*, the endemic species of Lake Ochrid, *Ochridaspongia rotunda*, and Lake Baikal, *Baikalospongia lacillifera*. Recent studies of thalassoid sponges of Lake Posso (Celebes) and lakes of the African Rift Valley (Racek & Harrison, unpublished data quoted in Racek, 1974) have demonstrated that all the globular, non-spongillids are closely related and descend from 'isolated marine ancestors'. Racek believes that all these thalassoid sponges have no relationship with the gemmule-producing Porifera and therefore proposes to elevate the subfamily of Globulospongillidae created by Brien (1973) to the rank of a separate family.

The discovery by Racek of *Cortispongilla barroisi* in the late Würm sediments of Lake Hula is rather surprising; this relict species was brought to the Central Jordan Valley by the last Pliocene transgression and it is difficult to explain how this thalassoid species could have reached Lake Hula.

Cortispongilla barroisi was reported (Por, 1968) to have been found for the last time in Lake Kinneret in 1938. The last sample found in the lake is kept in the Beth Gordon Museum at Deganya. The identification of the genus *Nudospongilla* with the genus *Cortispongilla* makes this account inexact. The endemic sponge of Lake Kinneret did not disappear. The only modification which seems to have taken place is a greater development of the shallow crust type colony which was once identified as *Nudospongilla* at the expense of the deeper globular type known as *Cortispongilla*.

C. Coelenterata

C. Serruya

Pelmatohydra sp. has been found in the profundal zone of the lake in winter when there is oxygen on the lake bottom (Por, 1968). The same author mentions also that nematocysts, probably belonging to this hydroid, have been found in the epidermis of the deep Rhabdocoela.

D. Turbellaria

C. Serruya

The Tricladida are found in the littoral zone of the eastern and southern coast and concentrate in the vicinity of shore spring. Their zonation extends to a depth of about 2 m. In 1913, Whitehouse described three new species of Planaria: *Planaria tiberiensis* n. sp., collected in shore springs at Migdal and Ein Tina, *Planaria salina*, collected in a saline spring near Tabgha, and *Planaria barroisi*.

In contrast to the previous littoral forms, some Rhabdocoela developed in winter on the oxygenated bottom. They show a typical pink-red granulation (Por, 1968). Other Rhabdocoela can be found at much shallower depths in *Nudospongilla*.

The types of *Dugesia* in Israel have been studied on the basis of their morphokaryological types by Bromley (1974). In Lake Kinneret and surrounding springs, this author found two types of *Dugesia*: (1) *Dugesia salina*, that she identifies with the *Planaria salina* described by Whitehouse in 1913. Like this latter specimen, *Dugesia salina* was sampled at Ein Sheva, a new name for the Tabgha springs. (2) *Dugesia biblica*, sampled on the northwestern and southeastern coasts of the lake. Whereas *Dugesia salina* has a regular karyological pattern ($2n = 16$ chromosomes), *Dugesia biblica*, a typical inhabitant of the Rift Valley freshwater springs, has an unusual karyological pattern ($3n = 27$ chromosomes). Moreover, 1 to 5 supernumerary chromosomes were observed and their number increases from south (Dead Sea shore) to north (Lake Kinneret shores and Dan springs). Generally, this species is fissiparous. When grown in the laboratory, it may develop a sexual pattern of reproduction but shows then a low fertility.

Bromley suggests that the triploidy and supernumeraries, characteristic of this species, are the expression of an adaptive process to the changing conditions of the Rift Valley. A similar interpretation has been given by White (1954) to account for supernumeraries found in insects, the other main group of animals besides the Turbellaria where this anomaly has been observed.

E. Trematoda

C. Serruya

In 1964, Paperna described twelve species of parasitic trematods found in various species of Kinneret fishes (List No. 7).

In 1971, Lengy & Wolff described three types of *Cercaria levantina* collected from snails in the vicinity of Lake Kinneret: a virgulate of the Paravirgulae subgroup hosted by *Bithynia sidoniensis* and *Bithynia saulcyi*, a *Gymnocephalus cercaria* of the Parapleurolophocerca sub-group collected from the same *Bithynia* species and a *Furcocercus cercaria* of the sub-group Strigeid found in *Bulinus truncatus*.

These trematods were found in Wadi Kursi (or Wadi Samakh) and in the Buteiha Valley, both located on the eastern side of the lake, and at the southern outlet of the lake.

List No. 7

List of Parasitic Trematods in Freshwater Fishes

(from Paperna, 1964)

Trematods	Hosts and location	Habitat
Dactylogyrus galilensis sp. nov.	*Barbus longiceps* Cuv. Val. Kinneret	gills
Dactylogyrus garrae sp. nov.	*Garra rufus* (Heckel) *Tylognathus steineitziorum* (Kosswig) Springs of Hula Valley Lower Jordan Valley Kinneret	gills
Cichlidogyrus cirratus sp. nov.	*Tilapia galilaea* (Artedi) Offshore waters of Kinneret	gills
Diplozoon minutum sp. nov.	*Phoxinellus kervillei* (Pellegrin *Tylognathus steineitziorum* (Kosswig) Western shores of Kinneret	gills)
Gyrodactylus cf. *medius*	*Cyprinus carpio* L. *Tilapia nilotica* L. *Tilapia zillii* (Gervais) *Tilapia galilaea* (Artedi) *Tristramella simonis* (Gunter) *Haphlochromis flavii-josephi* (Lortet) *Mirogrex terrae-sanctae* (Steinitz) Jordan Basin	gills skin fins

336

List No. 7 (*contd.*)

Gyrodactylus sp. B Paperna	*Tylognathus steinitziorum* (Kosswig) Shores of Kinneret	gills
Plagioporus biliaris biliaris ssp. nov.	*Tilapia zillii* (Gervais) *Haphlochromis flavii-josephi* (Lortet) *Tylognathus steinitziorum* (Kosswig) Kinneret Springs of Hula Basin	gall bladder
Podocotyle aphanii sp. nov.	*Aphanio mento* (Heckel) Western shore of Kinneret	intestine
Podocotyle lacustris sp. nov.	*Blennius vulgaris* Pollini *Garra rufus* (Heckel) Shores of Kinneret	intestine
Phlehniella dentata sp. nov.	*Clarias lazera* Cuv. Val. Kinneret, Hula Nature Reserve	intestine
Stictodora sclerogonocotyla sp. nov.	*Tilapia galilaea* (Artedi) Kinneret	muscles
Clinostomum sp.	*Tilapia zillii* (Gervais) *Tilapia nilotica* L. *Tilapia galilaea* (Artedi) *Tristramella simonis* (Gunter)	skin

F. Nemertea

C. Serruya

Several specimens of freshwater nemertines were collected by Reich (unpublished data) in 1941 in the northwest edge of Lake Hula and by Bromley (1972) in various locations of the Jordan system and Jerusalem area. All these specimens belong to the genus *Prostoma Dugès*.

Bromley found nemertine worms in four locations in the Jordan system: Nahal Snir and Hula Valley in the northern Jordan River, a few kilometers south of the exit of the lake in the southern Jordan River, and Wadi Faris, north of the Dead Sea. The same author also found three specimens in En Matta, west of Jerusalem, in a watershed draining in the Mediterranean Sea.

The occurrence of *Prostoma* in two distinct drainage systems suggests that it is widespread in freshwater bodies of the country. However, *Prostoma* has not been reported in the immediate vicinity of the lake.

G. Nematoda

C. Serruya

The fauna of Nematoda of Lake Kinneret has been investigated by Barrois at the end of the 19th Century and in the sixties by Andrassy (1963), Por (1968) and Por & Masry (1968). The non-exhaustive list of Nematoda published by Por (1968) includes:

Eudorylaimus andrassyi (Meyl)
Eudorylaimus attersbergensis (de Man)
Eudorylaimus obtusicaudatus (Bastian)
Dorylaimellus parvulus (Thorne)
Tobrilus aberrans (W. Schneider)
Tylenchus filiformis (Bütschli)
Aphelenchoides besseyi (Christie)
Anaplectus granulosus (Bastian)
Panagrolaimus rigidus (A. Schneider)

Many of these species live in the supra littoral area. In contrast with this general rule, *Eudorylaimus andrassyi* is found in large amounts in the profundal zone in summer in complete anaerobic conditions.

No further systematic work has been done on the Nematoda of Lake Kinneret.

Seven genera and nine species represent a poor fauna when compared to the seventeen genera and twenty-three species of Lake Ochrid.

H. Bryozoa

C. Serruya

Annandale (1913) reported two species of Bryozoa:

Fredericella sultana jordanica
Plumatella auricornis

Both belong to cosmopolitan genera. In 1915, Annandale found that both these species were represented in the Volga system together with *Plumatella casmiana* Oka, known until then only from Japan.

The Bryozoa of the lake would, therefore, possess eastern or cental Palaearctic affinities.

I. Oligochaeta

C. Serruya

The fauna of Oligochaeta in Lake Kinneret has been studied by Barrois (1892), Rosa (1893), Annandale (1912), Por (1968) and Gitay (1968). In presenting the list of species of Oligochaeta in the lake, Por (1968) notes that the supra littoral fauna is incompletely known and further studies might add new species.

Maididae	*Dero digitata* (Müller)
	Stephensoniana trivandrana (Aiyer)
Tubificidae	*Psammoryctides albicola* (Michaelson)
	Limnodrilus hoffmeisteri Claparede
	Euilyodrilus heuscheri (Bratscher)
	Euilyodrilus hammoniensis (Michaelson)
	Euilyodrilus bavaricus (Oschman)
Lumbricidae	*Eiseniella tetraedra* (Savigny)
	Allobophora patriarchalis Rosa
Glossoscolecidae	*Criodrillus lacuum* Hoffmeister
Haplotaxidae	*Haplotaxis gordioides* (Hartman)

This incomplete list includes five families, nine genera and eleven species and, as in the case of Nematoda, lies far behind the seventeen genera and twenty-three species described in Lake Ochrid.

J. Hirudinea

C. Serruya

The Hirudinean fauna of Lake Kinneret is a poor one, and three separate investigations have agreed on this point. The specimens collected by Barrois in 1890 and by Festa in 1893 were studied by Blanchard (1893, 1894). These authors reported that numerous specimens of *Dina blaisei* R. Bl. were found at Ein Tina, in the northern part of the lake, and in Lake Hula. Barrois also reported an abundant fauna of *Placobdella carinata* (Diesing) on turtles (*Emys caspica*) in the swamps of the Buteiha plain in the northeastern coast of the lake. This species is also known at Alep, living in the same turtles, and at Astrakhan.

In 1912, Annandale made a new collection and found a few specimens of *Placobdella catenigera* (Moq. Tand.) in Wadi Samakh and on the shore of Migdal. This species is widely distributed in Eastern Europe and Western Asia. Annandale agrees with his predecessors about *Dina blaisei* being the most common leech of Lake Kinneret. It is particularly abundant on the lower surface of stones and feeds mainly on small oligochaetes. This species is widely distributed in Europe, North and Central America, Madeira, Azores, Siberia and Mongolia.

The list of Hirudinea of Lake Kinneret is then restricted to:

Fam. Glossiphonidae
1. *Placobdella catenigera* (Moquin-Tandon)
2. *Placobdella carinata* (Diesing)
Fam. Herpobdellidae
3. *Herpobdella* (Dina) *lineata* (O. F. Müller)

In 1893, Blanchard described *Herpobdella lineata* as a new genus because it differs from the Herpobdellidae in that the third ring of the somite was enlarged and divided longitudinally by a superficial furrow. This feature was considered by Johansson (1913) to be of subgeneric value.

K. Crustacea

C. Serruya

1. Cladocera

The list of benthic Cladocera (Por, 1968) includes:

Moina dubia (Richard)
Macrothrix laticornis (Jurine)
Macrothrix goeldii (Richard)
Ilyocryptus sordidus (Lieven)
Ilyocryptus agilis (Kurz)
Leydigia leydigi (Schödler)
Leydigia acanthocercoides (Fisher)
Alona affinis (Leydig)
Alona rectangula (Sars)

No further work has been carried out on this group in Lake Kinneret.

2. Copepoda

a. Taxonomy

The copepods of Lake Kinneret have been studied by Richard (1893) who worked on the collection of Barrois and by Gurney (1913) who determined the samples collected by Annandale. Por (1964) studied the genus *Nitocra* Boeck in the Jordan Valley. The same author, in the framework of the survey sponsored by the Mekorot Water Company, which extended over the period 1963–1965, studied in detail the benthic copepods of the lake and of two inflowing springs. The following is based on his findings.

In comparison with other benthic animals, the copepod fauna is rich. It includes the following 19 species:

HARPACTICOIDA

Pseudobradya barroisi (Richard)
Horsiella brevicornis (Van Douwe)
Nitocra lacustris (Schmankewitsch)
Nitocra hibernica (Brady)
Nitocra incerta (Richard)
Schizopera taricheana n. sp.
Atheyella crassa (G.O. Sars)
Nannopus palustris tiberiadis n. ssp.
Onychocamptus mohammed (Blanchard & Richard)

343

Eucyclops serrulatus (Fischer)
Eucyclops macrurus (Fischer)
Afrocyclops gibsoni
Paracyclops fimbriatus (Fischer)
Paracyclops affinis (Sars)
Ectocyclops phaleratus (Koch)
Microcyclops bicolor (Sars)
Microcyclops varicans (Sars)
Acanthocyclops bicuspidatus (Claus)
Macrocyclops distinctus (Richard)

Among the above-mentioned species, *Horsiella brevicornis, Paracyclops affinis* and *Microcyclops bicolor* were reported for the first time in Israel by Por (1968). *Afrocyclops gibsoni* was already found by Por in Lake Kinneret in 1968. However, since this author found only one specimen of this species, he did not include it in his general list. In 1976, one of his students found *Afrocyclops gibsoni* in greater abundance in the Jordan River, south of Lake Kinneret (Por, personal communication). It seems then that the Kinneret – Jordan region is the northermost area of distribution of this African species.

Pseudobradya barroisi (Richard) was attributed by Richard to the genus *Ectinosoma*. In his monograph on the harpacticoids, Lang (1948) included it also in this same genus. In contrast, Por (1968) thinks that the form of the maxilla, of the mixillipede and of P V justifies its inclusion in the genus *Bradya*.

Schizopera taricheana (n. sp.) is related by Por, on the basis of similar form of P I and furca, to *Schizopera validior* Sars of Lake Tanganika.

(i) *Nannopus palustris tiberiadis* n. sp.
Although *Nannopus palustris* Brady is common, the Kinneret form is morphologically different from the common specimens. The main difference consists in 'the sexual dimorphism of the furcal armature: in the female, there is only one apical seta, very much inflated proximally, while the external apical spine is reduced to a mere chitinous prominence. In the male, there is a well-developed external seta, the internal seta being only slightly swollen' (Por, 1968).

It then follows that the isolation of this species caused certain morphological differences with the European population. The differences due to isolation are more marked in Lake Tanganika and led to the formation of another species, *Nannopus perplexus* Sars.

(ii) The genus *Nitocra*
Besides the three species of *Nitocra* found in Lake Kinneret and in its surrounding springs, a fourth species (*Nitocra balnearia* n. sp.) has been found in the hot springs of Hamei Zohar on the Dead Sea shores. It is interesting to note that these species of *Nitocra* are found in the Mediterranean littoral of Israel and the estuaries of some brooks and in the Jordan system but have never been found elsewhere in Israel. The harpacticoid fauna of the springs of the Judean Hills is composed of *Atheyella crassa* (Sars) and *Canthocamptus staphylinus* auct.

Nitocra hibernica is a species widespread in Europe. Variations from the typical species led Richard (1890) to describe the Lake Kinneret species as *Canthocamptus (Nitocra) hibernicus* var. *incertus*. Sars (1927) found both forms in the Caspian Sea, but Noodt (1954) was the first to mention, on the grounds of morphological differences, that the Near East populations of the variety *incerta* were phylogenetically older than the European *hibernica* population. The finding, in 1955, of *incerta* variety specimens in the interstitial waters of Lake Garda in northern Italy did not contradict this hypothesis; it is likely that the variety *incerta* of Lake Garda is a relict, as are many interstitial animals.

On the basis of morphological differences, Por (1964) considers these two forms as different species. Moreover, they do not appear in the same habitats: *Nitocra incerta* is found in the springs of Ein Gedi, in the Dead Sea area, and in some inflowing springs of Lake Kinneret whereas *Nitocra hibernica* lived in Lake Hula. In Lake Kinneret, where both forms are found, they occupy different niches: *Nitocra incerta* lives on the lake bottom whereas *Nitocra hibernica* is a periphytic species associated with the palustrine vegetation of the lake.

Por (1964) concludes that *Nitocra incerta* Richard is 'a geologically older and morphologically more primitive species of preglacial origin, which in Europe withdrew into subterranean waters but survived in the old water basins of the Near East: Lake Tiberias and the freshwater springs on the shores of the Dead Sea, Lake Egerdin and Lake Hazor, presumably remnants of the old Pleistocene Anatolian lake system (Kosswig, 1959), in Lake Manyas, a relict lake formed by a Pleistocene high eustatic level of the Mediterranean (Kosswig, 1955) and in the Caspian Sea, a well-known Neogene basin'. Conversely, *Nitocra hibernica* is a post-Pleistocene species which spread from the Sarmatic province to Europe and the Jordan system.

b. *Spatial distribution of the benthic copepods in Lake Kinneret* (Por, 1968)

(i) The spring fauna
Microcyclops bicolor, Microcyclops varicans, Ectocyclops phaleratus, Schizopera taricheana and *Atheyella crassa* are found exclusively in springs, whereas *Eucyclops serrulatus, Paracyclops fimbriatus* and *Nitocra incerta* are found in springs and in the lake. *Eucyclops serrulatus* is the only copepod found in the Ein Nur Spring from the Tabgha group where chloride reaches 2,500 mg Cl l^{-1}, indicating the large range of salinity tolerance of this animal.

(ii) The lacustrine fauna
The remaining ten species are found only in the lake. It appears that the cyclopoids *Paracyclops affinis, Eucyclops macrurus, Macrocyclops distinctus* and *Acanthocyclops bicuspidatus*, which are mainly found in the lake in winter and spring, are washed with the river floods but are not permanent inhabitants of the lake. It is worth noting that if we consider the nine permanent species, seven belong to the harpacticoids and only two to the cyclopoids.

Eucyclops serrulatus, Onychocamptus mohammed and *Nitocra incerta* were found down to 30 m when oxygen was present.

Nitocra hibernica, Nannopus palustris tiberiadis and *Pseudobradya barroisi* are littoral species never found below 5 meters. *Paracyclops fimbriatus* was found in the southern part of the lake down to 15 meters.

Nitocra lacustris lives in the interstitial water of the supralittoral and *Horsiella brevicornis* in the periphyton of the higher plants of the Jordan outflow.

c. Seasonal variations (Por, 1968)

Eucyclops serrulatus, Onychocamptus mohammed and *Nitocra incerta* are found at all depths in winter and frequently in the littoral and sublittoral throughout the year. The former species reproduces mainly from October to May; the two latter species reproduce throughout the year.

Nitocra hibernica and *Paracyclops fimbriatus* appear in December, culminate in April and decrease in summer. Egg-bearing females are mostly found from March to May.

Nannopus palustris tiberiadis occurs from March to July, whereas *Pseudobradya barroisi* is only found in April–May.

From a quantitative point of view, *Onychocamptus mohammed* is the predominant Copepoda in Lake Kinneret, especially below 2 meters. It seems abundant during all the seasons of the year. The absence of seasonal fluctuations also characterizes *Nitocra incerta*, but this species is rarer at great depths. Information concerning the relative importance of the different species and their area of distribution is included in Fig. 114 and 115.

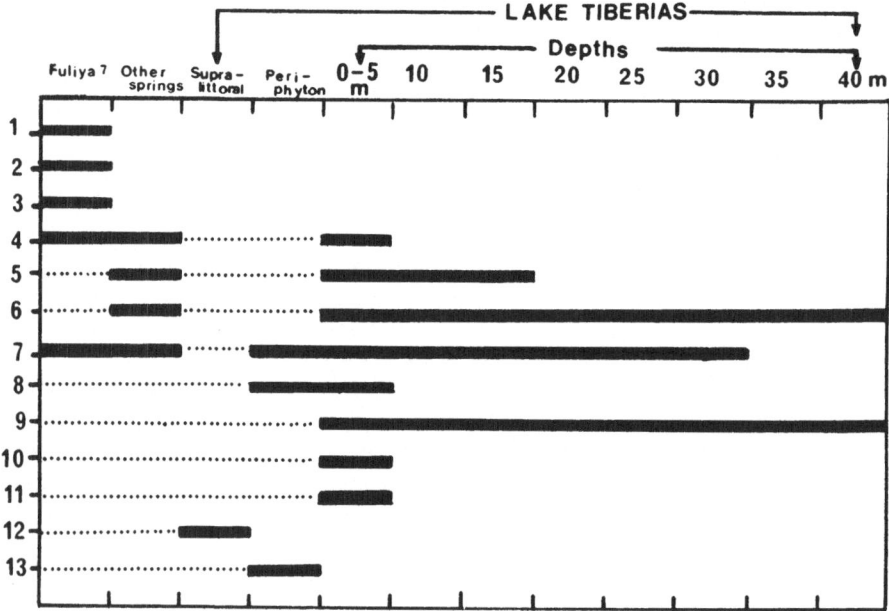

Fig. 114. The distribution of benthic Copepodes in Lake Tiberias and surrounding springs (from Por, 1968, Israel J. Zool., 17). 1 = *Ectocyclops phaleratus*, 2 = *Atheyella crassa*, 3 = *Microcyclops* spp, 4 = *Schizopera tarichaena*, 5 = *Paracyclops fimbriatus*, 6 = *Eucyclops serrulatus*, 7 = *Nitocra incerta*, 8 = *Nitocra hibernica*, 9 = *Onychocamptus mohammed*, 10 = *Pseudobradya barroisi*, 11 = *Nannopus palustris tiberiadis*, 12 = *Nitocra lacustris*, 13 = *Horsiella bisetosa*.

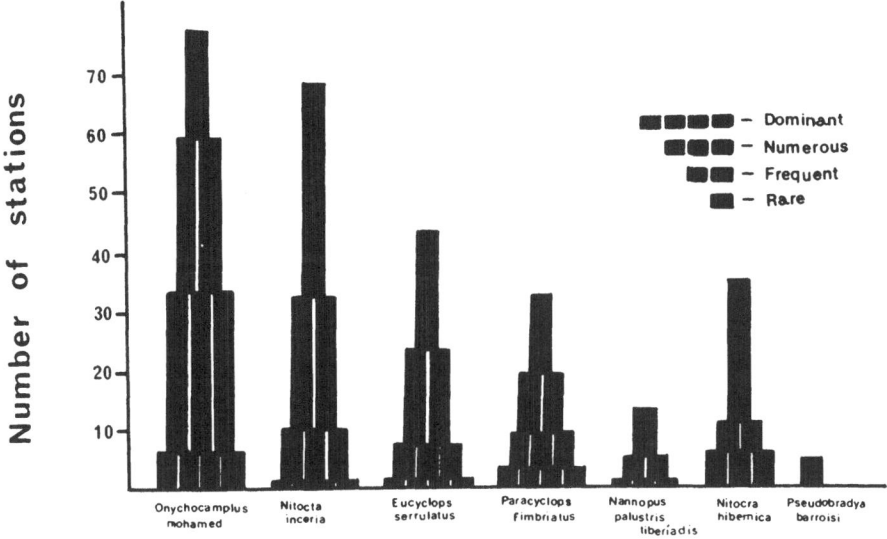

Fig. 115. Comparative frequency of benthic Copepoda in the littoral of Lake Tiberias (from Por, 1968, same source as Fig. 114).

d. *Origin of the benthic copepods of Lake Kinneret* (Por, 1968)

All the harpacticoids of the lake belong to the brackish fauna of marine origin. The Canthocamphidae, the widespread freshwater harpacticoids, are not represented at all. Moreover, the cyclopoids of the lake are euryhaline animals *par excellence*. These features underline the marine origin of this fauna, which is corroborated by the usual habitat where some of these harpacticoids are found elsewhere in the world; *Nannopus palustris*, for example, is found in brackish water still in contact with the open sea (Swedish Lake Mälaren). *Pseudobradya barroisi, Nannopus palustris tiberiadis* and *Nitocra incerta* are considered by Por (1968) as the oldest forms of copepods of the lake, relicts of the Pliocene Sea. This hypothesis is in good agreement with the fact that these species did not exist in the Hula, which came into existence only during the Pleistocene.

Por underlines the contradiction between the present oligohaline character of the Kinneret waters and the brackish type of its copepod fauna. This author considers that the stable 'osmotic climate' of the lake water is especially favourable to the acclimation of marine elements in oligohaline environments. Another possibility, mentioned by Gophen, is that these benthic species benefit from the slow but constant seepage of minerals from the interstitial waters of the sediments towards the lake water.

The zoogeographical distribution of the copepod species of the lake assigns them a dominant holarctic origin. However, the description by Borutzki (1952) of *Nannopus palustris* from the Black Sea and the Lake of Aral led Por to think that these forms are very near to the variety *tiberiadis*, which would mean that populations having the features of ssp. *tiberiadis* exist also in water bodies of Central Asia. Therefore, it is not excluded that besides holarctic species, Sarmatic

347

Pliocene elements may have contributed to the present copepod population of Lake Kinneret.

3. Ostracoda

The research on the Ostracoda of Lake Kinneret was, until recently, restricted to the paper of Barrois (1894) where he mentioned one species, *Limnocythere tiberiadis*, but did not describe it. The Ostracodes were studied in detail, in the framework of the study of benthos sponsored by the Mekorot Water Company, by Lerner-Seggev (1968). She found only one species of *Limnocythere, Limnocythere inopinate* (Baird), and assumed it to be identical to the species referred to by Barrois.

Lerner-Seggev found seven species of Ostracodes, three of which are new.

> *Cyprideis torosa* (Jones)
> *Limnocythere inopinata* (Baird)
> *Loxoconcha galilea* n. sp.
> *Darwinula stevensoni* (Brady and Robertson)
> *Ilyocypris hartmanni* n. sp.
> *Ilyocypris nitida* n. sp.
> *Potamocypris producta* (G. O. Sars)

The right and left valves of the three new species are of nearly similar size. The left valve is slightly smaller than the right one in *Loxoconcha galilea*; it slightly overlaps the right value in the case of *Ilyocypris hartmanni* and *Ilyocypris nitida*. The valves of *Ilyocypris hartmanni* are richly ornamented with protruding lobes, spines and marginal dentriculation.

Loxoconcha galilea can be found mainly during the spring period in the littoral but it is never found in large numbers. This species is probably a Pliocene marine relict. Its presence in the Caspian Sea strengthens this hypothesis.

Small populations of *Ilyocypris hartmanni* are found in the sublittoral zone of the lake throughout the year; in summer it spreads to the littoral.

Ilyocypris nitida is more abundant than the previous species and is mainly a spring form, especially in the sublittoral.

Cyprideis torosa dominates the littoral Ostracode fauna. It is represented by a smooth as well as a tuberculate form, the latter one being more frequent. The specimens of Lake Tiberias are generally of smaller dimensions than the specimens of the same species living in other water bodies. This species is known in Europe, Asia and Africa (plates 23, 24 and 25).

Darwinula stevensoni is the most abundant Ostracode in Lake Kinneret. It is found at all depths throughout the year. It is worth noting that only females were sampled. *Darwinula stevensoni* is also known in the freshwater springs of Ein Gedi in the Dead Sea area. This seems to indicate that there was in the past a continuity of freshwater bodies from Lake Kinneret to the Dead Sea (Por, personal communication).

Potamocypris producta is a rare species in the lake, and a spring form is mainly found in the southern part of the lake. Only female specimens were found. This

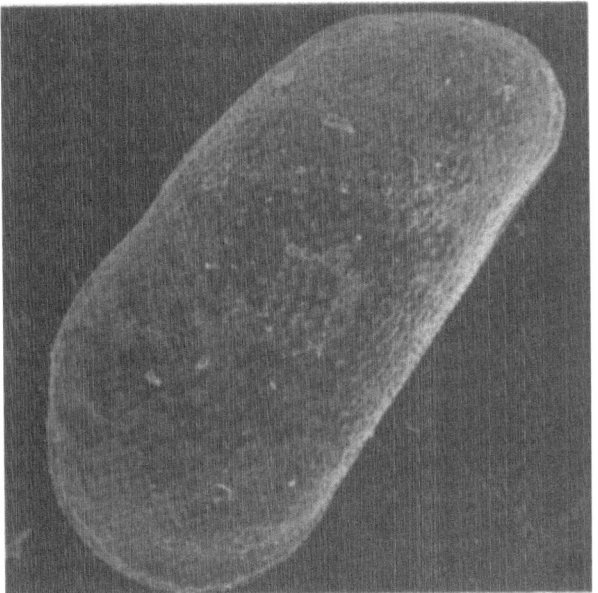

Plate 23. Cyprideis torosa (Jones), male specimen, external view of left valve. × 100 (Photo Cl. Guernet).

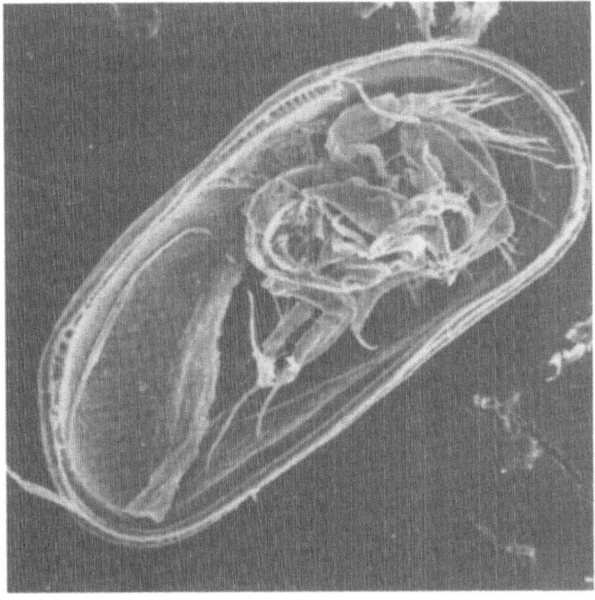

Plate 24. Cyprideis torosa (Jones), male specimen, internal view of left valve. × 100. (Photo Cl. Guernet).

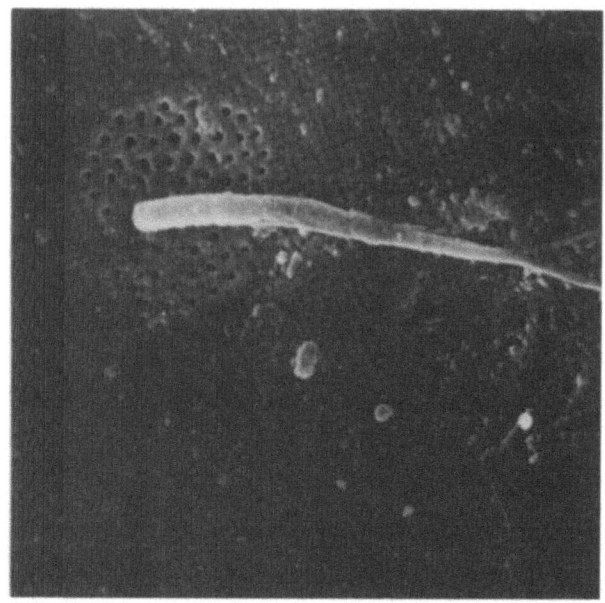

Plate 25. Cyprideis torosa (Jones), normal pore and subcentral seta × 3750. (Photo Cl. Guernet).

species has been reported in South, East, West and Central Africa (Hartman, 1957).

Limnocythere inopinata is also a rare species found mainly as female specimens in the southern part of the lake. This species is known in Europe, Asia and North America.

The fauna of Ostracode shows a clear seasonal variation. They proliferate in spring and their number decreases gradually towards summer. *Cyprideis torosa* and *Darwinula stevensoni* are perennial; *Ilyocypris hartmanni* is perennial in the sublittoral but in the littoral zone it is found only in spring and summer. The other species are only found in spring.

4. Amphipoda

Two species of Amphipoda are known in Lake Kinneret: *Orchestia platensis* Kröyer, *Echinogammarus veneris* (Heller).

In his revision of the species belonging to or confused with the 'Echinogammarus pungens' group, Stock (1968) examined the Gammarids collected on the shores of Lake Kinneret, at Tabgha springs and Fulya springs, and identified them as *Echinogammarus veneris* (Heller, 1865).

5. Isopoda

Por (1958) reports two species of Isopoda: Asellidae: *Proasellus coxalis* (Dollfuss), Oniscidae: *Halophiloscia couchii* (Kinahan).

This author indicates the presence of at least two species of unidentified Oniscidae in the supralittoral area.

Halophiloscia spp. are generally found in the marine littoral. The specific habitat of this animal strongly suggests a marine episode in the remote past of the lake (Por, 1975).

6. Decapoda

Two species of Decapoda are mentioned by Por (1968): *Athyaephyra desmaresti* (Millet), *Potamon potamios* (Olivier).

However, the most famous Decapoda of the Kinneret area, *Typhlocaris galilea*, is not found in the lake but in the springs of the Ein Sheva group on the northwest coast of the lake.

L. *Typhlocaris galilea*, the blind prawn of Galilee

M. Tsurnamal

The main population of the endemic blind and pigmentless palaemonid prawn, *Typhlocaris galilea*, considered to be a relict species of the Pliocenic Mediterranean fauna, recently inhabits some subterranean water-filled cavities in the Tabgha area (on the northern shore of Lake Tiberias). A small fraction of that population may, however, be observed also in the epigeic, rather saline water of two springs belonging to the Tabgha ('Heptapegon' or 'En-Sheva') complex, where the prawn is now regarded as a protected animal by the Israeli authorities.

The prawn has long been known to the Bedouins of the district as the 'white scorpion', but it was not until 1909 that it was first described by Dr. Calman of the British Museum (London). He suggested that because of several peculiar morphological features, this prawn, new to science, should be assigned to a new genus and also be regarded as the type of a distinct subfamily of the Palaemonidae, namely subfamily Typhlocaridinae. The description of the new species was based on two specimens which were collected, according to Calman, in 'a small pond near the town of Tiberias communicating with the lake (of Tiberias) and fed by a mineral spring'.

Annandale (1912) and Annandale & Kemp (1913) were the first to describe the small population of blind prawns in an octagonal pool at Tabgha, which they regarded as the only site at which *Typhlocaris galilea* may be found. They also added some information on the prawn's morphology, and made some field observations on its locomotion and feeding habits in the pool. According to these authors, the surface of the octagonal pool, known as 'Birket Ali-Ed-Dhahar', was at that time 'covered by a gigantic grass' and contained from six to ten feet of brackish and sulphurous water. Annandale collected several specimens of *Typhlocaris galilea*, some of which were later used by Gosh (1913) for his study on the anatomy of this prawn.

The presence of a large hypogeal population of *Typhlocaris* in the water tract of the En-Nur spring at Tabgha was first mentioned by Tsurnamal & Por (1968), who described this subterranean habitat and some organisms in it associated with the blind prawn. Tsurnamal (1978a, b) contributed more detailed information on the prawn's habitat and the distribution of the animals in it as well as on the behaviour, ecology, reproduction and temperature preference of *Typhlocaris galilea*.

A second species of *Typhlocaris*, *Typhlocaris lethaea*, was described by Parisi (1920, 1921) from a cave near Benghazi (Libya) and a third species, *Typhlocaris salentina*, was discovered and described by Caroli (1923, 1924) in caves near Otranto (S. Italy). Por (1963), in his discussion on the origin of the relict aquatic fauna of the Jordan Rift Valley, regarded *Typhlocaris galilea*, and also the two other congeneric species, as Pliocenic marine relicts that had succeeded in surviving to the present, perhaps via preadaptation. Their precursory form was already living in submarine springs and caves near the shore or in small rock pools, almost separated from the sea, all along the shores of the Pliocenic Mediterranean Sea. This sea also penetrated into the area of the recent Jordan Valley in late Pliocene times – about three million years ago – apparently through the Valley of Yizre'el (Neev & Emery, 1966). The sea later progressively withdrew, leaving behind a saline lake in the area of the recent Jordan Valley and also many other isolated water basins all around the Mediterranean Sea. The precursory species of *Typhlocaris* presumably died out later in the changing Pleistocenic Mediterranean, while some isolated populations in subterranean habitats succeeded in surviving and evolved simultaneously into isolated, different species.

Plate 26. Typhlocaris galilea Calman-Adult (body length 52 mm).

The blind prawn of Galilee (plate 26) is an opalescent, small white animal (length not exceeding 61 mm). Dark masses of the stomach and hepatopancreas, and sometimes also the white gonads, can be detected through the almost transparent integument. The flagella of the prawn's antennae are very long and thin but the eyes are degenerate and possess no pigment and no optical elements. The eyestocks have the form of short, slightly flattened scales, lying horizontally and nearly touching each other in the middle line, in front of the very short rostrum. From their internal structure, a neurosecretory function may be suggested. On each side of the smooth carapace runs a longitudinal suture-line or fine groove which Calman (1909) regards as a 'linea thalassinica' – a morphological feature which is found only in primitive decapod crustaceans (certain Penaeidae and Thalassinidea). This suture may indicate that *Typhlocaris* has been derived from some very ancient and primitive caridean stock.

354

There is a marked dissimilarity between the two large (second) chelae in many prawns. About 40% of the inspected males have been found to have one of the chelae (usually the left one) much broader and thicker than the other. The palm of that chelae is flattened and the 'fingers' are relatively short. This feature can, however, also be detected in about 10% of the female prawns. Thus, a broad-type chelae cannot be considered as a secondary sexual character of the male, as suggested by Annandale & Kemp (1913). The only genuinely masculine external sexual characters of *Typhlocaris galilea* are the appendix masculina on pleopod II and the special short and broadened spines (setae) on the posterior face of the protopodites of all pleopods (plate 27). These broadened spines are usually of a pale yellowish colour, but in males with ripe gonads they acquire a yellow to orange colour and are by then rather conspicuous. The colour is given to the special setae by a liquid substance stored in their broadened part and probably has the function of a pheromon.

The prawns are not sensitive to light but are highly sensitive to vibrations in the water or in the bottom in their vicinity. They may be observed to move slowly on the bottom of their habitat, while probing the substrate with their maxillipeds and occasionally also with the first, thin pair of pereiopods. While moving about, the prawns pick minute food material out of the sediment, chiefly with the aid of the delicate chelae of their first pair of pereiopods. Usually the second pair of pereiopods, bearing the large chelae, is not used in food collection. *Typhlocaris galilea* has been found to feed mainly on the small burrowing oligochaete, *Isochaeta israelis*, but some nematodes, filaments of sulphur bacteria, detritic material (rich in bacteria) and occasionally some juvenile *Theodoxus jordani* snails were also found in the stomach of collected prawns. In the epigeic habitat, the animals were observed to feed also on various organisms of terrestrial origin. *Typhlocaris* is rather sluggish and spends most of the time resting underneath stones, only emerging from the shelter for brief forages for food. If disturbed, however, the prawn would sprint backwards and upwards, swimming away quickly. The prawn was found to be active, albeit intermittently, throughout the diel period, showing no diurnal, nocturnal or seasonal cycle of activity. It seems that there is also no distinct seasonal cycle of molting and reproduction in this troglodyte prawn.

Despite many intensive and thorough observations during several years, egg-bearing females of *Typhlocaris galilea*, or freshly hatched, very young individuals (less than 10 mm long), were never observed at Tabgha, although some prawns of both sexes with maximally ripened gonads were seen there at all seasons of the year. Successful courting, copulation and oviposition has occurred in the laboratory after the water temperature in the aquaria was lowered from 28–30°C (similar to the temperature at En-Nur spring) to 23–25°C. The 35–42 eggs born on the pleopods of one female complete their development within 98–100 days and the juvenile prawns (5–7 mm long) hatch in a post-larva stage, already with degenerated eyes and devoid of any pigment. At the end of 15 months of development and growth in the laboratory, some juvenile *Typhlocaris* attained a length of 30–32 mm, and their almost ripe gonads were by then discernible through the almost transparent integument.

At Tabgha, *Typhlocaris galilea* was investigated at two habitats: the epigeic oc-

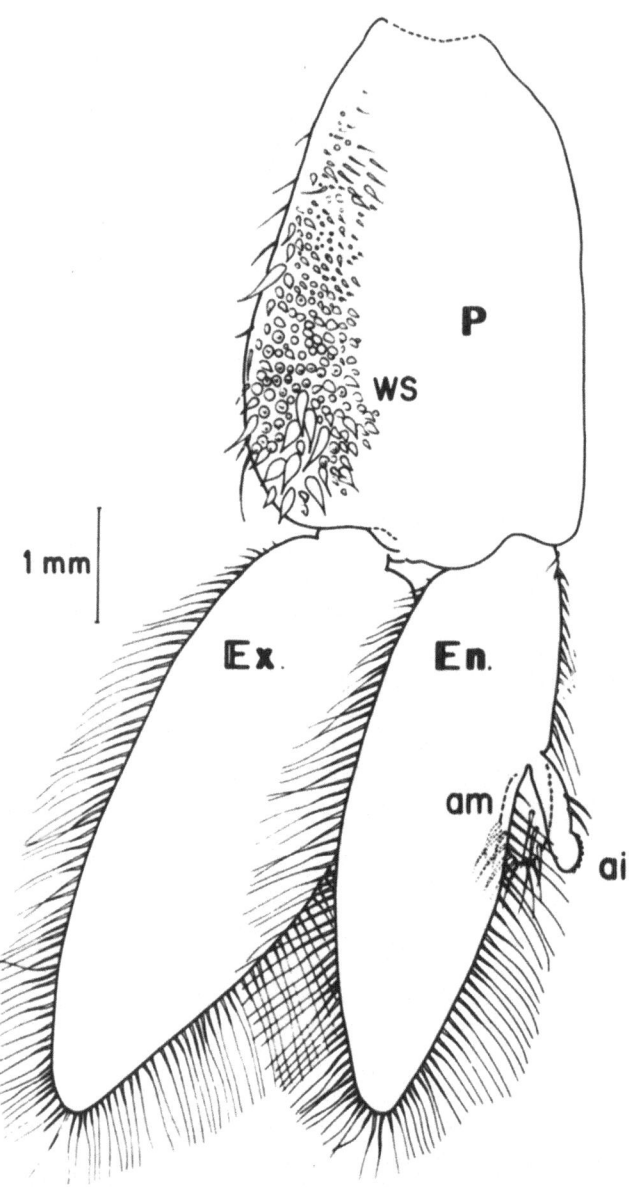

Plate 27. Typhlocaris galilea, left pleopod II of male, posterior view. ai = appendix interna; am = appendix masculina (dotted line); En, Ex = endo and exopodite; P = protopodite; ws = broadened short spines. (From Tsurnamal, Crustaceana 1978a).

tagonal pool of En-Nur and the subterranean water tract of a warm sulphur spring ('spring A') which drains into it.

The pool (about 20 m in diameter) assumed during many years the function of a cistern, and its water intermittently powered a water mill near the coast of the lake. The pool underwent many changes after it was visited and described by Annandale & Kemp (1913). The 'grasses' which then covered its water surface have

Plate 28. Frontal view of the building enclosing the octagonal pool of En-Nur at Tabgha. Lower arched aperture-entrance.

been cleared away and nowadays it is enclosed by a building made of basalt stones and a concrete roof (plate 28). The illumination in the pool is weak, as the light may penetrate into the building only through the entrance and several small apertures in the masonry.

Three different springs flow into the octagonal pool. The main spring ('spring A'), at the northern corner of the pool, is the most saline and its water chlorinity fluctuates seasonally in the range of 2,300–3,000 mg Cl 1^{-1}, while the temperature of the water is always 29–29.5°C. Oxygen is relatively low (0.9–2.0 cm^3 1^{-1}) and the pH is in the range of 6.8–7.0. The water also contains a considerable amount of dissolved hydrogen sulfide. The chlorinity and temperatures of the other two springs in the pool are lower.

The water level in the En-Nur pool, in recent years, has been kept constantly low (0.6–1.0 m deep) and the water is divided by elongated piles of stones into two, almost separate, masses.* One water mass, which is fed by the warmer and more saline 'spring A', has a temperature of 29.0–29.5°C, while the other water mass is cooler (27.0–27.5°C) but occupies the larger part of the pool. Both water masses intermix only at the outlet.

The population of *Typhlocaris galilea* in the pool is rather small, usually 35–40 individuals, and the prawns are confined to the warmer and more saline water mass originating from 'spring A'. Only rarely were prawns observed also in other regions of the pool, and in some of these cases, the prawns were observed to return quickly to the warmer water.

* In 1976, the stone piles were replaced by a solid concrete wall and the water level in the En-Nur pool was lowered since then to a depth of only 0.2 to 0.4 m.

Several individuals of the freshwater crab *Potamon potamios* (Olivier), and a few schools of the fish *Garra rufa* (Haeckel) (= *Discognathus rufus* H.), also inhabit the En-Nur pool, but so far only the latter have been observed to attack prawns. Adults of *Typhlocaris* are mostly ignored by the fish but small prawns are sometimes attacked, and remnants (limbs, antennae, etc.) of young prawns were detected in the stomachs of several fish. It would seem, however, that blind prawns are not an important constituent in the diet of *Garra*, as the guts of collected specimens were found to be filled mainly with detritic material and sulphur bacteria scraped off the stones in the pool.

The subterranean water tract of 'spring A' (plate 29), which was investigated while using SCUBA and underwater torches, consists of a 7 m long narrow inclined cave ('tunnel') through which the water ascends to the octagonal pool. The 'tunnel' widens into a rather oval karstic cavity ('hall') approximately 7.5 × 4.5 m and 2.5 m high. The bottom of the 'hall' lies 5 m beneath the bottom of the octagonal pool and is composed mainly of a yellow-brown sediment, rich in small gastropod (*Hydrobia* sp.) shells and detritic material. Like the 'tunnel', the 'hall' is entirely filled with water. About 2 m west of the entrance to the 'hall', just above the bottom, there is an opening to another subterranean water-filled cavity – the 'niche' (plate 29N). Apparently, water from the 'niche' flows directly into the octagonal pool, bypassing the main outflow through the 'tunnel'. An over 8 m deep vertical 'shaft' (plate 29S) opens in the northwestern part of the 'hall's' floor. At its deepest end, it narrows to such an extent that diving through it is impossible. Some narrow horizontal fissures open in the karstic rock at the extreme end of the 'shaft' conducting warm (29.5°C) water into the 'shaft' in a rapid up-flow. There is no light whatsoever in the subterranean water tract, except for the weak illumination which still penetrates the terminal section of the 'tunnel', near the outflow of 'spring A' into the pool. Water salinity, temperature, pH and dissolved O_2 are uniform throughout the subterranean tract and were always identical to those measured at the outflow of 'spring A'. The water of the spring, however, like the water of other springs in the vicinity, is known to be a mixture of waters from two main distinct sources: hot saline water seeping from very deep rock layers and colder freshwater (rain water) flowing rather horizontally from the upper rock layers of the surrounding hills. Both water masses intermix underground long before they reach the investigated section of the subterranean water tract. It should be noted that the uniformity of the water in the investigated section of the subterranean water tract of 'spring A', described above, does not mean that such conditions must also exist in all deep, yet unpenetrated cavities of the water tract system. On the contrary – a variety of conditions may be expected there as the water from both sources is expected to intermix in various proportions in different subterranean cavities of the local karst system.

A large population of *Typhlocaris* is present in the subterranean tract. A particularly large number of prawns is to be seen on the bottoms of the 'hall' and 'niche' where sometimes as many as ten prawns per m² could be seen foraging at one time. The animals also rest or crawl on the rocky walls of the subterranean cavities and feed on young snails from the local pigmentless hypogeal population of *Theodoxus jordani* (Sowerby) or on the filaments of sulphur bacteria protruding from crevices. Prawns were observed to enter or to leave narrow

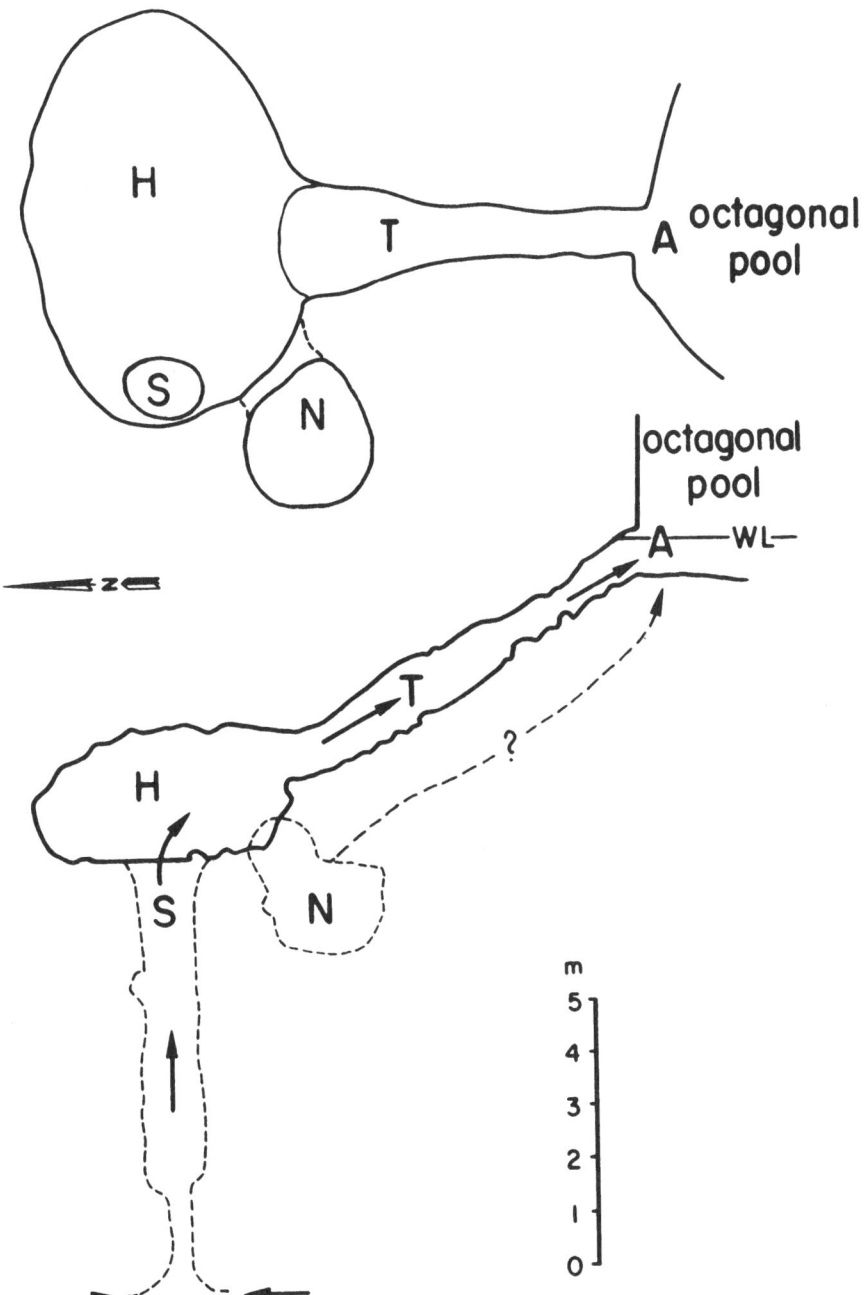

Plate 29. Subterranean water tract of the En-Nur "spring A". A = outflow of the spring into the pool; T = "tunnel"; H = "hall"; S = shaft; N = "niche"; WL = water level in the pool. Arrows indicate direction of water flow (same source as plate 27).

fissures in the walls, possibly reaching other, as yet uninvestigated cavities. No prawns were ever detected in the 'shaft' (plate 29S), but this may also be due to technical difficulties in making precise observations in that narrow site.

The subterranean cavities are also inhabited by several other peculiar organisms; most of these show some adaptations to subterranean life. Pigmentless *Theodoxus jordani* snails are quite common here and are confined to the hard substrate, where they browse on the filaments and films of bacteria covering the rocks. At the outflow of 'spring A' where some light is still available, the *Theodoxus* snails are somewhat darker, and in the more lighted sections of the pool they are almost black, similar to the usual colour of the *Theodoxus jordani* population in the nearby open springs (plate 30a). Another pigmentless but much smaller gastropod, *Hydrobia longiscata* (Bourtuignat), can also be found on rocky substrates. It is, however, much more common on the soft bottom sediments of the cavities. A dwarf variety of the black pigmented gastropod *Melanopsis praemorsa* (L.) (also inhabiting the octagonal pool) may be found among the above-mentioned snails. The shells of the En-Nur *Melanopsis* are only 6–9 mm long, while specimens from other springs nearby attain usually the length of 15–20 mm (plate 30b). A small, endemic tubified worm, *Isochaeta israelis* Brinkhurst (its other congeneric species live in brackish water or in old tectonic lakes, e.g. Baikal, Tahoe or Djoran), digs in the soft bottoms of the subterranean cavities and also in some areas in the octagonal pool. This oligochaete seems to be the main food of the blind prawn. It is accompanied by a few species of nematodes, among them *Punctodora ochridensis* W. Schneider, which was once considered as an endemic species of Lake Ochrid – a tectonic old lake in Macedonia (Yugoslavia). The aforementioned worms and individuals of *Theodoxus* constitute the main diet of *Typhlocaris*, but no *Hydrobia* or *Melanopsis* snails have ever been found in the stomach of collected prawns. The unique fauna of the hypogeal habitat also includes several species of tiny crustaceans. The most abundant is the cyclopoid copepod *Microcyclops minutus* (Claus), which was observed to swim just above the soft bottom in the subterranean cavities. The copepod shows a certain loss of its eye pigmentation. In addition, three species of malacostracean crustaceans were also discovered in samples collected at En-Nur using plankton nets.

A real surprise was the discovery of a small (1.3 mm) new species, *Cteniobathynella calmani* (Por, 1968), which is the first and the only syncarid crustacean ever found in Israel and in the whole Middle East area (plate 31a). *Cteniobathynella calmani* belongs to the ancient syncarid family Parabathynellidae, which today (apart from two relict species living in the old tectonic Lake Baikal) is adapted to life in subterranean waters, in interstitial spaces (between grains of sediment). This family evolved in the Carboniferous or even earlier in East Asia and spread from there, along two main routes, to other parts of the globe (Schminke, 1974). One led to the west in the direction of Europe, Africa and South America; the other, via Australia, also reached South America. On the western route, the Parabathynellidae apparently managed to reach Africa before the breakup of the old super-continent Pangaea in the Triassic. Here, a group of genera called the '*Cteniobathynella* group' started to expand, reaching Madagascar before its division from Africa and spreading to South America

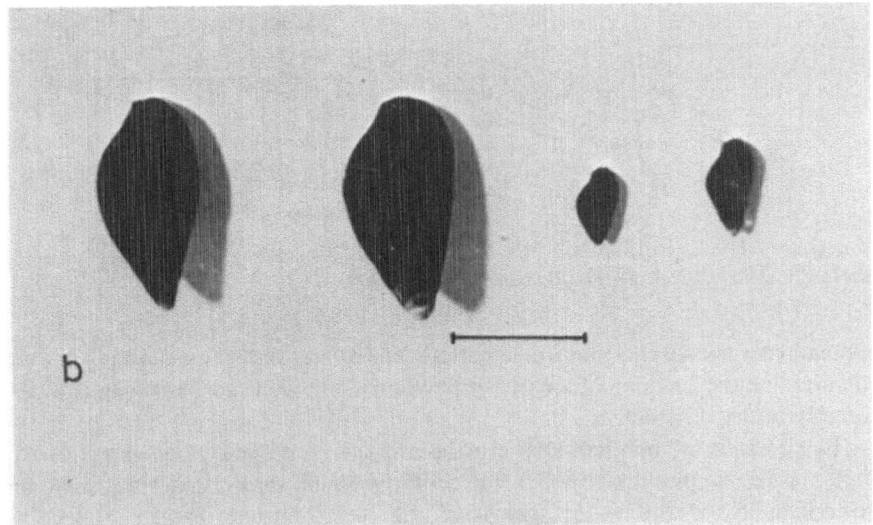

Plate 30. Gastropods from En-Nur. a, *Theodoxus jordani* (Sowerby). Pigmentation in individuals from various sites at En-Nur. Left to right: Two black specimens from slightly illuminated areas in octagonal pool; gray individual from the almost dark outflow of 'spring A'; pigmentless individuals from subterranean water tract. Pigmentless snails retain the ability to darken if maintained in illuminated habitats. b, *Melanopsis praemorsa* (L.). Specimens of the dwarf variety from En-Nur (right) as compared with specimens from neighboring springs. Scale = 1 cm.

before the opening of the southern Atlantic Ocean. Nothing is known on the ecology or biology of *Cteniobathynella* in the subterranean water tract of En-Nur.

Monodella relicta Por, belonging to the relict crustacean order Thermosbaenacea, was also observed and collected in En-Nur. This species (plate 31b) was first described from the warm mineral springs of En-Zohar, on the western coast of the Dead Sea (Por, 1962), and like *Typhlocaris galilea* has also been considered as a relict of the transgressive Pliocenic Mediterranean Sea. The presence of *Monodella relicta* in En-Nur and En-Zohar and the clear conspecificity of both

361

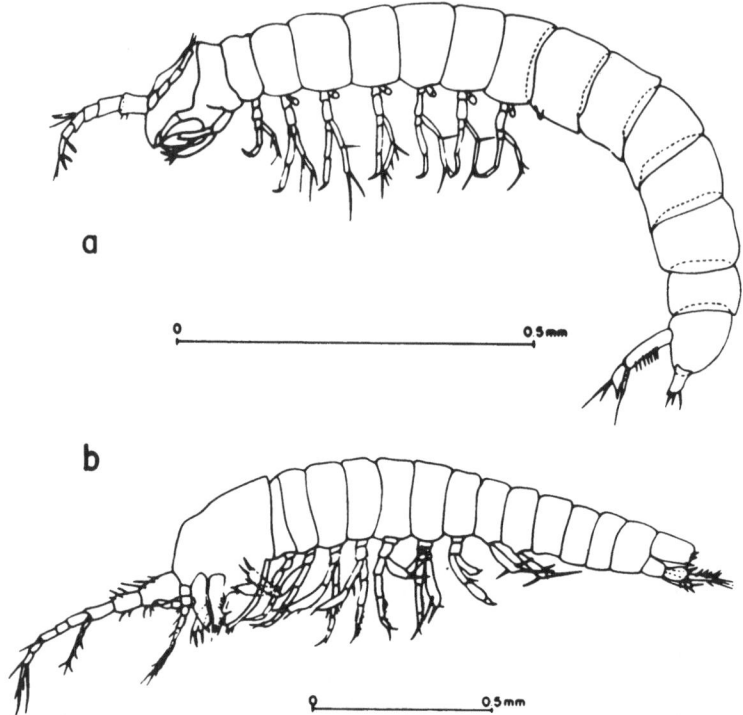

Plate 31. a, *Cteniobathynella calmani* (Por) and b, *Monodella relicta* Por from the hypogeal habitat of En-Nur (From Tsurnamal and Por, 1968, Intern. J. Speleology 3).

populations indicate that the whole Jordan Valley was once a continuous marine gulf and that the isolation of the two populations took place not very long ago, apparently in the Pleistocene.

The presence of another tiny crustacean, the amphipod *Bogdiella hebraea* Ruffo, in the samples taken in En-Nur confirms the aforementioned statement on the continuity of the water system of the late Pliocenic Jordan Valley, as *Bogdiella* was previously found in En-Hakikar springs, south of the Dead Sea (Ruffo, 1963).

In addition to the populations of *Typhlocaris galilea* in the epigeic and hypogeal habitats of En-Nur, a few adult prawns were also observed in the water of a small spring which is encased in a small ruin, located some 55 m southeast of the octagonal pool. The prawns were observed there on three separate occasions in the spring months of 1964 and 1965. However, the small spring dried up during the summer months and several years ago it had dried up completely, before any valuable information on its prawns could be gathered. The water of the spring (referred to as 'En-Sheva 6' by Mekorot, Water Company of Israel) has a temperature of 24°C and chlorinity of about 1,200 mg Cl l^{-1}. These values are much lower than those measured at 'spring A'.*

* On several occasions, a few blind prawns were also observed in the very shallow water of a small pool, about 25 m south of En-Nur. The pool is fed by two small springs (water temperature, 25°C; chloride content, 1000–1300 mg Cl l^{-1}).

The temperature preference of *Typhlocaris galilea* was experimentally studied by using a temperature gradient apparatus (Tsurnamal, 1978b). It has been found that prawns which, prior to the trials, were maintained at temperatures of 23–26.5°C clearly elected to stay longer in the section of the gradient where temperature was 26–27°C; animals (among them also individuals that had been used in the previous series of experiments) which, prior to the trials, were maintained at temperatures of 28–30°C spent most of the time in the 29–30°C section of the gradient. The only exception to this was an egg-bearing female prawn which was used in both experimental series. The ovigerous female showed a preference for the 26–27°C range even following prior maintenance at 28.5–29°C.

The results indicated that usually the acclimatization of the prawns to a certain range of temperature determines their future preferred range of temperatures. Such adaptive behaviour may explain why the blind prawns in the octagonal pool at Tabgha congregate almost exclusively in the warmer water mass. It would seem that some individuals from the main population of *Typhlocaris galilea* (which presumably lives regularly in some yet unpenetrated subterranean spaces) may occasionally enter other karstic cavities into which more hot and saline water is brought. Following prolonged stay in such relatively high temperatures, as actually prevail at the terminal (investigated) section of the water tract of 'spring A', the prawns become acclimatized to that temperature (29.5°C) and tend to refrain from entering the colder water mass in the pool. This may bring about an almost complete isolation of the populations of the pool and of the terminal section of the subterranean tract from the main *Typhlocaris* population, which is believed to live in slightly cooler water. The discovery of blind prawns in the outflow of the small 'En-Sheva 6' spring (water temperature: 24°C) brings some support to this belief.

The temperature range of 26–27°C may have special significance in the biology of *Typhlocaris galilea*, as this range was selected by all animals previously maintained at 23–26°C, but what seems to be more important is that the only ovigerous female used in the experiments showed a preference for the 26–27°C range even following prior maintenance at 28.5–29°C. In spite of the fact that these findings are based on a single test animal employed in a rather limited number of trials, they seem to have significance inasmuch as eggs of several ovigerous female prawns maintained in the laboratory at a 28–30°C range failed to complete their development, whereas eggs of the egg-bearing females maintained at 23–26°C always did so successfully (Tsurnamal, 1978a). Apparently, the optimal temperature for development of the embryos of *Typhlocaris galilea* is below that measured at the outflow of 'spring A' (29.5°C). It is possible, therefore, that the tendency of the egg-bearing female prawn to be located in the 26–27°C range is connected with a genetic-behavioural mechanism intended to ensure survival of the species.

If the egg-bearing females in nature also express a strong tendency to stay within such a limited temperature range, this may prevent them from leaving the suitable conditions of their hypogeal habitat, and they would be less likely to enter the warmer water which also flows into the octagonal pool at Tabgha. This may be at least one possible reason for the absence of ovigerous female prawns in 'spring A' and in the octagonal pool.

It thus seems that a temperature of 26–27°C or less exists in part of the water-filled, karstic subterranean spaces in the Tabgha area, and that these spaces are the primary breeding sites of *Typhlocaris galilea*, because the temperature of the water permits 'nuptial behaviour' by the prawns followed by oviposition, egg-bearing and embryo development like that observed in the laboratory under similar temperatures. The population of the blind prawns in the octagonal pool and in the investigated cavities of 'spring A' is a non-reproducing population which seems to be almost isolated from the main population of *Typhlocaris galilea* by becoming acclimatized to a higher water temperature in these habitats.

M. Insects

1. General

C. Serruya

The insect fauna of the Kinneret area has been investigated by Laidlaw (1913), Carpenter (1913), Edwards (1913), Brunetti (1913), Horvath (1913), Kieffer (1915) who worked on samples collected by Annandale. More recent studies include the work of Bodenheimer in the thirties, the unpublished work of Palmoni, the study of Chironomids by Kugler & Gasith in the late sixties and early seventies, the work on Trichoptera by Botosaneanu & Gasith (1971), the study of Mosquitoes by Margalit & Tahori in the seventies, the work of Matzliach (1972) on the Ephemeroptera, the study of Pleoptera by Zwick (1972) and the zoogeographical work of Botosaneanu (1973) concerning the distribution of Trichoptera.

In spite of these contributions, our knowledge of the insect fauna of the area is still incomplete. Among the most primitive insects (Apterygota), the springtail (Collembola), *Cyphoderus genneserae* was first described by Carpenter in 1913. Among the most common Ephemeridae, Palmoni found: *Cloeon dipterum, Centroptilum pennulatum, Baetis rhodani, Heptagenia longicauda* and *Caenis macrura*.

Trithmis annulata and *Brachytemis leucosticta* are the most common dragonflies (Odonata). However, Bodenheimer found, in the vicinity of the lake in July *Platycenemis dealbata, Orthetrum sabinae* and *O. chrysostigma*.

A common Neuropteron of Lake Kinneret, *Sisyra trilobata* lives preferentially in the sponges.

In the lake vicinity the most common flies are *Musca domestica*, the bloodsucking flies *Stomoxys calcitrans* and *Lyperosia minuta*. Hippoboscidae are represented by *Hippobosca equina* and the sand-flies (Psychodidae) by *Phlebotomus papatassi* and *P. minutus*. The most common crane-fly is *Geranomyia annandelei*.

The rhynchota are represented by semi-aquatic and true aquatic species as in the following list:

Semi-aquatic species:
   Hebrus pusillus, Fall
   Mesovelia vittigera, Horv.
   Dipsocoris alienus H. Sch.
   Hydrometra stagnorum L.
   Gerris paludum Fabr.
   Limnogonus aegyptiacus Put.
   Naboandelus bergevini Berg.
   Rhagovelia nigricans Burm.
   Microvelia pygmaea Duf.
   Patapius spinosus Rossi
   Erianotus lanosus Duf.
   Acanthia variabilis H.-Sch.

Ochterus strigicollis sp. nov.
Aquatic species
   Ranatra vicina Sign.
   Plea letourneuxi Sign.
   Anisops producta Fieb.
   Notonecta glauca L.
   Arctocorisa hieroglyphica Duf.
   Micronecta annandalei sp. nov.
   Micronecta isis Horv.
   Micronecta perparva sp. nov.
   Aphys nymphaeae

The Chironomidae, Trichoptera and Mosquitoes which have been studied in detail in the lake area are described separately.

2. Chironomidae and Trichoptera

J. Kugler

The first account of the Chironomidae of Lake Kinneret was given by Kieffer (1915). In his paper, Kieffer deals with 7 species collected by N. Annandale (Curator of the Indian Museum in Calcutta) in Tiberias in the autumn of 1912. Six of the seven species are described as new. The species given by Kieffer are:

Pelopia monilis (L.)
Pelopia cygnus
Trichotanypus tiberiadis
Polypedilum genesareth
Polypedilum tiberiadis
Tendipes bethseidae
Tendipes galilaeus

The same 7 species are mentioned in Bodenheimer's list of the lake's fauna in 'Animal Life in Palestine' (1935). In an unpublished report, 'Fishes of the genus *Barbus* in Lake Kinneret', Yashouv notes that while working on the benthos of the lake in 1948–1949, he found chironomid larvae at depths of 0–15 m all the year round. Larvae were absent at the depth of 24 m from June to December and at the depth of 30 m from June to January. Yashouv related the absence of chironomid larvae to the absence of oxygen in the hypolimnion during the summer stagnation period. By examining the stomach contents of *Barbus*, he found that *Barbus canis* C.V. is mainly vegetarian, while the stomach of *Barbus longiceps* C.V. was often filled with chironomid larvae.

From the end of 1963 until 1969, a survey of the chironomid fauna of Lake Kinneret was carried out, supplemented by a survey of the chironomids of the Hula Nature Preserve, 30 km north of Lake Kinneret. The results of these surveys were published (Kugler, 1966a, b, 1971; Kugler & Wool, 1968; Kugler & Chen, 1968; Wool & Kugler, 1969; Kugler & Reiss, 1973). There is a large number of abundant species and the chironomids are the most important insect group living in Lake Kinneret. They are the only insects which live not only in the littoral and sublittoral, but also in the profundal zone, reaching to the lake's deepest point at 42.5 m.

The adults of 37 species (List No. 8) were collected with the help of light traps posted on the western shore of the lake, in Kare-Deshe, and the eastern shore, in Ein-Gev. Of the 37 species, 28 are identified to the species level; some of the other 9 seem to be new species. Of the 7 species listed by Kieffer (1915), only 4 are included. The species *Pelopia cygnus* was not found; *Tendipes galilaeus* and *Tendipes bethseidae* cannot be recognized as they were insufficiently described from single females.

Adult chironomids emerge from Lake Kinneret throughout the year. Some of the species are very common, others are rare. The common species are: *Polypedilum tiberiadis, Procladius choreus, Tanypus punctipennis, Tanytarsus nigricornis, Cryptocladopelma virescens, Cladotanytarsus pseudomancus, Stictochironomus genesareth, Cricotopus silvestris, Dicrotendipes fusconotatus, Dicrotendipes pilosimanus, Xenochironomus xenolabis* and *Leptochironomus stilifer*. About 95% of the midges collected with the light trap belonged to these 12 species. They were collected in different seasons, indicating that they have several generations per year. Most of the chironomids of Lake Kinneret belong to the sub-family Chironominae (List No. 8). It is noteworthy that only a single species of the genus *Chironomus* was found, and from this relatively few larvae and imagines. This differs from eutrophic lakes in Europe which are characterized by the abundance of *Chironomus* larvae. Even in the Hula Nature Preserve, 30 km north of Lake Kinneret, larvae of 3 *Chironomus* species are abundant: the Holarctic species *Chironomus plumosus* L. and the two tropical species, *Chironomus calipterus* Kieff. and *Chironomus transvaalensis* Kieff.

The Tanypodinae are represented by 3 species; two of them, *Procladius choreus* and *Tanypus punctipennis*, are the characteristic species of the profundal zone of Lake Kinneret. A similar case where a *Procladius* (*Procladius umbrosus* Goetgh.) and a *Tanypus* (*Tanypus guttatipennis* Goetgh.) are the abundant species of the bottom is recorded by MacDonald (1953, 1956) for the Ekunu-Bay, Lake Victoria.

The Orthocladiinae are represented by 5 species. Larvae of an additional species, *Cricotopus ornatus* (Meigen), were found only in the salt water canal (which leads water from high salinity springs) along the western shore of the Kinneret to the Jordan River south of the lake. This species is known as a halobiont from Europe, Turkey, Egypt and maybe also from Canada and Formosa (Hirvenoja, 1973).

The Orthocladiinae are represented by more species in the swiftly flowing Jordan and its tributaries north of the Hula Nature Preserve.

The chironomid larval population differs quantitatively and qualitatively in different parts of the lake. All species and the greatest number per m² are found in the littoral. The number of larvae in the sandy littoral can reach 2,200 per m², not taking into account the very small larvae which passed through the mesh (0.6 mm) of the net which was used for straining. The peak number is reached in different months of different years, but always between April and June.

The number of species and specimens decreases in the sublittoral to a maximum of 1,000 larvae per m² and is lowest in the profundal, with a maximum of 180 per m². Although larvae are found in the littoral and sublittoral throughout the year, they are absent all summer in the profundal, where no measurable oxygen is present (Fig. 116).

From larvae collected on plant fragments of *Myriophyllum spicatum* L., *Phragmites australis* (Cav.) Trin. and *Typha australis* Schum & Thonn. in the littoral near Bet-Yerah at the southern border of the lake, the following adults were reared: *Dicrotendipes fusconotatus, Dicrotendipes pilosimanus, Cricotopus silvestris, Cricotopus vierriensis, Nilodorum brevibucca, Polypedilum tiberiadis, Polypedilum tropicum* and *Polypedilum* n. *laterale*. On stones of the littoral at Ras-el Burge at the southwest part of the lake, the larvae of the following species were collected: *Xenochironomus xenolabis, Polypedilum tiberiadis, Cladotanytarsus pseudomancus, Tanytarsus nigricornis, Chironomus calipterus, Cryptochironomus* sp., *Cricop-*

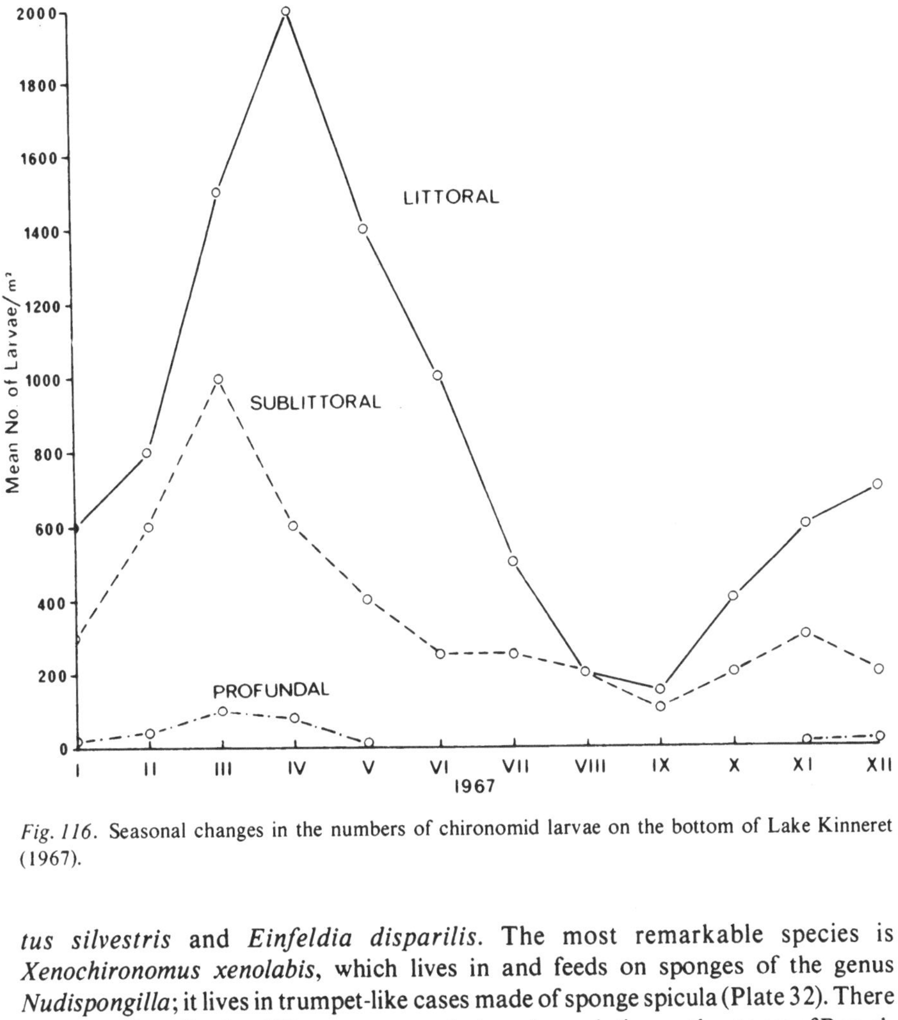

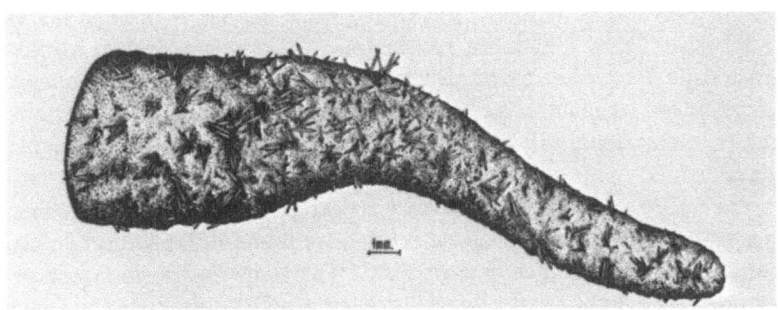

Fig. 116. Seasonal changes in the numbers of chironomid larvae on the bottom of Lake Kinneret (1967).

tus silvestris and *Einfeldia disparilis.* The most remarkable species is *Xenochironomus xenolabis*, which lives in and feeds on sponges of the genus *Nudispongilla*; it lives in trumpet-like cases made of sponge spicula (Plate 32). There was a small qualitative difference among the larval population on the stones of Ras-el-Burge in 1964–1965 and 1967–1968 (Fig. 117). A noteworthy decrease occurred in the number of *Xenochironomus* larvae in 1967–1968. This may have been due to the change in the water cover of the littoral stones as a result of the diversion of water from

Plate 32. Larval case of *Xenochironomus xenolabis* made of spiculae of *Nudispongilla.*

371

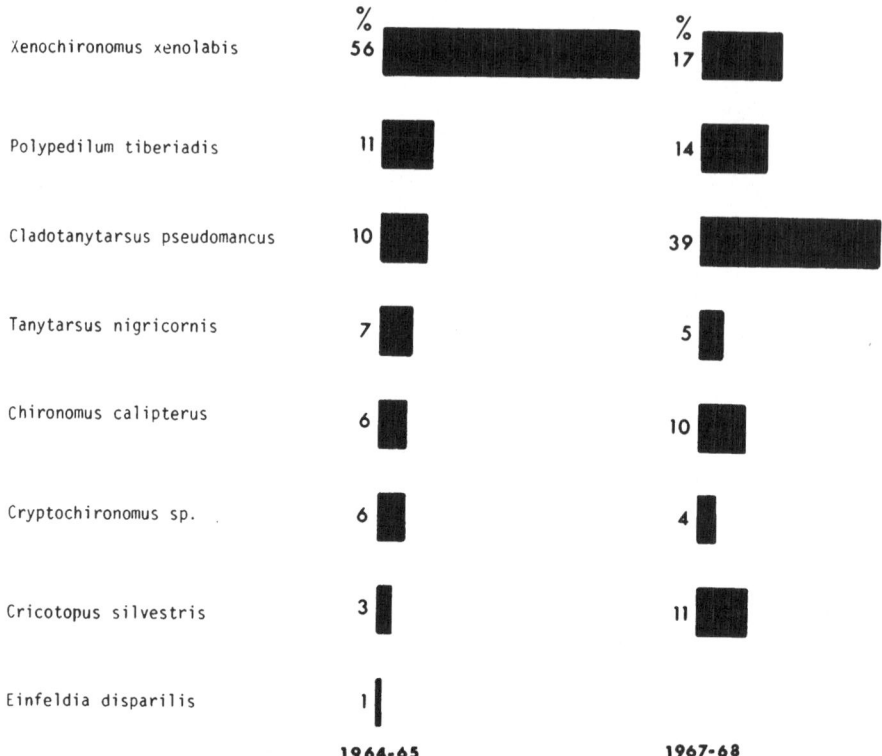

	%	1964-65	%	1967-68
Xenochironomus xenolabis	56		17	
Polypedilum tiberiadis	11		14	
Cladotanytarsus pseudomancus	10		39	
Tanytarsus nigricornis	7		5	
Chironomus calipterus	6		10	
Cryptochironomus sp.	6		4	
Cricotopus silvestris	3		11	
Einfeldia disparilis	1			

Fig. 117. Composition of chironomid larval population on stones and rocks at the Ras-El-Burge station in the years 1964–1965 and 1967–1968.

the lake by the National Water Carrier from 1964 onwards. The larvae of the very common *Cladotanytarsus pseudomancus* live in C-shaped tubes made of sand grains. In the sandy littoral, *Polypedilum tiberiadis* and *Cladotanytarsus pseudomancus* are the most common species, but *Xenochironomus xenolabis* is missing.

The sublittoral is poorer in species. In the winter months, *Stictochironomus genesareth* is the main element, while *Procladius choreus, Tanypus punctipennis* and *Cryptocladopelma virescens* are also better represented than in the littoral.

In the profundal, only the free-living white larvae of *Procladius choreus, Tanypus punctipennis* and the red larvae of *Cryptocladopelma virescens* are characteristic. According to the structure of the mouth parts, most larvae living in the littoral feed on plants, plankton or detritus. The number of predacious larvae increases in the sublittoral, and the three abundant species of the profundal are all predacious.

In Lake Kinneret, species of several zoogeographical regions are represented. Palaearctic, Holarctic and Tropical species are found in the littoral, in which the temperature in the summer reaches 30°C. The same composition occurs in the sublittoral, while in the relatively cold profundal, with a year-round temperature of about 15°C, only Holarctic and Palaearctic species are abundant.

a. Chironomids as fish food in Lake Kinneret

From November 1965 to December 1966, fish of different species were taken once a month from the night catch of the fishing boats. The exact locations where the fish were caught and the bottom depths of these places (measured with an echosounder) were obtained from the fishermen.

The fish were weighed, their stomachs were removed and stored in 70% alcohol. In the laboratory, the contents of the stomachs were examined. The chironomid larvae that were found were counted and identified.

A total of 322 fish of the following species were examined: *Barbus longiceps* C.V. (58 specimens), *Barbus canis* C.V. (17), *Cyprinus carpio* L. (an introduced species for fish culture) (40), *Mirogrex terraesanctae* Steinitz (66), *Capoeta damascina* (C.V.) (24), (Cyprinidae); *Tilapia zillii* (Gervais) (49), *Tilapia galilaea* (Arted.) (28), *Tristramella simonis* (Gunter) (15), *Tristramella sacra* (Gunter) (11), (Cichlidae); *Mugil cephalus* (L.) (an introduced species for fish culture) (14), (Mugilidae).

Chironomid larvae were found in stomachs of six of the ten fish species examined:

(i) *Barbus longiceps* and *Cyprinus carpio*

Both *Barbus longiceps* and *Cyprinus carpio* feed on chironomid larvae throughout the year. Larvae were found in 39 of the 58 specimens of *Barbus longiceps* and in 34 of the 40 of *Cyprinus carpio*. The number of larvae per stomach differed, reaching a maximum of almost 2,000 (1,930 in a specimen of *Cyprinus carpio* weighing 177 gr and 1,832 in a specimen of *Barbus longiceps* weighing 620 gr).

The same species of larvae appeared in both fishes. The larvae found in the largest proportions were: *Polypedilum tiberiadis*, *Cladotanytarsus pseudomancus*, *Cryptochironomus stilifer*, *Stictochironomus genesareth*, *Cryptocladopelma virescens*, *Tanypus punctipennis* and *Procladius choreus*.

In general, the larvae of the same species were found in the fish stomachs and on the bottom of the area where the fish were collected. Sometimes larvae from other areas were found in the fish stomachs, indicating that the fish obtained their food from a wider area.

Fish do not live in the profundal during the summer. In winter and spring, *Barbus longiceps* and *Cyprinus carpio* were caught in the profundal, and their stomachs were filled mainly with larvae of *Tanypus punctipennis*, *Procladius choreus* and *Cryptocladopelma virescens*.

(ii) *Mirogrex terraesanctae*

Mirogrex terraesanctae, better known as *Acanthobrama terraesanctae*, is the type species of the newly erected genus *Mirogrex* (Goren et al., 1975). From the 66 examined specimens of *Mirogrex terraesanctae*, chironomid larvae were found in 16 fish, collected in February, March and May. The fish were caught at a depth of 11–20 m. Most larvae found in the stomachs were of *Procladius choreus*, *Tanypus punctipennis* and *Cryptocladopelma virescens*. In May, larvae of *Polypedilum tiberiadis* were also common (26.9%). During all other months of the year, chironomid larvae were not found in the fish, but their stomachs contained large quantities of zooplankton. Taking into account that the breeding season of the fish in Lake Kinneret is in winter and spring, the change of food habits may be connected with the fish's reproduction.

(iii) *Capoeta damascina*; Synonym: *Varicorhinus damascinus*
The stomachs of *Capoeta damascina* contained mainly remains of plants, a great amount of gravel, sand, and sometimes snails. Only 5 of the 24 fish examined contained small quantities of chironomid larvae. The larvae of the following species were found in the stomachs: *Tanytarsus nigricornis*, *Cricotopus silvestris*, *Dicrotendipes pilosimanus*, *Cladotanytarsus pseudomancus* and *Polypedilum tiberiadis*. The finding of these larvae confirms that *Capoeta damascina* feeds by scraping plants, snails and chironomid larvae off the stones of the littoral.

(iv) *Mugil cephalus*
The fingerlings of *Mugil cephalus*, a Mediterranean breeding species, have been introduced into the lake for fish culture. Stomachs of only 5 of the 14 examined specimens contained small amounts of chironomid larvae (*Polypedilum tiberiadis*, *Cladotanytarsus pseudomancus* and *Cryptochironomus stilifer*).

(v) *Tilapia zillii*
From the 4 examined cichlid species, only *Tilapia zillii* feeds on Chironomidae. Thirty of the 49 specimens examined contained chironomids in their stomachs, mostly adults and pupae. Only in two specimens were small amounts of larvae found. In contrast to the other chironomid feeding fishes of Lake Kinneret, which feed on larvae from the bottom of the lake, *Tilapia zillii* collects chironomids mostly from the surface of the water. Adults were probably caught when hatching or floating as dead specimens. The pupae were probably eaten while ascending prior to hatching.

List No. 8

List of Chironomidae light-trapped on the shores of Lake Kinneret and their distribution, 1964–1969

Species	*Distribution*
Tanypodinae	
Ablabesmyia monilis (L.)	Holarctic
Tanypus punctipennis (Meig.)	Holarctic
Procladius choreus (Meig.)	Palaearctic
Synonym: *Trichotanypus tiberiadis* Kieff.	
Orthocladiinae	
Cricotopus silvestris (Fabr.)	Holarctic
Cricotopus vierriensis (Goetgh.)	Europe, Afghanistan, Israel
Psectrocladius sp.	
Limnophyes minimus (Meig.)	Europe, Israel
Smittia pratorum Goetgh.	Europe, Israel
Chironominae	
Chironomus calipterus Kieff.	Most parts of Africa, Israel
Einfeldia disparilis (Goetgh.)	Congo. Chad, S.W. Africa, Israel

374

Dicrotendipes pilosimanus Kieff.	Egypt, Sudan, Algeria, Ethiopia, Uganda, Kenya, Congo, Chad, Rhodesia, S. Africa, India, Australia, Israel
Dicrotendipes fusconotatus (Kieff.)	Egypt, Uganda, Chad, Congo, Israel
Dicrotendipes n. *ealae* Freeman	
Dicrotendipes sp.	
Nilodorum brevibucca Kieff.	Egypt, Sudan, Chad, Gabon, Uganda, Congo, Rhodesia, Israel
Xenochironomus xenolabis Kieff.	Holarctic
Cryptochironomus suplicans (Meig.)	Europe, Israel
Cryptochironomus acutus (Goetgh.)	Congo, Chad, Nigeria, S. Africa, Israel
Cryptochironomus sp. No. 1	
Cryptochironomus sp. No. 2	
Leptochironomus stilifer (Freeman)	Sudan, Uganda, Chad, S. Africa, Israel
Leptochironomus deribae (Freeman)	Sudan, Chad, Israel, Western Europe
Cryptocladopelma virescens (Meig.)	Holarctic
Harnischia nudiforceps (Kieff.)	Sudan, Congo, Chad, S. Africa, Israel
Stenochironomus sp.	
Polypedilum tropicum Kieff.	Sudan, Congo, Nigeria, S. Africa, Israel
Polypedilum n. *subovatum* Freeman	
Polypedilum n. *laterale* Goetgh.	
Polypedilum n. *dewulfi* Goetgh.	
Polypedilum nubifer Skuse	N. Africa, Iraq, Ceylon, Formosa, Australia, Israel
Polypedilum tiberiadis Kieff. indistinguishable from European species	
Polypedilum quadriguttatum Kieff.	Israel, Europe
Stictochironomus genesareth (Kieff.)	Egypt, Sudan, Ethiopia, Congo, Chad,
Synonym: *Stictochironomus caffrarius* (Kieff.)	Nigeria, Rhodesia, S. Africa
Tanytarsus nigricornis Goetgh.	Yemen, Ethiopia, Uganda, Rhodesia, S. Africa, Israel
Tanytarsus nocticolor Kieff.	Egypt, Sudan, Congo, Israel
Tanytarsus sp.	
Cladotanytarsus pseudomancus (Goetgh.)	Egypt, Sudan, Nigeria, Ghana, Congo, S.W. Africa, Israel

During the work on chironomids, a survey of the Trichoptera was also carried out (Gasith, 1969; Gasith & Kugler, 1971; Botosaneanu & Gasith, 1971). From more than 40 species of Trichoptera known from Israel and Sinai, only four live in Lake Kinneret: *Orthotrichria moselyi* Tjeder 1946 (Hydroptilidae), *Ecnomus galilaeus* Tjeder 1946 (Psychomyidae), *Ecnomus gedrosicus* Schmid 1959 (Psychomyidae) and *Pseudoneureclipsis palmoni* Flint 1967 (Polycentropodidae).

The Trichoptera are not a nuisance, as the number of adults emerging at the same time is not large. Their importance as fish food is also small.

The larvae of these Trichoptera live only on the stones of the littoral. Contrary to imagines of chironomids, which emerge at all seasons of the year, the imagines of Trichoptera do not emerge in winter.

a. *Orthotrichia moselyi*

Orthotrichia moselyi (2.5–3 mm long) is the smallest of the lake's Trichoptera. The wings are blackish, spotted with grey. Only the 5th instar larva secretes a seed-like case. At the end of its development, the larva glues the case to a stone, closes it and pupates inside. In winter, mainly 5th instar larvae are observed. The imagines emerge from mid-March until October. Two peaks in May and September–October show that there are at least two generations in a year. *Orthotrichia moselyi* is known only from Israel (Lake Kinneret, Ein Amal) and Sinai (W. Isla).

b. *Ecnomus galilaeus* and *Ecnomus gedrosicus*

It is easy to distinguish the adults of the two species. *Ecnomus galilaeus* is 4–6 mm long, uniformly colored light brown to nearly white. *Ecnomus gedrosicus* is 6–8 mm long, darker brown-grey with light spots on the forewings.

The larvae of the two species are very similar in appearance and habits. They usually live on the lower side of stones in self-secreted webs. They are both predacious. Before pupating in the cracks in the stone, the larvae build cases from small particles of stones and parts of mollusc shells. These species also have two generations per year. The imagines from the first generation emerge in spring, the second generation in summer.

Ecnomus galilaeus is also known from the Hula, Ein Amal, the water reservoirs of Zalmon and Beth-Netufa, and Birketh-Ram (Golan).

Ecnomus gedrosicus was described from Iran. In Israel, it is also known from Hula, the Jordan, the Yarmuck and from the springs of the Golan.

c. *Pseudoneureclipsis palmoni*

Pseudoneureclipsis palmoni is 5–6 mm long and uniformly black-grey. The larvae live in tube-like cases connected to the stones. It seems to have a single generation per year in Lake Kinneret. The adults emerge in the summer, mostly in September. This species is known only from Israel. Besides Lake Kinneret, it was also found in water canals of the Bet-She'an Valley and in Ein Amal.

It is noteworthy that part of the Trichoptera (*Ecnomus galilaeus* and *Pseudoneureclipsis palmoni*), like the chironomids, show affinity to the African fauna (Botosaneanu & Gasith, 1971).

376

3. Mosquitoes

J. Margalit & A. S. Tahori

a. History and present status

Little has been published about the mosquitoes of the Lake Kinneret area. The earliest record is that of Annandale (1915) who visited the area during the fall of 1912. In his comprehensive survey of the fauna of Lake Kinneret he lists the following mosquito species identified by Edwards (1913): *Anopheles palestinensis* Theobald, *A. culicifacies* Giles, *Stegomyia fasciata* Fabr., *Culex modestus* Ficalbi, *Uranotaenia unguiculata* Edwards and *Culex laticinctus* Edwards. The distribution of *A. culicifacies* is restricted to the parts of Asia from China to Iran and Oman (Stone *et al.*, 1959) and this species was possibly mistaken for *A. sergentii* (Theobald). *A. palestinensis* is a synonym for *A. superpictus* Grassi, and *Stegomyia fascatus* is a synonym for *Aedes aegypti* (Linnaeus). Other surveys were carried out by Barraud (1921), Buxton (1924), Theodor (1925) and during 1955–1958 by Margalit & Tahori (1970). The latter found the following species at Tabha: *Culex hortensis* Ficalbi, *Culiseta longiareolata* (Macquart), *Uranotaenia unguiculata* Edwards, *Anopheles claviger* (Meigen), *A. sacharovi* Favre, *A. sergentii* Theobald and *A. superpictus* Grassi. In 1976 during a survey for mosquito pathogens by Margalit (to be published) the following additional mosquito species were collected from the Lake Kinneret area especially from the Buteiha Valley: *Anopheles coustani tenebrosus* (Donitz), *A. dthali* Patton, *A. Multicolor* Cambouliu, *Aedes caspius* (Pallas), *Aedes detritus* (Haliday), *Culiseta annulata* (Schrank), *Culex pipiens complex*, *C. theileri* Theobald, *C. tritaeniorhynchus* Giles and *C. univittatus* Theobald.

After World War I intensive mosquito surveys were carried out first by the military authorities and later by Government personnel. The surroundings of Lake Kinneret were endemic malaria areas until the fifties.

Most of the earlier references mention the fauna of Lake Kinneret. It should, however, be stressed that, at least during the last 20 years, no mosquito larvae were ever found in the body of the lake, only in its surroundings. The drastic wave action in the littoral area of the lake, coupled with a predatory fish population, are probably detrimental to mosquito development.

It is interesting to note that the mosquito fauna around the Lake Kinneret, unlike in the rest of the country, has not drastically changed for the last 50 years. This may be due to the fact that very little anti-mosquito activity was carried out on the eastern banks of the lake until 1967, as many of the potential breeding sites in the north and north-east became accessible to the Israeli authorities only after 1967.

b. The mosquito species

The collection sites of the species are indicated in the map (Fig. 118).

Anopheles claviger was recorded from Tabha in 1956 by Margalit & Tahori

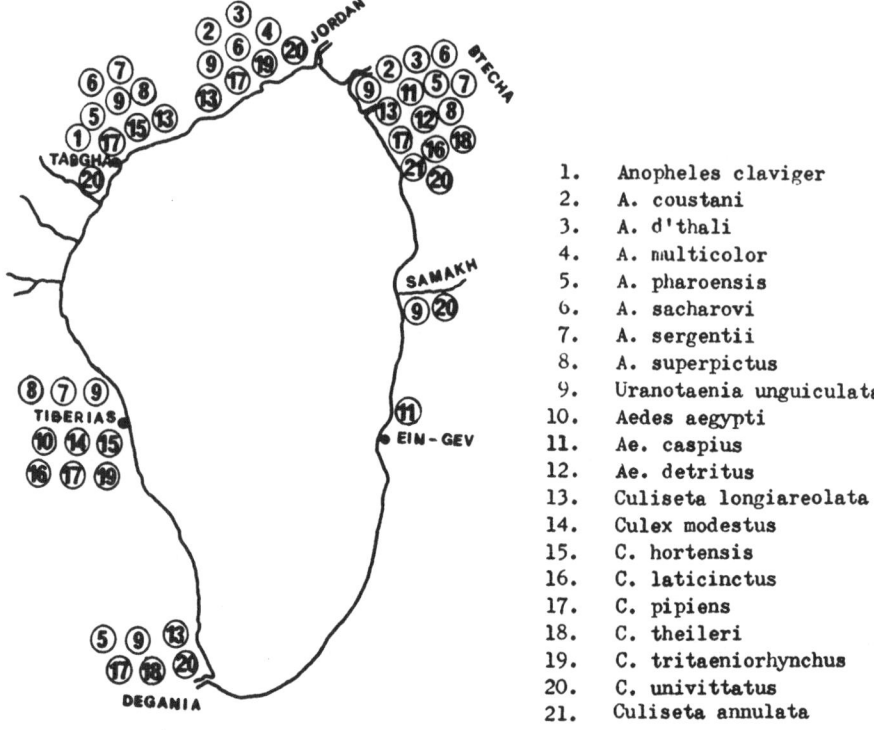

1. Anopheles claviger
2. A. coustani
3. A. d'thali
4. A. multicolor
5. A. pharoensis
6. A. sacharovi
7. A. sergentii
8. A. superpictus
9. Uranotaenia unguiculata
10. Aedes aegypti
11. Ae. caspius
12. Ae. detritus
13. Culiseta longiareolata
14. Culex modestus
15. C. hortensis
16. C. laticinctus
17. C. pipiens
18. C. theileri
19. C. tritaeniorhynchus
20. C. univittatus
21. Culiseta annulata

Fig. 118. Map of Lake Kinneret showing locations where mosquito species were collected. Numbers indicate species as listed.

(1970). This paleartic species which prefer cooler climates to that of Lake Kinneret is rather rare in this area. *Anopheles coustani tenebrosus* is currently breeding in the spring-fed swamps around Lake Kinneret at Buteiha Valley and in backwaters of Lake Kinneret near the entrance of the Jordan River. *Anopheles sacharovi* which was once the major vector of malaria in Israel (Saliternik, 1964) was no longer found in Israel in the fifties and sixties, probably because of stringent control measures and because of the drainage of most of the swamps of Israel. In 1969 it was recorded again from the Buteiha Valley (Ben-Dov, 1971) and has been collected since then by us many times in 1976. *Anopheles dthali* is an East African species extending into the Dead Sea area. In April 1976 we found it in the Buteiha Valley. This is the most northern record for this African species. Twenty larvae of *A. multicolor* were collected by Margalit in June 1976 in a swampy area near the Jordan River mouth. One female *Anopheles pharoensis* Theobald was recorded from Tabha in 1920 by Barraud (1921) and two from Deganya in 1923 by Theodor (1925). No further records from the Lake Kinneret area exist. *A. sergentii*, a potential vector of malaria, is very prevalent during April–October, especially in the spring-fed swamps of the Buteiha Valley. *A. superpictus*, also a potential vector of malaria, is now a little less common than *A. sergentii* and breeds under similar conditions, while Kligler (1924) found *A. sacharovi*, *A. superpictus* and *A. sergentii* to be equally common around Lake Kinneret.

378

Uranotaenia unguiculata is very common around Lake Kinneret. Its type locality is Tiberias (Annandale, 1915). *Aedes caspius*, a very common poikilohalinic species, breeds in periodically inundated areas around Lake Kinneret. *Aedes detritus* breeds under similar conditions as *Aedes caspius* and is found mainly during the winter months. *Aedes aegypti* was found by Annandale (1915) 'to be not uncommon' (Edwards, 1913), also by Barraud (1921) and in the twenties by Theodor (Margalit & Tahori, 1974). It has not been recorded since. Two larvae of *Culiseta annulata* were found in the Buteiha Valley by Margalit in August, 1976. *Culiseta longiareolata*, one of the most prevalent mosquito species in Israel, is also very common around Lake Kinneret.

Culex modestus was found by Annandale (1915), but has not been recorded since from this area. *Culex hortensis* was collected by Theodor from Tiberias in the twenties (Margalit & Tahori, 1974) and from Tabha in 1956 by Margalit & Tahori (1970). *Culex laticinctus* was found by Annandale (1915) and by Buxton (1924) from the Tiberias area (Type locality) and is very common there. *Culex pipiens* is the most abundant mosquito species in all of Israel (Margalit & Tahori, 1970); it was also reported by Edwards (1913) to be very common in the Lake Kinneret area. *Culex theileri* is quite common in the Tiberias area during the spring months. *Culex tritaeniorhynchus* was recorded from Tiberias by Barraud (1921) and since then was collected only once in the northern Kinneret area one kilometer west of the Jordan delta during August 1976 (Margalit, unpublished), although it can be found quite often in the Jordan Valley during summer months (Margalit & Tahori, 1974). *Culex univittatus* is the second commonest mosquito species in Israel and also in the Lake Kinneret area especially during the summer and fall months.

N. Hydracarina

C. Serruya

The freshwater mites of Lake Kinneret were studied by Koenicke (1894/1895). This group was revised in 1968 by Motas & Tanasachi (in litt.) on the samples collected by Por (1968). The following five species are recorded by these authors:

Unionicola (Pentatax) koenikei Viets
Unionicola (Pentatax) latilaminata Viets
Unionicola (Pentatax) uncata Viets
Atractides contemptus Lundblad
Torrenticola (T.) obtusidens Lundblad

In 1966, Petrova investigated the salty springs of Caphernaum and found two new species of Halacarina.

Limnohalacarus capernaumi sp. nov.

Four species belonging to the genus *Limnohalacarus* were previously known – *L. wackeri* (Walter) in Europe, *L. africanus* (Walter) in Africa, *L. cultellatus* Viets in South America and *L. fontinalis* (Walter and Bader). The species of this genus live generally in subterranean waters.

Lohmannella heptapegoni

This species is closely related to *L. stammeri*; whereas several specimens of *Limnohalacarus capernaumi* were found in the spring, only one specimen of *Lohmannella heptapegoni* was collected. According to Petrova, this may indicate that the spring does not represents its normal habitat.

O. Mollusca

C. Serruya

Bourguignat (1853), working on the collection of de Saulcy, was one of the first scientists who studied the molluscs of Lake Kinneret. Lartet (1878), Tristram (1884) and Barrois also contributed to our knowledge of the malacological fauna of the lake. Locard (1883) studied the samples brought by Chantre from Lake Homs and Lake Antioch and compared them with Bourguignat's collection. Blackenhorn (1897) described the fossil and living molluscs of Syria and Palestine. Preston (1914), working on the material collected by Annandale, split the known forms into numerous species. Pallary published in 1909 the results of his malacological research in Egypt; in 1929 and 1939 he published a similar inventory for the Syrian region. Germain (1921, 1922) worked on the collections of de Kerville and summarized the existing knowledge on the freshwater and terrestrial molluscs of Syria and Palestine. Bodenheimer, in his 'Animal Life in Palestine', reported 23 species of Mollusca in Lake Kinneret, based on the studies of the previously-mentioned authors. In 1963, in the framework of the project sponsored by the Mekorot Water Company, a detailed study of the present molluscan fauna of the lake, its distribution, life cycle and spatial and temporal connection with the fauna of the Near East was carried out by Tchernov (1975a, b). The following description is based on the results published by this author.

In this revision of the malacological taxonomy, Tchernov showed that 'some dozens of species were determined from a single population' (Tchernov, 1975a) and concluded that only eight species of molluscs presently inhabit Lake Kinneret and its shores according to the following list:

BIVALVIA

Unionidae
Unio (Psilunio) semirugatus Lamarck
Unio (Limnium) terminalis Bourguignat
Cyrenidae
Corbicula (Corbicula) fluminalis (Müller)

PROSOBRANCHIA

Neritidae
Theodoxus (Neritaea) jordani Sowerby
Truncatellidae
Pyrgula barroisi Dautzenberg
Bithynidae
Bithynia (Bithynia) hawaderiana Bourguignat
Thiaridae
Melanopsis (Melanopsis) praemorsum Linnaeus
Melanoides (Melanoides) tuberculata (Müller)

Special mention should be made of the species *Valvata saulcyi* Bourguignat. Although this species has been found in one instance at 31.5 m depth, it is not a permanent dweller of the lake. In the instance previously mentioned, its association with live plants and the fact that it has been found in the lake only in winter suggest that it is washed into the lake from brooks and rivers at the flood period but does not develop in the lake water (Tchernov, 1975a). This phenomenon is not restricted to the mollusc fauna; it seems that the abnormal ionic ratio of the Kinneret waters (Na > Ca) prevents the survival in the lake of many species living in the rivers surrounding the lake. (see p. 472.)

1. Main characteristics of each species

a. *Unio (Psilunio) semirugatus* Lamarck

Closely related species to *Unio (Psilunio) semirugatus* are found in Mesopotamia. Another close species, *Unio (Psilunio) homsensis* Lea has been found in Lake Homs in Syria, whereas *Unio (Psilunio) barroisi* Drouët is known in Lake Homs and the Orontes. In Israel, *Unio semirugatus* is not known outside Lake Kinneret, which seems to be located on the margin of the distribution of this genus.

b. *Unio (Limnium) terminalis* Bourguignat

Unio (Limnium) terminalis is known in most of the freshwater bodies of the Near East. In Israel, this species was common at the beginning of the 20th Century in the whole Jordan system and in the rivers of the coastal plain. The water pollution which developed with demographic development in most of the river basins explains why this species is now confined nearly exclusively to Lake Kinneret.

c. *Corbicula (Corbicula) fluminalis* (Müller)

At present, the distribution area of *Corbicula (Corbicula) fluminalis* includes the Near East, from the Sarmatic province in the north to the oriental areas in the east. In Africa, it is found only in the Nile. In previous periods of the Quaternary, it was widespread around the Mediterranean Sea. This species, like *Unio terminalis*, no longer inhabits the rivers of the coastal plain.

d. *Theodoxus (Neritaea) jordani* (Sowerby)

A common ancestor to the two subgenera *Theodoxus* and *Neritaea* inhabited the Tethys during the Oligo-Miocene period and divided later into two groups in the Sarmatic province. *Theodoxus (Theodoxus) fluviatilis* is found in western and northern Europe but is absent from the Danube system, and *Theodoxus (Neritaea) jordani* inhabits the Near East from Mesopotamia to Egypt. Many related species are found in Mesopotamia, but *Theodoxus jordani* is the only species found in Israel. Great intraspecific variability is observed in shell morphology, body anatomy and color pattern. Its area of distribution coincides with the areas of extension of the Neogene marine transgression.

Special mention should be made concerning the abundant population of *Theodoxus* found in the Ein Nur spring on the northwest shore of Lake Kinneret. This spring, which flows at a constant temperature of 28°C and 3,000 ppm chloride, is also inhabited by the marine relict *Typhlocaris galilaea*. The *Theodoxus* population of Ein Nur thrives in darkness and feeds on the algae growing in the hot, salty water of the spring. These specimens lack pigments in both shell and body and are smaller than the *Theodoxus* found in other areas of the lake. Exposure to light of the transparent *Theodoxus* during two or three months generates a nearly normal pigmentation of epithelium and shell.

e. *Pyrgula barroisi* Dautzenberg

The development of freshwater bodies which followed the regression of the Pliocene Sea was accompanied by the appearance of the genus *Pyrgula* in the Sarmatic province, and from there this genus colonized Europe, Russia and the Near East. It is probable that *Pyrgula* reached the Jordan Valley from the Orontes. In Israel, *Pyrgula* is restricted to Lake Kinneret.

There is a considerable variability in the shell morphology of *Pyrgula*, but highly carinated specimens are more frequent than smooth ones. The Syrian forms of this species are larger than those of Lake Kinneret.

f. *Bithynia hawaderiana* Bourguignat

The subgenus *Bithynia* invaded freshwater bodies in the Paleogene before *Theodoxus* or *Melanopsis*. *Bithynia* does not survive in water with salinity higher than 1,200 ppm chloride and requires well aerated water, which explains its disappearance from the coastal plain rivers. It is most common now along the eastern shore of the lake and in the River Jordan, where it is found under stones in shallow water. It seeks protections against storms by moving into deep water.

g. *Melanopsis praemorsum* Linnaeus

The area of distribution of *Melanopsis* coincides with the area of the Neogene marine transgression. This genus persisted in most oligohaline water bodies until the present. It is widespread in North Africa, Southern Europe (Italy, Spain, Greece), Turkey, Caucasia, Mesopotamia, Syria, Lebanon and Israel, but it is not known in Egypt. It has been able to survive in isolated springs in desert areas.

The species *Melanopsis praemorsum* has also been found in the Middle Pleistocene Ubediyya Formation (Picard, 1934) and in the lake deposits of Jisr-Banat-Yaqub (Picard, 1963).

h. *Melanoides tuberculata* (Müller)

Melanoides tuberculata is widespread in the inland water of the Near East, Sinai and North and East Africa. It is a mud dweller which can survive a salinity of 4,000 ppm of chloride. In contrast with *Melanopsis*, the hot, saline, nutrient-rich environments enhance the development of *Melanoides tuberculata*, which appears then in larger numbers and in bigger sizes. Like *Melanopsis, Melanoides* is a stable genus with very little variation of the radula teeth.

2. Distribution and ecology of molluscs in Lake Kinneret

The anaerobic conditions prevailing in Lake Kinneret during eight months prevent the development of a malacological fauna below 15 m. The molluscs of Lake Kinneret are rarely associated with higher plants which are rather scarce. Therefore, the nature of the bottom plays a greater role in their distribution. The rocky habitat, composed of gravel and cobble, is more developed on the western side of the lake, where it can reach a depth of 7 m. This habitat is populated mainly by *Melanopsis praemorsum, Theodoxus jordani* and *Bithynia hawaderiana*. The muddy habitat extending between the rocky habitat and the anaerobic zone is mainly occupied by two species of *Unio, Corbicula fluminalis, Melanoides tuberculata* and *Pyrgula barroisi* (Fig. 119).

The daily strong winds in summer and eastern storms in winter generate considerable wind action which can sweep away relatively large stones and their attached population. These hard physical conditions might partly explain the

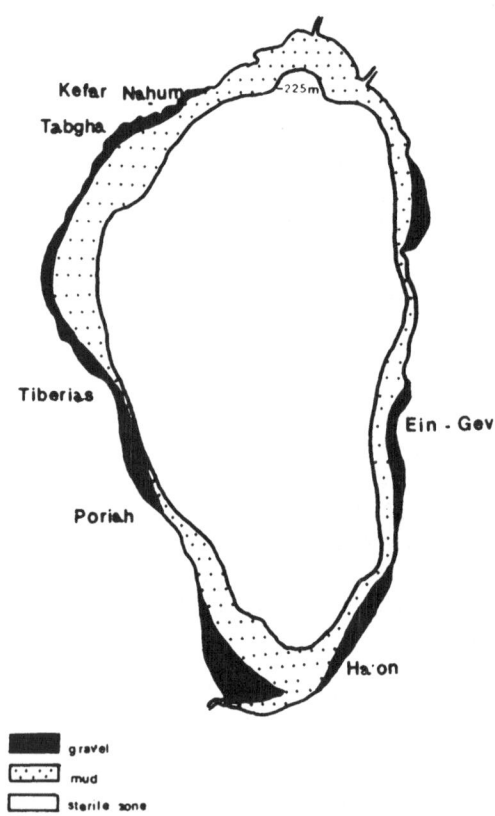

Fig. 119. Distribution of the rocky and muddy habitats of Molluscs. (From Tchernov, E., 1975, Malacologia, 15.)

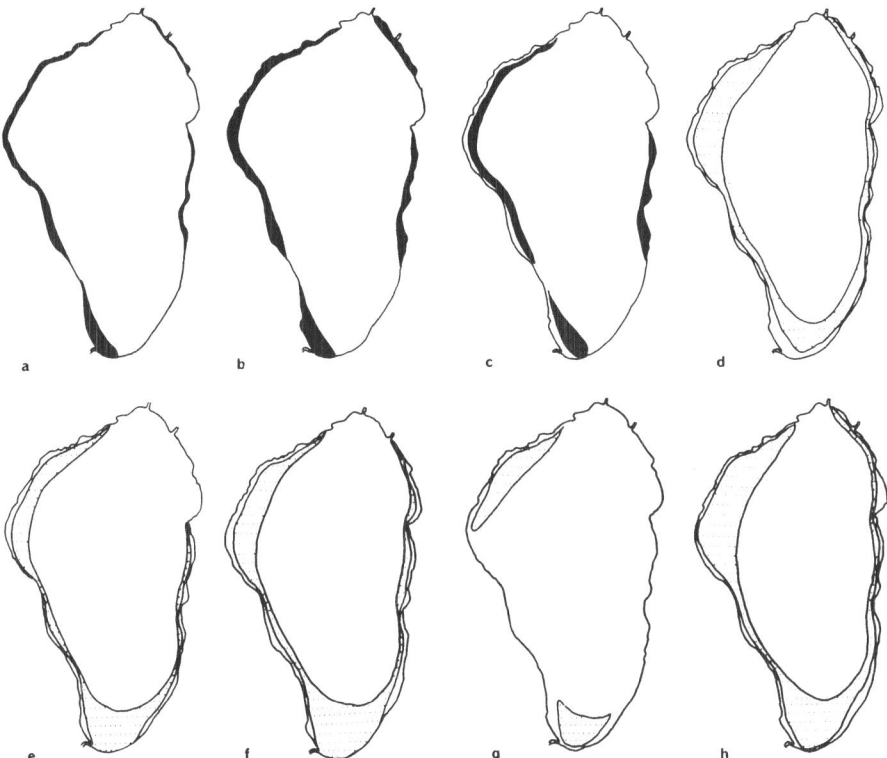

Fig. 120. Distribution of the various species of molluscs in Lake Kinneret (same source as Fig. 119). (a) *Melanopsis praemorsum*, (b) *Theodoxus jordani*, (c) *Bithynia hawaderiana*, (d) *Melanoides tuberculata*, (e) *Pyrgula barroisi*, (f) *Unio (Linmium) terminalis*, (g) *Unio (Psilunio) semirugatus*, (H) *Corbicula fluminalis*.

relatively low number of species of the rocky habitat, and might account for the fact that the species which could survive are all thick shell species. The distribution of the eight Kinneret species is shown in Fig. 120.

In general, the rock dwellers are more abundant on the eastern shore. Since the rock dwelling prosobranchs are obligatory herbivores, their abundance is likely to be in relation to the productivity of planktonic algae. As correctly mentioned by Tchernov, the dominant western winds sweep eastwards the upper layer of water, increasing the planktonic algal density on the eastern side. Measurements of empty shells, which give an idea of the total production of the population, indicate that the eastern shore has the highest number of empty shells of rock dwellers, especially of *Theodoxus*.

It is remarkable that the Gastropodes are far more numerous than Bivalves. The Tsemah area in the south and the Ginosar area on the western shore have unusually high amounts of Bivalves, especially of *Corbicula*. This species decreases rapidly from south to north. The mud dwelling Gastropodes are also more abundant in these two localities than elsewhere, probably because of the local enrichment of nutrients caused by the Ginosar streams and the agricultural drainage reaching the lake at Tsemah.

Table 46. Distribution of molluscs at different periods of the Pleistocene in the central and upper sections of the Jordan Valley (modified from Tchernov E., 1975, Isr. Acad. Sci. Hum. and Tchernov E., 1975, Malacologia, 15)

Species	Erq el Ahmar Early Pleistocene	Ubediyya Early-Middle Pleistocene	Lake Kinneret Post Würm	Gesher Benot Yaakov Early-Middle Pleistocene	Gesher Benot Yaakov Middle-Late Pleistocene	Lake Hula Post Würm
PROSOBRANCHIA						
Theodoxus jordani	+	+	+	+	+	+
Viviparus unicolor	+				+	
Viviparus apamae	+				+	
Valvata saulcyi	+	+		+	+	+
Hydrobia acuta	+					
Hydrobia longiscata					+	
Falsipyrgula barroisi	+		+			
Bithynia siriaca	+					
Bithynia hawaderiana		+	+	+	+	+
Bithynia multicostata	+					
Melanopsis praemorsum	+	+	+	+	+	+
Melanopsis doriae	+					
Melanopsis dadianus	+					
Melanoides tuberculata	+	+	+	+	+	+
Melanoides jordanicus	+					

PULMONATA

Lymnaea lagotis
Lymnaea palustris
Gyraulus piscinarum
Planorbarius planorbis
Anisus spirorbis
Segmentina nitida
Acroluxus lacustris
Succinea elegans
Succinea pfeifferi
Dreissena chantrei
Ancylus fluviatilis

VALVATA

Unio terminalis
Unio semirugatus
Unio subrectangularis
Leguminaia chantrei
Pisidium casertanum
Corbicula fluminalis
Sphaerium lacustre

3. The origin of the freshwater Mollusca

Information concerning the origin of the freshwater molluscs of the Jordan Valley can be found in Picard (1934), Schulman (1959), Dagan (1971), Por (1963), and Tchernov (1975a, b).

The origin of certain elements of the fauna of Lake Kinneret has been a subject of great controversy. Certain authors, such as Annandale, considered that this fauna is of recent and Palaeoarctic origin. However, he had to make exceptions for *Melanoides tuberculata*, an obviously 'old species', and for the fish *Tilapia* and *Clarias* of African origin. Arndt (1937) attributed an older origin to Lake Kinneret and its fauna. The detailed geological and faunistic work carried out in the last thirty years supports the thesis of Arndt. Freshwater molluscs are known in the Jordan Valley in the deposits of the early Pliocene, and their origin is a critical point since the area was, during the Miocene and part of the Pliocene, occupied by the Tethys. However, as described in the section on geology, the Neogene Tethys occupied only restricted areas of the Near East, and there have been continuously emerged lands and freshwater bodies in this part of the world since Neogene times until the present. Miocene and Pliocene lakes existed in the Golan Heights (where Neogene outcrops have supplied a fauna of *Theodoxus*), in the Damascus area, the Orontes Valley and Lake Homs. The freshwater fauna of these water bodies migrated to the Jordan system through hydrological connections with the Orontes system. After the last marine regression, in late Pliocene, this fauna invaded the Jordan Valley. It seems that in the Miocene, besides its connection to the Orontes system, our area was also in contact with the Ethiopian and Oriental units. The nilotic *Viviparus unicolor*, found in the early Pleistocene of the Jordan Valley, and other tropical forms such as the fishes *Tilapia galilaea*, *Clarias lazera*, and the vertebrates *Crocodilus niloticus*, *Trionix triunguis* and *Hypopotamus amphibius* are examples of Ethiopian influence in the fauna of the late Neogene and early Pleistocene.

It seems established that the molluscs which colonized the newly formed Jordan hydrographic net were old freshwater forms of various origin: European (*Bithynia*), Sarmatic (*Theodoxus*) and Ethiopian (*Viviparus unicolor*).

Rather detailed information is available on the evolution of the malacological fauna in the Pleistocene. Table 46 shows the distribution of molluscs at different periods of the Pleistocene in the central and upper sections of the Jordan Valley.

It is clear that the faunistic changes were minor in the upper part of the valley from the early Pleistocene until the present. Conversely, in the central area of the valley, the present fauna is extremely poor in comparison with the numerous species of the early Pleistocene. We note in particular the complete elimination of the Pulmonata in Lake Kinneret.

In the Middle Pleistocene, the Jordan Valley was occupied by a freshwater lake in which were deposited the Ubediyya beds and the oldest formations of Lake Hula. The post-Mindel development of a salty water body, the Lisan Lake, which extended from the southern end of the Dead Sea to Tiberias, eliminated a great number of the fresh-water molluscs. Conversely, the permanence of freshwaters in the northern part of the valley accounts for the stable populations in this area. The Lisan Lake came to an end about 20,000 years ago when minor tectonic movements created the two present lakes of the area: the Dead Sea and Lake Kinneret. In the post-Wurm period, the molluscs remaining in the rivers around Lake Kinneret began to repopulate the rapidly freshening Lake Kinneret. It follows that the molluscs of Lake Kinneret are the relict fauna which could survive the Lisan salty stage in rivers and freshwater springs in the same way that, at present, freshwater molluscs are found in freshwater springs on the shore of the Dead Sea.

III. Vertical distribution of benthic fauna

C. Serruya

The zonation of the benthic fauna has been described by Por (1968). This author distinguished five main faunal zones (Fig. 121).

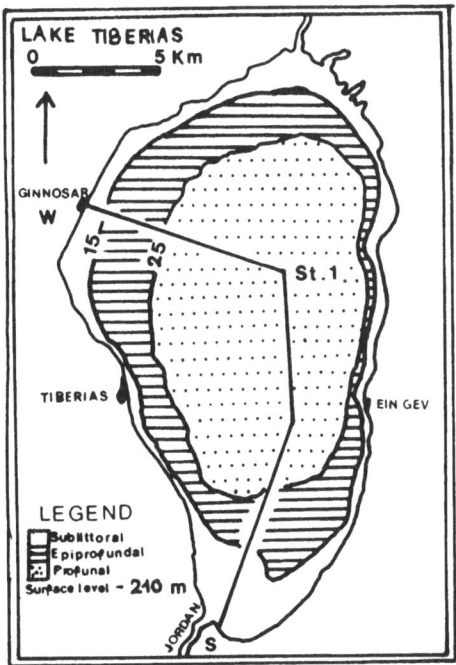

Fig. 121. The main faunal zones of Lake Kinneret. The sampling depths referred to in Fig. 123 are located on the full lines. (From Por, F. D. & Eitan, G., 1970, Isr. J. Zool., 19.)

A. The Supralittoral

The supralittoral belt refers to the wet but non saturated sediments which extend above lake level. This belt is composed generally of coarse gravel, covered with pebbles and stones. Sand and mud are exceptional since the wave action of the daily storms has a considerable sorting effect which transports lakewards all the fine grained particles. Muddy beaches are found in the bay of Ginosar less exposed to waves than the rest of the shores.

The fauna of the supralittoral belt is mainly composed of Nematodes: *Eudorylaimus* spp., *Panagrolaimus rigidus*, *Tylenchus filiformis*, *Aphelencoides besseyi*, *Anaplectus granulosus*, *Dorylaimellus parvulus*. The gravels also harbour a few oligochaetes such as *Haplotaxis gordioides*, *Criodrillus lacuum* and the Harpacticoid *Nitocra lacustris*. The Amphipoda *Orchestia platensis* and the Isopode

391

Halophiloscia couchii are also supralittoral dwellers.

B. The Littoral

The littoral belt is covered with boulders of basaltic origin in the northern part of the lake and of limestone near the Arbel cliffs. For a lake level of −209 m, the boulder zone extends from 2 to 5 m depth. During the last few years, the exceptionally low levels of the lake have uncovered part of the boulder belt.

The macrophyte vegetation of the littoral area is poor. The amphibious *Phragmites communis* is the dominant species with local development of *Typha* and *Arundo*. The submerged vegetation is limited to *Myriophyllum spicatum, Najas marina* and *Potamegon pectinatus* (Waisel, 1967). It follows that the associated fauna is also very limited. The Crustaceans *Athyaephyra desmaresti, Proasellus coxalis*, the leech *Placobdella carinata*, the freshwater crab *Potamon potamios* and the copepod *Horsiella bisetosa* are the main species bound to the macrophytic population.

Dense populations of *Bithynia badiella* and *Planaria barroisi* develop in the areas of freshwater seepage.

The organisms of the boulders show a clear zonation. The upper side of the rocks is covered with epilithic algae grazed by the snails *Theodoxus jordani* and *Melanopsis praemorsa*. The latter snail has highly ribbed and thick shells, probably an adaptation to the damaging wave action. Insect larvae occupy the sides of the boulders. The sides of the boulders are covered with colonies of *Fredericella sultanea jordanica* and of *Cortispongilla* spp. The lower part of the boulders is inhabited by large populations of *Echinogammarus veneris* and by *Proasellus coxalis*, planarians and trichopterans. In areas where the boulders are immersed in sediments, the leech *Dina blaisei* is present in the anaerobic area developing at the basis of the boulders.

The sponges of the lake harbour a specific fauna. The Hydracarina *Unionicola koenikei* Viets is found in large amounts in the spaces of the sponge tissue. The Neuropteran *Sysira trilobata* is also associated with the lake sponge. The genus *Sysira* is known to feed on sponges but this species is endemic in Lake Kinneret. The Chironomide *Xenochironomus xenolabis*, the nematode *Tobrillus aberrans*, several Oligochaetes and Rhabdocoelans are also found specifically associated with sponges. The Copepods *Nitocra hibernica* and *N. incerta* are non-specific inhabitants of the sponges.

The littoral level bottoms are inhabited by the Gasteropoda *Pyrgula barroisi* and *Melania tuberculata* and sporadically by *Echinogammarus veneris*. The littoral level bottoms have also a rich meiobenthic fauna consisting of the Ostracods *Loxoconcha galilea, Ilyocypris* spp., the harpacticoids *Nitocra hibernica, Nannopus palustris tiberiadis, Pseudobradya barroisi*, the Cladocera *Macrothrix goeldii*; the Oligochaetes *Psammoryctides albicola, Haplotaxis gordioides* and *Eiseniella tetraedra* are restricted to shallow depths. Freeliving Hydracarina (*Unionicola latilaminata* and *Torrenticola (T.) obtusidens*) are also found generally above 5 m. Crustaceans, such as *Moina dubia, Eucyclops macrurus* and *Acanthocyclops bicuspidatus*, found in winter in the littoral bottom, are 'accidental visitors' brought by the floods.

392

C. The Sublittoral

The sublittoral area extends between the boulder zone and the upper limit of the hypolimnion. It includes a wide belt from approximately 5 to 15 m depth. *Unio terminalis, Rhombunio rhomboides, Corbicula fluminalis* and the Ostracods *Darwinula stephensoni* and to a lesser degree *Cyprideis torosa* are the usual dwellers of the sublittoral belt together with a rich population of Oligochaetes. The southern part of the lake is especially rich in Lamellibranchs and Ostracods. The endemic sponge *Cortispongilla barroisi* was also found mostly in the southern sublittoral area.

In many other lakes, the sublittoral area is occupied by *Chara* meadows. The *Chara* meadows do not exist in Lake Kinneret which explains the absence of Amphipoda, Isopoda and Hydracarina from these depths. Between 10 and 15 m depth, in front of the building of the Regional Council of the Jordan Valley the lake bottom is covered by colonies of the blue green algae *Nostoc*. The *Nostoc* area extends eastwards but has not yet been exactly mapped.

D. The epiprofundal zone

In most lakes, the benthic fauna diminishes progressively with depth. In Lake Kinneret, Por (1968) and Gitay (1968) have found that the epiprofundal zone which extends from 15 to 30 m is poorer fauna than the profundal area. Por related this specific feature to seasonal fluctuations of the thermocline. More recent research on physical limnology suggests that this area corresponds to the area daily swept by the thermocline movements caused by the internal waves. (Serruya, 1975). This means that any given point in this area is, during a twenty-four hour period, located alternatively in the oxygenated and warm epilimnion and in the anoxic and cold hypolimnion. These rapid and extreme fluctuations of oxygen levels and temperature are probably very damageable to most organisms. Therefore during the stratified period, the Nematode *Eudorylaimus andrassyi* and the Oligochaete *Euilyodrilus* are the only dwellers of this area and the number of specimens found is much lower than that found in the profundal zone. It is more difficult to explain why the scarcity of fauna is maintained during the mixed period in winter.

E. The profundal area

This zone encompasses the bottom sediments located between 30 m and 42 m and corresponds to one third of the lake bottom. Anaerobic conditions prevail there during nine months of the year and the temperature ranges from 14 to 16°C.

During the stratified and anaerobic period, very abundant populations of the Nematoda *Eudorylaimus andrassyi* and of the oligochaete *Euilyodrilus heuscheri* are the main deep dwellers of the lake. Anaerobic bacteria mainly denitrifiers, sulfate reducers and methane bacteria are also present.

The oligochaetes are not only capable of surviving in anaerobic conditions but they reproduce in autumn when the concentrations of hydrogen sulfide reach their maximal values. The Nematodes seem to reproduce all through the year since young specimens are always present. The establishment of oxygenated conditions in winter does not seem to affect the Oligochaetes and Nematodes, the number of which remains comparable to the summer values. At this period, an additional population appears, superimposed on the anaerobic organisms; it includes *Pelmatohydra*, the Cladocera *Leydigia leydigi*, *L. acanthocercoides*, *Alona affinis*, *A. rectangulata* and *Ilyocryptus agilis*, the Copepods *Eucyclops serrulatus*, *Onychocamptus mohammed* and *Nitocra incerta*, the Chironomids *Tanypus punctipennis*, *Procladius choreus* and *Cryptocladopelma virescens* (Kugler, 1966).

It is not clear whether these transitory dwellers develop *in situ* from resting eggs or whether they come from shallow depths through active or passive processes. The high number of eggs of the two species of *Leydigia* (thousands per square meter) found by Por (1968) in the profundal zone indicates that these Cladocera develops *in situ*. This is not the case with the other Cladocera or the Copepods. The fact that very few Ostracods are found in winter in the profundal zone strengthens the hypothesis of an active downwards migration.

IV. Biomass of benthic organisms

C. Serruya

The information concerning the biomass of benthic organisms has been taken from Por & Eitan (1970).

The quantitative study of these authors emphasizes the peculiar drastic decrease of benthic population in the epiprofundal area from 15 to 25 m (Fig. 122). From thousands of specimens per square meter in the sublittoral zone, the benthic fauna is reduced to a few tens in the epiprofundal below which it rises again to a few thousand in the profundal.

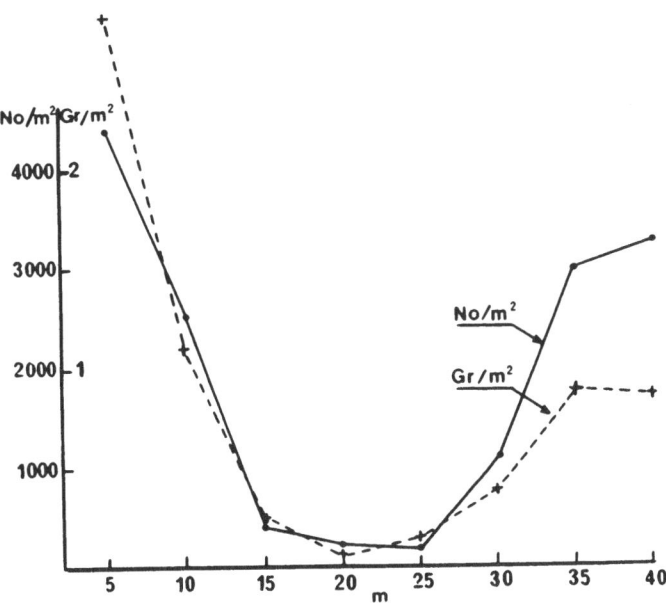

Fig. 122. Variation of meiobenthic fauna with depth. Number of specimens m^{-2} and weight biomass mg m^{-2} (same source as Fig. 121).

The quantitative changes with depth and seasons of the main groups of animals contributing to the benthic fauna are given in Fig. 123. This figure indicates that below 20 m the Entomostracans and Chironomids have no quantitative significance; the profundal is dominated by Nematodes with maximal numbers of 70,000 specimens per square meter and Oligochaetes (up to 2,000 specimens per square meter). Fig. 123 indicates also that there is a general decrease of populations in winter.

The maximum absolute values of benthic biomass are found in the sublittoral area with 7.6 g m^{-2} including 4.2 g m^{-2} Oligochaetes, and 2.8 g m^{-2} Nematodes.

395

Table 47. Annual average of number of specimens per m² calculated for the whole lake bottom (from Por F. D. & Eitan G. 1970, Isr. J. Zool. 19)

	Specimens m^{-2}	%
Oligochaeta	1,378	12.6
Nematoda	8,055	73.5
Chironomida	481	4.4
Copepoda	518	4.7
Cladocera	379	3.4
Ostracoda	152	1.4
Total	10,963	100.0

Table 48. Average biomass of the main faunistic belts of the lake (molluscs not included) (from Por F. D. & Eitan G. 1970, Isr. J. Zool. 19)

	Specimens m^{-2}	gm^{-2}	kgha^{-1}	Area in ha	Standing crop in tons
Sublittoral	2,421	1.35	13.5	3,952	53.4
Epiprofundal	472	0.18	1.8	3,214	5.8
Profundal	3,150	0.86	8.6	9,669	83.2
Total				16,835	142.4

In the profundal area, the maximum biomass value was 3.6 g m^{-2} including 3.4 g m^{-2} Nematodes and 0.2 g m^{-2} Oligochaetes.

The annual average of number of specimens for the whole lake bottom is shown in Table 47.

Table 48 shows the distribution of biomass for the three main bottom belts: 1.35 g m^{-2} in the sublittoral, 0.18 g m^{-2} in the epiprofundal and 0.86 g m^{-2} in the profundal.

The molluscs are not included in the previous table. A rough estimation of their biomass contribution, derived from two tows carried out in the southern part of the lake gave a value of 8.2 g m^{-2} (weight of soft parts only) or 320 tons for the whole lake. *Unio terminalis* appears to represent 75% of this biomass.

The values found by Por & Eitan are very low in comparison with values found in temperate lakes having an anoxic hypolimnion: Eddy (1963) gives an average biomass of 30 kg ha^{-1} for thirty-two lakes in North America and Kozhov (1963) reports 180 kg ha^{-1} for Lake Baikal. In 1972, Carmouze et al. measured, in Lake Tchad, biomass values ranging from 4 to 14 kg ha^{-1} in very shallow areas with a lake average of about 30 kg ha^{-1}. In Lake Tchad, 90% of the benthic population is contributed by molluscs and the biomass of worms rarely exceeds 2 kg ha^{-1}. However, Lake Tchad and Lake Kinneret cannot be compared directly since the dominance of molluscs in Lake Tchad might be explained by the fact that most of the bottom area of Lake Tchad corresponds to the sublittoral zone of Lake Kinneret where molluscs are also abundant.

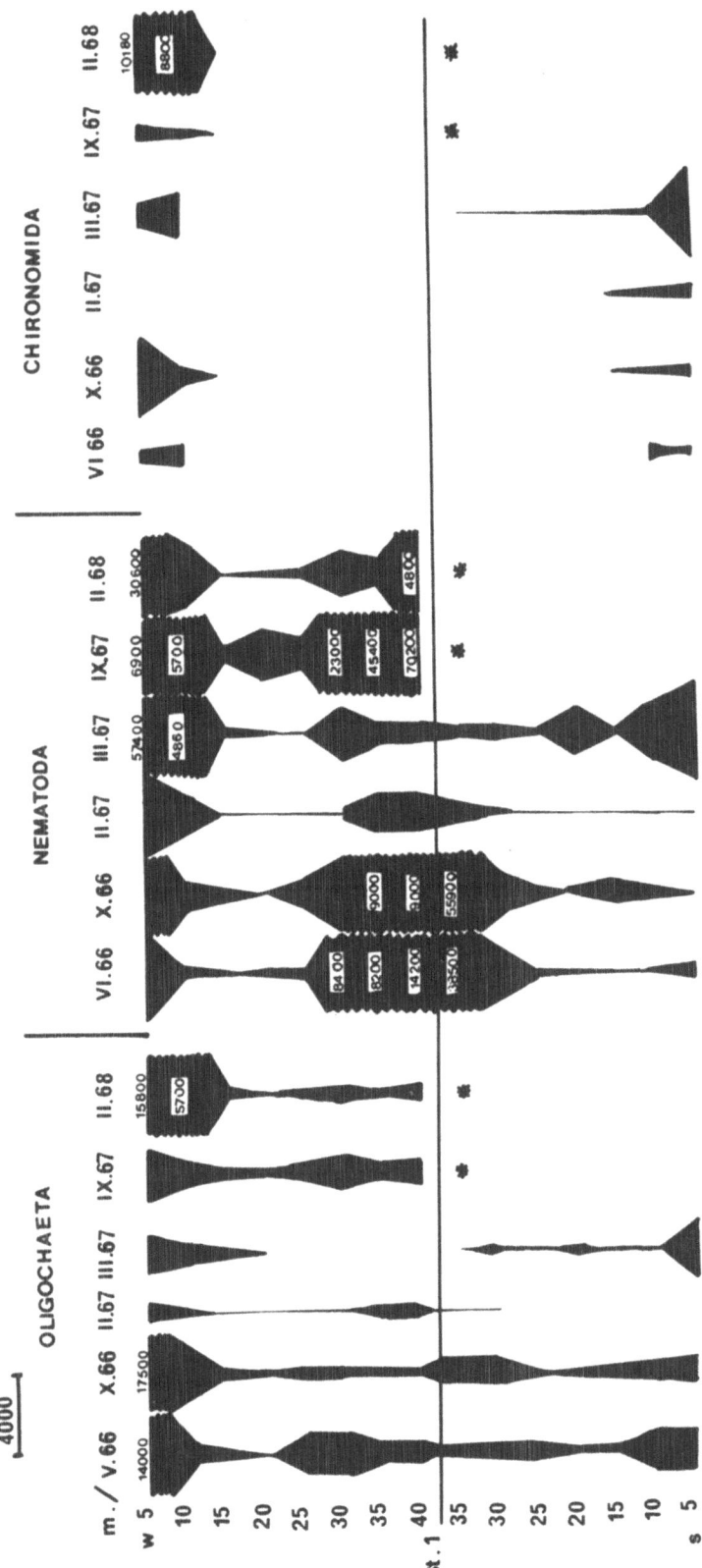

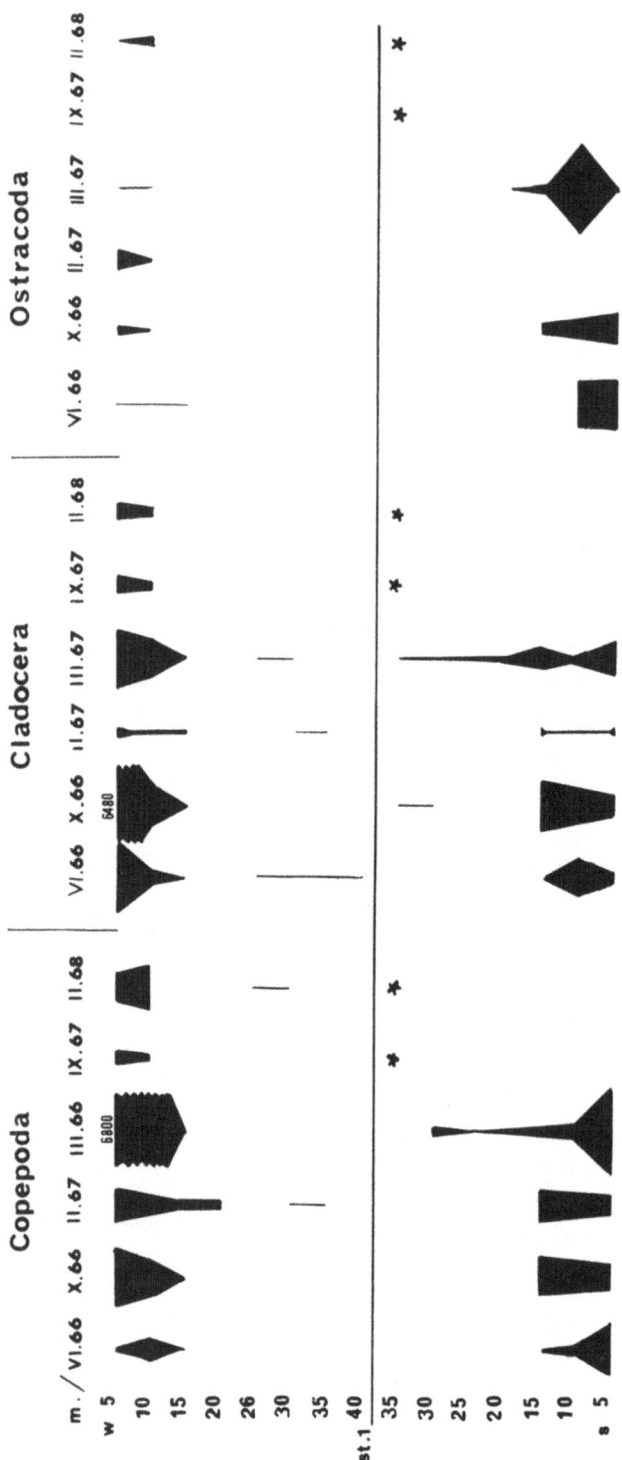

Fig. 123. Quantitative changes with depth of the main animal groups at various periods: w, st, s indicate the sampling profiles represented in Fig. 121. An asterisk indicates that no sample was taken at this date. The scale of 4,000 specimens is in the upper corner (same source as Fig. 121).

In Lake Ochrid, Stankovic (1960) found that the total average weight biomass does not go below 100 kg ha^{-1}.

Most of the lakes previously mentioned, have a profundal fauna, dominated by Oligochaetes and Chironomides. Lake Ohrid has been called an 'Oligochaete lake'. In contrast, Lake Kinneret is a 'Nematode lake'.

This unusual feature of Lake Kinneret may be due to the well known ability of the Nematodes to live without oxygen (Brand, 1946). Whereas most of the experiments on anaerobiosis of Nematodes were carried out at low temperatures of 2 to 6°C (Lindeman, 1942; Berg & Jonasson, 1965), the Nematode *Eudorylaimus andrassyi* was kept alive for six months at 15°C in complete anaerobic conditions (Por & Masry, 1968). The obvious ability of the animal to thrive during such a long period at the temperature observed in the hypolimnion confers on *Eudorylaimus* a considerable ecological advantage. However, the experiments of Por and Masry show that the Oligochaete *Euilyodrilus heuscheri* can also survive and even reproduce in similar conditions. This suggests that the dominant occupation of the anaerobic niche by Nematodes might be due to other factors, possibly, trophic ones.

References

Andrassy, L. 1963. Einige Nematoden aus der Umgebung des Toten Meeres. Israel J. Zool. 13:89–98.

Annandale, N. 1912. The blind prawn of Galilee. Nature (London) 90:251.

Annandale, N. 1913. The Polyzoa of the Lake of Tiberias. J. As. Soc. Gengal IX:223–228.

Annandale, N. 1913. An account of the sponges of the Lake of Tiberias. J. As. Soc. Bengal. IX:57–88.

Annandale, N. & S. Kemp. 1913. The Crustacea Decapoda of the Lake of Tiberias. J. As. Soc. Bengal. n. ser. 9:241–258.

Annandale, N. 1915. A report on the biology of the lake of Tiberias. J. & Proc. Asiatic Soc. of Bengal 11:435–476.

Arndt, W. 1937. *Ochridaspongia rotunda* n. gen., n. sp., ein nuer Süsswasserschwamm aus dem Ochridsee. Arch. Hydrobiol. 31:636–663.

Banarescu, P. 1970. Principii si probleme de zoogeografie. Ed. Acad. Repub. Soc. Rom. Bucuresti.

Barraud, P. S. 1921. Mosquitoes collected in Palestine and adjacent territories. Bull. Ent. Res. 11:387–395.

Barrois, T. 1894. Contribution a l'etude de quelques lacs de Syrie, Rev. biol. Nord. France. VI:224–314.

Ben-Dov, Y. 1971. Note on the occurrence of Anopheles sacharovi. Israel J. Entomol. 6:313.

Berg, K. & P. Jonasson. 1965. Oxygen consumption of profundal lake animals at low oxygen content of the water. Hydrob. 26:131–143.

Blanchard, R. 1893. Viaggio del Dr. E. Festa in Palestine nel libano e regioni vicine III. Hirudinees. Bull. Mus. Torino VIII, 161:1–3.

Blanchard, R. 1894. Voyage du Dr. Theodore Barrois en Syrie: Hirudinees. Rev. biol. Nord. France VI:41–46.

Blackenhorn, M. 1897. Zur Kenntniss der Süsswasserablagerungen und Mollusken Syriens. Paleontographica 44:71–144.

Bodenheimer, F. S. 1935. Animal Life in Palestine. L. Mayer, Jerusalem. 506 pp.

Botosaneanu, L. & A. Gasith. 1971. Contributions Taxonomiques et Ecologiques a la Connaissance des Trichopteres (Insecta) d'Israel. Israel J. Zool., 20:89–129.

Bourguignat, J. R. 1853. Catalogue raisonné des Mollusques terrestres et fluviatiles recueillis par M. F. Sauley pendant Son Voyage en Orient. Gide and Baudry, Paris. 96 p.

Brand, V. Th. 1946. Anaerobiosis in invertebrates. Biodynamic Monographs 4:1–328, Saint Louis, Missouri.

Brien, P. 1973. Malawispongia echinoides BRIEN. Etudes complémentaires, histologie, sexualité, embryologie, affinités systématiques. Rev. Zool. Bot. Afr. 50–76.

Bromley, H. 1972. A note on the occurrence of freshwater Nemertines of the genus Prostoma Duges (*Enopla, Monostylifera*) in Israel. Isr. J. Zool, 21:71–77.

Bromley, H. J. 1974. Morphokaryological types of Dugesia (*Turbellaria, Tricladida*) in Israel and their distribution patterns. Zoologica Scripta 3:239–242.

Buxton, P. A. 1924. Applied entomology of Palestine. Bull. Ent. Res. 14:289–340.

Calman, W. T. 1909. On a blind prawn from the Sea of Galilee (*Typhlocaris galilea* g. et sp. n.). Trans. Linn. Soc. London Zool., ser. 2, 11(5):93–97.

Carmouze, J. P., C. Dejoux, J. R. Durand, R. Gras, A. Iltis, L. Lauzanne, J. Lemoalle, C. Leveque, G. Louben & L. Saint-Jean. 1972. Grandes Zones ecologiques du lac Tchad. Cah. ORSTOM ser. Hydrobiol. VI 2:102–169.

Caroli, E. 1923. Di una specie italiana di Typhlocaris (*T. salentina* n. sp.) con osservazioni morfologiche e biologiche sul genere. Boll. Soc. Nat. Napoli 35:265–267.

Caroli, E. 1924. Sulla presenza della Typhlocaris (*T. salentina* n. sp.) in terra d'Otranto. Contributo alla conoscenza del genere. Ann. Mus. Zool. Univ. Napoli, n. ser. 5(9):1–20.

Cowgill, U. M. 1973. a. The waters of Merom: A study of Lake Huleh. II. The mineralogy of a 54 m core. Arch. Hydrobiol. 71:421–474.

Cowgill, U. M. 1973b. The waters of Merom: A study of Lake Hula III. The major chemical constituents of a 54 m core. Arch. Hydrobiol. 72:145–185.

Dagan, D. 1971. Taxonomic descrimination between certain species of the genus Theodoxus (*Gastropoda, Neritidae*). Isr. J. Zool. 20:223–230.

Eddy, S. 1963. Minnesota and the Dakotas. Chapter 10. In. Limnology of North America D. G. Frey Editor. University of Wisconsin Press, 734 pp.

Edwards, F. W. 1913. Tipulidae and Culicidae from the Lake of Tiberias and Damascus. J. & Proc. Asiat. Soc. of Bengal 9:48–51.

Etchecopar, R. D. & Hue, F. 1967. The Birds of North Africa (from the Canary Islands to the Red Sea). Oliver and Boyd, London.

Gasith, A. 1969. Trichoptera of Israel. M. Sc. Thesis submitted to the Tel Aviv University. 82 pp. (in Hebrew).

Gasith, A. & J. Kugler. 1971. Bionomics of the Trichoptera of Lake Tiberias (Kinneret). Israel J. Ent. 8:55–67.

Goren, M., L. Fishelson & E. Trewavas. 1973. The Cyprinid fishes of Acanthobrama Heckel and related genera. Bull. Br. Mus. (Nat. Hist.) Zool., 24:293–315.

Germain, L. 1921. Mollusques terrestres et fluviatiles de Syrie. Tome I. Introduction et Gasteropodes. Baillere et Fils, Paris. 523 p.

Germain, L. 1922. Mollusques terrestres et fluviatiles de Syrie. Tome II. Pelecypodes. Baillere et Fils, Paris. 242 p.

Ghosh, E. 1913. On the Internal Anatomy of the Blind Prawn of Galilee (*Typhlocaris galilea* Calman). J. As. Bengal, n. ser. 9:233–239.

Gitay, A. 1968. Preliminary data on the ecology of the level bottom fauna of Lake Tiberias. Israel J. Zool. 17:81–96.

Gurney, R. 1913. Entomostraca from the Lake of Tiberias. J. Ass. Soc. Bengal IX:231–232.

Haas, G. 1952. Remarks on the origin of the herpetofauna of Palestine. Istanb. Univ. Fen. Fak. Seri B 17:95–105.

Harrison, D. L. 1964–1968. The Mammals of Arabia I, II, Ernst. London.

Hartman, G. 1957. Asiatische Ostracoden, Akademie Verlag. Berlin:1–155.

Hirvenoja, M. 1973. Revision der gattung Cricotopus van der Wulp und ihrer Verwandten (*Diptera, Chironomidae*). Ann. Zool. Fennici. 10:1–363.

Hubault, E. 1938. *Sphaeromicola spheromidicola* n. sp. commensal de *Sphaeromides virei* Valle en Istrie et considerations sur l'origine des diverses especes cavernicoles perimediterraneennes. Arch. Zool. Exp. et Gen. 80, Notes et Revue:11–24.

Johansson, V. 1913. Zool. Auz. XLII, 2.

Kieffer, J. J. 1915. Chironomides du lac de Tiberiade. J. As. Soc. Bengal, 10:359–372.

Kligler, I. J. 1924. Breeding places and types of Anopheles. J. of Hygiene 23:298–307.

Koenike, F. 1895. Liste des Hydrachnides recucillies par le Dr. Theodore Barrois en Palestine, en Syrie et en Egypte. Rev. biol. Nord. France VII:139–147.

Kosswig, C. 1955. Beitrag zur zoogeographie der Seen im Marmaragebeit. Hidrobiologi Istanbul. Ser. B I(4).

Kosswig, C. 1955. Zoogeography of the Near East. Syst. Zool. 4:50–73.

Kosswig, C. 1959. Contribution to the knowledge of the zoogeographical situation in the Near and Middle East. Experientia 7:401–406.

Kozhov, M. 1963. Lake Baikal and its life, Monographiae Biologicae XI Dr. W. Junk Publishers. The Hague, 1963.

Kugler, J. 1966a. Vorlaeufige Mitteilung ueber die Chironomidenfauna des Tiberiasseess. Gewaesser und Abwaesser 41/42:70–84.

Kugler, J. 1966b. The Chironomids of the Kinneret. Mekorot Water Co. Ltd., Alon Techni. 2:30–32 (in Hebrew).

Kugler, J. 1971. The development stages of *Leptochironomus stilifer* (*Diptera: Chironomidae*) and the characters of the genus *Leptochironomus* Canad. Entom. 103:341–346.

Kugler, J. & H. Chen. 1968. The Distribution of chironomid Larvae in Lake Tiberias (Kinneret) and their Occurrence in Food of fish of the Lake. Israel J. Zool. 17:97–115.

Kugler, J. & D. Wool. 1968. Chironomidae (*Diptera*) from the Hula Nature Reserve. Ann. Zool. Fenn. 5:76–83.

Kugler, J. & F. Reiss. 1973. Die *triangularis* Gruppe der Gattung *Tanytarsus* v.d.W. (Chironomidae, Diptera). Ent. Tidskr. 94:59–82.

Lang, K. 1948. Monographie der Harpacticiden, I, II, Nordiska Bokhandeln, Stockholm. 1,682 pp.

Lartet, L. 1878. Exploration geologique de la Mer Morte, de la Palestine et de l'Idumée. Bertrand, Paris. 326 p.

Lengy, J. & Y. Wolff. 1971. Studies on larval stages of digenetic trematodes in aquatic molluscs in Israel. 3. On the cercariae encountered in the freshwater snails. Bithynia sidonensis Mousson 1861, B. Sauley; Bourguignat 1853 and Bulinus truncatus Audouin. Isr. J. Zool. 20:279–291.

Lerner-Seggev, R. 1968. The fauna of Ostracoda in Lake Tiberias. Israel J. Zool. 17:117–143.

Lindeman, R. L. 1942. Experimental stimulation of winter anaerobiosis in Senescent Lake. Ecology 23:1–13.

Locard, A. 1883. Malacologie des lacs de Tiberiade, d'Antioche et d'Homs. Arch. Mus. Hist. Nat. Lyon 3:195–293.

Macdonald, W. W. 1953. Lake flies. The Uganda Journal. 17(2):124–134.

Macdonald, W. W. 1956. Observations on the biology of chaoborids and chironomids in Lake Victoria, and on the feeding habits of the elephant snout fish (*Mormyrus kannume* Forsk). J. Animal Ecol. 25(1):36–53.

Margalit, J. & A. S. Tahori. 1970. Species of mosquitoes found in Israel during a survey 1955–58. Israel J. Entomol. 5:151–159.

Margalit, J. & A. S. Tahori. 1974. An annotated list of mosquitoes in Israel. Israel J. Entomol. 9:77–91.

Matzliah, S. 1972. Ephemeroptera of Israel. Thesis Tel Aviv University (in Hebrew).

Meinertzhagen, R. 1954. Birds of Arabia. Oliver and Boyd, London.

Neev, D. & K. O. Emery. 1966. The Dead Sea. Science Journal 2(12):50–55.

Noodt, W. 1954. Copepoda Harpacticoidea aus dem limnischen Mesopsammal der Turkei – Hidrobiologi, Istanbul. Ser. B, 2:27–40.

Pallary, P. 1909. Catalogue de la fauna Malacologique d'Egypte. Mem. Inst. Egyp. 6:1–92, 177–182.

Pallary, P. 1929. Premiere addition a la fauna Malacologique de la Syrie. Mem. Inst. Egypt. 12:1–43.

Pallary, P. 1939. Deuxieme addition a la fauna Malacologique de la Syrie. Mem. Inst. Egypt. 39:1–141.

Paperna, I. 1964. Parasitic Helminths of Inland Water Fishes of Israel. Isr. J. Zool. 13:1–26.

Parisi, B. 1920. Sulla presenza in Cirenaica della *Typhlocaris galilaea* Calman. Riv. Sci. nat. 'Natura', 11:101–104.

Parisi, B. 1921. Un nuovo Crostaceo cavernicolo: *Typhlocaris lethaea* n. sp. Atti. Soc. Ital. Sci. nat.

59:241:248.

Petrova, A. 1966. Deux Nouveaux Halacariens d'Israel. Intern. J. Speleology. II:354–362.

Picard, L. 1934. Mollusken der levantinischen Stufe Nordpalästinas (Jordantal) Arch. Molluskenk., 66:105–139.

Por, F. D. 1962. Un nouveau Thermosbaenace *Monodella relicta* n. sp. dans la depression de la Mer Morte. Crustaceana 3(4):304–310.

Por, F. D. 1963. The relict aquatic fauna of the Jordan Rift Valley (New contributions and review) Israel J. Zool. 12:47–58.

Por, F. D. 1964. The genus Nitocra, BOECK (Copepoda, Harpacticoide) in the Jordan Rift Valley. Isr. J. Zool. 13:78–88.

Por, F. D. 1968. *Parabathynella calmani* n. sp. (Syncarida, Bathynellacea) from Israel. Crustaceana 14(2):151–154.

Por, F. D. 1968. The invertebrate zoobenthos of Lake Tiberias I. Qualitative aspects. Isr. J. Zool. 17:51–79.

Por, F. D. 1975. An outline of the zoogeography of the Levant. Zoologica Scripta. 4:5–20.

Por, F. D. & D. Masry. 1968. Survival of a nematod and an oligochaete species in the anaerobic benthal of Lake Tiberias. Oikos 19:388–391.

Por, F. D. & G. Eitan. 1970. The invertebrate zoobenthos of Lake Tiberias II Quantitative data (level bottoms). Isr. J. Zool. 19:125–134.

Preston, H. B. 1914. A molluscan faunal list of the Lake Tiberias with description of new species. J. Proc. Asiat. Soc. Bengal. 9:465–475.

Racek, A. A. 1974. The waters of Merom: A study of Lake Huleh. IV. Spicular remains of freshwater sponges (Porifera). Arch. Hydrobiol. 74–2:137–158.

Richard, J. 1893. Copepodes recueillis par Mr. le Dr. Barrois en Egypt, en Syrie et en Palestine. Revue biol. Nord France V:400–405, 433–443, 458–475.

Rosa, D. 1893. Viaggio del Dr. E. Festa in Palestina, nel libano e regioni vicine. II Lumbricidi. Bull. Mus. Torino VIII, 160:1–14.

Ruffo, S. 1963. Studi sui Crostacei Anfipodi LVII. Una nuova specie di Bogidiella (Crustacea, Amphipoda) della depressione del Mar Morto. Bull. Res. Counc. Isr. II B (4):188–195.

Saliternik, Z. 1964. Malaria and its control in Israel. Hacohen, Jerusalem, 183 pp. (in Hebrew).

Sars, G. O. 1927. Notes on the crustacean fauna of the Caspian Sea, Festschrift für Prof. Knipowitsch. Moskwa.

Schminke, H. K. 1974. Mesozoic Intercontinental Relationships as evidenced by Bathynellid Crustacea (Syncarida: Malacostraca). Syst. Zool. 23(2):157–164.

Schulman, N. 1959. The geology of the Central Jordan Valley. Bull. Res. Counc. Isr. 8 G:63–90.

Stankovic, S. 1960. The Balkan Lake Ochrid and its living world. Monographiae Biologicae 356 pp. W. Jung, Publisher, Den Haag.

Steinitz, H. 1954. The distribution and evolution of fishes of Palestine. Istanb. Univ. Fen. Fak. Hidrobiologie B 1/4:225–275.

Stock, J. H. 1968. A revision of the European species of Echinogammarus pungens – group (Crustacea, Amphipoda). Beaufortia, Zool. Mus. Univ. Amsterdam. 211, 16.

Stone, A., K. L. Knight & H. Starke. 1959. A synoptic Catalog of the Mosquitoes of the world. The Thomas Say Foundation Vol. VI. 358 pp.

Tchernov, E. 1975. The Molluscs of the Sea of Galilee. Malacologia 15:147–184.

Tchernov, E. 1975. The early Pleistocene Molluscs of Erq el-Ahmar. The Isr. Acad. Sc. Hum.

Tchernov, E. (in press). Rodent faunas and environmental changes in the Pleistocene of Israel.

Theodor, O. 1925. Observations on Palestinian Anopheles. Bull. Ent. Res. 15:377–382.

Topsent, E. 1892. Sur une eponge du lac de Tiberiade. Rev. Biol. Nord. France. V:85–91.

Topsent, E. 1910. Description d'une variete Nouvelle d'Eponge d'eau douce. Bull. Soc. Amis. Sciences Rouen:1–5.

Tristram, H. B. 1884. Terrestrial and fluviatile Mollusca in 'Survey of Western Palestine'. Fauna and flora of Palestine. Comm. Palestine Explor:178–204.

Tsurnamal, M. 1978a. The biology and ecology of the blind prawn, *Typhlocaris galilea* Calman. Crustaceana, in press.

Tsurnamal, M. 1978b. Temperature preference of the blind prawn, *Typhlocaris galilea* Calman. Crustaceana, in press.

402

Tsurnamal, M. & F. D. Por. 1968. The subterranean fauna associated with the blind palaemonid prawn, *Typhlocaris galilea* Calman. Intern. J. Speleology 3(3 + 4):219–223.

Waisel, Y. 1967. A contribution to the knowledge of the Phanerogamous vegetation of Lake Tiberias. Bull. Sea Fish. Res. Stn. Haifa. 44:3–16.

White, M. J. D. 1954. Animal cytology and evolution. Cambridge University Press, Cambridge.

Whitehouse, R. H. 1913. The Planarians of the Lake of Tiberias. J. As. Soc. Bengal. IX:459–463.

Wool, D. & J. Kugler. 1969. Circardian rhythm in Chironomid species from Hula Nature Reserve. Israel. Ann. Zool. Fenn., 6:94–97.

Part Five

Vertebrates

I Fishes

A. Ben-Tuvia

A. Introduction

From ancient times, the fishes of Lake Kinneret have attracted the attention of scholars and travellers. Josephus Flavius explained the presence of *Clarias lazera*, referred to under the Roman name of Coracinus, by an underground arm of the Nile River. Old and New Testaments often refer to fish and fishermen of the lake (Tristram, 1911; Nun, 1964, 1977). Hasselquist (1757) was probably the first collector of fish and from his material Artedi described one of the most common cichlids of the lake, *Sarotherodon galilaeus*

Many later explorers and scientists collected and described the fish of the lake. The references to their work and publications can be found in the general bibliography of the lake by Steinitz & Oren (1968) which includes books and papers published until 31 May 1968. In the following, we refer mainly to the more recent publications pertinent to the fish and fisheries of the lake.

The majority of fresh water fishes of Israel can be found in Lake Kinneret (Goren, 1974). In this book, twenty-four species are listed as permanent inhabitants of the lake. They belong to eight families and are grouped in twenty genera (see the key in paragraph C). Among them, five species have been introduced within the last fifty years from other parts of the world: *Gambusia affinis* for its habit of feeding on mosquito larvae, and *Cyprinus carpio*, *Mugil cephalus*, *Liza ramada* and *Hypophthalmichthys molitrix* as additional commercial fish aimed at increasing the total catch.

From the point of view of the origin of the freshwater fish fauna, the families of the lake can be divided as follows: (1) primary fresh water fishes – Cyprinidae, Cobitidae, Clariidae; (2) secondary fresh water fishes – Cyprinodontidae, Poecilidae, Cichlidae, Blennidae; (3) euryhaline marine species introduced by man – Mugilidae; (4) incidental species, rare, introduced unintentionally into the lake as a result of man's activities; thus, for example, a large serranid fish, *Dicentrarchus labrax* has been found (Yashuv, 1969), apparently transferred from the Mediterranean Sea as a juvenile together with the fry of mugilids; the occurrence of one adult salmonid, most probably *Salmo gairdneri*, can be explained by the fact that this fish is being cultivated near the upper reaches of the Jordan River, at Kibbutz Dan.

The cyprinids are represented by ten species and show the greatest diversity in ecological adaptations. The cichlids are the second most numerous family, represented by seven species or subspecies. The mugilids are usually represented by two species. The remaining families of cobitids, clariids, cyprinodontids, poecilids and blennids are represented in the lake by one species only.

In ecological terms, the majority of species belong to benthic communities predominantly inhabiting the shallow, inshore waters. Pelagic species are fewer in numbers but dominate in terms of fish biomass: *Mirogrex terraesanctae*,

Hypophthalmichthys molitrix, mugilids and most of the cichlids. Among the cichlids, *Tilapia zillii* and *Haplochromis flaviijosephi* are bottom dwellers, usually feeding in shallow waters. The genera *Sarotherodon* and *Tristramella*, although basically pelagic, live in shallow waters close to the bottom during the spawning season. *Aphanius mento* and *Gambusia affinis* are found among submerged vegetation.

In the systematic survey of the fishes of the lake, I am following in general the recent taxonomic work of Goren (1974) although some of the names of genera and species need further considerations. As proposed by Trewavas (1973) the genus *Sarotherodon* is used for two species, formerly in the genus *Tilapia*, namely *S. galilaeus* and *S. aureus*.

Following Karaman (1971), the generic name *Tor* is used for *Barbus canis*, but for *Barbus longiceps* the old generic name is retained instead of the proposed *Bertinius* until additional data are provided to clarify the taxonomic status of this species. The same author (Karaman, 1971) placed *Tylognathus steinitziorum* as a junior synonym of *Hemigrammocapoeta nanus*. This change of nomenclature seems to receive confirmation from other studies (Banarescu, personal communication).

The knowledge concerning the fresh water fishes of the lake is far from being satisfactory. For most of them, the basic information on their life history and their place in the ecology of the lake is still lacking. Only a few commercial fishes have been investigated. Very little is known on the dynamics of the fish population in relation to the fishery.

B. Zoogeography

Nineteen native species or subspecies have been found in the ichthyofauna of Lake Kinneret; they represent 66% of the total 29* species known in the region of former Palestine and Golan Heights (Goren, 1974). This region is part of a wider zoogeographical unit, the Levant (Por, 1975) which includes Israel, Jordan, Lebanon, Western Syria and South-Eastern Anatolia of Turkey as far as Amanus and Toros (Taurus) mountains. It covers a distance of about 900 km in the North–South direction and gradual changes in the fish fauna composition can be observed along this axis. Palaearctic species are more numerous in the northern section whilst palaeotropic species prevail in the south.

According to the general division of zoogeographical regions, suggested as early as 1858 by Sclater (Beaufort, 1951), the border between palaearctic and palaeotropic regions crosses the Levant. The transitional character of fresh water fish fauna, which contains Ethiopian, Oriental and Palaearctic elements, has been stressed by several authors (Bodenheimer, 1935; Steinitz, 1954; Kosswig, 1952, 1955; Por, 1977). This is due to the fact that, since the Miocene, the Levant has formed a bridge between the Eurasian and African continents.

The exchange of fish fauna in both directions was possibly through the rivers of the Rift Valley, which, towards the north, are presently separated from the

---

* Among them is included *Garra tibanica* from the area south of the Dead Sea, identified recently by Banarescu (personal communication).

Euphrates by a narrow water divide. It is generally assumed (Kosswig, 1965, 1973 and others) that, during various periods of the late Pliocene and of the Pleistocene, the connections between the rivers of the Levant and the river system of South Anatolia and Mesopotamia were even closer. The geographical character of this area with the great differences in altitude between the low parts of the Jordan River System and the high mountains stretching along the Rift Valley, creates a whole range of climatological conditions in which both northern and southern species were able to find favourable ecological niches.

In Lake Kinneret, the most southern large fresh water body of the Levant, almost half of the total number of species are of Ethiopian origin, including all the cichlids, *Clarias lazera* and probably, as pointed out by Karaman (1971) *Tor canis*. According to the same author the subfamily Torinae which is also represented in Israel by *Garra* and *Hemigrammocapoeta*, originates from South Asia.

The Israeli representatives of the genera *Acanthobrama*, *Mirogrex*, *Pseudophoxinus*, *Capoeta* and *Barbus* are considered to belong to the southern part of the palaearctic region. They show close affinity to the Anatolian and Mesopotamian species and all of them reproduce during the cold winter months.

The distribution of *Aphanius mento* extends through Iran, Iraq and Levant reaching southern Anatolia. According to Kosswig (1943) and others (Steinitz, 1951; Villwock, 1972), the genus *Aphanius* should be considered as a relict of the Tethys, a distribution which covers the perimediterranean countries, the Middle East, the shores of the Red Sea and of the western Indian Ocean.

Table 49. The distribution and zoogeographical affinities of the Lake Kinneret fishes

Name	Distribution in Israel	General distribution	Zoogeographical affinities
Mirogrex t. terraesanctae	Kinneret	endemic	Palaearctic
Acanthobrama lissneri	Jordan R.,* Coastal	endemic	Palaearctic
Pseudophoxinus kervillei	North Kinneret, Hula	Levant	Palaearctic
Barbus longiceps	Jordan R.	Levant	Palaearctic
Capoeta damascina	Jordan R., Coastal	Levant	Palaearctic
Tor canis	Jordan R.	Levant, S. Anatolia	Ethiopian
Hemigrammocapoeta nana	Jordan R., Coastal	Levant	Oriental
Garra rufa	Jordan R.	Levant	Oriental
Noemacheilus tigris	Kinneret	Levant, Iraq	Palaearctic
Clarias lazera	Jordan R., Coastal	Levant, S. Anatolia, Africa	Ethiopian
Aphanius mento	Jordan R., Coastal	Levant, Anatolia, Iran	Tethys relict
Sarotherodon galileus	Jordan R.	Israel, Jordan, Africa	Ethiopian
Sarotherodon aureus	Jordan R., Coastal	Israel, Jordan, Africa	Ethiopian
Tilapia zilli	Jordan R., Coastal	Israel, Jordan, Africa	Ethiopian
Tristramella s. simonis	Kinneret	endemic	Ethiopian
Tristramella s. intermedia	North Kinneret, Hula	endemic	Ethiopian
Tristramella sacra	Kinneret	endemic	Ethiopian
Haplochromis flaviijosephi	Jordan R.	endemic	Ethiopian
Salaria fluviatilis	Jordan R., Coastal	Southern Europe, Algeria, Anatolia	Perimediterranean

* the Jordan River System.

Salaria fluviatilis is a marine relict with a characteristic perimediterranean distribution (Sasse, 1974). According to Ruggieri (1967) the penetration of this fish into the lakes of the Mediterranean countries and its speciation took place in the Miocene.

As shown in Table 49 the Ethiopian element is represented by nine species, the Palaearctic one by six species, the oriental one by two species and the Tethys relicts (or perimediterranean elements) amount to two species.

A high percentage of fresh water fishes are limited in their distribution to the southern section of the Levant (Table 49). *Mirogrex terraesanctae terraesanctae, Tristamella simonis simonis* and *Tristramella sacra* have not been found outside of the Kinneret. *Haplochromis flaviijosephi* and *Acanthobrama lissneri* are endemic to Israel, *Tristramella simonis intermedia* and *Mirogrex terraesanctae hulensis* were characteristic of Lake Hula and now are only rarely found in the northern shores of Lake Kinneret. All this endemism, expressed in distinct genera like *Tristramella* and *Mirogrex,* and in several endemic species and subspecies, shows that a high degree of speciation took place in the Jordan Valley system. Numerically the endemic forms amount to 25% of the total number of species in the Lake Kinneret.

C. Systematics

Key to Families of Fresh Water Fishes in Israel (Plate 33)

1a. Ventral fins absent; eel-like body shape . . . . . .ANGUILLIDAE*
1b. Ventral fins present; fish-like body 2
2a. Adipose fin present . . . . SALMONIDAE*
2b. Adipose fin absent 3
3a. Fins without true spines; ventral fins situated under dorsal fin 4
3b. Fins with true spines; ventral fins under or not far behind pectoral fins .8
4a. No teeth in jaws; anal fin starts behind the center of dorsal fin 5
4b. Teeth in jaws; anal fin starts in front of the center of dorsal fin . . .6
5a. Barbels absent or no more than 2 pairs . . . . . CYPRINIDAE (p. 411)
5b. Three pairs of barbels . . . . COBITIDAE† (p. 418)
6a. Four pairs of barbels . . . . CLARIIDAE† (p. 418)
6b. No barbels present .7
7a. Dorsal fin above anal fin . . . . . CYPRINODONTIDAE† (p. 419)
7b. Dorsal fin behind anal fin . . . . POECILIDAE† (p. 419)
8a. Lateral line divided into upper and lower sections; one pair of nasal openings . . . . CICHLIDAE (p. 419)
8b. Lateral line if present, continuous in one line; two pairs of nasal openings 9
9a. Ventral fins behind pectoral fins; scales present . MUGILIDAE (p. 425)
9b. Ventral fins in front of pectorals; scales absent . BLENNIDAE† (p. 426)

---

* These two families are not usually found in Lake Kinneret.

† These families are represented in Lake Kinneret by one species only.

SALMONIDAE

ANGUILLIDAE

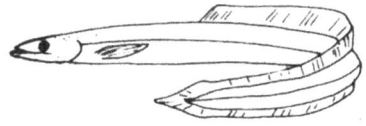

CYPRINIDAE

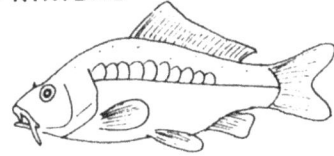

COBITIDAE

CLARIIDAE

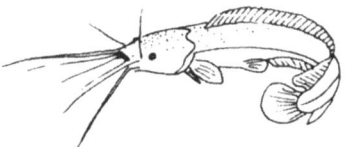

CYPRINODONTIDAE

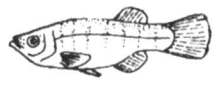

POECILIDAE

CICHLIDAE

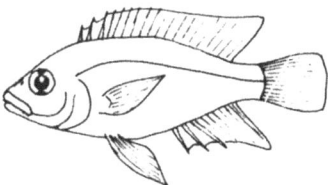

MUGILIDAE

BLENNIIDAE

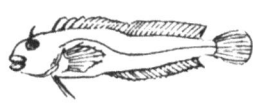

Plate 33. The families of freshwater fishes in Israel.

1. CYPRINIDAE

Key to Cyprinidae

1a. Sucking disc on lower jaw . . . . *Garra rufa*
1b. No sucking disc on lower jaw 2
2a. No barbels on chin 3
2b. Barbels on chin present 9

411

3a. Lower jaw protruding; eye below
mouth-line . . . . *Hypophthalmichtys molitrix*
3b. Lower jaw not protruding; eye clearly above mouth-line 4
4a. Base of anal fin equals dorsal fin;
lateral line incomplete . . . .*Pseudophoxinus kervillei*
4b. Base of anal fin longer than dorsal fin; lateral line incomplete 5
5a. Number of gill-rakers on the lower arm of first gill-arch less than 12 .6
5b. More than 12 gill-rakers on the lower arm 7
6a. Less than 58 scales in lateral line . . . *Acanthobrama telavivensis**
6b. More than 59 scales in lateral line . . . . . *Acanthobrama lissneri*
7a. Less than 15 gill-rakers on the lower arm *Mirogrex terraesanctae hulensis*†
7b. More than 15 gill-rakers on the
lower arm . . . . *Mirogrex terraesanctae terraesanctae*
8a. One pair of short barbels on lower jaw; scales small, each one smaller than
iris . . . . *Capoeta damascina*
8b. Two pairs of barbels around the mouth; scales larger, each one longer than
iris 9
9a. Dorsal fin extends as far back as end of anal fin . . .*Cyprinus carpio*
9b. Dorsal fin ends before the front of anal fin 10
10a. More than 15 scales in transversal line; front barbels longer than eye
diameter . . . . *Barbus longiceps*
10b. Less than 12 scales in transversal line; front barbels shorter than eye
diameter 11
11a. Mouth large and more or less terminal; front barbels shorter than rear
ones . . . . *Tor canis*
11b. Mouth small and inferior; front barbels as long as
rear ones . . . . *Hemigrammocapoeta nanus*

---

* Present only in the rivers of the coastal plain.

† Present only in the Hula Nature Reserve.

Plate 34. Garra rufa (from Heckell, VIII, 2).

Garra rufa (Heckel, 1843). Plate 34.
Synonym: *Discognathus rufus*.
Common name: agleset (H)*
Biology: no information available. Common on stones and gravel along the shores of the entire lake; often concentrating near thermal springs.
Size: usually 60–80 mm; attains 120 mm.
Distribution: throughout the lake; occurs in the whole of the Jordan River system.

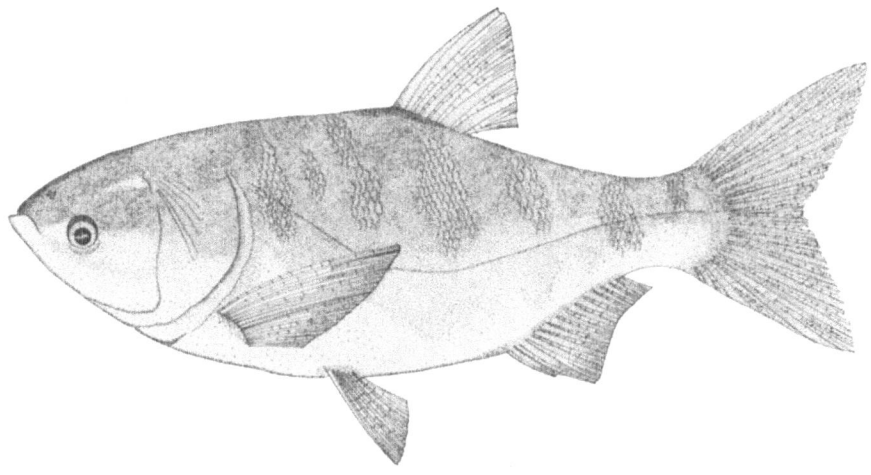

Plate 35. Hypophthalmichtys molitrix; 260 mm.

Hypophthalmichthys molitrix (Valenciennes, 1840). Plate 35.
Common name: kasif (H); silver carp (E).
Biology: introduced into the lake for the first time in 1969 and since then, hundreds of thousands of fingerlings are added each year. A native of Far East rivers, it does not reproduce in Israel in nature; reared artificially in hatcheries. A pelagic fish feeding on microscopic algae.
Size: usually 400–600 mm; attains 1,000 mm and 26 kg.
Distribution: throughout the lake.
Fishing: for the time being, in small numbers only; in 1975 the annual catch reached 23 tons.

* We have given the common names of species in Hebrew (H), Arabic (A) and English (E).

413

Pseudophoxinus kervillei (Pellegrin, 1911)
Common name: lavnunit hagalil (H)
Biology: no information available.
Size: 30–80 mm; attains 100 mm.
Distribution: rare, seems to occur only in the area of the inflow of the Jordan River into the lake and in the Batecha plain. Found in upper approaches of the Jordan River.

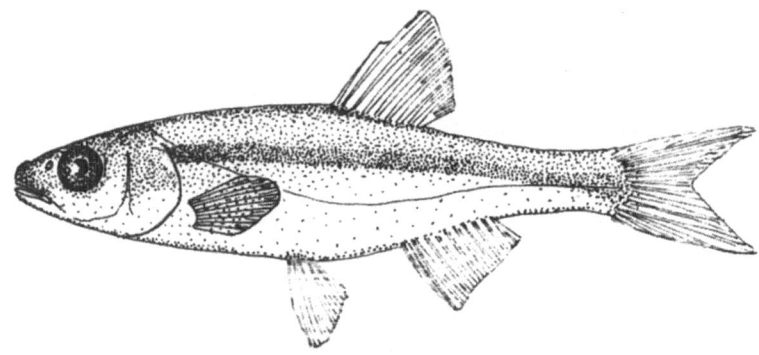

Plate 36. Acanthobrama lissneri; 82 mm.

Acanthobrama lissneri (Tortonese, 1952). Plate 36.
Common name: lavnum lissner (H).
Biology: no information available; known to inhabit shallow waters of the lake.
Size: usually 40–100 mm; attains 120 mm.
Distribution: in shallow coastal waters; Goren *et al.* (1973) reported it also from tributaries of the Jordan system, Lake Hula and Kishon River.

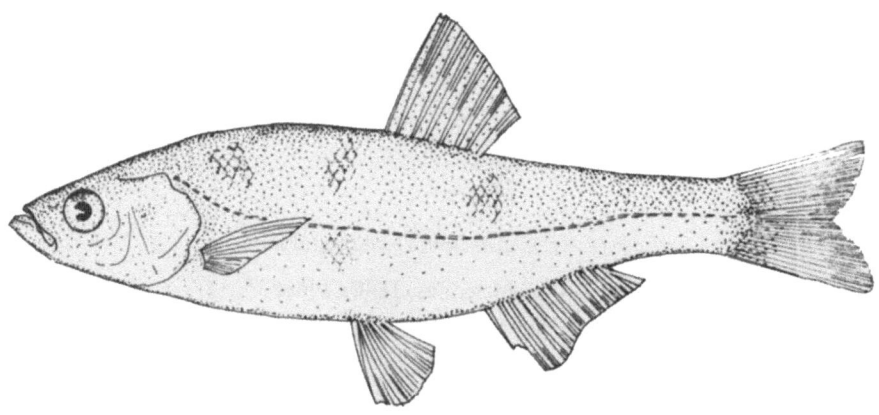

Plate 37. Mirogrex terraesanctae; 130 mm.

Mirogrex terraesanctae terraesanctae (Steinitz, 1952). Plate 37.
Synonym: *Acanthobrama terraesanctae*; *Alburnus sellal* (non Heckel).
Common name: lavnun hakineret (H); sardin tabarya (A).
Biology: one of the most common and commercially most important fish. Forms large schools which move close to the surface but occasionally descend to deep

layers. Feeds mostly on zooplankton, mainly Cladocera (*Bosmina longirostris*) and Copepoda (Komarovsky, 1952). Gophen & Landau (1977) found that it feeds preferentially on Cladocera; this is underlined by the positively electivity coefficients (Ivlev, 1961) found by Gophen for this group. Adult copepodes and copepodites are preyed more actively than nauplii and young copepodites. Reproductive season starts towards the end of November when the water temperature of the lake falls below 20°C, and continues until mid-March (Sivan, 1967). The minimum size of the spawners is 100 mm for males and 115 mm for females. Females grow faster than males and attain a larger size (Steinitz, 1959). Eggs are demersal and deposited over rocky bottoms, mostly along the eastern shores. One of the best known spawning places is the delta area of Wadi El-Kursi, north of Ein-Gev.

Size: usually 110–160 mm; attains 210 mm.

Distribution: endemic to Lake Kinneret.

Fishing: the annual catch varies between 700–1,100 tons, constituting about half of the total catch of the lake. The bulk of the catch is taken by purse-seine, usually in combination with light attraction. Utilized for canning industry and tins sold locally under the name of 'Lavnun', 'Sprat' or 'Sardine' of Lake Kinneret.

Capoeta damascina (Cuvier & Valenciennes, 1842).

Synonym: *Varicorhinus damascinus*.

Common name: hafaf (H); hafafi (A).

Biology: a common bottom fish in the lake and its tributaries. Feeds on bottom invertebrates and detritus and spawns in winter months, January–March, mostly in small streams flowing into the lake. Eggs demersal and deposited among gravel and pebbles.

Size: usually 250–350 mm; attains 500 mm.

Distribution: throughout the lake; occurs also in the Jordan River system; endemic to Israel and Syria.

Fishing: a commercial fish caught by trammel-net and gill-net; the annual catch amounts to approximately 10 tons.

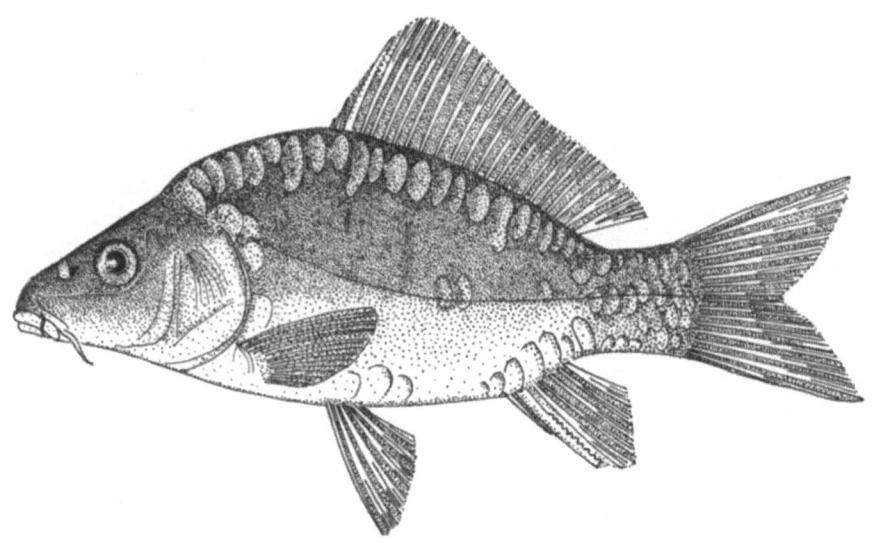

Plate 38. Cyprinus carpio; 137 mm.

Cyprinus carpio (Linnaeus, 1757). Plate 38.
Common name: karpyon mazui (H); common carp (E).
Biology: carp was introduced into Lake Kinneret during the years 1940–1941, 1948 and 1949, and unintentionally by occasional escapes from the adjacent fish culture ponds. Ripe gonads have been found in spring but it is not known whether the carp reproduces successfully in the lake. Feeds on bottom-living intertebrates.
Size: usually 300–500 mm; attains 1,000 mm and 20 kg.
Distribution: introduced in Israel in 1933 for fish farming and since then under cultivation in most of the regions of the country; also common in all fresh water streams and barrage lakes, particularly in Hula Nature Reserve.
Fishing: in small numbers only; the annual catch varies between 1–46 tons.

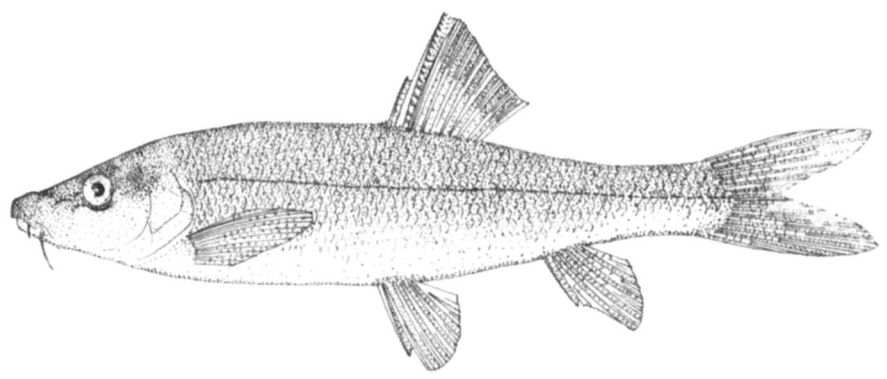

Plate 39. Barbus longiceps; 240 mm.

Barbus longiceps (Cuvier & Valenciennes, 1842). Plate 39.
Common name: binit arukat rosh (H); Kersin (A); barbel (E).
Biology: a common bottom fish found along all the shores of the entire lake.

416

Larger fish descend in winter to deeper waters. Feeds on chironomid larvae, Oligochaeta, Crustacea and Mollusca (Yashouv, unpublished; Kugler & Chen, 1968). Spawns in winter months, January–March, in shallow waters and in small streams flowing into the lake. Often in the same areas as *Capoeta damascina* with which it occasionally produces natural hybrids (Steinitz & Ben-Tuvia, 1957). Eggs are demersal and deposited among gravel and pebbles. The minimum size of spawners is 200 mm for males and 250 mm for females.

Size: usually 300–500 mm; males attain 500 mm and females, 750 mm.

Distribution: throughout the lake; occurs in the Jordan River System; endemic to Israel and Syria.

Fishing: an important commercial fish caught by trammel-net and gill-net; the annual average catch amounts to 80 tons.

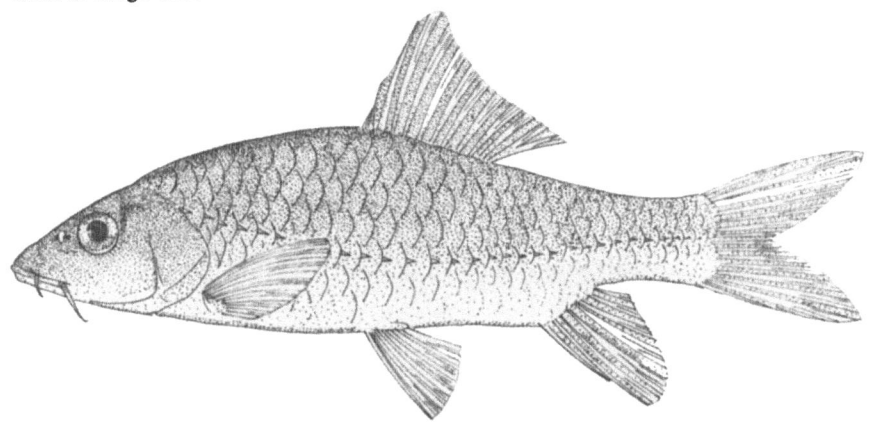

Plate 40. Tor canis; 134 mm.

Tor canis (Cuvier & Valenciennes, 1842). Plate 40.

Synonym: *Barbus canis*.

Common name: binit kishri (H); kishri (A).

Biology: a common fish found all over the lake and its tributaries. Feeds on small fish and invertebrates, both pelagic and demersal. Spawns in summer months, June–September, in shallow waters over gravel and pebbles. The minimum size of spawners 180 mm for males and 220 mm for females.

Size: usually 250–400 mm; males attain 450 mm and females 660 mm.

Distribution: throughout the lake; occurs also in the Jordan River system and its tributaries; endemic to Israel and Syria.

Fishing: an important commercial fish caught by trammel-net and gill-net; the annual average catch amounts to 53 tons.

Hemigrammocapoeta nanus (Heckel, 1843).

Synonym: *Tylognathus steinitziorum* (Kosswig, 1950).

Common name: yableset (H).

Biology: no information available; occurs in shallow waters, close to the bottom.

Size: usually 50–80 mm; attains 120 mm.

Distribution: Lake Kinneret and the Jordan River system but nowhere seems to occur in large numbers.

417

Plate 41. Noemacheilus tigris tigris (from Heckel, pl. XII. 4)

2. COBITIDAE

Noemacheilus tigris tigris (Heckel, 1843). Plate 41.
Common name: binun tigrisi (H); loach (E).
Biology: no information available.
Size: 30–60 mm; attains 70 mm.
Distribution: Lake Kinneret and Levant (Banarescu & Nalbant, 1964, 1966).

Plate 42. Clarias lazera (from Lortet, III, pl. XVII).

3. CLARIIDAE

Clarias lazera (Valenciennes, 1840). Plate 42.
Common name: sfamnun (H); barbut (A); cat-fish (E).
Biology: a common demersal fish found along the shores of entire lake. In winter, large fish often school near the hot springs as for example, the sublacustrine springs near Tabgha, known, among the fishermen, as the 'spring of cat-fish'. Omnivorous, feeding on invertebrates, fish and even seeds of plants. Spawns in spring in shallow waters. Juveniles are often found among stones and pebbles of small streams flowing into the lake. This fish is able to breathe air through an organ, formed as an outgrowth of two of the gill-arches and this enables movement some distances on land.

Size: usually 500–800 mm; attains 1,400 mm and 16 kg.

Distribution: throughout the lake; occurs in all the streams of the Jordan River system and of the coastal plain. Known from tropical Africa and from some lakes of Syria and Turkey.

Fishing: caught by trammel-net and gill-net; the average catch does not exceed 5 tons; a low-priced fish since it lacks scales and thus, does not conform to dietary laws and is not eaten by observant Jewish or Moslem people.

4. CYPRINODONTIDAE

Aphanius mento (Heckel, 1843).
Synonym: *Aphanius cypris*.
Common name: naavit khula (H); mento tooth-carp (E).
Biology: lives among water plants and submerged vegetation in protected bays and shallow pools along the shores. Feeds on crustaceans, insects and probably on some algae. Reproduces during the warm months of the year. Exhibits sexual dimorphism in the body shape and colour pattern.
Size: 25–40 mm; attains 60 mm.
Distribution: in the lake, and in many streams and springs throughout the country. Known from Iran and southern Anatolia.

5. POECILIDAE

Gambusia affinis (Baird & Girard, 1859).
Common name: gambusia (H); mosquito-fish (E).
Biology: a native of North America, introduced in the 1920's in attempts to control the larvae of malaria-carrying mosquitoes. Ovoviviparous, the female bears 2–100 young several times in a year. Reproduces during the warmer months of the year. Females larger and more numerous than males. Competes to a great extent with *Aphanius mento* for territory and food.
Size: males up to 30 mm, females up to 55 mm.
Distribution: protected bays and shallow waters among submerged vegetation.

6. CICHLIDAE

Key to the Family Cichlidae

1a. Dorsal fin with more than 27 spines and rays 2
1b. Dorsal fin with less than 26 spines and rays 4
2a. No more than 9 gill rakers on the lower arm of the first gill-arch; mouth horizontal . . . . *Tilapia zillii*
2b. More than 16 gill rakers on the lower arm; mouth oblique 3
3a. Depth of preorbital bone longer than length of the lower jaw and contained 3.5 to 4.3 in length of head; lower pharyngeal bone heart-shaped; no dark transversal streaks on the caudal fin . . . . . *Sarotherodon galilaeus*
3b. Depth of preorbital bone shorter than length of the lower jaw and contained 4.5 to 5.1 in length of head; lower pharyngeal bone of triangular shape;

caudal find with dark wavy transversal streaks . . *Sarotherodon aureus*
4a. Dorsal fin with less than 24 spines and rays; 7 gill rakers on the lower part of the anterior arch; 1 to 10 yellow spots on anal fin . . . . *Haplochromis flaviijosephi*
4b. Dorsal fin with 24 to 26 spines and rays; 9 to 12 gill rakers on the lower arm of the first gill-arch; no yellow spots on anal fin 5
5a. Teeth in jaw monocuspid; length of head longer than depth of body; lower jaw distinctly protruding; mouth very oblique . . .*Tristramella sacra*
5b. Teeth in jaw bicuspid or tricuspid; length of head shorter than depth of body; lower jaw not protruding or slightly protruding; mouth slightly oblique .6
6a. Jaws equal; interorbital width greater than length of lower jaw . . . . *Tristramella simonis simonis*
6b. Lower jaw slightly protruding; interorbital width shorter than length of lower jaw . . . . *Tristramella simonis intermedia*

Plate 43. Tilapia zillii (from Boulenger, pl. XCII); top left: young specimen; top right: jaws.

Tilapia zillii (Gervais, 1848). Plate 43.
Common name: amnun mazui (H); addadi (A); common St. Peter's fish (E).
Biology: spawning season lasts from April till end of August. Each fish spawns several times during the season. Males and females guard their eggs but are not mouthbreeders. Eggs are small and more numerous than in other cichlids. The minimum size at first maturity is 130–140 mm. Omnivorous, feeding on plankton, benthos and fragments of higher plants.
Size: usually 120–220 mm; attains 270 mm.

Distribution: very common throughout the lake and in the Jordan River system; one of the most common fishes in the coastal plain rivers. Known from African lakes and rivers.

Fishing: in small quantities only and of low commercial value.

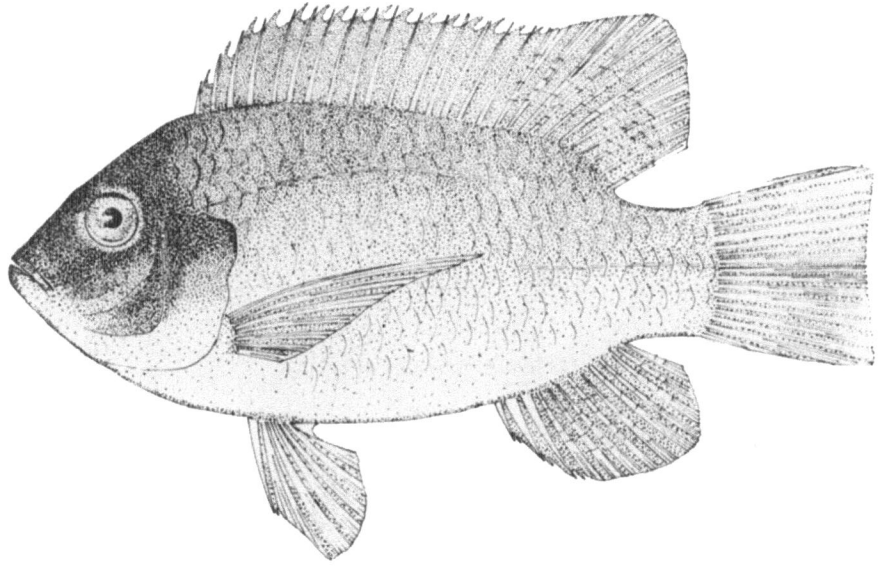

Plate 44. Sarotherodon galilaeus; 150 mm.

Sarotherodon galilaeus (Artedi, 1757). Plate 44.

Synonym: *Tilapia galilaea.*

Common name: amnun hagalil (H); musht abiad (A); galilee St. Peter's fish (E).

Biology: in winter months, behaves as a pelagic schooling fish. In March, schools disperse and pairs are formed. Spawning season lasts from end of March or beginning of April until mid-August. Both male and female participate in mouthbreeding and the number of eggs carried by each parent fish varies between 600 and 1,100. Each fish spawns more than once in a season. Juveniles leave mouth of parents after reaching the length of 12 mm. The minimum size at first maturity is 180–220 mm. Males grow faster than females and attain larger size. Feeds mostly on phytoplankton and *Peridinium cinctum fa. westii* (Dinophyceae, Pyrrophyta) is the most important component of its diet (Ben-Tuvia, 1959; Spataru, 1976). The population dynamics was studied by Landau (1975).

Size: usually 200–260 mm; attains 380 mm.

Distribution: throughout the lake and in the Jordan River system; found also in the Yarkon River. Known from African lakes and rivers.

Fishing: an important commercial fish renowned for its excellent taste and fetching high prices on local market. The average annual catch amounts to 113 tons (6.7% of the total catch). Caught by trammel-net and gill-net, seldom by purse seine.

421

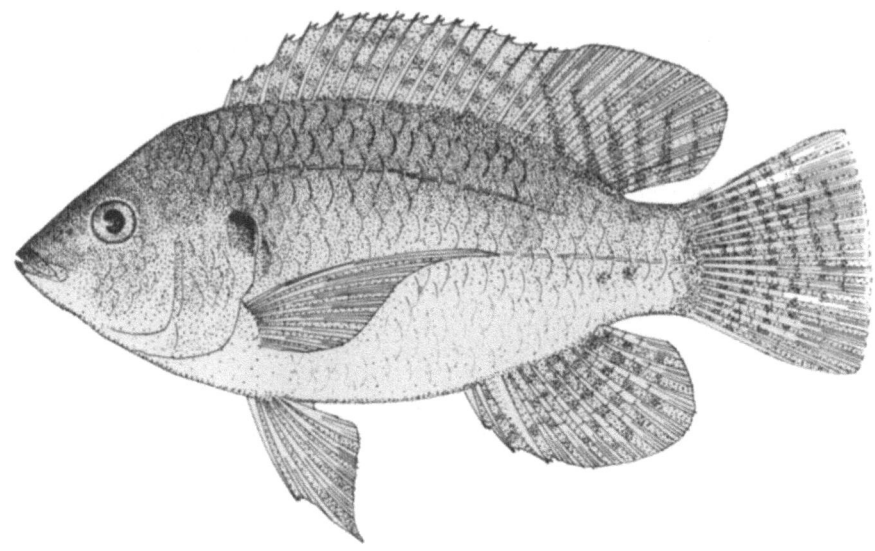

Plate 45. Sarotherodon aureus; 155 mm.

Sarotherodon aureus (Steindachner, 1864). Plate 45.

Synonym: *Tilapia aurea*; *Tilapia nilotica* (non *T. nilotica* Linné).

Common name: amnun hayarden (H); musht lubbad (A); Jordan St. Peter's fish (E).

Biology: spawning season lasts from April till August. Male digs a spawning pit but mouthbreeding is carried out by female only. Each fish spawns more than once in a season. Juveniles leave mother's mouth after reaching the length of 11 mm. The minimum size at first maturity is 180–200 mm. As in all cichlids, males grow faster than females and attain larger size. Feeds on plankton in which *Peridinium cinctum fa. westii* is an important component and on zoobenthos (Spataru, 1976).

Size: Usually 200–240 mm; attains 330 mm.

Distribution: through the lake and in the Jordan River system; common in coastal plain rivers. Known from African lakes and rivers.

Fishing: an important commercial fish caught by trammel-net and gill-net. The average catch amounts to 94 tons (5.7% of total catch). Since 1956, under cultivation in fish ponds in most regions of the country. The stock originated from Lake Hula, usually hybridized with *Tilapia nilotica* and *Tilapia vulcani* brought from Africa.

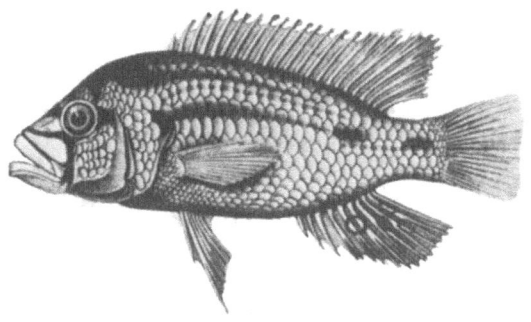

Plate 46. Haphlochromis flaviijosephi (from Lortet, III, pl. VIII).

Haplochromis flaviijosephi (Lortet, 1883). Plate 46.
Common name: amnunit josef (H); khanus (A).
Biology: spawning season lasts from April until at least July. Each fish spawns several times during the season. Mouthbreeding carried out by female only. Juvenile leaves mother's mouth at size of 8–9 mm (Werner, 1976). Presumably omnivorous, feeding also on small fish.
Size: usually 30–70 mm.
Distribution: throughout the lake and in the Jordan River system, in shallow water and among submerged vegetation. Endemic to Israel.

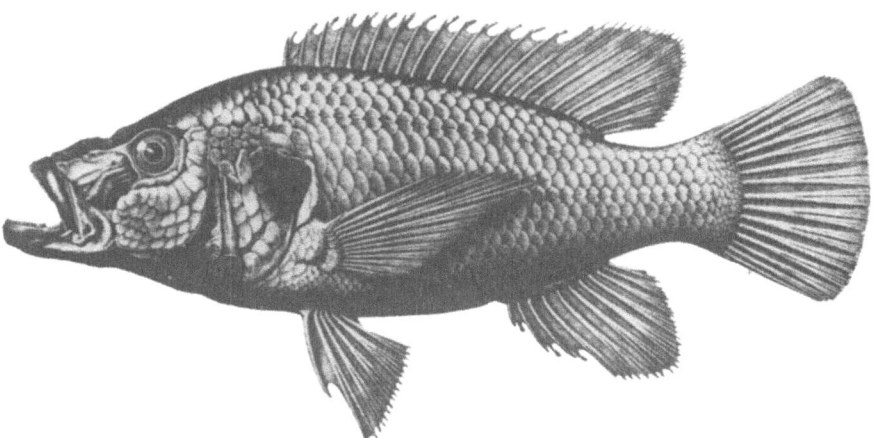

Plate 47. Tristramella sacra (from Lortet, III, pl. X)

Tristramella sacra (Gunther, 1864). Plate 47.
Common name: kdashnun kalbi (H); kelb (A); Tristram's St. Peter's fish (E).
Biology: spawning season lasts from April until July. Both male and female participate in mouthbreeding. Eggs large; 60 to 220 counted in the mouth of breeding parent fish; juvenile leave parents mouth at size of 14.5 mm. Omnivorous, feeds mostly on phytoplankton but occasionally predates on small fish.
Size: usually 210–240 mm; attains 280 mm.
Distribution: endemic to Lake Kinneret.
Fishing: in small quantities only.

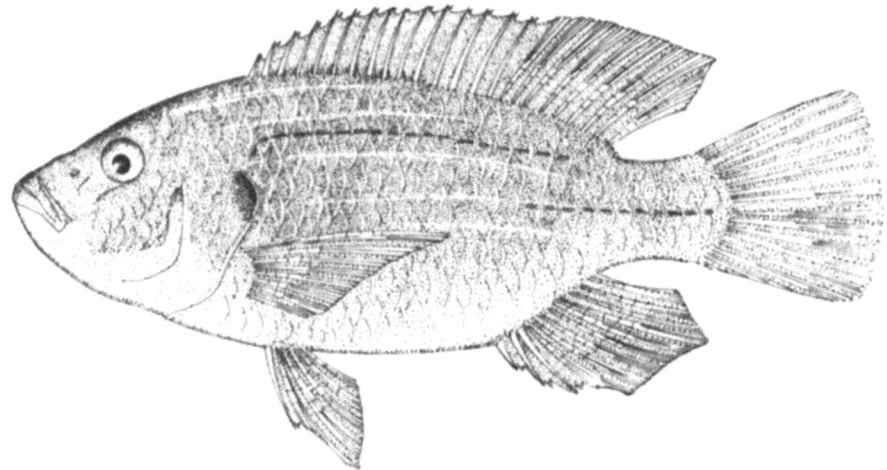

Plate 48. Tristramella simonis simonis; 150 mm.

Tristramella simonis simonis (Gunter, 1864). Plate 48.

Common name: kdashnun hakineret (H); marmur (A); Tristram's St. Peter's fish (E).

Biology: spawning season lasts from April until the end of August. Each fish spawns several times during the season. Mouthbreeding is carried out by female only which shelters 50–250 eggs. Juveniles leave mother's mouth at size of 14 mm. The minimum size at first maturity is 160 mm. Feeds on plankton and various benthic plants and invertebrates.

Size: usually 180–210 mm; attains 230 mm.

Distribution: endemic to Lake Kinneret.

Fishing: a commercial fish, caught by trammel-net and gill-net; the average annual catch amounts to 63 tons (3.8% of total catch).

Tristramella simonis intermedia (Steinitz & Ben-Tuvia, 1960).

Common name: kdashnun hahula (H); okar (A); Tristram's St. Peter fish (E).

Biology: spawning season lasts from April until end of June. Both male and female participate in mouthbreeding. Eggs large, as in *Tristramella simonis simonis*. The minimum size of spawner at first maturity 140–150 mm. Feeds on plankton and benthos.

Size: usually 180–220 mm; attains 240 mm.

Distribution: rare in the lake, occurs seldom in the area of Batecha. Before the drainage of the Hula Lake it was the most common and commercially exploited fish there.

Fishing: after the drainage of Hula, this fish is rare and therefore of no commercial value.

7. MUGILIDAE

Key to the Family Mugilidae

1a. Thick adipose lid covering most of the eye; 8 soft rays in anal fin; 2 pyloric
caeca . . . . *Mugil cephalus*
1b. No adipose lid or, if present, covers only edges of the eye; 9 soft rays in anal
fin; 7–9 pyloric caeca 2
2a. Lower lip high with fleshy papillae on its lower edge . *Chelon labrosus**
2b. Upper lip thin without fleshy papillae 3
3a. Scales on the top of head descending to the line of front nasal opening; teeth
minute, not visible with naked eye; dark axillary spot in the inner corner of
the pectoral fin . . . . *Liza ramada*
3b. Scales on the top of head descending only to the line of posterior nasal
opening; teeth small but clearly visible with naked eye;
no axillary spot . . . . *Liza aurata**

Plate 49. Mugil cephalus (from Boulenger, pl. LXXX)

Mugil cephalus (Linnaeus, 1758). Plate 49.
Common name: kifon buri (H); buri (A); grey mullet (E).
Biology: spawns along the Mediterranean coast in autumn. In
November–December juveniles 20–30 mm long concentrate near outlets of rivers,
in small bays and in lagoons. In this period they are collected by fishbreeders and
transferred gradually into fresh water to stock fish ponds and Lake Kinneret. Om-
nivorous, feeds on minute planktonic and benthic organisms, and by sieving
detritus.
Size: usually 300–400 mm; attains 800 mm and weight of 8 kg.
Distribution: throughout the lake; inshore-pelagic in habits, a fast swimmer, forming small schools. Euryhaline marine fish, circumtropical, known from both the
Mediterranean and the Red Sea (Ben-Tuvia, 1975).
Fishing: caught by trammel-net and gill-net; the average annual catch (together
with *Liza ramada*) amounts to 208 tons (12.4%) of the total.

* Rare fish in Lake Kinneret.

425

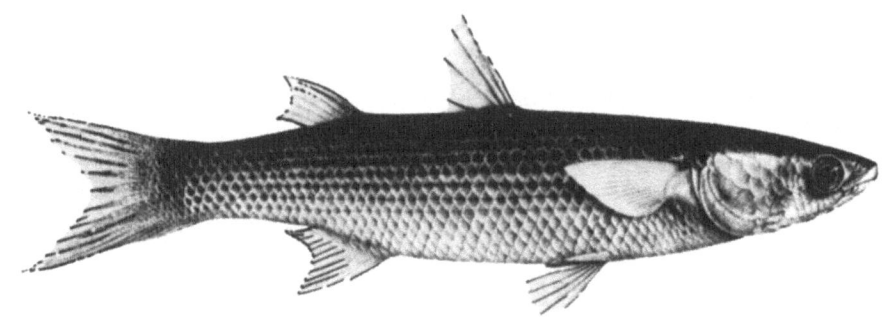

Plate 50. Liza ramada (from Boulenger, pl. LXXX)

Liza ramada (Risso, 1810). Plate 50.

Synonym: *Mugil capito.*

Common name: kifon tubar (H); tubara (A); grey mullet (E).

Biology: spawns along the Mediterranean coast in late autumn and winter. In December–March juvenile 20–30 mm long concentrate near outlets of rivers, in small bays and lagoons. In this period they are collected and used to stock fish ponds and Lake Kinneret. Omnivorus, feeds on minute planktonic and benthic organisms, and by sieving detritus.

Size: usually 300–400 mm; attains 500 mm.

Distribution: throughout the lake; inshore-pelagic in habits, a fast swimmer, forming small schools. An euryhaline marine fish known from the Mediterranean and eastern Atlantic.

Fishing: caught by trammel-net and gill-net; the average annual catch (together with *Mugil cephalus*) amounts to 208 tons (12.4%).

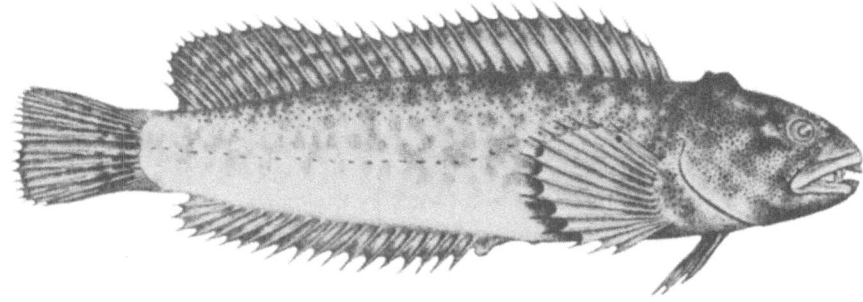

Plate 51. Salaria fluviatilis (from Lortet, III, pl. XVIII)

BLENNIDAE

Salaria fluviatilis (Asso, 1801). Plate 51.

Synonym: *Blennius fluviatilis.* (see Bath, 1977).

Common name: karnun naharoni; fresh water blenny (E).

Biology: spawning season lasts from April until July. Eggs deposited under stones and guarded by the male. Feeds on benthic invertebrates.

Size: usually 30–80 mm; attains 140 mm.

Distribution: throughout the lake in shallow waters among stones and pebbles; occurs in coastal plain rivers and rivulets; known from lakes and rivers of Southern Europe, Anatolia and Algeria (Sasse, 1974).

D. Development of Lake Kinneret Fisheries

During the period of the British Mandate which terminated in 1948, the annual catch varied between 200 and 500 tons. The fishery was seasonal, mostly confined to the winter months and the catch was taken by beach-seine, trammel-net, and cast-net. In 1940, a purse-seine of the Yugoslavian type 'saccoleva' for the lavnun (*Mirogrex terraesanctae*) fishing was introduced by the fishermen of the surrounding kibutzim, Ein-Gev & Ginnosar (Nun, 1977). As a result of recommendations by Hornell (1935) and Ricardo-Bertram (1944), steps were taken for rational management of the lake fisheries. The legal mesh size of commercial nets was enlarged to 10 mm from knot to knot for lavnun fishery (14 mm today), 35 mm for large cyprinids (37 mm today) and 45 mm in trammel-nets for cichlids. These measures are still valid for todays fishery regulations. The minimum legal size of fishes allowed for landings is 30 cm for the commercial cyprinids, 18 cm for cichlids and 12 cm for lavnun.

With the establishment of the State of Israel in 1948, a rapid development of the lake fishery took place. Catches increased considerably: 1,000 tons in 1957 and an all-time peak of over 2,000 tons in 1971. The average catch for the period 1971–1975 amounted to 1,682 tons (Sarid, 1972–1976). Derived from the same statistical data, the catch of lavnun for the same period constituted 61% of the total catch, cichlids 16%, grey-mullets 12% and all the others 11% (Table 50).

Table 50. Lake Kinneret fishery. Composition of the catch. Average of the period 1971–1975 (data compiled from Sarid, 1972–1976)

	1971	1972	1973	1974	1975	average	%
1. *Mirogrex terraesanctae*	1,209	1,016	969	949	1,017	1,032	61.3
2. *Mugilidae*	306	197	201	195	141	208	12.4
3. *Sarotherodon galilaeus*	153	115	101	73	121	112	6.7
4. *Sarotherodon aureus*	159	97	79	50	84	94	5.6
5. *Tilapia zillii*	7	1	—	9	—	3	0.2
6. *Tristramella* spp.	84	61	49	61	62	63	3.8
7. *Barbus longiceps*	111	79	69	71	70	80	4.8
8. *Tor canis*	65	72	58	37	35	53	3.2
9. *Copoeta damascina*	18	16	5	—	3	8	0.4
10. *Cyprinus carpio*	17	10	8	9	11	11	0.7
11. *Hypophthalmichthys molitrix*	—	—	—	11	23	7	0.4
12. *Clarias lazera*	4	7	4	3	5	4	0.3
13. Various	27	—	—	—	—	5	0.3
Total	2,160	1,671	1,543	1,468	1,572	1,680	100%

It is evident from Fig. 124 that the total catch of the lake depends entirely on the success of lavnun fishery. Catches of other fish contribute little to the overall picture of yields, however, their commercial value is comparatively high, exceeding the value of lavnun.

Cichlids and grey-mullets are the high-priced fish and an effort has been made during the last eighteen years to stock the lake with their fry. The first introduction of *Mugil cephalus* and *Liza ramada* originating from the Mediterranean Sea was

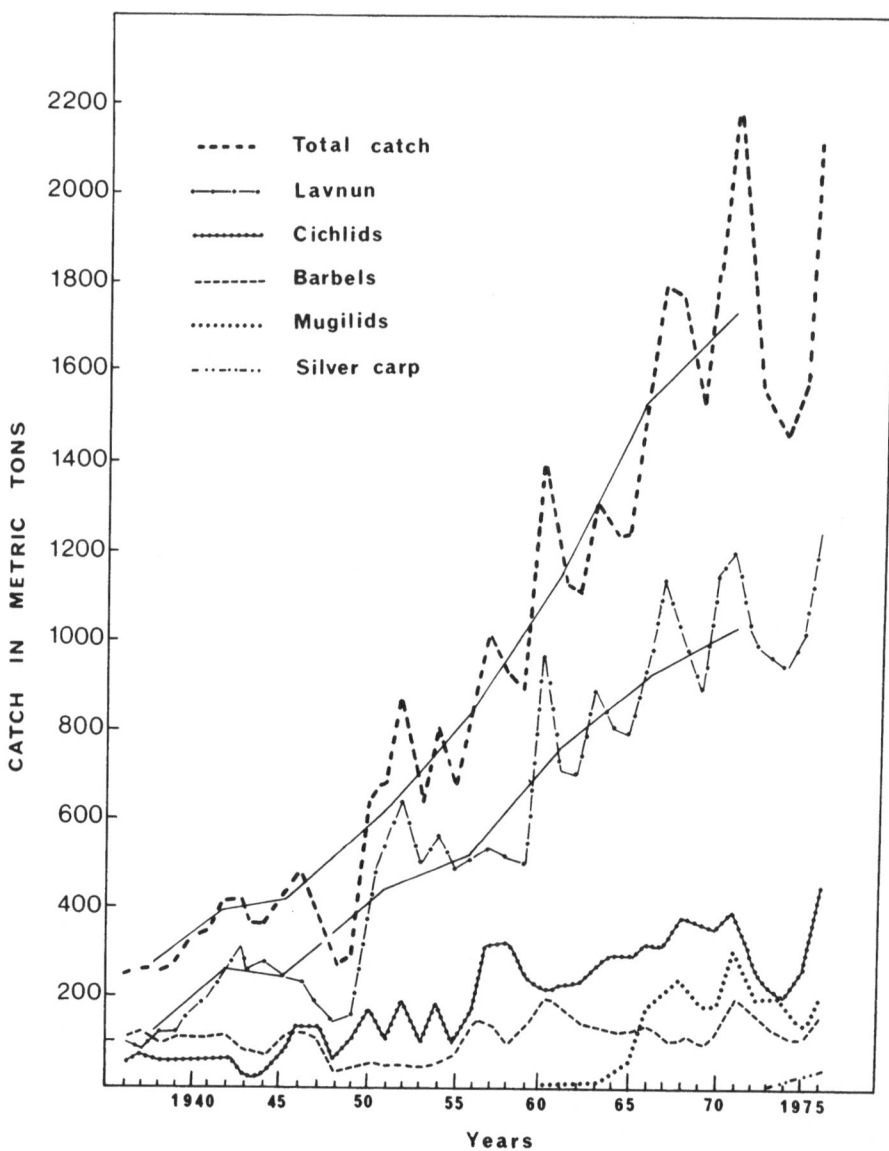

Fig. 124. Fluctuations of the annual catch in Lake Kinneret for the period 1935–1976; separate graphs represent the total catch and the catch of the main groups of fishes: lavnun (*Mirogrex terraesanctae*), cichlids, barbels (*Barbus longiceps, Tor canis* and *Capoeta damascina*), mugilids (*Mugil cephalus* and *Liza ramada*) and silver carp (*Hypophthalmichthys molitrix*). Trends are indicated by full lines drawn on the basis of five year averages (by courtesy of Dr. K. Reich, from various sources, unpublished).

made in small numbers in 1958. From 1962 onwards, the lake has been stocked annually with 1–3 million fingerlings, since grey-mullet does not reproduce in the lake. However, a sharp reduction in availability of grey-mullet fry along the Mediterranean coast of Israel has been experienced in recent years, thus limiting stocking activities.

428

In the same year, 1958, stocking of the lake with fingerlings of *Sarotherodon aureus* produced in ponds was initiated. The number of fry introduced varied from 350,000 in the first year to almost 2,500,000 in 1972 (Reich, 1976). According to data presented by the same author, the stocking had a direct impact on catches, about 60 tons per million of fingerlings, or 25–30% of the fingerlings reached a commercial size and contributed to the landings.

The catch of *Sarotherodon galilaeus* seems to be less dependent on stocking activities. In spite of considerable numbers of fingerlings released into the lake between 1952 and 1974 (Reich, 1976) there was no increase in landings that could be attributed to stocking. Further studies are needed in order to evaluate the year-to-year fluctuations in catches and to separate the influence of environmental factors from the influence of intensive fishing that took place in the last two decades.

Silver carp (*Hypophthalmichthys molitrix*) is the most recently introduced cyprinid fish into the lake. The first batch of 40,000 fingerlings was released in 1969, 25,000 in 1970 and over a million in 1974 (Shefler & Reich, 1977). A native of rivers of China and Amur, the silver carp does not reproduce in Lake Kinneret but is reared artificially in special hatcheries. According to Shefler & Reich (1977) it grows rapidly in the lake, reaching in some cases, a weight of 18 kg in five years. The catch in 1975 amounted to 23 tons.

References

Banarescu, P. & T. Nalbant. 1964. Cobitidae. Süsswasserfishe der Türkei, Teil 2. Mitt. Hamburg. Zool. Mus. Inst., 61 : 159–201.

Banarescu, P. & T. Nalbant. 1966. Zwei neue schmerlen der gattung *Noemacheilus* (Pisces, Cobitidae) aus Jordanien. Mitt. Hamburg. Zool. Mus. Inst., 63 : 329–336.

Bath, H. 1977. Revision der Blenniini (Pisces: Blenniidae) Senckenbergiana biol. 57 (4/6) : 167–234.

Beaufort de, L. F. 1951. Zoogeography of the land and inland waters. Sidgwick and Jackson, London. 208 pp.

Ben-Tuvia, A. 1959. The biology of the cichlid fishes of Lake Tiberias and Huleh. Bull. Res. Counc. of Israel, 8B(4) : 153–188.

Ben-Tuvia, A. 1975. Mugilid fishes of the Red Sea with a key to the Mediterranean and Red Sea species. Bamidgeh, Bull. Fish. Cult. Israel, 27(1) : 14–20.

Bodenheimer, F. S. 1935. Animal life of Palestine. L. Mayer, Jerusalem, 506 pp.

Gophen, M. & R. Landau. 1977. Trophic interactions between zooplankton and 'Sardines' (*Mirogrex terraesanctae*) populations in Lake Kinneret during the period 1973–1975. Oikos (in press).

Goren, M., L. Fishelson & E. Trewavas. 1973. The cyprinid fishes of *Acanthobrama* Heckel and related genera. Bull. Br. Mus. nat. Hist. (Zool.) 24(6) : 293–315.

Goren, M. 1974. The freshwater fishes of Israel. Israel J. of Zool., 23 : 67–118.

Hasselquist, F. 1757. Iter Palestinum, Stockholm.

Hornell, J. 1935. Report on the fisheries of Palestine. Crown Agents for the Colonies, London, 106 pp.

Ivlev, V. S. 1961. Experimental ecology of the feeding of fishes. Translated by D. Scott, Yale University Press, New Haven, Connecticut.

Karaman, M. S. 1971. Revision der Barben Europas, Vorderasiens und Nord-afrikas. Süsswasserfische der Turkei, Teil 8. Mitt. Hamburg. Zool. Mus. Inst., 67 : 175–254.

Komarovsky, B. 1952. An analysis of the stomach contents of *Acanthobrama terrae-sanctae* H. Steinitz, from Lake Tiberias. Bull. Sea Fish. Res. Stn, Haifa, (4) : 1–8.

Kosswig, C. 1943. Uber Tethysrelikte in der Türkischen Fauna. CR Soc. Türq. Sci. Phys. nat. Istanbul, 10 : 30–47.

Kosswig, C. 1952. Die Zoogeographie der turkischen Süsswasserfische. Istanbul Univ. Fen. Fak.

Hidrobologi, B, (2): 85–101.

Kosswig, C. 1955. Zoogeography of the Near East. Syst. Zool. 4: 50–73.

Kosswig, C. 1965. Zur historischen Zoogeographie der Ichthyofauna im Susswasser des sudlichen Kleinasiens. Zool. Jb. Syst. Bd., 92: 83–90.

Kosswig, C. 1969. New contributions to the zoogeography of fresh water fish of Asia Minor, based on collections made between 1964–1967. Israel J. of Zool., 18: 249–254.

Kosswig, C. 1973. Über die Ausbreitungswege sogenannter perimediterraner Süsswasserfische. Bonn. Zool. Beitr., 24: 165–177.

Kugler, J. & H. Chen. 1968. The distribution of chironomid larvae in Lake Tiberias and their occurrence in the food of fish of the lake. Israel J. of Zool., 17: 97–115.

Landau, R. 1975. The fish populations of Lake Kinneret. Israel Oceanogr. & Limnol. Res. Ltd., Interim Project Report, July, 1975.

Nun, M. 1964. Ancient Jewish fishery. Hakibbutz Hameuhad, 232 pp. (in Hebrew).

Nun, M. 1977. Sea of Kinneret; a monograph. Hakibbutz Hameuhad, 262 pp. (in Hebrew).

Por, F. D. 1975. An outline of the zoogeography of the Levant. Zool. Scr., 4(1): 5–20.

Reich, K. 1976. Problems of fisheries research in Lake Kineret. Bamidgeh, Bull. Fish Cult. Israel, 28: 3–11.

Ricardo-Bertram, C. K. 1944. Abridged report on the fish and fishery of Lake Tiberias. Dept. Agric. Fish., Palestine, 14 pp.

Ruggieri, G. 1967. The Miocene and later evolution of the Mediterranean Sea. Systematics Assoc. Publ. 7, Aspects of Tethyan Biogeography, 283–290, London.

Sarid, Z. (Ed.). 1972–1976. Israel fisheries in figures, 1971–1975. State of Israel, Ministry of Agriculture, Department of Fisheries, Tel-Aviv.

Sasse, H. 1974. *Blennius fluviatilis* Asso, 1784 (Blenniidae, Perciformes, Pisces). Süsswasserfische der Türkei, Teil 10. Mitt. Hamburg. Zool. Mus. Inst., 70: 267–275.

Shefler, D. & K. Reich. 1977. Growth of silver carp (*Hypophthalmichthys molitrix*) in Lake Kineret in 1969–1975. Bamidgeh, Bull. Fish Cult. Israel, 29 (1): 3–16.

Sivan, P. 1967. Seasonal changes in the gonads of *Acanthobrama terrae-sanctae* H. Steinitz. Bull. Sea Fish. Res. Stn, Haifa, (44): 22–41.

Spataru, P. 1976. The feeding habits of *Tilapia galilaea* (Artedi) in Lake Kinneret (Israel). Aquaculture, 9: 47–59.

Steinitz, H. 1951. On the distribution and evolution of the cyprinodont fishes of the Mediterranean region and the Near East. Bonner Zool. Beitr., 2 (1–2): 113–124.

Steinitz, H. 1954. The distribution and evolution of the fishes of Palestine. Istanb. Univ. Fen Fak. Hidrobiol., B1(4): 225–275.

Steinitz, H. 1959. Dr. Lissner's study of the biology of *Acanthobrama terrae-sanctae* in Lake Tiberias. Bull. Res. Counc. of Israel, 8B(2): 43–64.

Steinitz, H. & A. Ben-Tuvia. 1957. The hybrid of *Barbus longiceps* C.V. and *Varicorhinus damascinus* C.V. (Cyprinidae, Teleostei). Bull. Res. Counc. Israel, 6B (3–4): 176–188.

Steinitz, H. & O. H. Oren. 1968. Bibliography on Lake Kinneret (Lake Tiberias). Bull. Sea Fish. Res. Stn, (53): 1–48.

Tortonese, E. 1952. Viaggio del dott. Enrico Festa in Palestina e Siria (1893): Pesci. Boll. Mus. Zool. Anat. Comp. Torino, 46(85): 3–48.

Trewavas, E. 1973. I. On the cichlid fishes of the genus *Pelmatochromis* with proposal of a new genus for *P. congicus*; on the relationship between *Pelmatochromis* and *Tilapia* and the recognition of *Sarotherodon* as a distinct genus. Bull. Br. Mus. nat. Hist. (Zool.), 25(1): 1–26.

Tristram, H. B. 1911. The natural history of the Bible. London, 10th edition, 520 pp.

Villwock, W. 1972. Beitrag zur Kenntnis der Zahnentwicklung bie Zahnkarpfen der Tribus Aphaniini. Mitt. Hamb. Zool. Mus. Inst., 68: 135–176.

Werner, Y. L. 1976. Notes on the reproduction in the mouth-brooding fish *Haplochromis flaviijosephi* (Teleostei: Cichlidae) in the aquarium. J. nat. Hist., 10: 669–680.

Yashouv, A. 1950. Report on the *Barbus* fishes and their environment (in Hebrew, unpublished).

Yashouv, A. 1969. The bass, *Dicentrarchus punctatus* (Bloch), fish for culture in fresh and brackish waters. Fisheries and fish breeding in Israel, 4(3): 27–29 (in Hebrew).

430

II Amphibians

S. Lulav

Three species of Amphibians are known on the Kinneret shores and other water bodies of the Jordan Valley: the green toad (*Bufo viridis*), the lemon yellow tree frog (*Hyla arborea*) and the edible frog (*Rana ridibunda*). This latter lays eggs on the beaches of the lake but only in very quiet environments since this animal is very much disturbed by water turbulence. It also lays eggs on irrigation canals. *Bufo viridis* lays eggs in pools lacking vegetation whereas *Hyla arborea* lays its eggs amongst the vegetation of shallow pools.

III Reptiles

S. Lulav

The Reptiles of Lake Kinneret and surrounding area include thirty-one species according to the following list:

Name	Common name	Observations
A. Testudines		
Mauremys (Clemmys) caspica	Caspian terrapin	
Testudo graeca	Greek tortoise	
Trionyx triunguis	Nile soft-shelled turtle	(A fossil Trionyx sp. was found in the Pleistocene deposits of 'Ubeidiya)
B. Sauria:		
Agama stellio	Starred lizard (hardun)	
Ophisaurus apodus	Scheltopusik	
Chamaeleo chamaeleon	Common chamaeleon	
Hemidactylus turcicus	Turkish gecko	
Ptyodactylus hasselquisti puiseuxi	Fan-footed gecko	
Ophisops elegans	Menetries's lizard	
Ablepharus kitaibelii	Lidless skink	
Chalcides guentheri	One-toed skink	
Chalcides ocellatus	Eyed skink	
Eumeces schneideri pavimentatus	Gold skink	
Mabuya vittata	Bridled skink	
Ophiomorus latastii	Sand skink	
C. Ophidia:		
Eryx jaculus	Javelin sand-boa	
Psammophis schokari	Schokari sand-snake	
Coluber jugularis	Syrian black-snake	
Coluber najadum	Dahl's whip-snake	
Coluber ravergieri	Coin-marked snake	
Eirenis lineomaculata	Crowned peace-snake	
Eirenis decemlineata	Lined peace-snake	
Eirenis rothi	Jan's peace-snake	
Malpolon monspessulanus	Montpellier snake	
Micrelaps muelleri	Falce coral-snake	
Natrix tessellata	Dice water-snake	
Rhynchocalamus melanocephalus	Palestine black-headed snake	
Telescopus fallax	European cat-snake	
Typhlops simoni	Blind snake	
Typhlops vermicularis	Greek blind-snake	
Vipera palaestina	Palestine viper	

From an ecological point of view, three species are closely associated with the aquatic habitat. *Mauremys caspica* is found in Lake Kinneret, in the Buteiha swamps and in the River Jordan (plate 52). *Natrix tesselata* lives in the water; it is a very active

Plate 52. Meal of the caspian
terrapin (*Mauremys caspica*)
(Photo A. Shuv).

Plate 53. Hatching of the dice
water snake (*Natrix tessellata*)
(Photo A. Shuv).

Plate 54. The
Palestine viper (*Vipera
palaestina*)
(Photo A. Shuv).

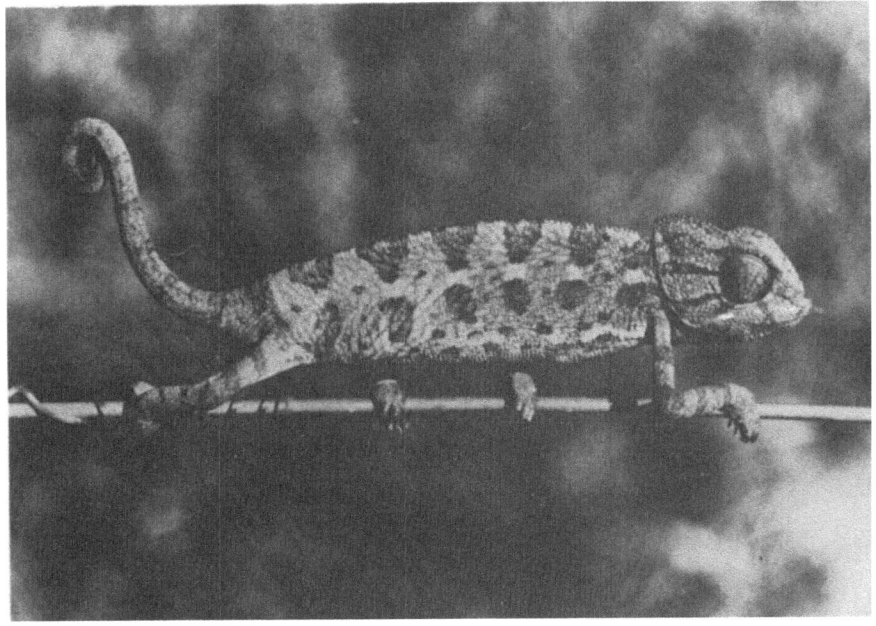

Plate 55. The common chamaeleon
(*Chamaeleo chamaeleon*)
(Photo A. Shuv).

Plate 56. The Syrian black snake
(*Coluber jugularis*)
(Photo A. Shuv).

Plate 57. The starred lizard
(*Agama stellio*) (Photo A. Shuv).

Plate 58. The fan footed gecko
(*Ptyodactylus hasselquisti*) (Photo A. Shuv).

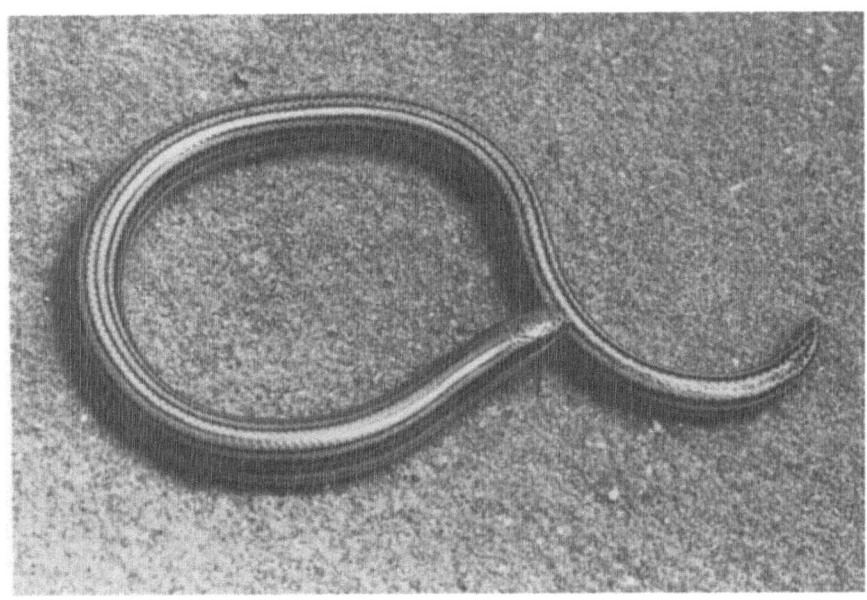

Plate 59. The sand skink
(*Ophiomorus latastii*) (Photo A. Shuv).

predator feeding on fish and frogs (Mendelssohn, personal communication) (plate 53). On one occasion (20 March 1975) a Nile soft-shelled turtle (*Trionyx triunguis*) was caught in nets of fishermen on Lake Kinneret. It was probably introduced artificially by Man. We note however, that a fossil soft shelled turtle of the same genus was found in the Pleistocene deposits of Ubeidiya, 4 km SSW of Lake Kinneret. The present habitat of *Trionyx* in Israel is limited to the few unpolluted streams of the coastal plain flowing westwards to the Mediterranean Sea.

It is worthwhile mentioning that the viper (*Vipera palaestina*) is the only venomous snake in the Kinneret area (plate 54). The coin-marked snake (*Coluber ravergieri*) is sometimes mistaken for a viper because of its viper-like pattern and colour. A few other Reptiles of the Kinneret area are shown in plates 55, 56, 57, 58 and 59.

IV Birds

S. Lulav

The Kinneret area belongs to the Irano–Turanian Steppe separating the Mediterranean vegetation from the desert. This area represents the northernmost intrusion of the steppe vegetation into Israel through the Jordan Valley because of the relative scarcity of precipitation. In contrast to the Mediterranean region where forests of pine trees can develop and olive trees can grow without irrigation, trees cannot grow naturally in the steppe area and its climax plant is the shrub *Zisyphus lotus*. During the past sixty years, the Jewish settlements transformed this semi-arid zone into a green plain where irrigated fields, gardens, lawns and the creation of fish ponds encouraged the influx of animals and especially of birds which had previously found the terrain unsuitable.

A. The Water Fowl

The kingfishers are very common around the lake and the fish ponds. Two species, the pied kingfisher (*Ceryle rudis*) and the smyrna kingfisher (*Halcyon smyrnensis*) are permanent residents whereas the European kingfisher (*Alcedo atthis*) is a winter visitor. The pied kingfisher feeds on fish only. The smyrna kingfisher is less dependent on fish for its diet which is composed mostly of insects, small reptiles and amphibians. Both species are nesting in the high banks of the lake and nearby rivers and wadis.

From September to April, Lake Kinneret serves as a winter hostel for a large gull population, mostly – about 80% – black-headed gulls (*Larus ridibundus*), herring gulls (*Larus argentatus*) and smaller numbers of black-headed gulls (*Larcus fuscus*) and little gulls (*Larus minutus*).

Usually, the gulls spend their nights on the lake. Every morning, at dawn, most of them leave the lake to visit the fish ponds and garbage heaps in the Kinneret Valley. At sunset these birds return to the Kinneret, the flocks coming in one after another: the bigger herring gulls in groups of up to 20, the black-headed gulls numering up to 200 per flock – all flying in broad, wavy arrow-head formations.

The number of gulls wintering on the Kinneret is estimated at about 10,000. The most cautious estimates are that 2,000–3,000 gulls migrate north to the Hula valley, at least 1,000 remaining on the lake during the day. According to several observers, the gulls return from the Hula valley along the slopes of the Naftali ridge, flying low along the upper Jordan to the Kinneret. Before Lake Hula was drained in the 1950's, it also served as a winter hostel for hundreds of black-headed gulls and for some of the other species mentioned above. Great flocks of black-headed gulls could be seen in the afternoon hours flying south from the swamps.

The distances covered every day are up to 40–50 kilometers and may well be even longer.

Watching these daily migrations of gulls, especially of the returning flocks at sunset, is one of the typical winter experiences in the Kinneret region.

The number of species and number of individuals of winter Water Fowl increased in the Kinneret–Jordan Valley area after the drying of Lake Hula and the development of fishponds. The most common species are the Coot (*Fulica atra*), the tufted duck (*Aythya fuligula*), the crested grebe (*Podiceps cristatus*) and the grey heron (*Ardea cinerea*).

In winter, a few specimens of Osprey (*Pandion haliaetus*) are observed from time to time as well as the marsh harrier (*Circus aeruginosus*) and the winter eagles (*Aquila pomarina* and *A. clanga*). The white pelican (*Pelecanus onocrotalus*) and the flamingo (*Phoenicopterus ruber*) are winter visitors to the lake and fishponds.

Among the summer breeders it is worthwhile mentioning the moorhen (*Gallinula chloropus*), the little grebe (*Tachybaptus* (*Podiceps*) *ruficollis*), the purple heron (*Ardea purpurea*), the night heron (*Nycticorax nycticorax*), the squacco heron (*Ardeola ralloides*) and the cattle egret (*Bubulcus ibis*).

B. Typical birds of the lake surroundings

In 1860, Tristram mentioned the existence of the francolin (*Francolinus francolinus*) in the plain of Acco, the Valley of Jesreel, the Hula plain and the western bank of Lake Kinneret. The drainage of the swamps restricted the distribution of this 'black partridge' to the Kinneret area and the Bet Shean Valley. His loud, far reaching, high pitched call during the breeding season is a familiar sound in the Kinneret area and competes with the whip sound of the quail (*Coturnix coturnix*) which also breeds in the fields of the area and which served as an important food source to the Israelites during their wandering in the Sinai Desert (Exodus XVI, 12–13).

The rock partridge (*Alectoris graeca*) is a very common bird on the hills bordering the lake. It has been noted recently that its breeding territory penetrates into the flat areas which generally constitute the territory of the Francolin.

The sun-bird (*Cynniris* osea*), which was known in the Dead Sea area, colonized the gardens of Israel during the last forty years, including the gardens of the Jordan Valley where it is now very abundant. From the gardens the sun bird spread to the orchards.

The Dead Sea sparrow (*Passer moabiticus*) was discovered on the southern bank of the Dead Sea and described by Tristram in 1864. In the thirties, Mendelssohn (1955), studied this bird in the Allenby Bridge area. In the forties, the Dead Sea sparrow was observed in the Bet Shean Valley; it reached the Kinneret area in the late fifties and since then, nests in the valley. The large egg-shaped nests are always open to direct sun radiation which is partly responsible for the egg maturation.

The fish owl (*Ketupa ceylonensis Semenovi*) was reported in 1863 by Tristram in Western Galilee. Since 1969, this rare bird has been permanently recorded in the wadis of the eastern shores of Lake Kinneret. This is the westernmost area where this Indian bird is recorded. Moreover, there is a discontinuity in its global area of distribution and in this respect the Kinneret area is a 'Ketupa Island'.

* Synonym: Nectarinia osea.

Although in summer, the turtle dove (*Streptopelia turtur*) which is a summer breeder in our region, fills the air with its characteristic cooing just as in Biblical times (Cant. cant. II, 12), the collared dove (*Streptopelia decaocta*) is the characteristic species of the Jordan banks. It is a permanent resident.

The palm dove (*Streptopelia senegalensis*) was brought from Africa by the Moslems in the seventh century to the mosque-yards of this country and was called until recently the 'City dove'. This species reached the Kinneret area in the sixties; it is now a widespread permanent resident.

Until the sixties, the blackbird (*Turdus merula*) was only a winter visitor in the lake area. In the sixties, it appeared in the Yarmoukh Valley and nests now in the gardens of the kibbutzim. It is one of the best singers of the country. Its song competes with that of the native common bulbul (*Pycnonotus barbatus xanthopygos*), which is known as the 'Nightingale of Palestine'.

The white wagtail (*Motacilla alba*) has always been considered as the winter announcer in our country but recently a few couples have been observed nesting near the fishponds.

From the winter 1972–1973 we can observe, in the southern part of the lake the mass flight of hundreds of thousands of starlings (*Sturnus vulgaris*). In winter evenings, the starlings of the Lower Galilee, Kinneret region and Southern Golan, converge on the orchards of Hamat Gader in the Yarmoukh Valley. They spend the night there and disperse again the next morning; the number is estimated at several millions. In winter, during the late hours of the day, huge flocks of starlings cross our area eastwards, whereas the gulls, returning to the lake, move northwards above them.

During the last twenty years, swans (*Cygnus*) were observed at least twice in the Jordan Valley. On one occasion it was *Cygnus bewickii* (one of the northernmost breeders on earth) presumably on its way to Egypt.

Visitors came to our area also from the south, such as the marabou (*Leptoptylos cruminiferus*) which joined a flock of white storks (*Ciconia ciconia*).

The Jordan Valley in general is one of the highways of the massive white stork migrations from Eastern Europe to Africa. The relatively constant warm air rising from the Jordan Valley is utilized by these gliding birds in spring and autumn. A few birds, probably immature or too old, remain in the Jordan Valley in summer but do not breed.

The imperial eagle (*Aquila heliaca*) has been observed a few times. The large summer flocks of the Egyptian vultures (*Neophron percnopterus*) and the winter flocks of black kites (*Milvus migrans*) have disappeared. Only a few birds belonging to these species are observed at present. The buzzards (*Buteo* spp.) are still observed in winter and during migration. The short toed eagle (*Circaetus gallicus*) feeding on reptiles, is the only diurnal bird of prey which was not affected by the poisoning of rodents by the farmers. It is breeding in the area. The cliffs of Wadi Amud, located north-west of the lake, and the Golan Heights are the nesting area of the griffon vulture (*Gyps fulvus*).

Among the nocturnal birds of prey, the barn owl (*Tyto alba*) and the scops owl (*Otus scops*) are the most common. The eagle owl (*Bubo bubo*) is also observed on the hill slopes.

In 1948, Marcus (1969) reported a total number of 108 bird species in the

Table 51. Number of species of various types of birds observed in the Jordan Valley in 1948 and 1969. The 1969 figures are based on observations carried out during the period 1959–1969 for vagrant species and during the period 1964–1969 for other species. In brackets, percentage of increase

Types	Number of species	
	1948	1969
Resident	14	36 (257)
Summer resident (breeders)	7	30 (443)
Winter visitors	50	65 (130)
Migrants	27	35 (130)
Vagrants	10	27 (270)
Total	108	193 (179)

Kinneret area. The number of species observed in 1969 reached 193 (Table 51). Even if we take into account that the observations have been more frequent in the last ten years, it is undeniable that there is a general trend towards an increase in the number of species and in the number of specimens in the lake area. In spite of the presence of birds which are specific to the Kinneret Jordan Valley, there is a general tendency to uniformity of species with the Mediterranean region of the country.

References

Tristram, H. B. 1888. Passer moabiticus. In: The fauna and flora of Palestine. Published by the Committee of the Palestine Exploration Fund. London.
Mendelssohn, H. 1955. Biology and ethology of the Dead Sea sparrow. Sal'it I : 37–36 (in Hebrew).
Marcus, E. 1969. Quoted from an unpublished mimeographed report in 'Changes in avifauna of the Kinneret Valley' by S. Lulav, Teva Vaaretz 11(6):264–268, 1969.

V Mammals

S. Lulav

The mammal fauna of the Kinneret area includes elements from Palaearctic, Ethiopian, Oriental and Mediterranean origin. Ethiopian species advanced northwards in the Rift Valley whereas Palaearctic elements extended southwards in the mountainous areas bordering the depression. Thirty species are represented.

The Egyptian fruit bat (*Rousettus aegyptiacus*) invades gardens and plantations at night to feed on fruit bearing trees. The rat-tailed bat (*Rhinopoma hardwickei*) is mainly found in caves near Tiberias. Kuhl's pipistrelle bat (*Pipistrellus kuhli*) is abundant in settlements all through the year, but its activity decreases in winter.

The striped hyaena (*Hyaena striata syriaca*) is found mainly in the area East of the lake and more specifically in the Buteiha plain. It is however more common to observe its tracks and faeces than to meet the animal itself. The Egyptian mongoose (*Herpestes ichneumon*) is very common, even near human settlements (Plate 60). Since modern poultry-sheds are properly closed, they feed on snakes (including vipers) and rodents only. It is very common to see mongooses crossing the highways in the vicinity of the lake. A caracal lynx (*Felis caracal smitzi*) was caught on the Plateau of Sirin, some 10 km SW of the lake. Its main habitat in Israel is the Arava

Plate 60. The Egyptian mongoose
(*Herpestes ichneumon*).

Name	Common name	Observations
A. Insectivora:		
Erinaceus europaeus concolor	Short-eared hedgehog	
Crocidura russula judaica	White-toothed shrew	
Crocidura suaveolens portali	Lesser white-toothed shrew	
Suncus etruscus	Pygmy shrew	
B. Chiroptera:		
Roussetus aegyptiacus	Egyptian fruit-bat	
Rhinopoma hardwickei	Lesser rat-tailed bat	
Pipistrellus p. kuhlii	Kuhl's pipistrelle	
C. Carnivora:		
Hyaena striata syriaca	Striped hyaena	
Herpestes ichneumon	Egyptian mongoose	
Felis lybica (ocreata)	African wild-cat	
Felis caracal schmitzi	Caracal lynx	
Canis aureus syriacus	Golden jackal	
Vulpes vulpes palaestina	Red fox	
Vormela peregusna syriaca	Marbled polecat	
Meles meles canescens	Badger	
Lutra lutra seistanica	Otter	
D. Hyracoidea:		
Procavia capensis syriaca	Hyrax	('Coney of the Bible')
E. Artiodactyla:		
Sus scrofa libycus	Wild boar	
Gazella gazella	Mountain gazella	
F. Lagomorpha:		
Lepus europaeus syriacus	European hare	
G. Rodentia:		
Hystrix indica	Indian crested porcupine	
Spalax ehrenbergi	Palestine mole-rat	
Rattus rattus alexandrinus	House rat	
Rattus rattus frugivorus	(Tree) rat	
Mus musculus praetextus	House mouse	
Acomys cahirinus (dimitiatus)	Cairo spiny mouse	
Meriones tristrami	Tristram's jird	
Microtus guentheri (M. socialis guentheri)	Gunther's vole	
Myocastor coypu	Nutria (swamp beaver)	
Arvicola terrestris		

Valley, south of the Dead Sea. The jackal (*canis aureus syriacus*) was once very abundant. Its habitat extended to the cultivated areas, plantations and vineyards. It was nearly exterminated in the sixties by intentional poisoning. Its population is now gradually recovering. It is rather frequent in the area east of the Lake Kinneret. The jackal is the 'fox' of Samson who caught three hundred of them and sent them to the fields of the Philistines with their tails set on fire (Judicum XV, 4–5). The fox (*Vulpes*

vulpes palaestina) is found mainly on the border of the valley. The marbled polecat (*Vormela peregusna syriaca*) is a small and alert carnivore dwelling on the hill slopes facing the lake. It is also found in the Valley, in areas of dense vegetation and sometimes near or inside the settlements. The badger (*Meles meles canescens*) is a carnivore which does not despise vegetables and other vegetarian food. The otter (*Lutra lutra seistanica*) was very common until the fifties all over the country from the north to Nahal Soreq, in the Rehovot area. This animal needs running water and it was specially common in the streams flowing into Lake Kinneret. The hunting for fur decimated the otter population in this area. In the coastal plain, water pollution was the main cause of its decline. The otter found shelter in fishponds where its diet is mainly composed of fish but also includes snakes, crabs and water-fowls. However, the otter causes relatively minor damage since it requires a rather wide territory (1 km² per animal). This characteristic has generally been ignored and the otter has been driven out of this last refuge (Mendelssohn, personal communication). In the Kinneret area, it is found near the water but many specimens are killed on the nearby highways. (plate 61). The hyrax (*Procavia capensis syriaca*) has multiplied in the last years. It inhabits the caves and cliffs of Wadi Amud and Wadi Arbel, west of the lake. This is the 'coney' of the Bible (Leviticus XI, 5; Psalmi CIV, 18; Proverbia XXX, 26). The wild boar (*Sus scrofa libycus*) left the Hula Valley after the reclamation of the swamps and emigrated to the Western Galilee and the Kinneret area. They are found in the Buteiha plain and the Yarmouk Valley. They cause damage in the banana plantations of this area. In the area east of Lake Kinneret, the gazelles (*Gazella gazella*) belong to the scenery; they are also common in plantations in the Jordan Valley south of Lake Kinneret and on the hill slopes up to Yavneel. The hares (*Lepus europaeus syriacus*) are abundant in the fields of the plateaux bordering the Kinneret area.

Plate 61. Lutra lutra seistanica This specimen was found dead on the highway near the lake
(photo M. Gophen).

Before the reclamation of the Hula (Mendelssohn, personal communication), *Rattus rattus* used to nest on the water among the stems of *Phragmites* and *Papyrus* and prey on water fowl eggs. The specimens living on the water were more blackish than the terrestrial ones.

The water vole (*Arvicola terrestris*) is a mysterious animal which has never been observed in our region (Mendelssohn, personal communication). However, skulls and other bones of this rodent were found in the regurgitated pellets of *Tyto alba* in the Hula area. A nest of this animal can be observed in the museum of the A. D. Gordon Agriculture and Nature Study Institute at Deganya, located on the southern shore of the lake.

The nutria or swamp beaver (*Myocastor coypus*) was brought from Argentina in the fifties for the fur industry. The experiment did not succeed because, as a result of the warm climate, the fur was not of high quality. The animals escaped to the natural environment and settled around the lake and the fish ponds where they are known and appreciated as effective weed-cleaners. They may also cause damage to vegetables. As their number increased, they became 'partners' for the utilization of food delivered to the fish ponds.

VI Man – an outline of the Prehistory of the Kinneret area

O. Bar-Yosef

1. History of Research

The interest of archaeologists and geologists in the Jordan Valley during the nineteenth and the early decades of the twentieth century was stimulated by the discovery of Biblical remains. Several surveys such as those by Karge, Zumoffen and others demonstrated the presence of prehistorical stone tools and megalithic structures. It is therefore not surprising that F. Turville Petre in 1925–1926 chose the caves of Wadi Amud as the main site of his pioneer work in the area. These excavations established the existence of Stone Age assemblages in this region which became famous with the discovery of a fragmentary human skull in the Zuttiyeh cave (Turville Petre, 1927).

Partial surveys mainly south of the lake and along the Jordan River were conducted first by L. Picard and later by M. Stekelis & P. Solomonica (Picard, 1932). The intensive survey of the triangular area which lies between Lake Kinneret, the Jordan River and the Yarmuk River by Stekelis, Meisler & Yeivin led to the discovery, among others, of the Neolithic site of Sha'ar Hagolan. However, this site located on the right bank of the Yarmuk River was partially excavated only in 1949–1951 (Stekelis, 1973). Other Neolithic sites have been excavated by M. W. Prausnitz at Tell Eli (1955–1959) and by J. Perrot at Munhata from 1963–1967 (Prausnitz, 1970; Perrot, 1966, 1967).

A major discovery was made in 1959, on the western flanks of the Jordan Valley, when surface finds of bones and stone tools were found in the proximity of the Tell 'Ubeidiya. The suggested Lower Pleistocene age of these remains stimulated a long term project of excavations together with geological and faunistic studies (Stekelis et al., 1960; Stekelis, 1966; Picard & Baida, 1966a, 1966b; Haas, 1966, 1968; Tchernov, 1968, 1973, 1975; Stekelis et al., 1969; Bar-Yosef & Tchernov, 1972; Bar-Yosef, 1975a, 1975b).

During the same period two expeditions returned to Wadi Amud. One, under the direction of H. Suzuki and H. Watanabe, worked in 1961–1962 in the cave of Amud and the other, headed by S. Binford dug at the cave of Shovakh (Suzuki & Takai, 1970; Binford, 1966). In 1973, the caves of Zuttiyeh and Emireh were examined again by I. Gisis and the author and a brief report that clarifies the stratigraphy of Zuttiyeh is already available (Gisis & Bar-Yosef, 1974).

Besides the controversial assemblage from the Emireh cave, attributed to an early phase of the Upper Palaeolithic, a Late Aurignacian site containing a human grave was excavated in Nahal Ein Gev (Bar-Yosef, 1973; Arensburg, 1977). In this area near Ein Gev, several in situ Epi-Palaeolithic sites were also encountered. Four sites, attributed to the Kebaran and Geometric Kebaran A were excavated in 1963–1964, 1968, 1973–1977 (Stekelis & Bar-Yosef, 1965; Bar-Yosef, 1970,

1975c; Arensburg & Bar-Yosef, 1973; Davis, 1974; Martin & Bar-Yosef, in press).

During the past fifteen years, further surveys were carried out in the Kinneret basin but none can be considered as an intensive survey. The most extensive ones were made by D. Ben-Ami on the east bank of the lake and the late B. Rabani on the west side.

The map of Fig. 125 is mainly based on the record of excavations and partially on the known surveys. The following is a brief resumé of the prehistoric sequence as can be established from the published reports.

2. Lower Palaeolithic

The site of 'Ubeidiya, located 3 km south of Lake Kinneret is one of the oldest known sites in the Near East. The fourteen archaeological horizons uncovered were found stratigraphically within the folded and faulted 'Ubeidiya formation (Picard, 1965; Picard & Baida, 1966a, 1966b). According to its sediment facies the 'Ubeidiya formation has been divided into four members, representing transgressions or regressions of a fresh water lake (Fig. 126).

Most of the Palaeolithic stone tools were found within the Fi member; they were generally incorporated in beach deposits and, more rarely, sunken in silty-clayey layers, or mixed with gravel accumulations. These lithic assemblages are characterized by the abundance of core-choppers and flakes along with spheroids and hand-axes. Four types of assemblages were recognized in the following order from the lower to the upper archaeological horizons (Stekelis, 1966; Stekelis et al., 1969; Bar-Yosef & Tchernov, 1972; Bar-Yosef, 1975b): (a) Core-choppers with spheroids. (b) Core-choppers, spheroids and a low percentage of hand-axes. (c) Abundant handaxes with core-choppers and no spheroids. (d) Numerous core-choppers and rare hand-axes (Fig. 127).

Except for the third variety which was found in a wadi infilling in rolled condition and was therefore transported from the hills, all the other three varieties were made and embedded either on the beaches of the Lake of Ubeidiya or in its immediate proximity (Plates 62 and 63). Since the material necessary for the manufacture of tools (flint for core-choppers, limestone for spheroids and basalt for most of the hand-axes) was available in wadi gravels or pebbly beaches, it is likely that these quantitative typological differences are of cultural origin and not of a functional nature.

The age of the site of 'Ubeidiya is determined mainly on the basis of its faunal assemblages which are generally related to the Cromerian and Biharian faunas (Haas, 1968; Maglio, 1975) and by the preliminary palaeomagnetic checks that showed a transition from reversal to normal from the Li to the Lu member hinting perhaps to the Bruhnes–Matuyama borderline. This suggested date is supported by a K/Ar dating of a lava flow in the upper part of the Mishmar Hayarden formation (contemporary with the 'Ubeidiya formation on the basis of malacological evidence) which yielded an age of 640,000 ± 120,000 years (Horowitz et al., 1973; Siedner & Horowitz, 1974; Tchernov, 1973).

The tectonic lowering of the valley floor put an end to the deposition of the 'Ubeidiya sediments in the Kinneret basin. A period of wadi aggradation followed, and large alluvial fans were deposited in that area. Little is known about

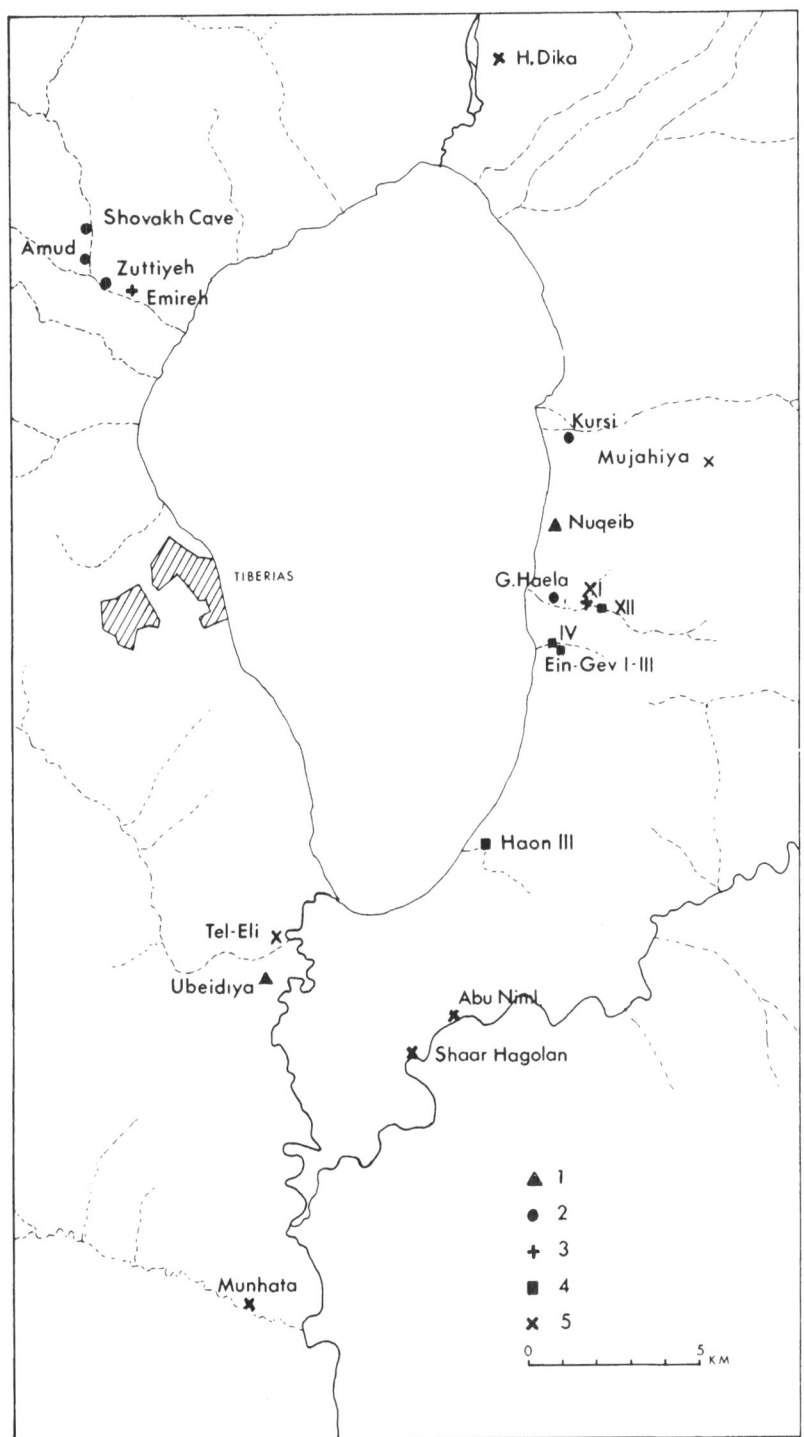

Fig. 125. Map of the Kinneret basin showing the location of major sites. Legend: 1. Lower Palaeolithic. 2. Middle Palaeolithic. 3. Upper Palaeolithic. 4. Epi-Palaeolithic. 5. Neolithic.

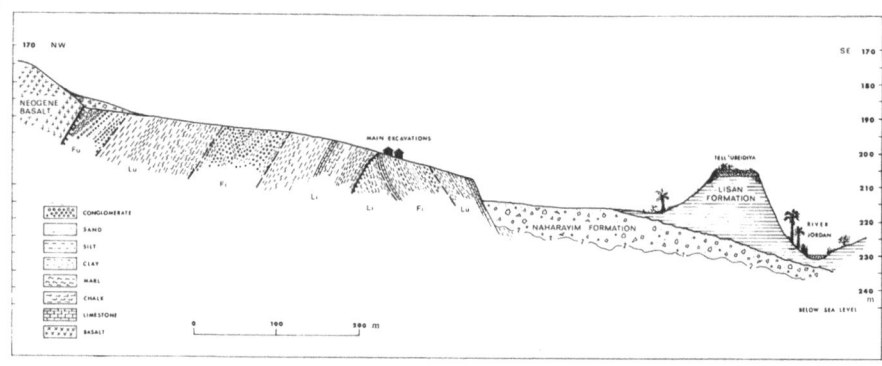

Fig. 126. A cross-section at 'Ubeidiya showing the 'Ubeidiya Formation, its four members and the overlying formations (from Bar-Yosef and Tchernov, 1972).

Acheulian hunters and food-gatherers who roamed around, as most of the area is presently covered by the Upper Pleistocene Lisan formation. Besides scattered hand-axes of Upper Acheulian types (ovaloids, cordiforms, etc.) found on the basaltic plateaux of the Eastern Galilee and the Golan, a derived Acheulian assemblage was collected by the author from a wadi infill near Nukeib, north of Ein Gev. This wadi infill, today within inversed topography, contained, beside flint hand-axes and flakes, some cleavers similar to those of Gesher Benot Ya'acov. Typologically the small assemblage of Nukeib, still unpublished in details, seems to be younger than the famous assemblage of Gesher Benot Ya'acov V which is made mainly of basalt (Stekelis, 1960; Gilead, 1970).

Plate 62. 'Ubeidiya, the trench of layers I–26b and I–26c. in the background, trenches I–15 and II–23—II–24 are visible.

450

Plate 63. 'Ubeidiya, trench I–26 showing the stratigraphy along the strike of layers I–26b and I–26c.

4. Middle Palaeolithic

During the Upper Pleistocene, the Kinneret basin was partially covered by the Lisan Lake and its northern part was an alluvial plain (Horowitz, 1971). Middle Palaeolithic sites are therefore known only from the margins of the area or as derived artefacts in conglomerates incorporated within the Lisan marly deposits near Haon and near Gesher (Bar-Yosef, 1975d). The major sites representing this period are therefore located in the wadis that descend to the Jordan Valley either from the Golan Heights or the Eastern Galilee. A few sites are presently known from the surroundings of Ein Gev and are considered to be mainly open-air workshops; their lithic assemblages are dominated by the Levallois technique. As many sites of later age were found to be buried in redeposited Ein Gev sands (a Miocene formation, see Michelson, 1972), Middle Palaeolithic habitations should come to light in the future.

The principal source of information remains the caves of Wadi Amud, three of which provided a large body of data (Fig. 128). The early assemblages excavated by Turville Petre at Zuttiyeh cave, are now called Acheulo-Yabrudian or Acheulian of Yabrudian facies and Yabrudian (Gisis, 1976). The lower levels at Zuttiyeh are rich in Yabrudian scrapers (these resemble Quina-type racloirs) and are accompanied by a few hand-axes (Fig. 129). It is difficult to ascertain the character of the industry in every level as the samples of the 1973 controlled excavations are too small and the original collections were mixed during the dig. Several Micoquian hand-axes perhaps hint at the existence of a Late Acheulian industry.

The upper levels at Zuttiyeh contained Mousterian of Levallois facies quite

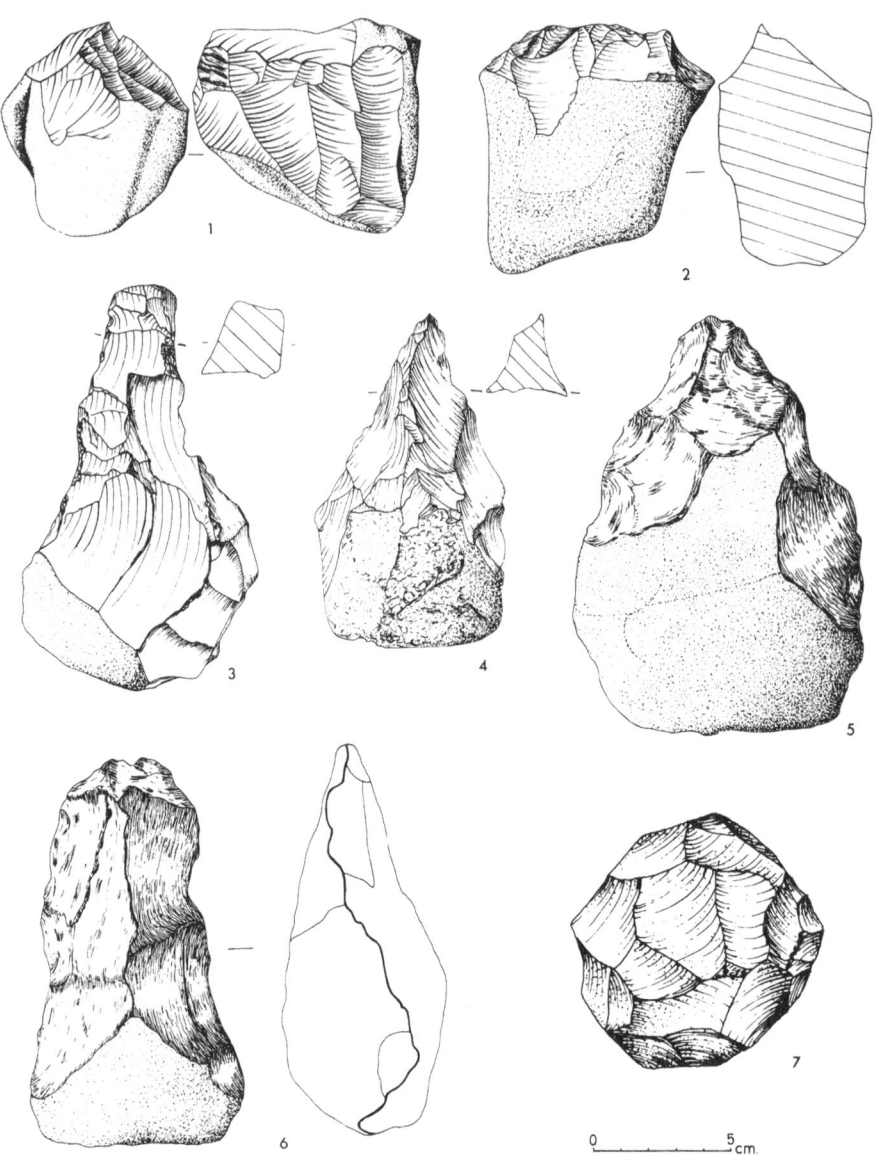

Fig. 127. Stone-tools from 'Ubeidiya: 1–2 Core-choppers made of flint; 3–4 Quadrihedral and trihedral made of flint; 5–6 Handaxes made of basalt; 6 Limestone spheroid (after Stekelis, 1966).

similar to those exposed in Amud cave and Shovakh cave but perhaps with more Lower Mousterian affinities (i.e. such as Tabun D).

The hominid remains known as the 'Galilee Man' were found beneath the lower most levels. Therefore they should be ascribed to the same time stratigraphic unit as Tabun E, i.e. the Last Interpluvial period (Farrand & Goldberg in Jelinek et al., 1973). This fragmentary skull, assigned to the local Neanderthals group, has some

452

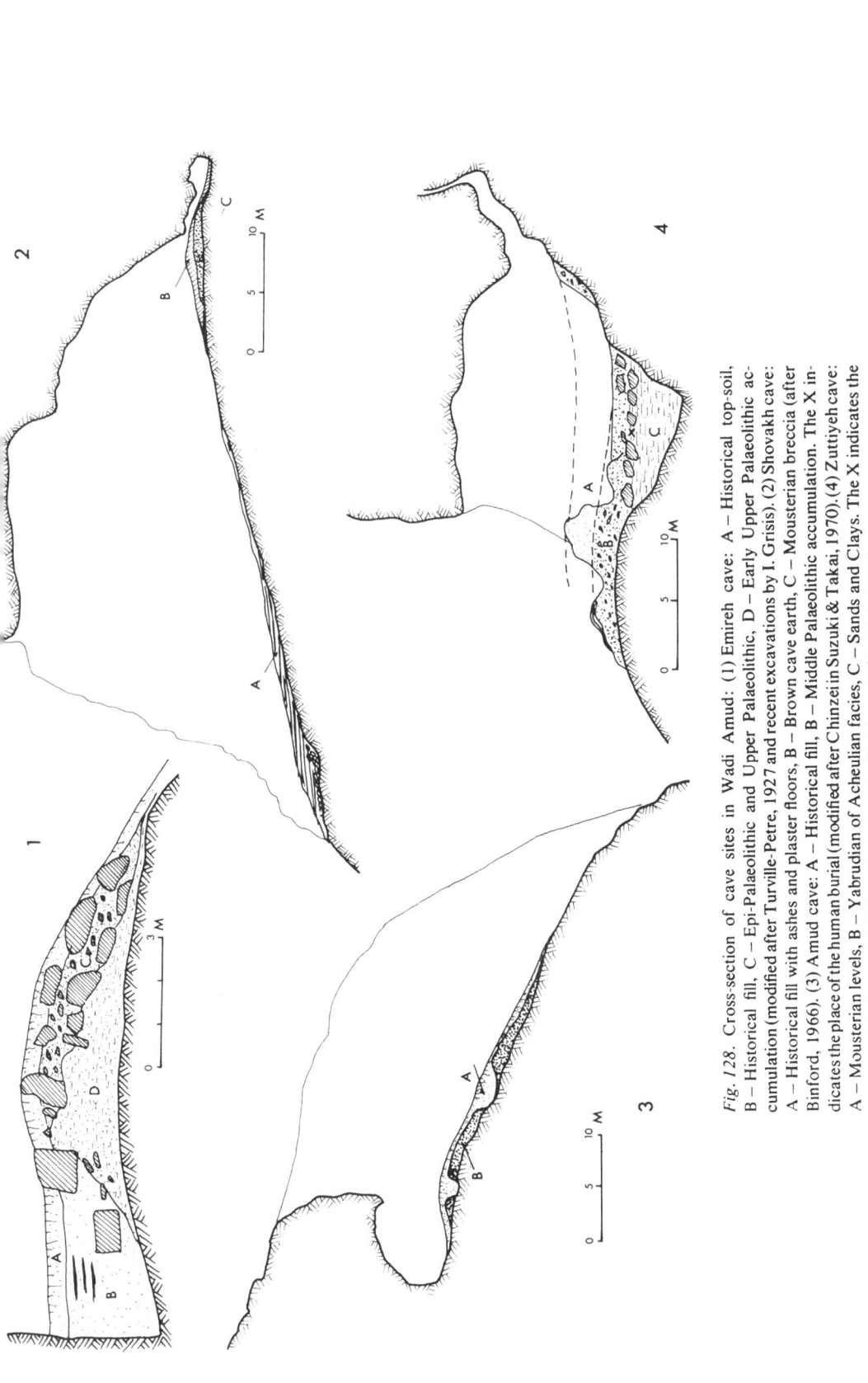

Fig. 128. Cross-section of cave sites in Wadi Amud: (1) Emireh cave: A – Historical top-soil, B – Historical fill, C – Epi-Palaeolithic and Upper Palaeolithic, D – Early Upper Palaeolithic ac-cumulation (modified after Turville-Petre, 1927 and recent excavations by I. Grisis). (2) Shovakh cave: A – Historical fill with ashes and plaster floors, B – Brown cave earth, C – Mousterian breccia (after Binford, 1966). (3) Amud cave: A – Historical fill, B – Middle Palaeolithic accumulation. The X in-dicates the place of the human burial (modified after Chinzei in Suzuki & Takai, 1970). (4) Zuttiyeh cave: A – Mousterian levels, B – Yabrudian of Acheulian facies, C – Sands and Clays. The X indicates the location of the human skull (modified after Turville-Petre, 1927; Gisis, 1976).

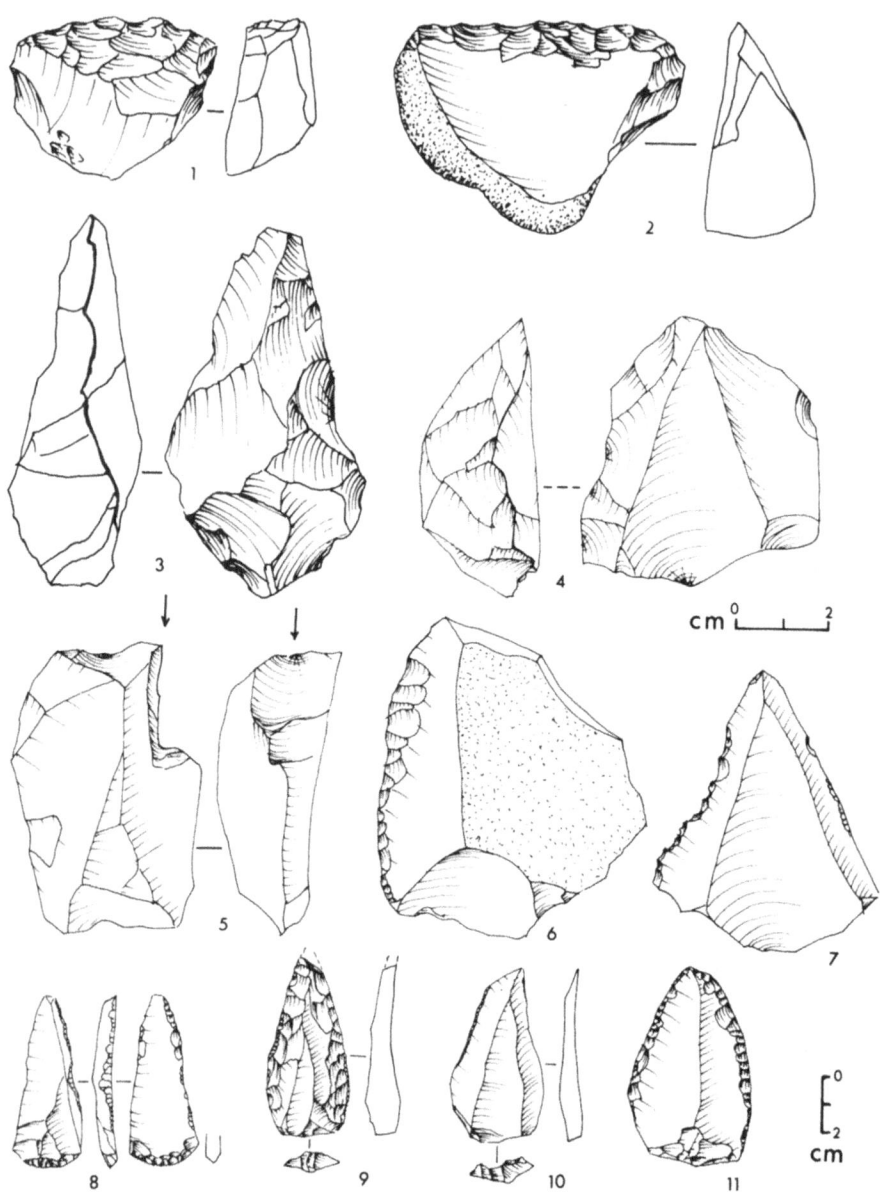

Fig. 129. Stone tools from Zuttiyeh (nos. 1–3), Amud (nos. 4–7) and Shovakh (nos. 8–11) caves (after Gisis & Bar-Yosef, 1974; Watanabe, in Suzuki & Takai, 1970; Binford, 1966). Note the different scale of the fourth row. 1–2 Transverse racloirs; 3 Handaxe; 4 Levallois core; 5 Burin; 6 Convex racloir; 7 Denticulate Levallois point; 8 Emireh point; 9 Mousterian point; 10 Retouched Levallois point; 11 Convergent scraper.

454

features which place it between *Homo erectus* and *Homo sapiens neanderthalensis*.

Shovakh cave provided us with an undisturbed Mousterian layer only from the back, dark portion of the cave (Binford, 1966). The rest of the cave was used for temporary habitation during historical times, but the lithic material confirmed the general aspect of the industry as typical Mousterian of Levallois facies (Fig. 129).

A similar industry has been unearthed by the Tokyo Expedition in Amud cave, along with a complete human grave and the fragments of three other hominids (Suzuki & Takai, 1970). The preliminary reports claimed for a Transitional industry (i.e., an early Upper Palaeolithic assemblage) but the drawings and the descriptions do not support it (Watanabe, 1964, 1968 and in Suzuki & Takai, 1970) (Fig. 129). The fully detailed report of the lithics is urgently needed.

The faunal remains uncovered in these sites demonstrate that the most frequently hunted animals were the fallow deer and gazelle. However slight differences exist between the caves: in Zuttiyeh some animal species indicate a wetter climate, while Amud cave shows evidence of a drier period. Worth mentioning is the large number of carnivore remains found in the Zuttiyeh cave. It is impossible to ascertain whether the three cave sites served as base camps or only as seasonal habitations as suggested by S. Binford (1968).

5. The Upper Palaeolithic and Epi-Palaeolithic

The cultural transition from the Middle to the Upper Palaeolithic in Israel is still unclear although new data have been obtained from the Negev and Mount Carmel. However, the large assemblage collected from the main archaeological level at Emireh cave occupies a special place and was described by Garrod as a Transitional industry but its exact position is still controversial (Plate 64). In 1925 Turville Petre uncovered there what seemed to be a relatively clear stratigraphy with a Palaeolithic layer at the bottom covered by a layer of fallen blocks and historical layers. The lithic assemblage from this layer had the characteristics of both Mousterian and Upper Palaeolithic tool-kits containing points, racloirs, endscrapers, etc. Garrod, who first believed it to be a mixed industry later rehabilitated it (Garrod, 1951, 1955). A special point, with a bifacial flaked proximal end, was found in this assemblage and has since been named as 'Emireh point' (Fig. 129). This point was considered to be a 'guide fossil' of the Emiran industry or the Upper Palaeolithic Phase I (Neuville, 1951; Garrod, 1955). Its chronological importance is unfortunately questioned as sporadic Emireh points were found in Mousterian assemblages such as Shovakh cave (Binford, 1966) or Kebarah cave (Schick & Stekelis, 1977).

Even if the controversial Emiran assemblage belongs to the beginnings of the Upper Palaeolithic sequence, most of its time span is still not represented in the Kinneret basin.

The end of the Aurignacian tradition (of estimated age *ca.* 20–18000 B.C.) is known from the site Ein Gev XI, previously named Nahal Ein Gev I (Bar-Yosef, 1973). This open-air site is located on the right bank of Nahal Ein Gev and although a large portion has been destroyed, a salvage excavation succeeded in revealing a Palaeolithic level that contained numerous burins, denticulates, a few nosed scrapers,

Plate 64. Emireh Cave.

Plate 65. Ein Gev XI, the woman's burial as exposed in the excavations.

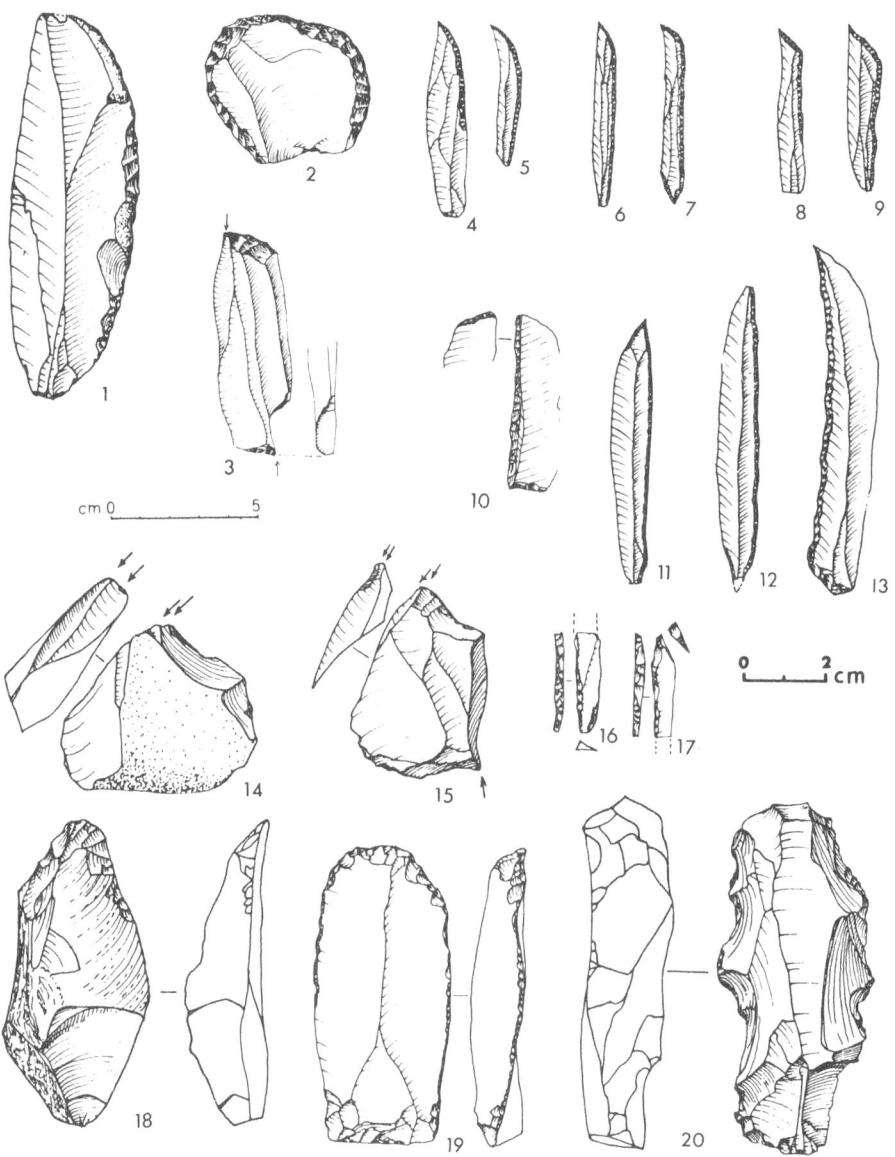

Fig. 130. Stone tools from Ein Gev I (nos. 1–13) and Ein Gev XI (after Stekelis & Bar-Yosef, 1965; Bar-Yosef, 1972). (1) Backed blade; (2) Rounded scraper; (3) Double angle burin on truncations; (4–9) Microliths, mainly obliquely truncated; (10–13) Falita points (note the different scale); (14–15) Burins on concave truncations; (16–17) Backed bladelets; (18) Nosed scraper; (19) Endscraper; (20) Denticulate.

backed blades and bladelets (Fig. 130). Under the archaeological horizon a grave of a 30–35 year old woman in a flexed position was uncovered (Plate 65). It has Proto-Mediterranean features basically similar to those of the Kebaran woman, known from Ein Gev I and the Natufian population of Israel (Arensburg, 1973, 1977).

457

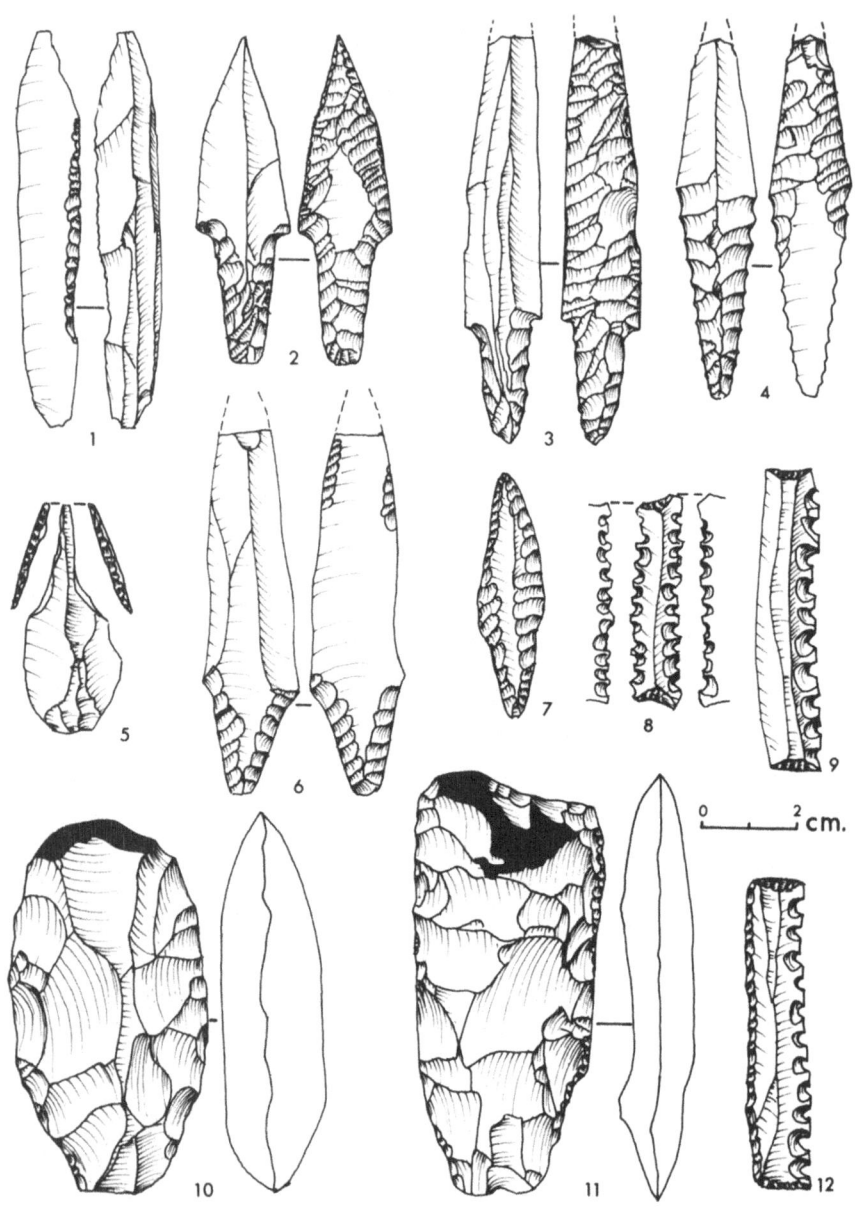

Fig. 131. Neolithic stone tools from Munhata (nos. 1–4) and Sha'ar Hagolan (after Perrot, 1966 and Stekelis, 1973): (1) Sickle knife; (2, 4) Arrowheads; (3) Spear-head; (5) Awl; (6) Spear-head; (7) Arrow-head; (8–9, 12) Denticulated sickle-blades; (10–11) Axes (the black areas are polished).

A flint industry similar to the one of Ein Gev XI was encountered in the lower levels of Ein Gev III. This site is one of a cluster of sites situated on a sandy hill south of Susita. The four partially excavated Epi-Palaeolithic sites are located in a decreasing topographic order from −155 m to −180 m below sea level, and are of Kebaran and Geometric Kebaran sequence (Stekelis & Bar-Yosef, 1965;

Bar-Yosef, 1970; Arensburg & Bar-Yosef, 1973; Davis, 1974; Martin & Bar-Yosef, in press).

Ein Gev I was the main target of the project and contained the remains of a Kebaran open-air habitation dug into the sandy slope. It had an oval form 5–7 m in diameter, and about 0.30–0.50 m deep. It had been successively inhabited and deserted and as a result the sandy-silty accumulated deposits attain a thickness of *ca.* 1.20 m. Several layers of occupation were observed and the semi-flexed burial of a woman 30–35 years old was unearthed under one of these levels (Plate 66). Burned aminal bones collected near her skull were dated to 13,750 ± 415 yrs. B.C. (GrN −5,576). The lithic industry of this site and of Ein Gev II is classified as Kebaran, typified by a high content of backed and retouched bladelets (dominated by the obliquely truncated ones, see Fig. 6). Scrapers, burins and backed blades were part of the tool-kit but worth-noting are the Falita points, defined first at Yabrud Rock-shelter III (Rust, 1950) and which have only been found in Syrian or Trans-jordanian sites (Bar-Yosef, 1970, 1975c).

Plate 66. Ein Gev I, the women's burial during the excavations.

The industry of Ein Gev III seems to be slightly younger as demonstrated by the appearance of small quantities of trapeze-rectangles. The site of Ein Gev IV tested only in a very small trench, has special lithic features. The microburin technique was used to shape triangles and lunates, and the microgravette points seem to be a miniature modified version of the Falita point.

To this Geometric Kebaran complex another site, near Haon should be added. Again a salvage excavation uncovered a portion of a habitation, close to the

Plate 67. Ein Gev I, mortar and pestle found within the hut remains.

shores of the lake, at an elevation of about −195 m below sea level. The lithics are dominated by a high blade content, some modified into scrapers and burins, along with backed bladelets, mainly trapeze-rectangles. Similar Geometric Kebaran A remains, but with a much more microlithic blade technology and trapeze-rectangles, were identified in Emireh cave (I. Gisis, personal communication).

In every Kebaran or Geometric Kebaran site mortars or bowls and pestles were found, made of basalt or limestone (Plate 67). These pounding tools indicate the consumption of vegetal food, although flotation used in some of the excavations failed to retrieve carbonized seed or plant remains.

The meat diet was based on hunting gazelle and deer along with some wild goats, wild oxen, hare and tortoise. In a few cases wild horses and wild boar were found, but except for Ein Gev I all bone samples should be considered as too small (Davis, 1972, 1974). Evidence for fishing in Lake Kinneret has only been found at Ein Gev IV (a site that might be dated *ca.* 12,000–10,000 B.C.). The

460

question whether these sites were occupied during winter, summer or all year round is yet unsolved, however, rare swan and goose-bones in Ein Gev I suggest at least a winter occupancy from the presence of these migratory birds.

Dentalium shells as well as *Connus* and *Columbella* originating in the Mediterranean coast point out to some sort of exchange between the people around Lake Kinneret and those living seasonally near the sea coast during the Epi-Palaeolithic period.

The Epi-Palaeolithic sequence is terminated with Ein Gev XII, a Late Natufian site, situated on the right bank of Nahal Ein Gev, north of Susita. It is a large site 2–3,000 m² in area and its industry contains large quantities of sickle blades, borers, some lunates, fishing weights and imported *Dentalium* shells both from the Mediterranean Sea and the Red Sea (Nir & Bar-Yosef, 1976).

6. The Neolithic

The number of Neolithic sites in the proximity of the Kinneret Lake is very small, although the hilly flanks around are rich in wild wheat stands (Harlan & Zohary, 1966). The main Neolithic settlements known and studied, either through small or large scale excavation, are situated near wadi courses and are buried within the alluvial fans of these wadis. This is the case of Sha'ar Hagolan, Tell Eli and Munhata (Stekelis, 1973; Prausnitz, 1970; Perrot, 1967). While the archaeological levels at Sha'ar Hagolan are covered by layers of silts, sands and gravels deposited by the Yarmuk River, the Neolithic and Chalcolithic levels of Tell Eli and Munhata are part of the natural accumulation of clays caused by continued alluviation or by shallow muddy lake deposits.

The cultural sequence starts with Pre-Pottery Neolithic levels at both Munhata and Tel Eli. The rectangular buildings, as traced along their stony foundations, are the common architectural features, but the rounded courtyard house that maintains the local tradition, was also uncovered in Munhata.

The lithic assemblages characterized by tanged and winged arrow-heads, axes (commonly sharpened by transversal blows) and sickle-blades, mainly dominated by the straight elongated sickle knife hafted probably only at the proximal end.

Obsidian trade from Anatolia is evidenced in the few blades found from this period, and this may also be the route through which Einkorn wheat and sheep reached the Central Jordan Valley. Domesticated goat seems to be present but hunted animals remained an important source of the meat diet.

The period of the Pre-Pottery B terminated in the desertion of Munhata and Tel Eli, and of many other sites in the country. The Pottery Neolithic period, in the Jordan Valley, started according to several authorities after a long gap (Perrot, 1968; Moore, 1973). This hiatus is said to have been caused by the onset of the Atlantic period, a time stratigraphic unit well defined on the basis of palynological evidence in North Western Europe (Butzer, 1971). The local record from the Hula Valley and the Kinneret Basin indicate that it was relatively a warmer and wetter period (Horowitz, 1971, 1973, 1974). It therefore seems bizarre that a gap of 1,500 years (approximately from 6000 to 4500 B.C.) is assumed to exist in these Neolithic sites. It is true that the nature of the Pottery Neolithic occupations differ from previous levels. These are pit-dwellings rarely accompanied by additional structures or burials. Two phases have been recognized the 'Sha'ar Hagolan' phase (named also the 'Yarmukian') and the 'Munhata phase' (Perrot, 1968). The later Wadi Rabbah phase demonstrates generally regular architectural remains, although not in every site.

There are many features common to the Sha'ar Hagolan and the Munhata phases, both with regard to the pottery and the lithic assemblages as well as in the female clay figurines and schematic human engravings on pebbles (Stekelis, 1973; Perrot, 1966, 1967). Long distance comparisons with Byblos (mainly with the 'Néolithique Moyen' of Cauvin, 1968) led archaeologists to conclude that the Yarmukian should be dated to the middle of the fifth millenium B.C. However, closer detailed comparisons indicate that the Yarmukian might be earlier and even earlier than Byblos 'Néolithique Ancien'. It seems that a date in the middle of the sixth millenium B.C. may be more appropriate, thus decreasing the visible stratigraphic gap between the PPNB and the PN to a few centuries.

Apparently the reasons for the special nature of the PN settlements should be sought in their socio-economic structure; details on daily activities such as herding, farming etc. might be obtained through careful collection of bone and plant remains during excavations. The available information from bones (Ducos, 1968, 1969) indicates husbandry while farming can be inferred from the flint and ground stone tools.

At the close of the fifth millenium and early fourth millenium B.C. Chalcolithic settlements spread over all parts of the Kinneret basin, but the description of their site size, site location, material culture etc., is beyond the scope of this contribution.

On the basis of our present knowledge from other regions of Israel, it is clear that the Stone Age sequence in the Kinneret basin remains incomplete. But contrary to the arid zones, during the Quaternary this area remained within the Mediterranean climatic zone and in spite of both climatic fluctuations and geomorphic changes it could have been inhabited during every time period. It seems that the completion of the cultural sequence as well as the inter and intra-site variability is a matter of intensive surveys and excavations in this part of the Jordan Rift Valley.

References

Arensburg, B. 1973. The People in the Land of Israel from the Epi-Palaeolithic to Present times. Unpublished Ph.D. thesis, Tel Aviv University.

Arensburg, B. 1977. New Upper Palaeolithic Human Remains from Israel. Eretz-Israel 13:208–215.

Arensburg, B. & O. Bar-Yosef. 1973. Human Remains from Ein Gev I, Jordan Valley, Israel. Paleorient 1:201–206.

Bar-Yosef, O. 1970. The Epi-Palaeolithic Cultures of Palestine. Unpublished Ph.D. thesis, Hebrew University, Jerusalem.

Bar-Yosef, O. 1973. Nahal Ein Gev I, Mitequfat Haeven 11:1–7 (in Hebrew).

Bar-Yosef, O. 1975a. 'Ubeidiya-Early Man Site in the Jordan Valley. Archaeology 28:30–37.

Bar-Yosef, O. 1975b. Archaeological Occurrences in the Middle Pleistocene of Israel, in: Butzer, K. W. & G. L. Isaac (eds.) After the Australopithecines, Mouton, The Hague. pp. 571–604.

Bar-Yosef, O. 1975c. The Epi-Palaeolithic in Palestine and Sinai. In: Wendorf, F. & A. E. Marks (eds.) Problems in Prehistory: Northwest Africa and the Levant, SMU Press, Dallas, pp. 363–378.

Bar-Yosef, O. 1975d. Les gisements 'Kebarien Géometrique A' d'Haon, Vallé du Jourdain, Israël.

Bull. Soc. Préhist. Franc. C.R.S.M. 72:10–14.

Bar-Yosef, O. & E. Tchernov. 1972. On the Palaeoecological History of the Site of 'Ubeidiya. Publications of the Israel Academy of Sciences and Humanities, Jerusalem.

Binford, S. R. 1966. Me'arat Shovakh (Mugharet esh-Shubbabiq). I.E.J. 16:18–32, 96–103.

Binford, S. R. 1968. Early Pleistocene Adaptations in the Levant. American Anthropologist 70:707–717.

Butzer, K. W. 1971. Environment and Archaeology. 2nd edition, Aldine-Atherton, Chicago.

Cauvin, J. 1968. Les Outillages Néolithiques du Byblos et du Littoral Libanaïs. Fouilles de Byblos IV. Librairie d'Amerique et d'Orient, Paris.

Davis, S. 1972. Faunal Remains of Upper Palaeolithic Sites at Ein Gev (Israel). Unpublished M.Sc. thesis, Hebrew University, Jerusalem.

Davis, S. 1974. Animal Remains from the Kebaran Site of Ein Gev I, Jordan Valley, Israel. Palaeorient 2:453–462.

Ducos, P. 1968. L'origine des Animaux Domestiques en Palestine, Delmas, Bordeaux.

Ducos, P. 1969. Methodology and Results of the study of the Earliest Domesticated Animals in the Near East (Palestine). In: Ucko, P. J. & G. W. Dimbleby (eds.). The Domestication and Exploitation of Plants and Animals, Duckworth, London, pp. 247–264.

Garrod, D. A. E. 1951. A Transitional Industry from the Base of the Upper Palaeolithic in Palestine and Syria. Journal Roy. Anthrop. Inst. 81:121–129.

Garrod, D. A. E. 1955. The Mugharet el Emireh in Lower Galilee; Type Station of the Emiran Industry. Journ. Roy. Anthrop. Inst. 85:143–161.

Gilead, D. 1970. Early Palaeolithic Cultures in Israel and the Near East. Unpublished Ph.D. thesis, Hebrew University, Jerusalem.

Gisis, I. & O. Bar-Yosef. 1974. New Excavations in Zuttiyeh Cave, Wadi Amud, Israel. Paleorient 2:175–180.

Gisis, I. 1976. Zuttiyeh Cave and its Place in the Middle Palaeolithic of the Levant. Unpublished M.A. thesis, Hebrew University, Jerusalem (in Hebrew).

Haas, G. 1966. On the Vertebrate Fauna of the Lower Pleistocene Site 'Ubeidiya. Publications of the Israel Academy of Sciences and Humanities, Jerusalem.

Haas, G. 1968. On the fauna of 'Ubeidiya. Proc. Israel Acad. Sc. Hum. (Section of Science) No. 7.

Harlan, J. & D. Zohary. 1966. Distribution of Wild Wheats and Barley. Science 153:1074–1080.

Horowitz, A. 1971. Climate and Vegetational Developments in Northeastern Israel during Upper Pleistocene–Holocene Times. Pollen et Spores 13:255–278.

Horowitz, A. 1973. Development of the Hula Basin, Israel. Israel Journ. Earth Sci. 22:107–139.

Horowitz, A. 1974. Preliminary Palynological Indications as to the Climate of Israel during the Last 6,000 Years. Paleorient 2:407–414.

Horowitz, A., G. Siedner & O. Bar-Yosef. 1973. Radiometric Dating of the 'Ubeidiya Formation, Jordan Valley, Israel. Nature 242:186–187.

Jelinek, A., W. R. Farrand, G. Haas, A. Horowitz & P. Goldberg. 1973. New Excavations at the Tabun Cave, Mount Carmel, Israel, 1967–1972; a preliminary report. Paleorient 1:151–183.

Maglio, V. J. 1975. The Pleistocene Faunal Evolution in Africa and Eurasia. in: K. W. Butzer and G. L. Isaac (eds.) After the Australopithecines, Mouton, The Hague, pp. 419–476.

Martin, G. & O. Bar-Yosef. (In press). Excavations at Ein Gev III, Jordan Valley, Israel. Paleorient 3.

Michelson, H. 1972. The Geohydrology of the Southern Golan Plateau. Tahal, Tel Aviv (in Hebrew).

Moore, A. 1973. The Late Neolithic in Palestine, Levant 5:36–68.

Neuville, R. 1951. Le Paléolithique et le Mésolithique du Desert de Judée. Archives de l'Institut de Paléontologie Humaine, Paris. Vol. 24.

Nir, D. & O. Bar-Yosef. 1976. Quaternary Environment and Man in Israel (in Hebrew). Society for the Protection of Nature and the Israel Exploration Society, Tel Aviv.

Perrot, J. 1966. La Troisième Campagne de Fouilles à Munhata (1964). Syria XLIII:49–63.

Perrot, J. 1967. Munhata. Bible et Terre Sainte 93:4–16.

Perrot, J. 1968. La Prehistoire Palestinienne, Supplement au Dictionaire de La Bible, Paris.

Picard, L. 1932. Zur Geologie, des Mittleren Jordantales, Z.D.P.V. LV:169–236.

Picard, L. 1965. The Geological Evolution of the Quaternary in the Central-Northern Jordan

Graben, Israel. in: Wright, H. Jr. and D. Frey (eds.) International Studies on the Quaternary, Boulder, Colorado. pp. 337–366.

Picard, L. & U. Baida. 1966a. Geological Report on the Lower Pleistocene Deposits of the 'Ubeidiya Excavations. Publications of the Israel Academy of Sciences and Humanities, Jerusalem.

Picard, L. & U. Baida. 1966b. Stratigraphic Position of the 'Ubeidiya Formation. Proc. Israel Acad. Sci. Hum. (Section of Science) No. 4.

Prausnitz, M. W. 1970. From Hunter to Farmer and Trader, Jerusalem.

Rust, A. 1950. Die Höhlenfunde von Jabrud (Syrien). Karl Wacholtz, Neumunster.

Schick, T. & M. Stekelis. 1977. The lithic Industries of Kebarah Cave, Eretz Israel 13 (in press).

Siedner, G. & A. Horowitz. 1974. Radiometric Ages of the Late Cenozoic Basalts from Northern Israel. Chronostratigraphic Implications. Nature 250 : 23–26.

Stekelis, M. 1960. The Palaeolithic Deposits of Jisr. Banat Yaqub. Bull. Res. Counc. Israel. 9G : 61–87.

Stekelis, M. 1966. Archaeological Excavations at 'Ubeidiya, 1960–1963. Publications of the Israel Academy of Sciences and Humanities, Jerusalem.

Stekelis, M. 1973. The Yarmukian Culture. Magness Press, Jerusalem.

Stekelis, M., L. Picard, N. Schulman & O. Haas. 1960. Villafranchian Deposits near 'Ubeidiya in the Central Jordan Valley, Israel (preliminary report). Bull. Res. Counc. Israel 9G : 175–183.

Stekelis, M. & O. Bar-Yosef. 1965. Un habitat du Paléolithique Superieur à Ein Gev (Israël), Note preliminaire. L'Anthropologie 69 : 176–183.

Stekelis, M., O. Bar-Yosef & E. Tchernov. 1966. Prehistoric Settlement near Ein Gev. Yediot Behagirat Eretz Israel Weatigotheha 30 : 5–21. (in Hebrew).

Stekelis, M., O. Bar-Yosef & T. Schick. 1969. Archaeological Excavations at 'Ubeidiya 1964–1966. Publications of the Israel Academy of Sciences and Humanities, Jerusalem.

Suzuki, H. & F. Takai. 1970. The Amud Man and his Cave. University of Tokyo, Tokyo.

Tchernov, E. 1968. A Preliminary Investigation of the Birds in the Pleistocene Deposits of 'Ubeidiya. Publications of the Israel Academy of Sciences and Humanities, Jerusalem.

Tchernov, E. 1973. On the Pleistocene Molluscs of the Jordan Valley. Publications of the Israel Academy of Sciences and Humanities, Jerusalem.

Tchernov, E. 1975. The Early Pleistocene Molluscs of 'Erq el-Ahmar. Publications of the Israel Academy of Sciences and Humanities, Jerusalem.

Turville-Petre, F. 1927. Researches in Prehistoric Galilee (1925–1926). Bull. British School of Archaeology, Jerusalem. No. 14.

Watanabe, H. 1964. Les Éclats et Lames à Chanfrein et la Technique de fabrication Transversale dans un Horizon Paleolithique en Palestine. Bull. Soc. Préhist. Franc. 61 : 84–88.

Watanabe, H. 1968. Flake Production in a transitional Industry from the Amud Cave, Israel. in: Bordes, F. (ed.) La Prehistoire: Problemes et Tendances. Editions du Centre National de la Recherche Scientifique. Paris. pp. 499–509.

VII The origin of the Kinneret fauna

C. Serruya

A. History

Tristram, in 1888, stated that the overwhelming majority of species in Lake Kinneret are of Palaearctic origin, but that there are, in each class, exceptions which can be accounted for only by the geological history of the Rift area. Annandale, in 1915, found that 'the most remarkable point in the distribution of aquatic animals of the Jordan system is the fact that, whereas there is an unmistakable Ethiopian element among the fish, no such element can be detected with certainty among the invertebrates'. Annandale found an answer to this anomaly in the theories of the geographer Suess (1906). According to these theories, the Jordan Valley was covered, in the Pliocene by the 'Jordan Lake' which was drained at its southern end by an 'Erythrean River' which reached the Red Sea at the level of Aden. Upward tectonic movements and climatic changes cut off the Jordan system from the Erythrean Valley, which was then filled with the Red Sea waters. In a later stage, high salinities developed in the now closed system. As long as the Erythrean River connected the Jordan system to Africa, there was a constant colonization of the Jordan waters by African species which were trapped when the connection was cut off, and, with the further increase of salinities, most of the African species perished. The only animals which were able to survive were the fish from two families, Clariidae and Cichlidae. Annandale insists on the remarkable power of resistance of this latter family and indicates that fish of this family although belonging to fresh water species, could be grown successfully in sea water. Consequently, most of the invertebrates which are presently found in Lake Kinneret belong to a new fauna which, after the saline period, migrated into the lake from the Palaearctic and Oriental districts.

In this context, Annandale described *Typhlocaris galilea* in the Tabgha spring, as 'an inhabitant of the waters under the earth' ejected from its normal habitat by earth movements. Hubault (1938), was among the first scientists who mentioned that certain crustaceans found in subterranean water around the Mediterranean Sea are the remnants of a neogene littoral marine fauna, which, after regression of the sea, survived in this limited environment. During the same period, Bodenheimer (1935, 1937) attempted the first comprehensive zoogeographical analysis of the Levantine Province. Later zoogeographic analysis were carried out by Haas (1952) on reptiles, by Steinitz (1954) on freshwater fish, by Meinertzhagen (1954) on birds, by Harrison (1964, 1968) on mammals, by Etchecopar & Hüe (1967) on birds, and by Tchernov (in press) on rodents. In 1975, Por presented a review and zoogeographical synthesis of the existing knowledge concerning the marine, freshwater and terrestrial fauna of the Levant.

B. General background

From a zoogeographical point of view, the Levantine province is a meeting place for the Palaearctic, Ethiopian and Oriental fauna, although the connections between these various districts were greatly affected by the regional geological history.

In late Oligo–Miocene, the alpine orogenic movements which added to the old continents the southern part of Europe and northern part of Africa, brought also into existence most of the Levant province. However, whereas the general alpine tectonic lines created east–west zoogeographical belts, the main tectonic lines of the Levant province are north–south oriented. These features conferred on the Levant the properties of a transition–communication area where terrestrial and freshwater faunas could easily pass from the Palaearctic and Oriental regions to the Ethiopian one and vice-versa through north–south valleys and rivers.

These easy communications were altered in late Miocene–Pliocene, when the Sahara and Arabo–Syrian deserts developed. The terrestrial connections between the Ethiopian and Oriental regions suffered an additional disturbance in late Pliocene with the formation of the Red Sea. Conversely, this event brought the marine Indo–Pacific fauna deep into the Levant province. The last marine transgression of the Jordan Valley came from the Mediterranean Sea through the Esdraelon Valley and left interesting relicts. The ultimate form of the Rift Valley in Plio–Pleistocene times created more or less permanent water bodies in the down-faulted area and a north–south hydrographic net composed of the Orontes–Litani–Jordan Valleys with 'steeple-chase' barriers over which the organisms could be transported. A similar connection was established between an affluent of the Orontes River and an affluent of the Euphrates River. The Levant province thus developed easy connections with the Palaearctic and Oriental regions, but was cut off from the Mediterranean Sea.

C. Where did the Rift fauna come from?

For most of the freshwater fauna of the Levant, the sources of the Jordan constitute the southern limit of extension of a typical temperate Palaearctic fauna. The freshwater fish seem to be an exception to this general rule. The southern limit of the Palaearctic fish species lies some 500 km further north in the Taurus mountains (Banarescu, 1970). The fish fauna of the Levant is thus mainly composed of Oriental and Ethiopian species. The genera *Varicorhinus*, *Barynotus*, *Tylognathus*, *Acanthobrama*, *Mastacembellus* and various species of *Barbus* belong to the Oriental province and have reached the Rift via the Euphrates–Orontes water connection. According to Por (1975), the Ethiopian elements did not reach the Jordan system through the southern part of the Rift Valley as Annandale suggested but crossed the desert belt with the Nile, reached the Levant coast and penetrated into the Rift through coastal rivers (Fig. 132). The fish of African origin belong to euryhaline species of the Cichlidae (*Tilapia*, *Tristramella*, *Haplochromis*) and of the Clariidae (*Clarias*). Ethiopian elements also dominate the populations of various groups of aquatic insects. This applies to Culicidae, Odonata, Trichoptera (Botosaneanu & Gasith, 1971) and

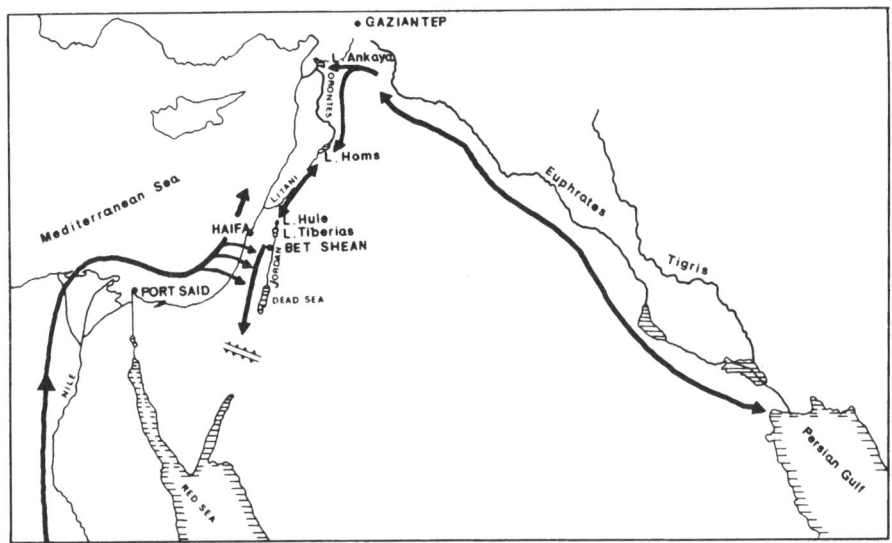

Fig. 132. Hypothetical distribution ways of the fresh-water fishes of the Levant (from Por, 1975, Zoologica scripta, 4 : 5–20).

Chironomidae (Kugler, 1966).

Most of the other species of the northern freshwater fauna have a completely different distribution; they penetrate as far south as the sources of the Jordan but pulmonate snails, cold water copepods, stone flies (Plecoptera), Rhyacophila among the Trichoptera are not found south of the Jordan head waters. South of this line, the Palaearctic freshwater fauna is restricted to a limited number of genera as if this area had 'filtered' the original fauna. For example, south of the Jordan head waters, the Ephemeroptera are mainly represented by two species, *Cloeon dipterum* and *Caenis macrura* (Matzliah, 1972) the Mollusca are dominated by *Melanopsis praemorsa* and *Pisidium casernatum*, the Crustacea by *Proasellus coxalis*, *Echinogammarus syriacus*, *Eucyclops serrulatus* and the Hirudinea by *Dina lineata*.

The study of the distribution of the various species of *Dugesia* by Bromley (1974) and of *Nitocra* by Por (1964) shows that Israel can be divided into five subprovinces: the area north of the Jordan sources, the Rift Valley, the mountainous central area, the coastal plain and the southern desertic area (Fig. 133). The distribution of species indicated that there have been connections between the coastal plain and the Rift Valley possibly as indicated on Fig. 132. The few perennial springs of the desert subprovinces are populated only by insects or passively transported invertebrates but fish are completely absent from their waters.

D. When did the various faunal elements colonize the Jordan Valley?

The lake fauna is composed of Pliocene elements both of marine and freshwater origin, of euryhalines species of freshwater origin which reached the Jordan Valley at different periods of the Pleistocene and Post Wurm, and of tropical elements. The period of penetration of the tropical species remains a controversial point. In

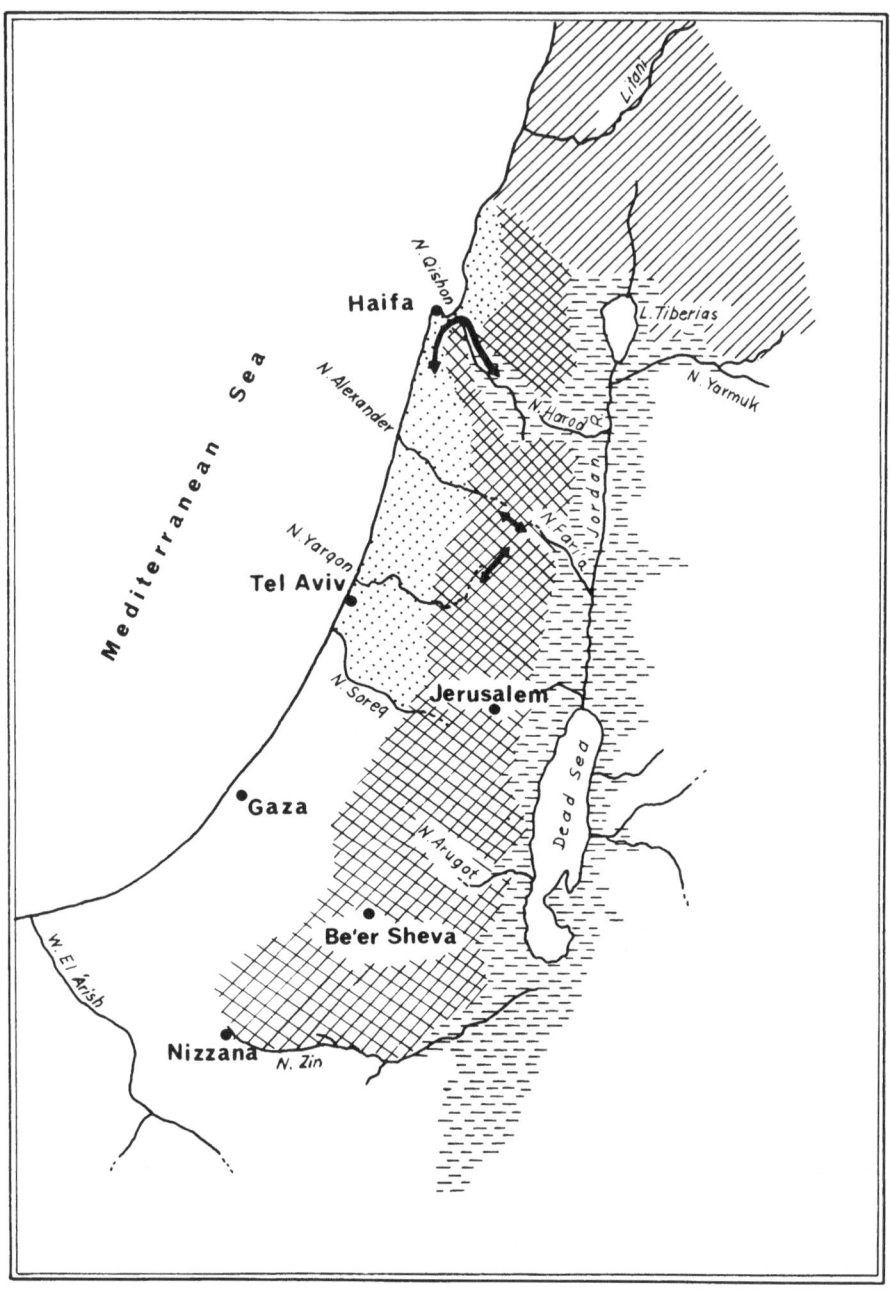

Fig. 133. Subprovinces of the fresh-water fauna of Israel and surrounding countries (from Por, 1975, Zoologica scripta, 4 : 5–20).

1963, Por suggested that the tropical elements colonized the Jordan Valley during the 'wet pluvial period'. Tchernov (1975) objected that tropical elements have existed in the Levant since the Miocene.

1. The Marine Pliocene Relicts

In late Pliocene, the Mediterranean Sea flooded the Rift Valley through the Valley of Ysreel, putting an end to the lacustrine phase which had already developed. Primitive Crustaceans occupied subterranean shelters on the sea shores. The regression of the Pliocene Sea left these populations in their limited habitats and they became the oldest relicts of the Rift. In the Kinneret area the most famous of these relicts is the prawn *Typhlocaris galilea* living in the salty and warm spring of Ein Nur on the northwest side of the lake. Other primitive relict crustaceans were also found in Ein Nur and in various springs near the Dead Sea; *Monodella relicta, Bogidiella hebraea*. The Kinneret Ostracods *Pseudobradya barroisi* and *Loxoconcha galilea* are also remnants of this old fauna. The existence of the relict isopod *Typhlocirolana steinitzi* in the Bay of Haifa is additional evidence of the Pliocene connection between the Mediterranean Sea and the Rift.

The fish *Blennius vulgaris* and *Aphanius mento*, the harpacticoids *Nannopus palustris tiberiadis* and *Onychocamptus mohammed* have also penetrated into the Rift in late Pliocene. Por (1968) suggested a similar origin for the snail *Theodoxus jordani* and the nematode *Eudorylaimus andrassyi*; this opinion is supported by the distribution of *Theodoxus* which is absent from the areas untouched by the Pliocene sea (Tchernov, 1975). However, this latter author thinks that '*Theodoxus*, although of marine origin, invaded this area much later from an open body of freshwater'.

2. The Freshwater Pliocene Relicts

Arndt (1937) was the first to point out that Lake Kinneret is harbouring preglacial lacustrine relicts; among them the most famous is certainly the sponge *Cortispongilla barroisi* – Arndt showed that the Kinneret species is closely related to the endemic sponge of Lake Ochrid, *Ochridaspongia rotundata* Arndt and concluded that they both belong to the 'Pliocene freshwater fauna of Eurasia'.

The snail *Pyrgula barroisi* is also an endemic species, closely related to *Pyrgula dybowskii polinski* from Lake Ochrid and to *Pyrgula pfeifferi* Weber from Lake Egerdir (Anatolia). Fossil species of the subgenus *Xesto pyrgula* are also known from Pliocene layers in Hungary and Yugoslavia.

The Haparcticoid Copepod *Nitocra incerta* seems to be the remnant of a population widely spread in Pliocene times in the Anatolian lake system. It is now found in subterranean waters in Europe and in some permanent water bodies of the Near East such as Lake Kinneret, Lake Egerdir and Lake Hazor in Turkey; it is also in the Caspian Sea which is of Neogene origin. The two species of Ostracods *Ilyocypris* spp. and the Bryozoan *Fredericella sultana jordanica* have a similar origin.

3. The Euryhaline species of Freshwater Origin

These animals have numerous representatives among the Oligochaetes and Cladocerans; the Ostracod *Cypreidis torosa*, the Copepods *Eucyclops serrulatus* and *Nitocra hibernica*, the Oligochaete *Euilyodrilus* spp., the Amphipod *Echinogammarus venus* and the Gastropod *Melanopsis praemorsa* belong also to this group. Although certain species might be newcomers in the Jordan Valley as suggested by Por (1963), many euryhalines of Palaearctic origin occurred already in the Ubeidiya formation (Haas, 1968). This indicates a hydrographic connection of the Jordan Valley with the Orontes system at least as early as Mindel times.

E. The Plio–Pleistocene sequence of events

During the Miocene and the Pliocene the Jordan Valley was connected with the open sea but numerous freshwater lakes existed in the surrounding areas. The fauna of these lakes constituted the stock of Pliocene species which invaded the graben during its formation. After the regression of the Pliocene Sea the marine elements of the Jordan Valley survived in underground waters or became adapted to the lacustrine conditions which developed in the valley.

The Mollusc fauna found in the Early Pleistocene sediments of the limnic formation of Erq el Ahmar located a few kilometers south of the lake includes many fossil species some of which are typical of Neogene formations. Out of eighteen species of Mollusc found in the Erq el Ahmar deposits, only seven were found in the 'Ubeidiya formation (Tchernov, 1975). In contrast, the 'Ubeidiya malacological fauna is very similar to the present fauna, indicating that a drastic faunal disruption occurred between the Early and the Middle Pleistocene. The freshwater Lake 'Ubeidiya which covered a much greater area than the present lake, was present throughout the Mindel and Mindel–Riss periods. During the Würm, the salty Lisan Lake replaced the freshwater. This increase in salinity considerably impoverished the freshwater fauna but a small number of species survived in the surrounding streams. Tectonic movements and climatic changes put an end to the Lisan stage some 20,000 years ago. The fauna which had remained alive in the lake vicinity now colonized the newly formed Lake Kinneret.

This brief review of our present knowledge of the fauna of Lake Kinneret and its connections with neighbouring provinces emphasizes certain features and poses numerous questions.

The benthic fauna of the lake includes no more than one hundred known species and Por (1968) estimated that complementary studies might bring this number to approximately one hundred and fifty species. This is a surprisingly poor fauna e.g., Lake Ochrid, another relict lake, has more than four hundred known species. The scarcity of species is accompanied by the great abundance of a few species particularly of oligochetes, nematodes and copepodes. A similar feature has been found by F. E. Round (personal communication) in the communities of benthic diatoms.

Although the salinity of the lake water is only slightly above the salinity range of freshwater, the benthic fauna of the lake includes a large number of species of marine–brackish origin. A typical example is given by the Mollusc fauna where the elements of marine origin predominate. This is another distinct difference from Lake Ochrid where according to Stankovic (1960), the fauna is typically of freshwater origin. However, even in Lake Ochrid, various exceptions can be found such as the Harpacticoids (*Ectinosoma abrau* and *Nitocrella hibernica*, the

470

nematode *Theristus dubius*, the ostracod *Leptocythere* etc.), but these species have no quantitative importance. In Lake Kinneret, the animals of marine origin represent a larger fraction of the total biomass than in Lake Ochrid, especially if we accept the marine origin of the nematode *Eudorylaimus andrassyi*.

These unusual features of the benthic fauna of Lake Kinneret have been explained in various ways. Tchernov (1975) underlined the destructive effect of the Lisan salty stage on the fauna. In contrast Por (1968) claimed that 'during the warm and dry interglacial (Lisan stage) the salinity of the lake did not exceed by much the present values'. For this author the presence of freshwater sponges of Pliocene origin, unable to produce gemmules is a definite evidence of the permanence of a freshwater body in the Jordan Valley during the whole Pleistocene. However, the discovery by Racek (1974) of *Cortispongilla barroisi* in the sediments of Lake Hula allows a different interpretation. In Lake Hula, *Cortispongilla* reached its greatest abundance between 30,000 and 10,000 years B.C. (Racek, 1974) and then disappeared completely during the main lacustrine period of the Hula.

Racek believes that *Cortispongilla* sp. are of marine origin and could survive in Lake Hula until the establishment of 'unsuitable conditions for its survival at the peak of the extreme limnetic character of this lake'. According to this interpretation, it is quite possible that the relict sponges survived a salty phase; then it is no longer necessary to assume the permanence of freshwater conditions during the Lisan stage. This is in good agreement with the fact that the types of sediments and flora assemblages found in the Lisan formations indicate high salinity levels (Begin *et al.*, 1974).

We know then that the freshwater fauna of the Jordan Valley has undergone dramatic changes at the end of the Early Pleistocene, since the fauna assemblages of the Erq el Ahmar and Ubeidiya formations are very different and it is likely that the Lisan salty phase caused also the disappearance of many species. Since the present freshwater cycle started approximately 20,000 years ago, one would expect to find in the lake a large variety of newcomers and observe a systematic colonization of the lake by the species common in the Northern Jordan River. Instead, the faunistic assemblage of the lake (planktic and benthic) is rather different from that of the Jordan River. For example, planktic Rotifers constitute the bulk of the zooplankton of the river. In contrast, in the lake, Rotifers represent no more than 7% of the total zooplankton biomass. Moreover, the lake Rotifers are mainly composed of winter visitors brought by the winter floods. They disappear completely in summer. An extreme case is presented by *Asplanchna* which is one of the few zooplankters feeding on the alga *Peridinium*. It is brought into the lake with the spring floods when the alga is extremely abundant in the lake. In spite of this rich food at its disposal, *Asplanchna* does not develop in the lake water. Other Rotifers such as *Filinia longiseta*, *Brachionus calyciflorus*, *Brachionus caudatus* are also brought with the Jordan waters and are considered by Gophen (1972) as good tracers of the Jordan flow in the lake.

Most of the fauna of copepods reaching the lake with the Jordan inflow is composed of *Microcyclops minutus*. This species is found in large amounts at the mouth of the Jordan but is absent from the lake; in spite of the frequent sampling carried out during the last eight years, only a few specimens have been found occasionally.

All these examples emphasize the fact that penetration of certain species into the lake does not mean necessarily survival and colonization. Various physical factors such as waves and storms have been invoked. Although these physical disturbances may affect the benthos of the littoral zone it does not account for the non-survival of planktic species in the lake waters. Trophic factors are obviously not the reason for the disappearance of animals such as *Asplanchna*. The difference of salinity between the river and the lake is within the range of salinity tolerance of most freshwater animals.

In contrast, it seems that the considerable difference in ionic composition existing between the river and lake water may represent a serious obstacle to the adaptation of 'river species' to lake conditions. Droop (1958) has shown that marine algae are very sensitive to ionic ratios. Pora *et al.* (1958) have shown that in the Mollusc *Anodonta cygnaea*, the physiological functions are greatly affected by the Na:Ca ratio of the external medium. These authors think that the difference in ionic ratios existing between the Black Sea and the Mediterranean Sea may explain why only 21% of the Mediterranean species are found in the Black Sea. More recently Nkudu & Harrison (1976) working on the Gastropode *Biomphalaria pfeifferi* (Krauss), reported that a decrease of the Ca:Mg weight ratio in the culture medium from 4 to 1 causes a dramatic decrease in the productivity of this animal. Similar effects, although less drastic, were obtained by a decrease of the Ca:Na and Ca:K ratios.

In the Jordan–Kinneret system, the Ca:Mg ratio varies from 8 to 10 in the river water but ranges from 1.2 to 1.5 in the lake water. The Ca:Na ratio fluctuates around 6 in the river water but do not exceed 0.4 in the lake. It is very likely that a difference of nearly one order of magnitude in the ratio of the major ions may greatly affect the main physiological functions of the river species. The chemical differences existing between the lake water (NaCl type) and the river water (calcium bicarbonate type) may prove to be a very serious 'fauna filter' which prevented and still prevents the colonization of the lake by river species.

It appears then that the distribution of the fauna along the Jordan Valley is presently regulated by a series of filters which cause a progressive impoverishment in species from north to south. The rapid increase of temperature and decrease in humidity occurring from the headwaters of the Jordan southwards constitute a first filter of climatic origin and only a relatively small percentage of the species living north of the Jordan springs can get adapted to the semi-arid conditions of the valley. A second 'fauna filter', of chemical origin, prevents the development in the lake water of numerous river species carried mechanically into the lake. This would also presumably apply to species transported by other means also. The sensitivity of many animals to the ionic composition of the water is, in my opinion, the main factor which has prevented a more successful colonization of the lake by the freshwater species of the watershed. Such a chemical faunal filter would certainly limit the number of species in the lake. It would also allow the perpetuation of old assemblages which are preserved from competition. Moreover, since the ionic ratios found in Lake Kinneret are very far from the ionic rations of most inland waters, the unusual chemical composition of this lake may have played a role in the conservation of species of marine origin and explain their relative abundance.

References

Annandale, N. 1915. A report on the biology of the lake of Tiberias. J. and Proc. Asiatic Soc. of Bengal 11:435–476.

Arndt, W. 1937. *Ochridaspongia rotunda* n. gen., n. sp. ein nuer Süsswasserschwamm aus dem Ochridsee. Arch. Hydrobiol. 31:636–663.

Banarescu, P. 1970. Principii si probleme de zoogeografie. Ed. Acad. Repub. Soc. Rom. Bucuresti.

Begin, Z. B., Ehrlich, A. & Y. Nathan, 1974. Lake Lisan: The Pleistocene precursor of the Dead Sea. Geological Survey of Israel Bull. No. 63.

Bodenheimer, F. S. 1935. Animal life in Palestine. L. Mayer, Jerusalem. 506 pp.

Bromley, H. J. 1974. Morphokaryological types of Dugesia (*Turbellaria Frielida*) in Israel and their distribution patterns. Zoologica Scripta 3:239–242.

Botosaneanu, L. & A. Gasith, 1971. Contributions taxonomiques et ecologiques a la connaissance des Trichopteres (Insecta) d'Israel. Israel J. Zool. 20:89–129.

Droop, R. 1958. Optimum, relative and actual ionic concentrations for growth of some euryhaline algae. Ver. Int. Ver. Limnol. 13:722.

Etchecopar, R. D. & F. Hue, 1967. The birds of North Africa (from the Canary Islands to the Red Sea). Oliver and Boyd, London.

Gophen, M. 1972. Zooplankton in Lake Kinneret. Internal Report. Israel Oceanographic & Limnological Research Ltd., Haifa, Israel.

Haas, G. 1952. Remarks on the origin of the herpetofauna of Palestine. Istanb. Univ. Fen. Fak. Seri B 17:95–105.

Haas, G. 1968. On the fauna of Ubeidiya. Proc. Is. Acad. Sci. and Human (section Science) No. 7.

Harrison, D. L. 1964. The mammals of Arabia. I. Ernst. London.

Harrison, D. L. 1968. The mammals of Arabia. II. Ernst. London.

Hubault, E. 1938. *Sphaeromicola sphaeromidicola* n. sp. commensal de *Sphaeromides virei* Valle en Istrie et considerations sur l'origine des especes cavernicoles perimediterraneennes. Arch. Zool. Exp. et Gen., 80, Notes et Revue, 11–24.

Kugler, J. 1966. Vorlaeufige Mitteilungen uber die Chironomi denfauna des Tiberiasseess. Gewaesser und Abwasser 41/42:70–84.

Matzliah, S. 1972. Ephemeroptera of Israel, Thesis, Tel Aviv University (in Hebrew).

Meinertzhagen, R. 1954. Birds of Arabia. Oliver and Boyd, London.

Nduku, W. K. & A. D. Harrison, 1976. Calcium as a limiting factor on the biology of *Biomphalaria pfeifferi* (Krauss) (Gastropoda-Planorbidae). Hydrobiol. 49(2):143–170.

Pora, A. E., D. Rusdea, Fl. Stoicovici, C. Wittenberger, H. Kolasovith & D. I. Rosca. 1958. La modification de la composition ionique du milieu interieur et de la respiration tissulaire chez *Anondota cygnaea* en fonction du rapport ionique du milieu exterieur. Reunion Comm. int. Explor. Sci. Mer. Medit. Monaco.

Por, F. D. 1963. The relict aquatic fauna of the Jordan Rift Valley (New contributions and review). Israel J. Zool. 12:47–58.

Por, F. D. 1964. The genus Nitroca, BOECK (Copepoda, Harpacticoida) in the Jordan Rift Valley. Israel J. Zool. 13:78–88.

Por, F. D. 1968. The intertebrate zoobenthos of Lake Tiberias I. Qualitative aspects. Israel J. Zool. 17:51–79.

Por, F. D. 1975. An outline of the zoogeography of the Levant. Zoologica Scripta. 4:5–20.

Racek, A. A. 1974. The waters of Merom: A study of Lake Huleh. IV Spicular remains of freshwater sponges (Porifera). Arch. Hydrobiol. 74–2:137–158.

Stankovic, S. 1960. The Balkan Lake Ochrid and its living world. Monographiae biologicae, W. Junk. Den Haag.

Steinitz, H. 1954. The distribution and evolution of fishes of Palestine. Istanb. Univ. Fen. Fak. Hidrobiologie B 1/4:225–275.

Suess, M. 1906. The face of the Earth. Eng. Ed. II.

Tchernov, E. 1975. The molluscs of the Sea of Galilee. Malacologia. 15:147–184.

Tchernov, E. 1975. The early Pleistocene Molluscs of Erq El Ahmar. The Isr. Acad. Sc. Hum.

Tristram, H. B. 1888. 'Reptilia, Batrachia and freshwater Fishes'. Survey of Western Palestine: Fauna and Flora.

473

Part Six

History of research
Present and future developments

History of research

C. Serruya

Josephus Flavius was probably the first author who formulated hypothesis concerning various physical and biological features of lake Kinneret. Describing the path of the Jordan in Lake Kinneret, J. Flavius notes that the river 'goes straight in the middle of the Lake Gennesareth'. In his 'Jewish War' (75) J. Flavius who knew of the existence of the fish *Clarias lazera* in the Nile tried to explain its presence in Lake Kinneret by assuming a subterranean connection between both water bodies.

The systematic investigation of the lake was initiated, in 1757, by Hasselquist, a student of Linnaeus who encouraged the exploration of the Holy Land. It is on the faunal specimens collected by Hasselquist that Linnaeus based his description of *Tilapia galilea*. The increasing interest in the Holy Land brought to the Levant famous hydrobiologists such as Tristram (1862), Lortet (1883) who worked mainly on the fauna of the lake and its surroundings. Barrois (1894) was the first scientist to study the lake from a broad limnological point of view. He worked simultaneously on the bathymetry, water temperature, benthic and planktic organisms. In 1913, Annandale who was Curator of the Bombay Museum of Natural History carried out a diversified study of the lake which included systematic classification, hydrographic observations, zoogeographic correlations and the first chemical analysis of the Kinneret waters.

The first world war reduced the stream of foreign scientists and slowed down the investigations which were resumed as the Jewish immigration increased. The project of draining the Lake Hula made many scientists realize that this water body was of unique importance from zoogeographical point of view and that it had not been sufficiently investigated. In 1936, the British Percy Sladen Memorial Fund financed an expedition aimed at studying the living world of Lake Hula (Washbourn, 1936; Washbourn & Jones, 1938). A team of scientists of the Hebrew University also organized field trips and scientific investigations on Lake Hula (1939–1944).

With the economic development which preceded the creation of the State of Israel, the scientific community became more interested in practical problems of water productivity and fisheries (Hornell, 1934; Ricardo-Bertram, 1944). It was also during this period that Ashbel extended routine meteorological measurements to stations on the lake shore (1936–1937).

After the creation of the state, numerous investigations were carried out by various institutions. In 1948, Lissner, Director of the Sea Fisheries Research Station of Haifa, initiated the biological investigation of Lake Kinneret and worked particularly on the Kinneret sardine. Kimor (1959), studied the plankton of the lake and Oren (1962) investigated its physical and chemical features. Steinitz (1951–1959), Yashuv (1957–1960) and Ben Tuvia (1959) continued Lissner's work on fish and fish production within the framework of the Ministry of Agriculture.

In the early sixties, with the completion of the National Water System, carrying water from the lake and distributing it as far south as the Negev, Lake Kinneret became the main reservoir of freshwater of Israel. This new role again modified the direction of research. Tahal, the Government Company responsible for water resource planning in Israel, invited J. Shapiro to collect all the existing data and prepare a global limnological picture of the lake. Shapiro's conclusions (1962) constituted the first comprehensive report on the limnology of the lake and his recommendations served as a stimulus for a series of studies made during the 1960's under the auspices of Mekorot Water Company. These studies were carried out by numerous scientists belonging to the Hebrew University, Jerusalem, Tel Aviv University, The Sea Fisheries Research Station of Haifa, the Fish Culture Research Station of Dor, the Technion (Technical University in Haifa), the Geological Survey, Tahal, Mekorat Water Company, Weizmann Institute, etc. The research fields included geology, hydrology, meteorology, climatology, productivity, plankton studies, etc.

However, the dispersion of the scientific efforts between various institutions proved to be an obstacle to co-ordination. In 1966, upon the suggestion of the Oceanographic and Limnological Committee of the Israel National Council for Research and Development chaired by Professor M. Shilo, the Mekorot Water Company invited Professor W. Rodhe from the University of Uppsala (Sweden) for consultations concerning the orientation of future limnological research. In his report, Rodhe emphasized the advantage of concentrating research efforts in one lakeside laboratory; this recommendation was accepted in 1967 and materialized in 1968 with the newly created Israel Oceanographic and Limnological Research Company. This institution, headed by Admiral Y. Ben-Nunn and working in cooperation with Mekorot, undertook the task of building and staffing the Kinneret Limnological Laboratory (KLL) which was opened in November 1968, at Tabgha, near the main pumping station of the National Water Carrier. The watershed research team, created in 1965 by the Mekorot Water Company also moved into the new building of the Kinneret Limnological Laboratory although it continued to function within the administrative framework of the Water Company.

The Kinneret Limnological Laboratory (KLL)

C. Serruya & Y. Avnimelech

I. The Lake Team (plate 68)

The circumstances of its creation, ascribed to the Kinneret Limnological Laboratory a mission orientated character. The problems of water quality which arose with the operation of the National Water Carrier made it a matter of urgency to know whether the spring algal bloom, partly responsible for the bad taste of the water, was a constant feature of the lake or whether the lake was undergoing a process of rapid eutrophication.

Plate 68. 'Hermona', the research vessel of the Kinneret Limnological Laboratory and her Skipper, Moshe Hatab.

In spite of the numerous and useful data collected by foreign and Israeli scientists, the information concerning the dynamic processes of the ecosystem were very scanty. In particular, little attention has been given to the causes of the algal bloom. Since blooms are the integration of meteorological, physical, chemical and biological factors, a ten year research program was established on an interdisciplinary basis. Meteorology, hydrodynamics, water chemistry, phycology, zoology, microbiology, biochemistry and sedimentology were the main disciplines

479

involved. This program of research was organized on an integrated basis and data collection was organized according to the structural pattern of the ecosystem. From a consideration of the three types of energy input to the lake, mechanical (wind), chemical (dissolved and particulate substances) and electromagnetic (radiation), the research aims were defined as the study of the utilization and transformations of these energy inputs within the system.

The watershed of Lake Kinneret is characterized by its low demographic density (30 inhabitants km^{-2}) and by its nearly complete absence of industry. The deforestation of the mountain areas, the draining of Lake Hula and the development of agriculture and aquiculture are the main factors which contribute to the nutrient load of the lake. In the lake itself, denitrification of nitrogen and precipitation–sedimentation of phosphorus decrease considerably the available nutrient load. In spite of these apparently unfavorable conditions, the algal blooms in Lake Kinneret reaches biomass values which compare with the most productive of world lakes.

The integrative research carried out at the Kinneret Limnological Laboratory has shown the prominant role of wind energy in the formation of blooms. In particular, the winter storms occurring during the homothermal period increase the water turbulence, allow the resuspension of sediments and the re-circulation of the internal nutrient load. This mechanism accounts for the normal appearance of algal blooms in dry years when the external load is minimal. Moreover, the cessation of storms in early spring and the consequent decrease of turbulence allows the motile algae to concentrate and remain in the upper layers of the lake. This purely physical modification results in a dramatic increase of the division rate of the algae which is the immediate trigger of the bloom. However, a high division rate of algae does not necessarily produce a bloom and a bloom can develop even at moderate rates of division depending on the grazing pressure exerted by zooplankton.

In Lake Kinneret, the grazing pressure on the alga *Peridinium* is extremely low because, in winter, zooplankton finds more easily assimilable sources of food, for example, bacteria. It follows that the organic matter produced at the primary producer level accumulates as detritus instead of flowing through the grazing pathway.

These are only a few examples of non-trophic factors which deeply affect the blooms in Lake Kinneret. In many lakes, cultural eutrophication has magnified the role of trophic factors in the occurrences of blooms and masked the essential role of many non-trophic factors. Thus the quantitative models based on the assumption that the quantity of algae produced in a lake is a direct function of the amount of nutrients supplied by the watershed may give good results in industrial areas where the artificial input of nutrients is so high that it has reduced all other factors to a minor role. This approach cannot be applied to lacustrine ecosystems which are highly productive despite the fact that they are not subjected to intensive cultural eutrophication. Lake Kinneret and probably many other tropical and subtropical water bodies belong to this latter group; these ecosystems require a different approach which is why, at the Kinneret Limnological Laboratory, the integration of trophic and non-trophic factors in the production of algal bloom has been developed.

Part of the results were presented at a discussion organized as an International

Symposium on Warm Lakes held in September 1973 at Ginosar on the shores of Lake Kinneret by the Israel National Research Council.

II. The Watershed Team

The classical approach to prevention of lake pollution advocates the reduction, to a minimum, of the watershed development and severe restriction on its existing activities. Such an approach is valid when other alternatives exist. The size of Israel, her social and economic features and the fact that most of the country is arid require more elaborate programs. For example, the high rate of immigration does not allow the country to neglect arable soils simply because they are located in the lake watershed. These specific features have led to a different conception of watershed management where economic development is considered as an intrinsic constraint on the system. The role of the watershed team then becomes the development of watershed management along with economic projects at the same time avoiding damaging effects on the quality of the lake water.

The adequate management of the water system is the cornerstone of watershed policy. The water system performs three main functions: it conveys the water to the lake, it is the water source for local consumption and it is a drainage system for waste-water (fishpond drainage, sewage and agricultural drainage). It is essential to separate these three different functions in the following way: the amount of water necessary for domestic supply is tapped at the water sources, a sufficient amount of water is allowed to reach the lake free of pollutants. To achieve this, the drainage and sewage waters are utilized for irrigation and fishponds and if necessary stored in special reservoirs.

The implementation of such a policy makes it imperative to clarify critical points concerning the mineralization capacity of soils, the behaviour of nitrogen and phosphorus in the soil column, the effect of utilization of waste-water on agriculture and aquiculture, the self purification potential of streams, etc. These determine the research program of the team.

This relatively liberal concept has interesting social consequences. Since its successful implementation requires the cooperation of the local population, a dialogue is established between the scientific team and the local people who thus become more and more involved in problems of preservation of water quality, land management, etc.

An example of the above mentioned approach is the work done by the watershed team in the drained Hula Lake system. The organic soils exposed after the drainage, underwent rapid decomposition and oxidation which led to the accumulation of nitrates. During some years, about 50% of the nitrogen flowing into the lake originated from this area which covers less than 1% of the watershed. A multi-disciplinary research program was initiated. The different aspects of the problem, from hydrology to soil microbiology were studied. Four years after the initiation of the research the results indicated that several procedures should be implemented: (a) Water flow through the affected area was minimized through the use of improved water conduits, dams and gates. (b) Water level was adjusted to minimize oxidation and nitrification during the summer and to maximize water storage capacity during the rainy winter. (c) The irrigation system was changed to sprinkling systems in order to induce denitrification in the soil.

The implementation of these recommendations have drastically reduced the leaching of nitrates from the Hula Valley and has made possible the coexistence of intensive agricultural practices with reasonable water quality standards in the lake.

The Kinneret Authority

A. Harpaz

The economic development of the watershed and the expanding tourist industry exert and will exert in the future an increasing ecological pressure on Lake Kinneret. The protection of the lake water quality required the creation of an appropriate administrative framework allowing the coordination of scientific research, the planning and execution of engineering projects and the supervision of polluting activities.

To this end, the fifth amendment to the Water Law, passed by the Knesset (Parliament) on 29 November 1971, granted the Water Commission of the Ministry of Agriculture special financial and legal authority to prevent the development of pollution in the Kinneret watershed. The Water Commission created the Kinneret Administration and delegated to it part of its authority. In March 1975 the Kinneret Administration was transformed into the Kinneret Authority with a significant extension of its legal powers.

Composition and organization

A Board of Directors representing the Government Ministries, the local authorities and public institutions, decides the long term policy of the Authority and confirms the budget proposed by the Authority.

The Kinneret Authority is headed by a representative of the Water Commissioner and includes members of the National Water Commission Authority and of the Water and Drainage Authorities of the Upper Galilee and Jordan Valley districts.

The Authority is assisted by a Research Council which coordinates the activities of the various research institutions working in the Kinneret watershed.

Activities

(1) Interdepartment coordination: The composition of the Authority allows the regional bodies, responsible for the actual management of their water resources to be kept permanently informed of the overall policies of the Authority.

(2) Coordination between research teams and circulation of results: A considerable amount of continuous data concerning the lake and its watershed has been collected by the Kinneret Limnological Laboratory, and the Mekorot Water Company Research team over the past seven years. This knowledge represents the basis for the policy decisions of the Authority. A research coordinator is responsible for the circulation of scientific information through reports and meetings.

(3) Planning of engineering projects designed to control pollution: The Directoral Committee of the Kinneret Authority appointed a Steering Committee which prepared the outlines of a comprehensive long term plan for the development of

the Kinneret watershed. This plan is drawn up by a team from Water Planning for Israel (Tahal) and aimed at determining optimal sites and size of urban and rural settlements, optimization of the number of inhabitants, occupations, land use and building categories. It will include plans for shore development around Lake Kinneret and will outline a regional infrastructure of roads, air transport, sewage and drainage systems. A plan for comprehensive local sewage treatment of all the settlements in the Kinneret watershed area had already been approved and should be completed within the next few years.

(4) Implementation: The Kinneret Authority has been active in projects such as care and maintenance of bathing areas, shoring up of beaches, fire prevention and soil conservation. The Authority collaborates with the Department of Soil Conservation of the Ministry of Agriculture in controlling the flow of the Jordan River and its tributaries to prevent flooding of the Hula Valley and minimize the flow of silt and nutrients to the lake.

(5) Supervision: The Kinneret Authority maintains a field team to pinpoint potential sources of pollution. This team reports on infringement of ordinances (quarrying of beach stones, illegal constructions, etc.). The operation of sewage facilities, agricultural crop spraying, irrigation and drainage methods are also controlled by this team which works in cooperation with various governmental and public agencies in the region.

References

Annandale, N. 1913. Introduction to a report on the biology of the Lake of Tiberias. J. As. Soc. Bengal. 9:17–23.

Barrois, Th. 1894. Contribution a l'etude de quelques lacs de Syrie. Rev. Biol. Nord. France. 6:224–312.

Ben-Tuvia, A. 1959. The biology of the Cichlid fishes of Lake Tiberias and Huleh. Bull. Sea Fish. Res. Stn., Haifa (27).

Hornell, J. 1934. Report on the Fisheries of Palestine. Crown agent for the Colonies, London, 71, 11.

Josephus Flavius. 75. The Jewish war, translated by G. A. Williamson, 1959, Penguin Books Ltd.

Kimor (Komarovsky), B. 1959. The plankton of Lake Tiberias. Bull. Sea Fish. Res. Stn. Haifa, 25.

Lortet, L. 1883. Poissons et reptiles du lac de Tiberiade et de quelques autres parties de la Syrie. Arch. Mus. Hist. Nat. Lyon 2:99–189.

Ricardo-Bertram, C. K. 1944. Abridged report on the fish and fishery of Lake Tiberias. Dept. Agric. Fish. Palestine. 1–44.

Steinitz, H. 1954. The freshwater fishes of Palestine. Publ. Hydrobiol. Res. Inst. Univ. Istanbul 3, 1 (4):225–275.

Steinitz, H. 1959. Dr. Lissner's study on the biology of *Acanthobrama terraesanctae* in Lake Tiberias. Bull. Sea Fish. Res. Stn. Haifa, 24.

Tristram, H. B. 1862. Notes on the birds of Palestine. Ibis 4:277–279.

Washbourn, R. 1936. The Percy Sladen Expedition to Lake Huleh. 1936. Pal. Expl. Fund. Quart. St. 204–210.

Washbourn, R. & R. F. Jones. 1938. Report on the Percy Sladen Expedition to Lake Huleh. A contribution to the study of the freshwaters of Palestine. Ann. Mag. Nat. Hist., 2(1):517–560.

Yashuv, A. 1957. Contribution to the knowledge of Lake Tiberias carps. Fishermen's Bull., Haifa 2(11):4–7 (in Hebrew).

Yashuv, A. 1960. Fishing management in Lake Tiberias. Fishermen's Bull., Haifa, 3(26):24–27 (in Hebrew).

Taxonomic index

487

489

General index

Absorbed, 153
Air temperature 47, 53, 60–62
Algae, 223–225, 234, 252, 257, 309, 315
Algal cell division, 269, 271, 272, 279
Algal cell division rate, 271, 273, 274, 287, 288, 290
Algal groups, 245, 247
Algal pigments, 258, 270
Algal succession, 243
Alkaline phosphatases, 269
Alkalinity 187, 189
Ammonia, 186, 193, 197, 200, 201, 315, 316
Amphibians, 439
Amphipod, 350, 362, 391, 393, 470
Assimilation numbers, 255–257
Atmospheric nitrogen, 317
Available radiation, 258

Bacteria, 303, 309, 313, 348, 355, 360, 393
Barbutim, 47, 105, 109, 113–115, 120
Barbutim group, 97
Barometric pressure, 7, 47, 53, 60, 61, 63, 64
Batecha, 424
Bathymetric, 126, 128, 130, 136
Bathymetry, 125
Benthic algae, 322, 323
Benthic fauna, 322, 323, 391
Benthic flora, 322
Benthic organisms, 395
Bicarbonate, 109, 189
Biomass of algae, 151, 201, 203, 204, 223–225, 244, 245, 247, 248, 252, 253, 255–257, 285, 287–290, 301, 315, 328, 395, 399
Biomass of Zooplankton, 297, 299, 301-303
Biomass of benthos (of benthic organised) 395, 471
Birds, 439, 465
Bivalves, 385
Blennids, 407
Bloom, 153, 154, 160, 186, 189, 193, 201, 212, 231, 236, 243, 246, 248, 269–271, 280, 286, 288, 292, 314, 315
Breezes, 52, 53
Bromide, 105, 107, 109, 113–117, 130, 185, 187

Btecha valley, 378
Buteiha, 12, 132, 134, 433
Buteiha plain, 13, 104, 445
Buteiha valley, 377, 389

C:N atomic ratio, 209
Calcium, 113, 116, 185, 187, 189–191, 201, 234
Calcium carbonates, 180, 189, 205, 210–212
Calcium phosphate, 202
Capernaum, 1, 138, 382
Caphernaum, 380
Carbon, 161, 180, 187, 189, 191, 201, 269, 278, 281, 315
Carbon: Chlorophyll ratios, 247, 248, 270
Carbonates 183, 208, 278
Catch, 422, 427–429
Cells in division 274
Chalcolithic, 461, 462
Chironomid, 305, 322–5, 366, 367, 369, 370, 375, 395–397, 399, 417, 467
Chloride, 96, 97, 104, 109, 185, 186, 187, 188, 189, 385
Chlorinity, 359, 364
Chlorophyll, 151, 203, 204, 249, 251, 252, 255–258, 270
Chlorophyll concentrations, 253
Chlorophyll a, 152
Cichlids, 407–409, 420, 422, 427, 428, 465, 466
Cl:Br, 113, 114
Cladocera, 302, 303, 343, 392, 394, 396, 398, 415
Clariids, 407, 465, 466
Clobitids, 407, 410, 411, 418
Clockwise, 170
Clockwise circulation, 169
Cloud, 49, 50, 53, 59, 62
Composition, 205
Conductivity, 137, 186, 188, 189, 276
Contents of the stomachs, 373
Copepod, 252, 302, 303, 305, 322, 343, 345–347, 360, 392, 394, 396, 398, 415, 467, 467, 469, 471
Counterclockwise, 169, 170